Zooplankton of the Atlantic and Gulf Coasts

ZOOPLANKTON of the ATLANTIC and GULF COASTS

A Guide to Their Identification and Ecology

SECOND EDITION

William S. Johnson

Biological Sciences
Goucher College
Baltimore, Maryland

and

Dennis M. Allen

Belle W. Baruch Institute for Marine & Coastal Sciences
University of South Carolina
Columbia, South Carolina

With Illustrations by

Marni Fylling

THE JOHNS HOPKINS UNIVERSITY PRESS
Baltimore

Printed in the United States of America on acid-free paper
2 4 6 8 9 7 5 3 1

The Johns Hopkins University Press
2715 North Charles Street
Baltimore, Maryland 21218-4363
www.press.jhu.edu

Library of Congress Cataloging-in-Publication Data

Johnson, William S., 1946–
Zooplankton of the Atlantic and Gulf coasts : a guide to their identification and ecology / William S. Johnson and Dennis M. Allen. — 2nd ed.
p. cm.
Includes bibliographical references and index.
ISBN-13: 978-1-4214-0618-3 (pbk. : acid-free paper)
ISBN-10: 1-4214-0618-7 (pbk. : acid-free paper)
ISBN-13: 978-1-4214-0746-3 (electronic)
ISBN-10: 1-4214-0746-9 (electronic)
1. Marine zooplankton—Atlantic States—Identification. 2. Marine zooplankton—Gulf States—Identification. 3. Marine zooplankton—Ecology—Atlantic States. 4. Marine zooplankton—Ecology—Gulf States. I. Allen, Dennis M. II. Title.
QL123.J649 2012
592.177'6—dc23 2012002473

A catalog record for this book is available from the British Library.

Special discounts are available for bulk purchases of this book. For more information, please contact Special Sales at 410-516-6936 or specialsales@press.jhu.edu.

The Johns Hopkins University Press uses environmentally friendly book materials, including recycled text paper that is composed of at least 30 percent post-consumer waste, whenever possible.

Contents

IDENTIFICATION and BIOLOGY of COMMON ZOOPLANKTON

APPENDIXES

Preface

I often towed astern a net made of bunting, and thus caught many curious animals.
—Charles Darwin, *Notes,* December 1833, made during the voyage of the *Beagle*

Marine zooplankton can be appreciated by anyone with access to the shore. Tow a net of fine mesh for five minutes in virtually any estuarine or nearshore water, and you will collect hundreds or, more likely, thousands of organisms. Most will be barely large enough to see if you put them in a jar. Slight magnification will reveal the fantastic world of zooplankton. These unlikely looking creatures astound both students and seasoned biologists with their bizarre shapes, strange movements, and rare beauty. The initial fascination and excitement generated by even a brief look at zooplankton soon leads to curiosity about what these organisms are, how they live, and how they fit into the ecological complex.

Zooplankton seldom get more than a cursory examination in most marine biology courses despite their abundance and importance in marine ecosystems. This is largely because identifying these often-unfamiliar animals is difficult, especially for beginners. We have designed this guide to facilitate students' and biologists' identification of zooplankton without previous training in zooplankton systematics. Our first aim is a guide that blends scientific accuracy with ease of use. Our second aim is to provide an understanding of and appreciation for the zooplankton of the Atlantic and Gulf Coasts. In addition to developing a general introduction to plankton biology and ecology, we have collected and synthesized the biological and ecological data relevant to each taxon. While this book is not intended for specialists, we have provided references to technical literature for more advanced users.

Geographic Scope. We have included taxa commonly found in estuarine and nearshore waters from Cape Cod to Florida on the Atlantic Coast and from Florida to Texas along the Gulf of Mexico. This guide includes zooplankton that occur from the low-salinity reaches of rivers to the coastal ocean within 10 km of shore. While the zooplankton assemblage is not homogeneous through this entire range, a surprising number of the most common taxa occur throughout. Deep-ocean plankton and tropical plankton found in the Caribbean Sea and carried northward with the Gulf Stream represent distinct faunal assemblages that are beyond the scope of this guide.

Taxonomic Coverage. We included the taxa most frequently encountered and most characteristic of coastal or estuarine zooplankton collections, including the common large phytoplankton and the fish larvae often retained in zooplankton nets. Most rare species are not treated here. The level of taxonomic treatment varies with the difficulty of identification. To avoid undue frustration, only taxa that can be distinguished with reasonable ease using a dissecting microscope are included. However, we provide technical references to less common species and specialized life stages.

Approach to Identification. The illustrations call attention to the critical features useful for identification. Arrows enable the user to focus on the key characteristics at a glance. We emphasize features easily observed by the novice more so than those used by specialists in classification. This approach eliminates many specialized anatomical terms. We devote special attention to identification of fresh or living specimens. Often the mode of movement or the color of a zooplankton specimen provides instant clues to identification. Since size or relative size may readily distinguish similar taxa, we have included a size range with most illustrations.

Biology and Ecology. Identification is the gateway to appreciating the diversity, biology, and ecology of the organisms collected. We include general background information in the "Introduction to Zooplankton" and more detailed notes on individual groups or species in the "Identification and Biology of Common Zooplankton" section. Still, this is only an introduction. Readers are encouraged to pursue areas of interest using the general texts and scientific literature cited.

We strongly encourage users to start the process of identifying zooplankton with this guide by reading the section entitled "How to Use This Book to Identify Zooplankton" and then to read carefully the "Identification Hints" for each group before attempting identification.

The illustrations in this book, all drawn by Marni Fylling, were inspired by a number of sources. They were usually drawn using a combination of photographs, specimens collected in the wild, experience of the authors, advice from external reviewers, and descriptions and illustrations from published literature as cited in the "Sources of Models for Illustrations" following the Index at the back of this book.

Acknowledgments

The first of many thanks go to our wives, Valerie Chase and Wendy Allen, without whose understanding, encouragement, and support this book would probably not have materialized. As professional marine science educators with a passion for advancing the world's understanding for our oceans and coastal systems, they inspired and energized our effort.

We also give special recognition and thanks to Marni Fylling, whose talents as an artist and personal commitment to excellence throughout the project are deeply appreciated. In addition to carefully preparing hundreds of illustrations, Marni played an active role in formatting the figures and identification plates. Thousands of e-mails and files later, we can still say that we enjoyed it all. We also thank Tammy Clark, whose initial illustrations in the early drafts helped us to refine our approach.

Many of our colleagues who are experts in their respective fields of biology reviewed chapters in this book. We were overwhelmed with the generosity with which they contributed their time and expertise. *Zooplankton of the Atlantic and Gulf Coasts* is far richer for their insights and firsthand knowledge of the local fauna and flora. We also appreciate the timely encouragement they provided along with their comments. The contributions of the following individuals are gratefully acknowledged: Julie Ambler (Introduction and Copepoda), O. Roger Anderson (Protozooplankton), Stephan Bullard (Decapoda), Edward Buskey (Introduction, Copepoda, and Protozooplankton), Nancy Butler (Introduction), Dale Calder (Cnidaria and Ctenophora), Wayne Coats (Microzooplankton), Jack Costello (Cnidaria and Ctenophora), Jim Ditty (Fishes), John R. Dolan (Rotifera), Michael Fahay (Fishes), Paul Fofonoff (Cladocera), Richard Fox (Pericarida), Harold G. Marshall (Phytoplankton), Charles Gallegos (Phytoplankton), Sharyn Hedrick (Phytoplankton), Matthew Johnson (Protozooplankton), Richard Lacouture (Phytoplankton and Protozooplankton), Mike Mallin (Introduction, Phytoplankton, and Copepoda), Nancy Marcus (Copepoda), Jerry A. McLelland (Chaetognatha), Cecily Natunewicz (Introduction and Decapoda), Wayne Price (Peracarida), Jennifer Purcell (Cnidaria and Ctenophora), Edward Ruppert (Introduction), Katherine Schaefer (Phytoplankton), Kevin Sellner (Phytoplankton), Richard Shaw (Fishes), Barbara Sullivan (Copepoda, Cnidaria, and Ctenophora), Howard L. Taylor (Rotifera), Erik Thuesen (Chaetognatha), and Jennifer Wolny (Phytoplankton).

We gratefully acknowledge the following experts who reviewed changes to the second edition: Julie Ambler (Cladocera and Lower Chordates), Scott Burghart (Decapoda), Edward Buskey (Copepoda), Dale Calder (Cnidaria), Jose Cuesta (Decapoda), Michael Fahay (Fishes), Darryl Felder (Decapoda), Richard Heard (Peracarida), Richard Lacouture (Phytoplankton), Sara LeCroy (Peracarida), Steve Morton (Phytoplankton), Michael Parsons (Phytoplankton), Ernst Peebles (Peracarida), Mark Peterson (Fishes), Gerhard Pohle (Decapoda), Patricia Ramy (Annelida), Jan Reid (Copepoda), Kevin Sellner (Phytoplankton), Cecily Steppe (Decapoda), and Gregory Tolley (Fishes).

Others who provided valuable comments, materials, and/or encouragement as the manuscript took form include Charles Barans, Donna Lynn Barker, Bruce C. Coull, John Dindo, David Egloff, Elizabeth Fensin, Ray Gerber, Barbara Henderson, David Knott, Alan Lewitus, Stephen Stancyk, Jack McGovern, Rick Tinnin, and Elizabeth Wenner. The following colleagues provided information and suggestions that helped guide revisions for the second edition: Jack Costello, Elisabeth Fensin, Frank Ferrari, Terri Kirby Hathaway, Matt Johnson, John McDermott, Richard H. Moore, Christian Sardet, and Mare Timmons.

Sincere thanks go to the office and technical staff and colleagues at Goucher College and the University of South Carolina who provided many kind services and extra efforts that made it possible for us to complete the project. We are particularly grateful to the Johns Hopkins University Press for encouraging and enabling the production of this edition. Many thanks go to our editor Vincent J. Burke and our copyeditor Andre Barnett, whose trained eye, thorough readings, and patience resulted in a more properly formatted, accurate, and readable book. We also thank all of the other members of the Press's production and marketing team.

Among our many mentors and instructors, the late Professor Sidney S. Herman stands out as a great champion for the appreciation of zooplankton. We thank him greatly for the knowledge he shared and the enthusiasm he generated during our years together (1972–78) at the Wetlands Institute, Stone Harbor, New Jersey. A second inspiration was *The Field Book of Seashore Life* by Roy Waldo Miner, published in 1950. This field guide profoundly affected both of us in our early years as undergraduate biologists. More than a guide, it both fed and stimulated our curiosity about the marine life that we were each discovering for the first time. Miner's book was both an inspiration and a reference as we worked on this guide to zooplankton.

This book was developed in response to (1) our professional interests in zooplankton ecology and (2) the need for a single reference source for introducing students and teachers to the subject for the first time and for training advanced students and technical staff to identify local forms. Throughout the project, we marveled at the meticulous and mostly forgotten studies of early workers, ongoing controversies about taxonomic status, and the diversity of zooplankton in our area. Early on, we recognized the great challenge we faced in trying to synthesize and compose an accurate and useful product. In the end, we share equal responsibility for our successes and failures. Although Bill developed most of the text and Dennis was responsible for the identification sections, the final product reflects an equal and collaborative effort that was both instructional and fun. In the absence of a more reasonable standard for making a decision about the order of authors, age emerged as the criterion of choice, with Bill having provided guidance as a member of Dennis's dissertation committee all those years ago.

In preparing the second edition, we added more than 50 new taxa (and illustrations) to the identification sections. We also adjusted and improved illustrations and/or features for identification for more than 60 other taxa. Much new information about the biology, ecology, and taxonomic status of dozens of species became available since the first edition was prepared in 2004, so many entries have been updated and more than 420 new and recent references have been added. Scientific names of some familiar species have been changed since the first edition. The "Introduction to Zooplankton" and other sections of the book have also been updated.

This second edition has benefited from the suggestions of many users, whose input has resulted in improvements, corrections, and the addition of many new species. We know that this second edition is still not as correct or complete as all would hope. We, and perhaps future users, will benefit from observations and suggestions provided to us by readers at all levels.

Funding for the excellent line drawings developed by Marni Fylling for the first edition originated from the McClane Endowment and other funds from Goucher College and the Franklin H. and Virginia M. Spivey Endowment for environmental education, Baruch Institute, University of South Carolina. The Spivey Endowment provided funding for Marni Fylling to prepare new, original illustrations and to adjust many existing illustrations and figures for the second edition.

Zooplankton of the Atlantic and Gulf Coasts

INTRODUCTION TO ZOOPLANKTON

Plankton (from the Greek word for wanderer) are the small organisms suspended in the water, neither attached to the bottom (**benthos**) nor able to swim effectively against most currents (**nekton**). **Phytoplankton** include the photosynthetic unicellular protozoans and bacteria. **Zooplankton** include both unicellular and multicellular organisms. Most zooplankton are only a few millimeters in size, but some are much larger. The traditional definition of plankton evokes an image of organisms drifting passively as they are carried to and fro with the currents. Although they are at the mercy of currents, many zooplankton are accomplished swimmers capable of complex feeding and evasive maneuvers.

Most zooplankton samples include a diverse mixture of **holoplankton**, organisms that spend their entire lives in the plankton, and **meroplankton**, zooplankton that spend only a part of their life cycle in the water column, usually as larval stages (Fig. 1). **Demersal zooplankton** spend much of their time on or near the bottom but periodically swim upward into the water column, especially at night. The **neuston** consist of organisms specifically associated with the uppermost layer of the water column, either at or just below the surface. The Portuguese man-of-war (*Physalia*) is an example of a surface drifter, but many other planktonic animals swim just below the surface.

Plankton span a size range of more than eight orders of magnitude, as shown in Tables 1 and 2. Because **pico-** and **nanoplankton** pass through the finest plankton nets, we were long unaware of their abundance and ecological importance. In fact, these organisms may equal or exceed the more visible forms in productivity and energy flow. The term **net phytoplankton** refers to phytoplankton retained by a 20 μm mesh, primarily larger diatoms and dinoflagellates. **Microzooplankton** includes a mixed assemblage of organisms in the 20- to 200-μm-size range. Most organisms in these groups will pass through zooplankton nets or be too small to be noticed, except at high magnification. The predominant microzooplankton are ciliate, flagellate, and amoeboid protozoans and copepod nauplii. **Mesozooplankton** in the 0.2 to 20 mm size range dominate most samples and are the focus of this book. Copepods alone often represent 50%–90% of the catch, with densities as high as 1 million per cubic meter (m^{-3}). Rotifers, larval barnacles, crab zoeae, and mollusc veligers can also reach impressive densities, but their presence is more sporadic. **Macrozooplankton** (2–20 cm) shrimps, larval fishes, and other large, mobile animals, though common, tend to be less abundant and less susceptible to capture. The largest zooplankton, primarily jellyfishes and ctenophores, are easily seen from the surface and may be collected with a bucket or dip net.

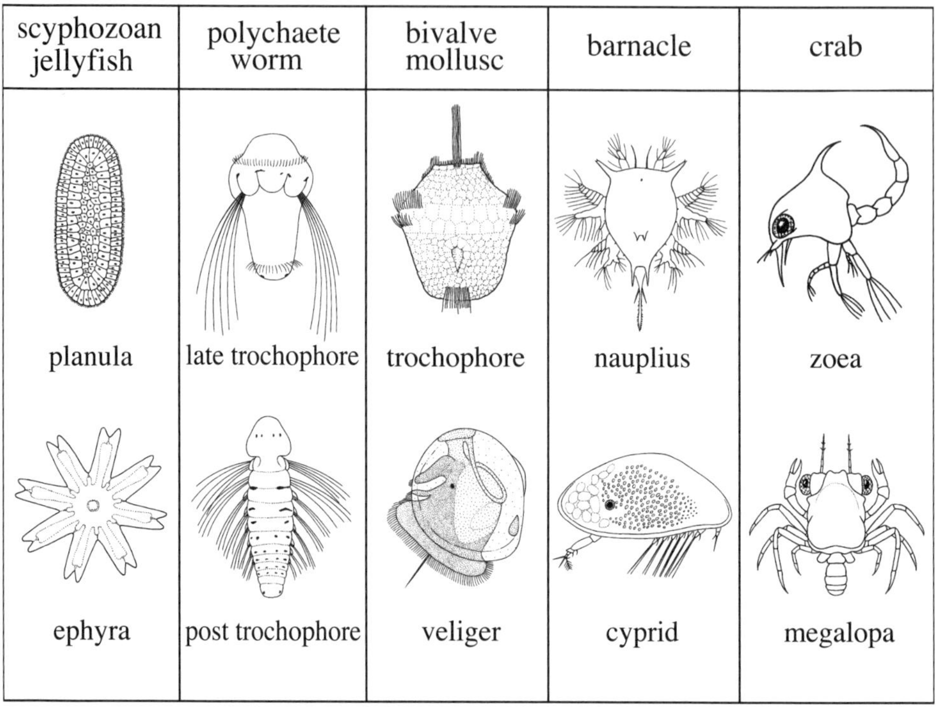

scyphozoan jellyfish	polychaete worm	bivalve mollusc	barnacle	crab
planula	late trochophore	trochophore	nauplius	zoea
ephyra	post trochophore	veliger	cyprid	megalopa

Fig. 1. Representatives of some common larvae, illustrating only a hint of the great diversity of planktonic larval forms released by benthic invertebrate phyla. Note how several different larval forms may occur in the development of single species and how different these are from the benthic adult.

ADAPTATIONS TO PLANKTONIC LIFE

Life in the plankton poses many challenges. Zooplankton live in a viscous liquid medium in which locomotion is difficult. There is no place to hide from drifting or swimming predators. Survival in the plankton requires specific adaptations to remain in the water column, to secure suitable food, and to avoid predators. Solutions to these challenges have evolved separately in different groups of zooplankton.

Feeding

While some benthic invertebrates release **lecithotrophic** larvae with sufficient yolk reserves to complete their planktonic phase without feeding, all of the permanent zooplankton and most of the temporary larval stages feed actively in the water column and are thus **planktotrophic**. In contrast to terrestrial ecosystems, most of the primary productivity in the water column is in the form of small particles. Phytoplankton and suspended bits of organic matter (and associated microbes) called **detritus** are usually less than 1 mm long. Most coastal zooplankton graze on these particles. Their task is to collect or to concentrate

Table 1. Size classes of planktonic organisms illustrating the extreme range of sizes of planktonic life forms

Planktonic group	Femto-plankton (0.02–0.2 μm)	Pico-plankton (0.2–2.0 μm)	Nano-plankton (2.0–20 μm)	Micro-plankton (20–200 μm)	Mesoplankton (0.2–20 mm)	Macro-plankton (2–20 cm)	Mega-plankton (>20 cm)
Viruses							
Bacteria							
Phytoplankton: cyanobacteria, dinoflagellates, and diatoms							
Protozooplankton: ciliates, forams, and radiolarians							
Copepods, crab zoeae, etc.[a]							
Decapod shrimps, euphausids, salps, and chaetognaths							
Scyphomedusae and siphonophores							

Source: Modified after Sieburth et al. 1978.
Notes: Any single collection usually targets only a small fraction of this spectrum. The thicker bars indicate the typical size range for the group and the lines the extremes although authorities differ on exactly where to draw the lines. Note the log scale: each category contains a 10-fold size range except the 100-fold range given for mesoplankton.
[a] See Table 2.

the food. Most of these particle grazers are loosely termed **suspension feeders**. Until recently, grazing zooplankton were thought to be filter feeders, but studies now show that, rather than sieving food particles, they actively harvest them.

With the major exception of crustaceans, cilia represent the tool of choice for most suspension feeders. Cilia create feeding currents to move water and food over a mucus-covered surface. As small particles become stuck, cilia move the mucus and trapped food to the mouth. Feeding using cilia and mucus is well suited to removing nanoplankton and other particles too small to be caught by crustacean setae. Lacking cilia, the crustacean grazers sweep their setae through the water and remove phytoplankton and detritus particles. The active, mobile predators in the nearshore plankton, such as larger crustaceans, chaetognaths (arrow worms), and larval fishes, detect and attack individual prey. Medusae and ctenophores are more passive "ambush" or "entanglement" feeders but equally efficient predators. They drift or cruise slowly through the water as their trailing tentacles subdue and collect zooplankton.

Table 2. Size ranges of the mesozooplankton groups commonly caught in estuarine nearshore waters of the Gulf and Atlantic coasts.

Taxon	Size range (mm) log scale: 0.1 · 0.5 · 1 · 5 · 10 · 20
Holoplankton	
Hydromedusae	
Scyphomedusase	
Ctenophores	
Cumaceans	
Copepods	
Ostracods and Cladocera	
Mysids	
Siphonophores	
Chaetognaths (arrow worms)	
Larvaceans (without their "houses")	
Meroplankton (larval stages)	
Mollusc veligers	
Copepod nauplii	
Barnacle nauplii	
Barnacle cyprids	
Polychaete larvae	
Crab zoeae and megalopae	
Decapod shrimp larvae	
Fish eggs	
Fish larvae	

Notes: Zooplankton span over 2 orders of magnitude as indicated here using a log scale.

A complete description of the complexity and elegance of planktonic feeding adaptations is beyond the scope of this introduction, but detailed descriptions of feeding and diets are included with each taxon in the chapters on identification and biology. Ongoing research in this area may soon fill the remaining gaps in our understanding of zooplankton feeding.

Locomotion

By definition, plankton are swept to and fro by currents. However, it would be a mistake to think of them as passive drifters. Many are accomplished swimmers capable of rapid escape maneuvers and extensive vertical excursions. While overall swimming speeds seldom exceed 5 cm per second, these speeds are more impressive when we consider the animals' sizes. For example, some ciliates and rotifers, though propelled by cilia, can attain speeds of more than 10 body lengths per second. The fastest copepods swim at more than 100 body lengths per second when in escape mode. By contrast, human Olympic swimmers reach only reach about 1 body length per second. These high relative speeds become even more remarkable when we consider that the viscosity of water has a far greater impact on motion as size decreases. Swimming resistance due to viscosity may be minimal for a fish but becomes the dominant form of resistance (expressed as a low Reynolds number) for smaller plankton of a millimeter or less in length. Some people suggest that for smaller plankton swimming might be like swimming in syrup. Zooplankton use widely varying mechanisms to move. Cilia, jointed appendages, or whole-body contractions generally accomplish swimming, but each group has its distinctive form of locomotion. For more detailed descriptions, refer to the introductory sections of specific zooplankton groups in the "Identification and Biology of Common Zooplankton" section.

Vertical Migrations and Selective Tidal Transport

When gazing at the ocean's flat surface, it is easy to forget that the sea is a three-dimensional world. Waters at different depths differ in light level, temperature, salinity, oxygen availability, and concentrations of food and predators. Most zooplankton can actively control their vertical position through directed swimming or slight changes in buoyancy. Extensive vertical migrations of entire planktonic communities are common in the open ocean but are seldom seen in shallow coastal and estuarine areas where dramatic vertical excursions by individual species occur. We see two primary patterns of vertical migration in our area.

Many species aggregate near the bottom during the day and rise into the water column at night. Restricting time in the water column to nocturnal excursions may reduce predation by visual feeders. Epibenthic crustaceans, including some mysids, copepods, and larval shrimps, engage in time-sharing between benthic and planktonic lifestyles. Their nocturnal forays are often associated with feeding. Other primarily benthic animals leave the bottom specifically for mating. Nereid and other polychaetes gather at the surface to mate or to release eggs, and many amphipod and cumacean crustaceans enter the plankton to find mates. Many zooplankton have photoreceptors and apparently use light intensity to coordinate vertical migrations. Even moonlight may provide enough illumination to affect vertical migrations or to permit some visual predators to feed near the surface. Surprisingly, some zooplankton are attracted to strong lights at night. Samples collected after dark using spotlights or in lighted areas often result in unusual catches.

Both holoplankton and meroplankton use vertical positioning to take advantage of favorable currents for migration or to retain their position within estuaries, a process called selective tidal transport. For example, larval fishes produced in offshore spawning areas undergo vertical movements that place them in onshore currents that transport them into estuaries.

Defense

Various planktonic predators feed on herbivorous zooplankton and on one another. The other major threat to zooplankton comes from planktivorous fishes. In fact, the most abundant fishes in coastal and nearshore waters (anchovies, sand lances, herrings, and silversides) are zooplanktivores. Most are selective visual feeders on zooplankton. Despite the absence of hiding places, zooplankton reduce predation in a variety of ways. Many are transparent. Some zooplankton have impressive spines or contain distasteful or toxic chemicals to reduce predation. Others have well-developed predator-detection capability and escape responses. Last, some zooplankton have so little nutrient value that most selective predators ignore them. Table 3 illustrates the diversity and distribution of antipredator defenses found in coastal zooplankton.

Table 3. Summary of adaptations to deter predators

Antipredator tactics	Representative taxa
Transparency	Arrow worms, hydromedusae, comb jellies, fish eggs and larvae
Protective spines	Crab zoeae, polychaete bristles
Detection and escape. Detection of minute pressure waves followed by evasive maneuvers	Copepods and some fish larvae
Distasteful or toxic	Some crab larvae
Diel swimming rhythms, usually with nocturnal excursions into the plankton	Late shrimp and crab larvae, mysids, gammaridean amphipods, estuarine calanoid copepods
Synchronized mass spawning; timing of larval release; limited planktonic phase	Many bivalves, echinoderms, some polychaetes; many crabs and shrimps; sea squirts
Bioluminescence; sudden flashes may startle predators, but this hypothesis needs confirmation	Ctenophores and some dinoflagellates

Note: For specific information, see the "Suggested Readings" at the end of this chapter, including Ohman (1988), for an overview.

ADAPTATIONS OF PLANKTONIC LARVAE

Most meroplankton are larval forms of benthic invertebrates. The larval phase may last only a few hours or a year or more. Regardless of its duration, the larval stage plays a critical role in the overall life cycle. Dispersal by means of planktonic larvae reduces overcrowding, allows colonization of new and potentially unexploited habitats, and promotes genetic exchange between populations. Other potential advantages conferred by having planktonic larvae are listed in Table 4, but there are considerable risks as well. Very few survive to settle in a favorable habitat and assume the adult form. Perhaps this is why many benthic species are so prolific.

Sometimes larval development takes place at a considerable distance from the adult habitat. The life cycles of fishes and invertebrates often exhibit movements between coastal ocean and estuarine systems. The successful transition between larval and adult habitats requires precise timing and, often, long-distance transport. Because larvae cannot swim against most tidal currents, behavioral responses to advantageous tidal and wind-driven currents are keys to successful transport and settlement. Table 5 gives a glimpse of the diversity of strategies involved.

Table 4. Benefits and risks of producing planktonic larvae

	Benefits
Dispersal	Larvae may disperse to new areas not previously colonized. This is a major benefit since the adults are more or less sedentary. Disturbances may render old habitats unsuitable and create new ones. Recruitment into these areas is by larval settlement and may be crucial to survival of the population or species.
Genetic diversity	When gametes are shed into the water, zygotes may be formed from eggs and sperm produced by parents some distance apart, perhaps even from separate populations. This outcrossing is a special benefit for species with sedentary adults.
Avoiding competition with adults	Planktonic larvae typically exploit entirely different food resources in a different location.
Lower predation	Mass spawnings produce huge pulses in larval abundance that may overwhelm predators and thus lower predation risk for individuals, at least for short periods.
	Risks
Finding food	Many planktonic larvae begin with limited yolk reserves and must find food or starve.
Predation	Larvae are at risk while in the plankton.
Finding a home	Larvae are at the mercy of the currents and may be swept many miles from suitable adult habitat. Adults may have specific habitat requirements. Larvae that survive the planktonic stage but settle in inappropriate locations die.

Table 5. Larval transport and retention.

Developmental pattern	Example
Estuarine adults with oceanic larvae	Blue crabs Adults spawn at the mouths of estuaries. Larval (zoeal) stages develop offshore over the continental shelf. Late larvae approach shore as they grow and enter inlets to complete larval and juvenile stages.
Intertidal adults with coastal oceanic larvae	Fiddler crabs Adults spawn in salt marshes or high intertidal shorelines. Zoeae develop outside of nearby inlets. Megalopae return to marshes.
Estuarine adults with estuarine larvae	Mud crabs (some), gobies Adults spawn within estuaries. All larval stages are retained within the estuary and not exported to the ocean.
Ocean spawners with part-time estuarine larvae	Brown shrimp, spot, menhaden Larvae begin development in the ocean. Late larvae approach shore as they grow and enter inlets to complete larval and juvenile stages before returning to the sea.
Oceanic/estuarine adults with fresh/estuarine larvae (anadromous fishes)	Shads, herrings, striped bass Adults migrate up rivers to spawn in freshwater or at the freshwater/brackish interface. Larvae gradually move toward the ocean as they grow.

How different larvae accomplish their return or entry into estuarine systems has long been a mystery, but recent investigations provide some insights. The example of blue crabs in the Chesapeake Bay (Fig. 2) is remarkable for the long-distance transport involved, but it illustrates many of the adaptations associated with larval transport. More information on life cycles and planktonic larval stages can be found in the notes on specific taxa in the identification section.

For benthic invertebrates, the planktonic larval phase ends as the larvae settle and take up the adult lifestyle. Because most invertebrates have specific habitat requirements, the location of the final settlement is crucial for adult survival. A few decades ago, we assumed that larval settlement was largely a chance event where only those fortunate enough to be in the right spot at the right time survived. Although serendipity is indeed a factor, many larvae have surprisingly sophisticated physical and chemical sensory capabilities employed solely for selecting a favorable habitat. Often larvae approaching settlement will explore and evaluate several prospective substrates before selecting one. If no suitable substrate is found, some larvae can postpone the final metamorphosis to the adult form, thus giving a wider temporal window to finding a suitable spot.

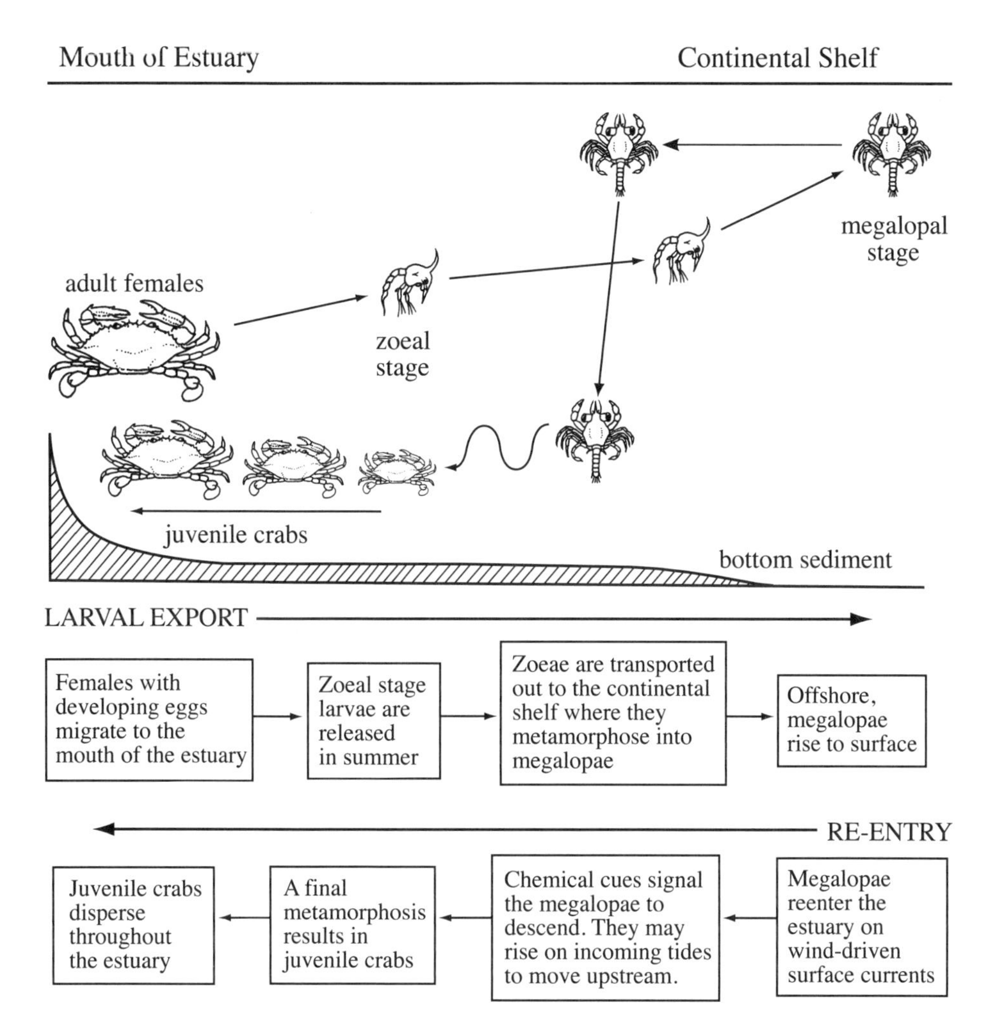

Fig. 2. Working model of the blue crab life cycle in Mid-Atlantic estuaries illustrates long-distance transport of larvae out of the estuary and their return. The use of currents and the cues involved in vertical migration and habitat selection are still under investigation. (Modeled after Dame and Allen 1996.)

FACTORS AFFECTING ZOOPLANKTON DISTRIBUTION

Factors acting on many scales influence zooplankton distribution. Water temperatures primarily determine geographic ranges, a function of both latitude and major ocean currents. Within a given temperature range, salinity is probably the largest factor affecting distribution. Our nearshore habitats include freshwater, brackish water, seawater, and even hypersaline areas. Within each of these salinity divisions, many species show preferences for specific habitats or hydrographic conditions. Differences in depth, current velocity, wave energy, and turbidity affect local distributions. Proximity of certain habitats such as salt marshes, oyster beds, and submerged vegetation also affects plankton distribution. Although we cannot treat all of these factors in detail, this summary provides a general background for understanding the distribution of individual species. The distribution of individual planktonic taxa change over time as abiotic and biotic factors change. These factors include the distribution of food, abundance of predators, and changes in temperature or salinity. Different larval or other life stages of a taxon can have different distributions within the same water body

Long ago, zoogeographers divided the globe into "geographic provinces" according to the distribution of flora and fauna. Most oceanographers today use slightly different terminology when talking about specific regions of the ocean. Listed below is the terminology we will use to designate faunal distributions associated with nearshore distributions.

New England (Cape Cod, Massachusetts, to Sandy Hook, New Jersey)
Mid-Atlantic (Sandy Hook to Cape Hatteras, North Carolina)
Southeast (Cape Hatteras to Cape Canaveral, Florida)
Gulf of Mexico (Fort Myers, Florida, through Texas)

Figure 3 shows these regions and the major ocean currents influencing zooplankton distribution. "Suggested Readings" at the end of this chapter lists references describing the detailed hydrological aspects of each region.

Plankton Distribution and Ocean Currents

Coastal surface currents flow westward from Cape Cod and continue past Sandy Hook and then south or southwest to the Virginia capes. This is a slow net flow produced by prevailing winds with periodic reversals when southerly winds blow for several weeks. To the south of Cape Hatteras, the nearshore flow is also southward—a countercurrent to the north-flowing Gulf Stream (Fig. 3). Even though it ranges from tens to hundreds of kilometers offshore, the Gulf Stream influences plankton along the entire Atlantic Coast. The Gulf Stream transports tropical fauna from the Caribbean Sea northward along the southern Florida coast and then curves seaward as it passes northern Florida, Georgia, and South Carolina. At Cape Hatteras, the Gulf Stream comes within 20–40 km of the coast before swinging northeast toward Great Britain. As the Gulf Stream moves northward, it sheds eddies, often 100–300 km in diameter, which enclose masses of Gulf Stream water. These eddies drift toward the southwest with the prevailing coastal current and transport offshore and tropical zooplankton into nearshore waters.

The nearshore currents in the Gulf of Mexico are generally weak, with a slow net westward flow of northern coastal Gulf water from Mississippi toward Texas. The Loop Current (Fig. 3) is the major current in the Gulf. Even at its northernmost extent, the Loop Current remains offshore. As the Loop Current bends, it sheds eddies at approximately 8–10

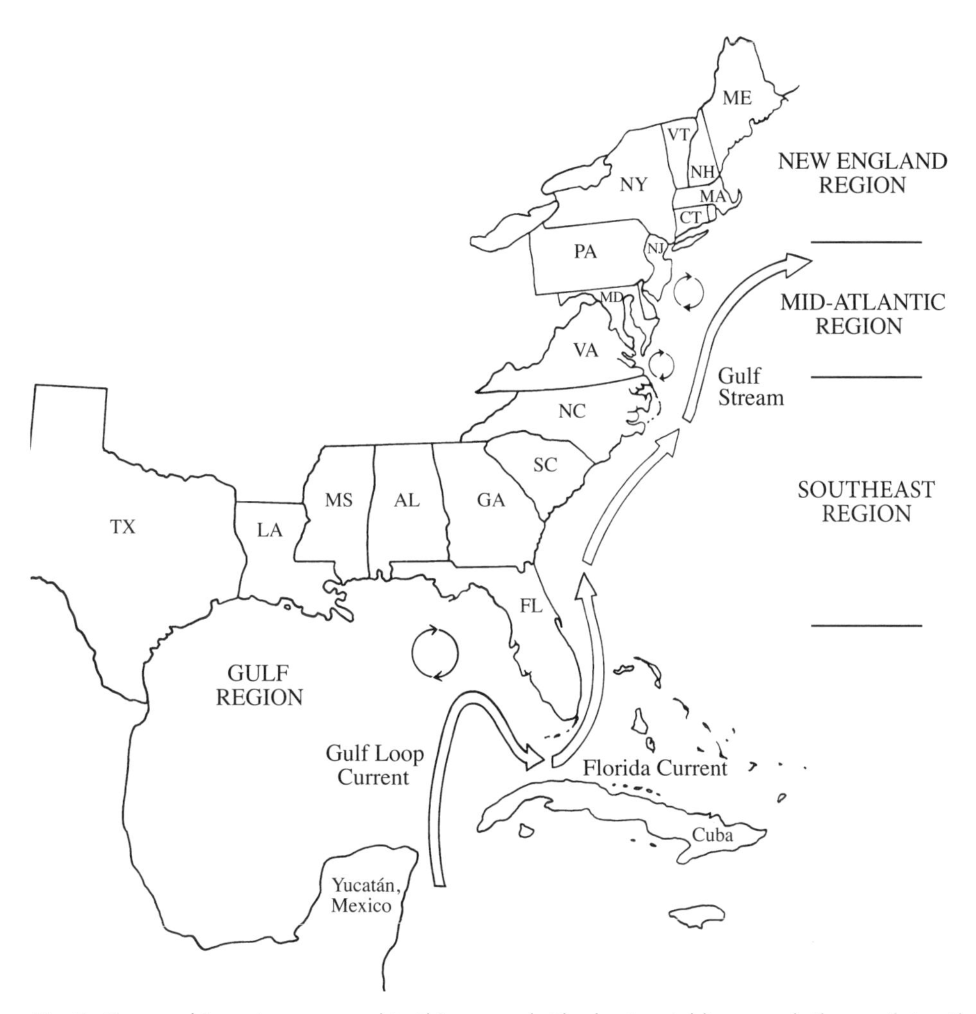

Fig. 3. Geographic regions covered in this manual. The horizontal lines mark the north/south boundaries of the Atlantic regions but note that our coverage only extends out to 10 km from shore. Eddies, routinely spinning off from the Gulf Stream and the Gulf Loop Current, carry tropical, oceanic waters nearshore.

month intervals. These continue northward and sometimes reach the shoreline, especially in the eastern Gulf, bringing species typically associated with open Gulf waters.

Coastal Waters

South of New England the bottom slopes gently across the broad continental shelf. Even comparatively shallow depths of 50 m usually lie many kilometers offshore. Many complex shallow-water phenomena affect zooplankton in coastal waters. Here we introduce some of the basic aspects of the nearshore ocean environment. Consult Able and Fahay

(1998) and Garland, Zimmer, and Lentz (2002) in the "Suggested Readings" for details to supplement this brief overview.

As the prevailing ocean currents flow past the coastline, they mix with brackish water from numerous bays and estuaries to produce a narrow band of slightly, but measurably, lower-salinity water, extending roughly 5–10 km from shore. Zooplankton characteristic of this coastal boundary layer are seldom found much farther offshore. Open-ocean plankton over the continental shelf are periodically transported into this nearshore region by vagaries of wind or tide to mix with the more coastal fauna, but only some of these species persist in the narrow coastal zone.

Plumes of estuarine water often extend well beyond the shoreline. The interface where estuarine discharge meets ocean water produces a sharp boundary, or front, often marked by foam or by color discontinuity. The edges of frontal plumes exhibit enhanced plankton productivity and biomass. Most of the plumes along the Atlantic Coast affect local conditions within the mouths of estuaries to several kilometers offshore, depending on tidal stage, wind conditions, and discharge volume. In contrast, two-thirds of the entire freshwater inflow into the Gulf enters via a single source: the Mississippi-Atchafalaya discharge—the only river system in our geographic range with sufficient flow to consistently push the brackish water–mixing area tens of kilometers offshore. This Mississippi Plume of nutrient-laden freshwater mixes with Gulf water and increases productivity. The mixing area at the edge of the plume is especially rich with up to 10-fold more zooplankton at this interface than in the waters to either side. The Gulf water and the freshwater from the Mississippi mix slowly as they drift westward. This brackish-water mass influences the salinity and character of Gulf waters—and its zooplankton—as far west as Port Aransas, Texas.

Thermoclines and Coastal Stratification: Invisible Boundaries

Oceanographers use the term **water column** to refer to the vertical dimension of the oceans and to describe specific physical conditions or biotic distributions associated with depth. Profound differences in temperature, salinity, light penetration, oxygen, or nutrients may occur within a few meters of depth, depending on how well the water column is mixed. Currents, winds, and waves can keep the water completely mixed from top to bottom, especially in shallow waters. Particularly during calm weather, the summer sun warms surface waters, making them less dense. In absence of strong mixing, a sharp boundary or temperature gradient, the **thermocline,** develops between warm surface water and cooler and denser bottom water. Waters with different densities do not mix readily, thus the water column becomes stratified into two distinct layers.

If **stratification** persists, the bottom waters, cut off from oxygen-rich surface waters, may suffer severe oxygen depletion (**hypoxia**). While most zooplankton can swim up to the oxygen-rich surface waters, this may increase the risk of predation or position them in currents that transport them into unsuitable areas or adverse conditions. Along the Atlantic Coast in summer, stratification occurs occasionally during prolonged periods of calm weather. Coastal stratification and hypoxia are more prevalent along the Gulf Coast, particularly near the Mississippi River discharge.

Prolonged stratification may alter the overall dynamics of the system and thus indirectly influence zooplankton assemblages. Thermoclines prevent nutrients regenerated in bottom waters from reaching the surface where they would support phytoplankton blooms. Reduced nutrient availability influences both the quantity of phytoplankton available and the phytoplankton species composition. At times, these changes lead to cascading effects that impact entire planktonic food web.

Cross-Shelf Transport to Nearshore and Estuarine Areas

Many nearshore fishes and invertebrates spawn offshore (Table 5). Their planktonic larvae then move to nearshore or estuarine nursery areas. How do these feeble swimmers traverse the often-considerable distances from spawning to nursery areas? We still don't know, but recent research suggests that various complex mechanisms may be involved.

Density-Driven Currents. Along the southeastern United States, winter cold snaps can chill the shallow nearshore waters. In theory, this denser cold water would flow slowly offshore, downslope along the bottom, and then be replaced by a shoreward flow of warmer shelf waters. These currents could be a primary mechanism for transporting larvae of the winter-spawning fishes, such as spot, croaker, pinfish, menhaden, and mullet, to inshore nursery grounds.

Subsurface Transport. Along the Mid-Atlantic Coast, the warm coastal boundary layer is buoyant and often flows over the colder, denser bottom waters from the open ocean. In fact, the surface and bottom waters can flow in different directions. As the surface layer moves offshore, it is replaced by a slow shoreward flow of denser bottom water. This mechanism can inject salty ocean water, including larvae, into the deeper estuarine basins.

Upwelling and Downwelling. Coastal **upwelling** and **downwelling** result from winds blowing up and down the coast, often aided by Ekman circulation (see "Alongshore Wind Stress (Gulf)" section). When winds move surface waters offshore, this water is replaced by deeper ocean water moving shoreward. Downwelling occurs when surface waters pile up along the coasts, sink, and then move offshore. Recent evidence from automated plankton pumps moored along the North Carolina coast suggests that both upwelling and downwelling can transport larvae shoreward across the shelf, depending on whether the animal is in surface or bottom water. On the Atlantic Coast, upwelling and downwelling events typically last only a few days or weeks because of a lack of sustained winds.

Internal Waves. Underwater waves form as tidal currents flow over rough bottom topography, such as the shelf break. Internal waves also flow along the density discontinuities created by stratification within the water column. Although internal waves cause little water displacement in deep water, surface slicks (convergence zones) associated with internal waves move shoreward and can transport neustonic larvae and other zooplankton. In shallow water, internal waves break and cause advection (net shoreward movement) similar to the advection caused by surface waves when they break in the surf zone. Breaking internal waves can thus bring larvae ashore. Internal waves are present along most coastlines and have been identified in the Mississippi Sound and Massachusetts Bay, but their significance in larval transport along the Atlantic and Gulf Coasts requires further study.

Alongshore Wind Stress (Gulf). Westerly winds along the northern Gulf coastline are deflected northward by Coriolis "force" and transmitted to deeper layers (Ekman circulation). This pushes Gulf water northward, piling it up along the coast. The resulting increase in water levels along the coastline forces water (and plankton) through inlets along the Gulf Coast. Winter storms at 3 to 10 day intervals and tropical storms in summer and fall may also propel plankton into nearshore areas along the Gulf.

Gulf Stream and Loop Current Eddies. Spin-offs from major offshore currents occur along both Atlantic and Gulf Coasts. They produce significant, episodic influxes of offshore plankton into nearshore waters. Eddy shedding by the Loop Current in the Gulf may be a significant factor in onshore movement of larvae, including blue crabs in the northern Gulf.

Tides and Tidal Currents

Tides, the alternating rise and fall of sea level produced by the combined gravitational attraction of the moon and sun, are quite variable within our geographic range. Atlantic tides are semidiurnal with two high and two low tides each day. The average tidal range is 1–2 m in most of our region, higher in New England and Georgia and lower in the Chesapeake Bay. The heights and timing of tides vary daily as the relative positions of earth, moon, and sun change. During full and new moon phases, the sun and moon are aligned, and their combined gravitational pull produces the maximum tidal amplitude with especially high and low tides; these are known as spring tides. Neap tides, with less than average tidal amplitude, occur when the sun and moon are not aligned and their gravitational forces are partially canceled. Tables of predicted tidal heights are available for most areas, but strong coastal winds and offshore storms may affect both the amplitude and timing of tides on any given day, sometimes dramatically. In much of the Gulf of Mexico, there is only one high and one low tide per day (diurnal tides), and these are 1 m or less. Some areas of the Gulf have very low amplitude gravitational tides, but wind-driven changes in water level can be significant. Narrow inlets and channels concentrate tidal flow to produce relatively strong currents, even in areas of moderate tides.

Knowledge of the times and heights of tides and the strength of tidal currents is important when planning zooplankton collections. Access to some areas may be limited to specific tides or current conditions. Depending on the type of collecting techniques employed, tidal flow can either facilitate or prevent successful collection. Furthermore, the catch may differ dramatically, depending on both current velocities and tidal stages. Because tides are frequently the dominant force influencing both physical and biological dynamics of the coastal and estuarine systems, be sure to note the stage of the tide whenever samples are collected.

Coastal Embayments

With the exception of the rocky coasts of New England, the Atlantic and Gulf Coasts of North America share a number of physical features. Along the coastline, the gentle slope of sandy beaches continues subtidally. Shifting barrier islands (Fig. 4) occur at intervals along much of the shoreline from Long Island through Texas, creating "bar-built" estuaries connected to the sea by narrow inlets, or passes. Where these islands lie close to shore, extensive systems of salt marshes, channels, and embayments lie between barrier islands and the mainland. These are typically high-salinity estuaries because they seldom have large freshwater input. Shallow, open waters enclosed by barrier islands are often called lagoons. Along the southern Texas coast, high evaporation, lack of summer rain, and limited seawater exchange produce hypersaline coastal lagoons that sometimes reach 150% of normal ocean salinity (e.g., Laguna Madre, TX). Where the barrier islands lie further offshore, the larger, shallow embayments are termed sounds. The most notable are the Mississippi Sound and the Albemarle and Pamlico Sounds that separate the Outer Banks of North Carolina from the mainland.

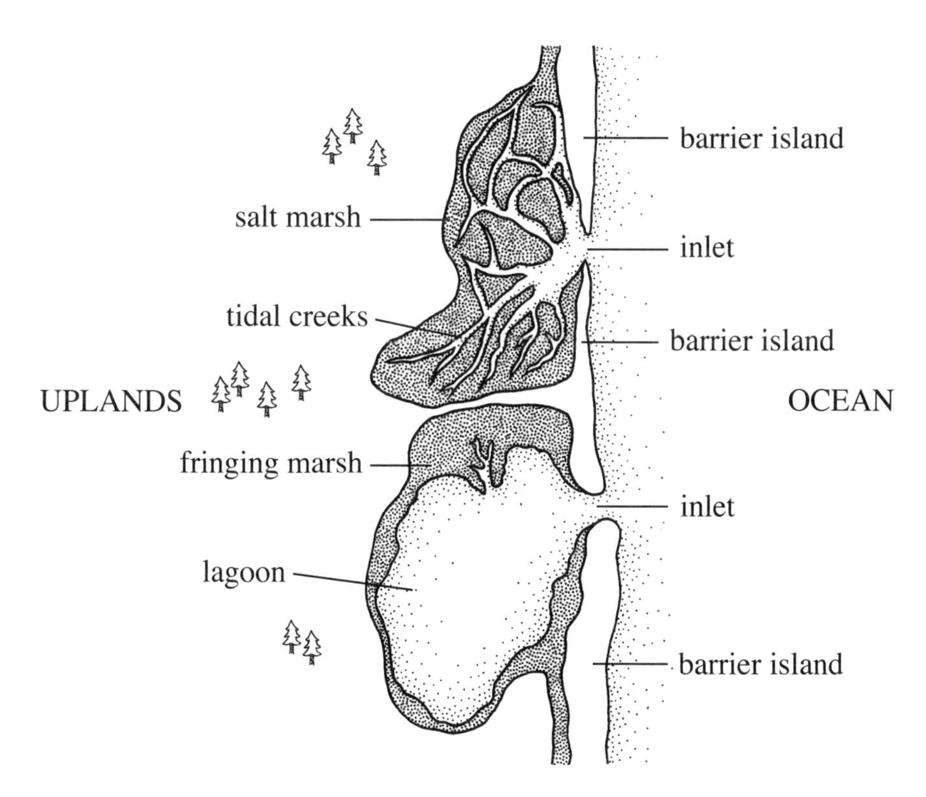

Fig. 4. Overhead views of two types of ocean-dominated barrier island–type estuaries. The upper inlet opens into an extensive salt marsh drained by a complex system of creeks. The lower lagoonal type has larger expanses of open water with a fringe of salt marsh around the margins.

The Atlantic Coast is punctuated by a series of large coastal plain estuaries with extensive reaches of brackish water. Chesapeake, Delaware, Winyah, and Mobile Bays and the Hudson-Raritan Estuary are premier examples.

Salinity and Estuarine Stratification

In contrast to the relatively uniform salinities of the coastal ocean, salinities in estuaries range from near zero in almost fresh headwaters to the full seawater concentration of 34–35 psu[1] near the coast. Salinity, more than any other factor, determines the diversity and distribution of zooplankton within estuaries. Both lowered salinity and the variable conditions within estuaries reduce species diversity. Many coastal zooplankton are **euhaline**, restricted to areas with salinities approaching that of seawater (30–35 psu) and occur only in the ocean or in mouths of estuaries. As salinity decreases, species diversity declines because fewer coastal zooplankton tolerate less saline waters. The relatively few species tolerant of intermediate salinities often have broad (**euryhaline**) distributions within estuaries. Some **brackish** water (0.5–30 psu) species tolerate very low (**oligohaline**) conditions, where they are joined by a few salt-tolerant freshwater zooplankton. Thus, brackish waters contain a mixture of typically brackish zooplankton plus some especially hardy marine and freshwater species with more restricted estuarine distributions. Figure 5 illustrates the "Venice" system for classifying brackish waters. These divisions are widely used by estuarine biologists, although few species adhere strictly to the ranges listed.

The spatial and temporal salinity variability within estuaries affects zooplankton distribution. Salinities are not static but change gradually over weeks or months because of seasonal or episodic rainfall. Dramatic salinity changes of up to 12 psu may occur in a single tidal cycle. Zooplankton sampled at any one location at different tidal stages may represent entirely different communities, which characterize different water masses as they move to and fro with the tides.

In some deeper estuaries such as the Chesapeake Bay and Long Island Sound, a periodic layering of water, called estuarine stratification, results from the combined effects of temperature and salinity differences in the water column. The sharp density boundary, or **pycnocline**, between warmer and fresher surface water and the cooler, saltier bottom water, restricts vertical mixing. The heavier water from the ocean tends to move into the estuary along the bottom, while the fresher river water "floats" overhead toward the sea. This two-layered circulation facilitates selective tidal transport via vertical positioning used by many estuarine zooplankton. Even in estuaries where stratification is common, it may not occur in all seasons or in all years, depending on the amount of freshwater input and on the strength of wind and tidal mixing. Some "moderately stratified" Chesapeake tributaries may become destratified because of strong spring tides and thus alternate between stratified and destratified conditions at biweekly intervals. As in coastal ocean areas, prolonged stratification causes low oxygen conditions in deeper bottom waters.

---

1. See **salinity** and **psu** in the glossary for details on salinity measurement and units.

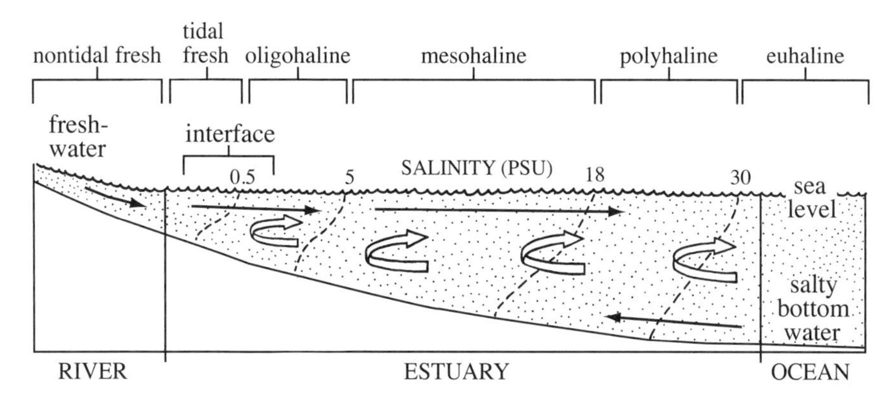

Fig. 5. A cross section of an estuary, showing the Venice salinity zones (oligo-, meso-, poly-, and euhaline) of a partially stratified estuary with a two-layered circulation. A net flow of saltier bottom water moves upstream beneath the outflowing fresher surface water. Moderate mixing between the two layers (open arrows) results in a gradual increase in salinity (dashed lines) toward the ocean with bottom waters slightly more salty.

Benthic Habitats and Zooplankton Distribution

Thus far, critical aspects of the physical factors of the water column and their effects on zooplankton have been discussed. However, the nature of the bottom, its fauna, and rooted aquatic vegetation all affect plankton composition in shallow waters.

Along vast areas of both Atlantic and Gulf Coasts, rooted aquatic vegetation forms two distinctive habitats: grass beds and salt marshes. Subtidal grass beds are typical of broad, protected shallow areas where sufficient sunlight reaches the bottom to support submerged aquatic vegetation (SAV). The once extensive eelgrass (*Zostera*) beds of the Chesapeake and many northern areas are now severely reduced, but remaining grassy areas still provide critical habitat throughout our region. In the Gulf, meadows of turtle grass (*Thalassia testudinum*) and shoal grass (*Halodule wrightii*) cover broad areas of the shallow coastal lagoons. Grass beds can influence zooplankton distribution with some zooplankton being 10-fold more abundant in grass beds than in adjacent open channels.

Salt marshes dominated by grasses (primarily *Spartina* spp.) cover protected low-lying coastal and estuarine areas that are only submerged at high tide. The channels and creeks that drain the marshes abound with planktonic larvae of marsh-dwelling invertebrates, as well as copepods and other holoplankton, taking advantage of these highly productive habitats. Tidal rhythms have an especially pronounced effect on availability of zooplankton tidal creeks, with greatly differing responses shown by different species. For example, fiddler crabs (*Uca* spp.) time the release of their larvae to coincide with the extra high tides of the full moon. These larvae are then swept into the tidal creeks and into open water. After a period of development in open water, the late stage *Uca* larvae (megalopae) return to the marsh, using high tides to carry them back onto the marsh surface where they metamorphose into juvenile crabs. Such tidally linked life cycles, varying in details among species, explain why zooplankton samples in marsh creeks are particularly variable with respect to lunar and tidal cycles.

With the exception of the rocky areas of southern New England, soft sediment dominates estuarine and coastal ocean bottoms. These are known as soft bottoms, and they can be composed of sand, mud, or any combination of the two. Most soft-bottom invertebrates are selective about the bottom texture, resulting in distinct benthic communities in sand bottoms compared with nearby communities in mud. These sediment preferences affect zooplankton assemblages in two ways. First, demersal or epibenthic zooplankton spend part of their time on the bottom and are more likely to be collected in waters overlying their preferred sediment type than elsewhere. Second, because many benthic invertebrates produce planktonic larvae, early larval stages may be more common near centers of adult populations.

Artificial hard substrates are widespread in nearshore areas from New England to Texas and support diverse communities of animals not found on soft bottoms. For example, oyster beds are a primary habitat for many fishes, especially gobies, and a host of invertebrates. Thus, areas with extensive oyster beds may yield a different suite of invertebrate larvae compared with areas that lack oysters. Rocky structures, such as breakwaters and jetties, provide additional hard surfaces and are usually densely covered with a diverse community quite unlike that in the surrounding sandy or muddy bottom. Animals requiring a hard substrate are not confined to using rocks. Pilings and floating docks provide additional hard substrate. A plankton net hung downstream from a large marina may yield larvae from invertebrates found only on these substrates.

ESTUARINE AND COASTAL PLANKTONIC FOOD WEBS

Estuarine and nearshore areas are among the most productive marine environments. By virtue of their abundance and biomass, zooplankton form a key link in transferring the primary production from photosynthesis to higher trophic levels. Nearshore food webs are complex and dynamic. Within a given area, the species composition and the feeding relationships within the water column show pronounced short-term, seasonal, and annual variations. This is an area of active research, and major modifications to the information presented next are likely.

Primary Productivity

Nearshore primary productivity originates from higher plants, seaweeds, and unicellular algae, especially phytoplankton. Extensive tidal wetlands are extremely productive; however, direct consumption of marsh grasses and other plants by terrestrial or aquatic herbivores is minimal. Each year, old stems and leaves left on the marsh surface are broken up and washed into the tidal creeks as **detritus**, small particles of dead and decomposing organic material. Detrital particles are usually covered with bacteria and fungi. These particles, including microbes, are food for many primarily suspension-feeding zooplankton. Algae growing on the marshes and shorelines contribute to primary production and add more detritus. Leaf litter and other wetland debris from upland vegetation provide yet more detritus. The amount of detritus produced and exported from inland areas is considerable. Beds of eelgrass (*Zostera*), turtle grass (*Thalassia*), and other submerged aquatic vegetation produce even more detritus. A variety of zooplankton feed on detritus, but whether detritus is an essential direct food source for zooplankton is unknown.

Phytoplankton provide the primary production in the water column, and phytoplankton cycles are tightly linked to zooplankton dynamics. Dinoflagellates and diatoms are the two most important groups of the larger nearshore phytoplankton, but various taxa of the smaller nano- and picoplankton often contribute more to overall productivity. When conditions are optimal, rapid cell divisions produce impressive blooms, with cell densities exceeding a million cells per milliliter. Each species of phytoplankton has its own optimal range of light, temperature, and nutrient conditions so that plankton blooms may be dominated by a succession of new species as conditions change. In shallow areas, benthic diatoms and other microflora are commonly resuspended and thus contribute an unknown and perhaps significant amount of primary productivity to planktonic food webs.

Planktonic Food Webs

The "Classic" Phytoplankton → Herbivore Food Web

It was thought that phytoplankton productivity directly leads to consumption by herbivorous zooplankton, primarily copepods, which then support carnivores. Although this "classic" view is only part of the story, it is a good place to begin. Planktonic food webs are dynamic and change seasonally in response to both temperature and nutrient conditions. The seasonal patterns of productivity and their effect on the food webs outlined next are common from southern New England to the Mid-Atlantic Coasts. Similar patterns occur widely in other regions, but the timing is variable, depending on local conditions.

Winter/Spring. The classic pattern of coastal phytoplankton dynamics begins with high nutrient conditions and increasing daylight in spring that trigger blooms of large diatoms, especially *Skeletonema costatum*. Herbivorous copepods graze preferentially on these diatoms. A time lag between the onset of the spring phytoplankton bloom and increases in copepods and other water column grazers allows a significant fraction of the total productivity to sink to the bottom or to be assimilated by water column decomposers. The huge, if belated, increase in copepods follows several weeks later. The dominant zooplanktivores of summer (ctenophores, jellyfishes, and anchovies) are largely absent in the spring, but larval fishes may partially fill this trophic void. Whether their feeding has an appreciable effect on copepod populations is uncertain. Researchers are trying to determine how much copepod productivity is passed on to higher trophic levels.

Summer. As summer arrives and nutrients decline, smaller diatoms, dinoflagellates, and microflagellates (nanoplankton and picoplankton, many <5 μm) often supplant the larger diatoms. This change benefits smaller grazers that can better use the smaller phytoplankton. Microzooplankton, including ciliates, rotifers, and copepod nauplii, remove a considerable fraction of these small phytoplankton. Smaller copepod species and benthic invertebrate larvae (especially bivalves) often peak at this time. In summer, the comb jelly

(*Mnemiopsis leidyi*), and the sea nettle (*Chrysaora quinquecirrha*), are dominant estuarine predators, consuming vast numbers of copepods and other small zooplankton. In high-salinity coastal areas, arrow worms, comb jellies, and the moon jelly (*Aurelia aurita*), are important zooplankton predators. Planktivorous fishes, including anchovies, silversides, and herrings, form a direct link between zooplanktonic productivity and larger fishes.

The Microbial Loop

One of the more important discoveries of recent decades is that a large fraction of the primary productivity may not go directly to herbivorous consumers as outlined in the "classic" pathway discussed earlier. Instead, detritus and dead phytoplankton (particulate organic carbon, or POC) and dissolved organic carbon (DOC) support a rich **heterotrophic** bacterial community with cell densities sometimes reaching 10 million per milliliter. In turn, these bacteria support a largely separate planktonic food web. Now widely termed the **microbial loop**, this trophic pathway consists of bacteria, ciliates, and heterotrophic flagellates. Even when the classic food web is dominant, a substantial fraction of the total energy flow in nearshore ecosystems is through the less obvious microbial loop. Recently, we have identified some important cross-links between the microbial loop and the classic food web. For example, copepods feed on large ciliates that feed on bactivorous flagellates.

Food Web Dynamics

Each geographic region and each specific body of water has its own particular physical features, including size, depth, tidal range, and freshwater input, resulting in different patterns of species dominance and different trophic connections. Furthermore, our present ideas about nearshore food webs will surely change as new research fills in some missing links, especially concerning the role of planktonic larvae. Nevertheless, the basic features of the estuarine food web shown in Figure 6 probably apply to most Atlantic and Gulf estuarine areas, although details will vary. The cycle begins with terrestrial nutrient input fueling phytoplankton productivity. This supports dual microbial and herbivorous food webs connected by rotifers (especially in low to mid salinities) and by the copepod *Acartia tonsa*. The ctenophore *Mnemiopsis leidyi* is often a dominant copepod grazer in the nearshore zooplankton, and it can substantially reduce copepod and meroplankton densities. Thus, overall plankton dynamics are influenced by **bottom-up** control dependent on nutrient availability and by **top-down** control from predators.

As Figure 6 shows, planktonic food webs are connected to benthic, nektonic, and terrestrial systems. Runoff from the land contains dissolved nutrients, detritus, and sediment that determine much of the character of both estuarine and coastal waters, especially primary productivity. Planktonic production, in turn, affects the dynamics of the water column and the benthos. Because grazers do not consume all productivity, the excess often sinks to the lower part of the water column or to the bottom, where it is an important food source for the benthic consumers and decomposers. Zooplanktivorous fishes, especially the bay

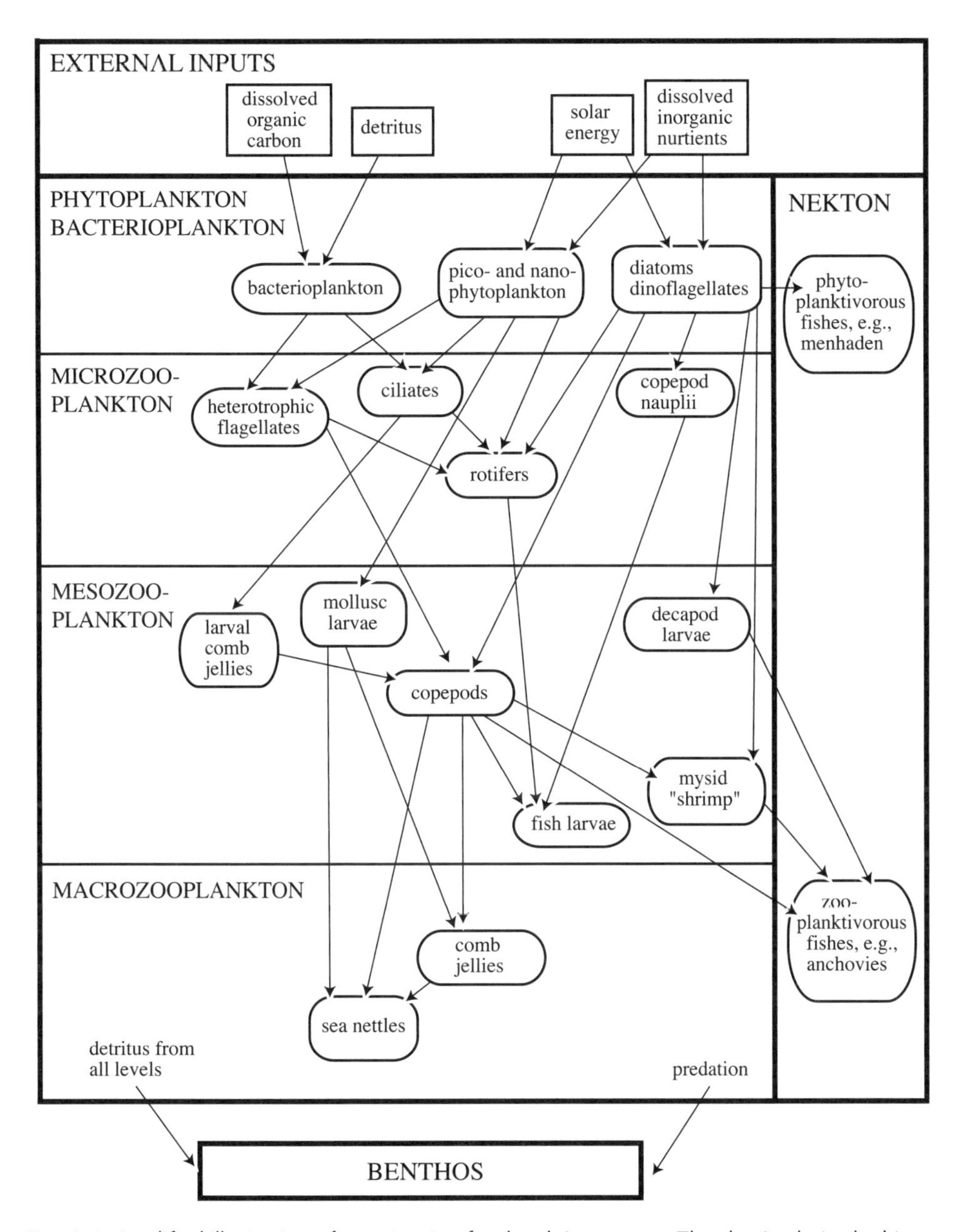

Fig. 6. A simplified illustration of an estuarine food web in summer. The classic phytoplankton-based connections begin at the upper right with the microbial loop on the left. Not all connections are illustrated. (Based on data from the Chesapeake Bay in Baird and Ulanowicz 1989).

anchovy, provide a trophic link to many important nearshore recreational and commercial fishes and to birds such as terns, ospreys, bald eagles, herons, and kingfishers.

Although Figure 6 appears complex, it still does not convey the dynamic nature of the system, integrated over time and space. A pulse of freshwater input or even a temporary thermocline can have an immediate effect. Phytoplankton and microzooplankton numbers can change significantly in just a few days, depending on local conditions. Likewise, copepod communities can respond with shifts in species dominance in just a few weeks. Even more dramatic are the sudden pulses of planktonic fish or invertebrate larvae as individual species spawn or migrate. Currents and eddies produce a small-scale patchiness so that water masses just 10–100 m apart can differ significantly in physical, chemical, and biological properties. Our understanding of how these communities work is changing rapidly as new connections between the physical, biological, and temporal components of estuarine and nearshore ecosystems are made. The complexity and dynamic nature of planktonic systems makes plankton research exciting—and challenging.

A BRIEF HISTORY OF ZOOPLANKTON RESEARCH

Pioneers in Zooplankton Research[2]

Antony van Leeuwenhoek, using his newly developed microscope, observed oyster veligers as early as 1722. Another Dutchman, Martinus Slabber, produced some of the first published illustrations of invertebrate larvae in 1778. The use of nets to collect and concentrate plankton during the early nineteenth century, pioneered by J. V. Thompson and Johannes Müller in the mid-1800s, ushered in a new era of discovery. Müller said, "I have used this method for many years with the best results . . . It is quite indispensable, and in no way to be replaced" (Haeckel 1893, 567). This statement is still valid, and the basic plankton net is still the most widely used plankton-collection device, although the silk gauze of the early nets is now nylon. Haeckel documented the many new and exciting lines of inquiry that ensued:

> In the beginning of the second half of our century [nineteenth century], the astonishing wealth of interesting and instructive forms of life which the surface of the sea offers to the naturalist first became known, and that long series of important discoveries began which in the last forty years have filled so many volumes of our rapidly increasing zoölogical literature. A new and inexhaustibly rich field was thus opened to zoötomical and microscopical investigation, and anatomy and physiology, organology and systematic zoölogy have been advanced to a surprising degree. The investigation of the lower animals has since been recognized as a wide field of work, whose exploration is of great significance for all branches of science. (Haeckel 1893, 567)

2. This section is a compilation drawn in part from Ernst Haeckel's (1893) first-person account of the early years of zooplanktology, supplemented by C. M. Young's (1990) review of the history of larval ecology.

Johannes Müller's systematic collections in the Baltic Sea were instrumental in founding the discipline of plankton research. He illustrated many new life-forms, including the first larval echinoderms, ribbon worms, acorn worms, flatworms, polychaetes, and phoronids, collected from plankton. In early studies, these strange organisms were not recognized as larvae and were assigned to new phyla. By the end of the nineteenth century, most of these larval stages were linked to adults and assigned to the proper phylum. Still, the geographic, or three-dimensional, extent of plankton distribution was not appreciated until the monumental global explorations of the British *Challenger* expedition (1873–76) led by Sir Charles Wyville Thompson and Dr. John Murray. The HMS *Challenger* towed nets to depths below 3,000 m and made the then astonishing discovery of a rich plankton fauna in the deep sea. This finding caused widespread consternation among leading scientists of the day, including Alexander Agassiz, who believed that the deep sea was devoid of life. The latter part of the nineteenth century also provided the basic vocabulary of plankton research. Victor Hensen's first use of the term *plankton* was rapidly adopted and extended by Haeckel (1893, 583): "Numerous organisms pass their whole life and whole cycle of development hovering in the oceans, while with others, this is not the case. These, rather, pass a part of their life in the benthos . . . This first group we call holoplanktonic and the second meroplanktonic."

Sir Alister Hardy's invention of the Continuous Plankton Recorder (CPR) proved to be one of the most notable developments in plankton research of the early twentieth century. This relatively simple device is towed at a depth of 10 m and collects plankton on moving bands of silk mesh. Hardy first deployed his invention while on the *Discovery* expedition (1925–27) to the Antarctic. He then arranged to have the recorders towed by merchant vessels as they traversed the oceans. Routine collections began in 1932 in the North Sea and then spread to routes across the entire North Atlantic. These collections continue today under the auspices of the Sir Alister Hardy Foundation for Ocean Science (SAFOS). This long-term dataset of more than 200,000 samples is unparalleled in both scope and duration. More than 400 scientific publications are based on these collections.

In the United States, early plankton research began with the establishment of the US Commission of Fish and Fisheries Laboratory in 1871 and the Marine Biological Laboratory (MBL) in 1887, both at Woods Hole, Massachusetts. Because virtually nothing was known about the local fauna, initial work focused on collection and classification. William M. Wheeler published the first list of copepods from the area in 1899, with descriptions of 30 species. Charles B. Wilson expanded on Wheeler's first copepod accounts with two monographs in 1932 that contained descriptions of more than 300 copepod species from Woods Hole and the Chesapeake Bay. Other notable early planktologists based in Woods Hole from 1912 through the 1930s were Charles J. Fish and Henry B. Bigelow. In contrast, zooplankton of the southeastern United States and Gulf Coasts remained relatively unexplored until early contributions by Joseph King (1950), Robert A. Woodmansee (1958), and George D. Grice (1956, 1960), primarily from Florida. The founding of many new marine labs in the southeastern United States and the Gulf of Mexico was instrumental in the recent upsurge in zooplankton research and publications along these coasts.

While collections continue to add to our knowledge of fauna, research in the latter part of the twentieth century expanded into new areas. During the 1960s and 1970s, John D. Costlow Jr., C. G. Bookhout, and Austin B. Williams raised many Mid-Atlantic crustacean larvae in the laboratory, providing definitive life history information for the first time. Fish's publications in the 1930s followed by Gordon A. Riley and Georgiana B. Deevey in the 1940s and 1950s initiated work on planktonic food webs that led to studies on productivity, energy flow, and trophic efficiency. More recently, mathematical models of food webs and energy flow provide new insights that link structure and function within planktonic communities and more accurately capture the dynamic nature of planktonic systems. Much new research is devoted to understanding the structure and dynamics of deep-sea plankton. Meanwhile modern techniques provide increasingly detailed analyses of behavioral adaptations peculiar to life in the plankton.

As the twenty-first century begins, the impacts of DNA technology are increasing on many fronts. Heretofore, the only way to link planktonic larvae with a particular species was to have the adults spawn in the laboratory and then raise their larvae through successive stages that were carefully drawn and described. This process is difficult and painstaking. As a result, the larvae of many species are unknown, and, conversely, many planktonic larvae cannot be identified to species. DNA sequence databases, accessible to the public, promise relatively rapid identification of field-caught larvae. Shipboard facilities for PCR (polymerase chain reaction) are now being used to identify pelagic fish larvae. In addition to species identification, DNA sequences can distinguish genetic differences within species to facilitate studies of larval dispersal and genetic connectivity. Last, advances in molecular biology and cladistics have evolved into useful tools for unraveling phylogenetic relationships of some enigmatic planktonic groups as well as revelations regarding evolutionary relationships at all levels resulting in a flurry of reclassifications and name changes.

Collection Techniques: Old and New

Over the years, many techniques have been used to collect zooplankton, but the plankton net is by far the most widely used for mesozooplankton. More recently developed techniques with specific applications include plankton traps, plankton pumps, and video surveys.

Plankton Nets

A plankton net is still the easiest and least expensive way to collect plankton. In its simplest form, a plankton net is nothing more than a cone made of fine mesh towed through the water. A rigid frame keeps the mouth open, and there is usually a jar or container at the small end where the plankton are retained (Fig. 7). Nets 20–50 cm across with a mesh of about 150–200 μm are used for many routine nearshore collections and will catch most taxa. Smaller mesh sizes retain smaller plankton, but their increased resistance to the water flow through the meshes means that they have small diameters and must be towed slowly. Many larger and more mobile zooplankton can detect pressure waves from approaching

A. Simple Plankton Net

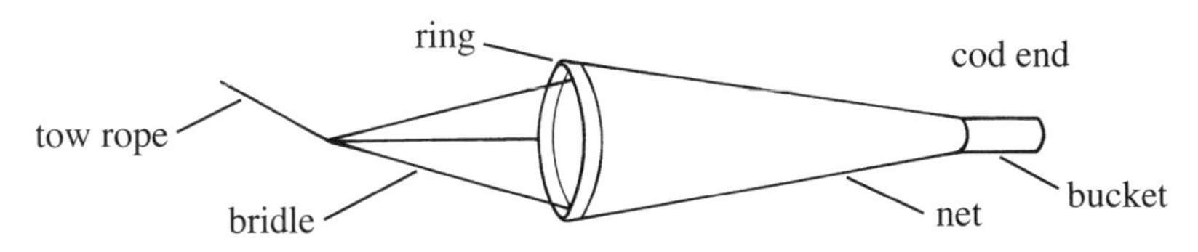

B. Vertically Suspended Closing Net

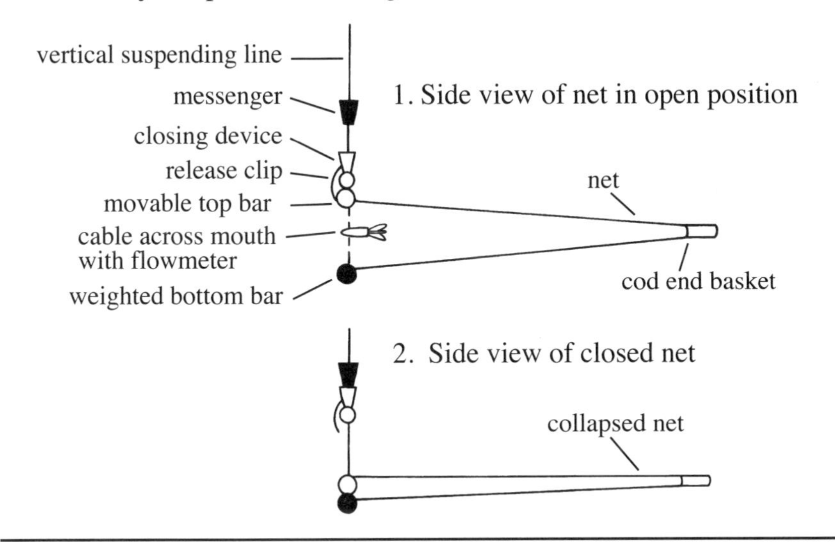

C. Epibenthic Sled

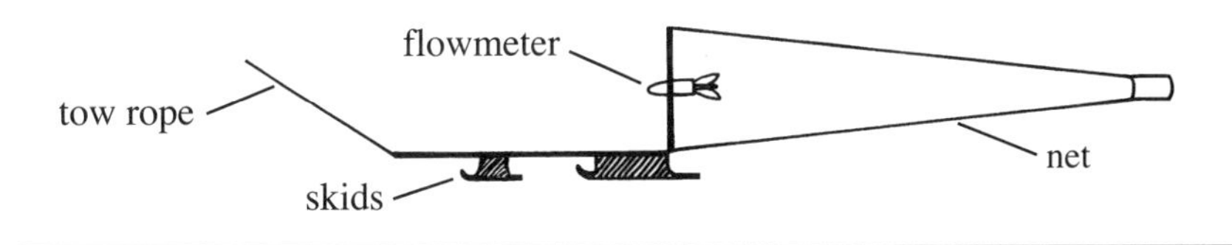

Fig. 7. Three types of plankton sampling nets. *A*, The simple plankton net is usually towed by boat or by hand. *B*, The vertically suspended closing net (side view) is usually hung from an anchored boat to sample discrete depths by taking advantage of tidal currents. A spring-loaded closing device collapses the mouth of the net. *C*, The epibenthic sled is a metal frame device towed to collect zooplankton close to the bottom. All of these net devices can be fitted with flowmeters to estimate the volume of water filtered in a tow.

nets and often avoid capture by simply swimming away. Large nets, up to 1 m in diameter, with mesh openings of 0.5–1.0 mm may be towed rapidly to reduce this "net avoidance." However, large meshes allow small plankton to pass through. In addition, large nets are expensive and difficult to handle. Because all plankton nets are selective, each net must be carefully matched to its intended use. Appendix 1 discusses the bias and application of different nets for specific applications.

A startling finding of the *Challenger* expedition was the abundance of life below 1,000 m. These results were immediately challenged by eminent scientists of the day, who said that the animals caught were merely the result of contamination of the sample with plankton from surface waters. This led to one of the first major advances in plankton net technology—a way to control nets so that they would open and then close at a specific depth and thus bring back a sample from a discrete depth interval. The first crude "opening-closing" nets soon proved the existence of deep-sea plankton, ended the controversy about life in the deep, and paved the way for continued studies of deep-sea plankton. This was the first in an ongoing series of modifications and improvements in plankton nets and trawls for specific applications.

The epibenthic sled is a plankton net fitted to a sled-like contraption that slides along the bottom (Fig. 7). The sled samples within a few centimeters of the substrate, collecting many species seldom caught using standard plankton nets—some present in surprising abundance in nearshore and estuarine waters.

Plankton Pumps, Traps, and New Technologies

Plankton pumps use a hose attached to a pump to deliver a water sample to the surface, where it is then strained through a mesh screen or plankton net. Surprisingly, many taxa pass through plankton pumps unscathed. The main advantage of pumps is that they can take samples from discrete positions in the water column or close to structures such as reefs and drilling platforms. However, they are poorly suited to deep sampling because of the long length of unwieldy hose and because the resistance of pumping water through a hose increases with its length. Zooplankton larger than 2 mm can usually avoid pump intakes.

Three very different types of traps provide specialized, selective information on plankton distributions. (1) Freshwater planktologists have used volumetric plankton traps in lakes for decades. The traps snap shut to enclose a known volume of water that is then brought to the surface. Although researchers in the Baltic Sea have had success using a 26-L trap, American marine biologists have seldom used this approach. (2) Emergence traps are specifically designed to catch demersal plankton—animals that leave the benthos for periodic sojourns into the plankton. These traps, often simple funnels, are placed on the bottom to catch plankton as they return to the substrate. (3) Light traps use artificial illumination to lure and capture many species of plankton drawn to lights at night.

Automated Plankton Samplers, Acoustic Technology, and Video Techniques

Three new techniques are still in development but are already providing unique data, unavailable from other means. (1) On the Atlantic Coasts off North Carolina and New Jersey, automated plankton samplers moored to fixed stations several kilometers from shore are using pumps to collect plankton from discrete depths at preselected intervals. The aim is to correlate distribution and recruitment with currents and hydrologic events in nearshore waters. (2) Acoustic technology (sonar) has been refined to determine the location and density of zooplankton in oceanic and estuarine systems. An acoustic Doppler current profiler (ADCP) can be towed from a boat to record spatial distribution or mounted in a fixed location to monitor zooplankton changes over time. Acoustic techniques provide useful general inferences about the density of zooplankton that produce acoustic signals, but the technology is not refined enough to replace more traditional collection techniques for quantifying individual zooplankton groups. (3) Towed video cameras can record zooplankton passing through a narrow field. Computer software programmed to "recognize," or to identify, particles based on size, shape, and density counts the zooplankton. These image analysis systems cannot distinguish between similar taxa and have difficulty identifying asymmetrical species. Video recordings and digital image analysis software are not yet able to supplant collecting with nets, especially in turbid coastal waters. Reliable species identification remains elusive, but video can provide useful estimates of biomass and abundance. In addition, video provides unique glimpses of behavior of larger zooplankton. Recordings of fragile species and deep-sea fauna from manned submersibles and unmanned ROVs (remotely operated vehicles) are especially valuable.

ZOOPLANKTON AND ENVIRONMENTAL QUALITY

Most coastal and estuarine areas from Cape Cod to Texas are environmentally challenged by nutrient and pollutant loading and by other changes caused by human alterations. We would be remiss not to mention some of the major environmental challenges and their relationships to zooplankton.

Hypoxia

Water fully saturated with dissolved oxygen (DO) contains 6–14 mg of oxygen per liter (mg L^{-1}), depending on temperature and salinity. One milligram per liter represents 1 ppm (part per million). In areas where productivity greatly exceeds its utilization by water column or benthic macroconsumers, microbial decomposition increases. Increased decomposition may result in lowered oxygen levels in the water column, especially in bottom waters in locations where the water column is stratified (Fig. 8). Areas with seriously depleted oxygen levels are termed **hypoxic** if oxygen drops below 2–3 ppm and **anoxic** when no oxygen is present. Evidence from bottom sediments shows that hypoxia in the waters of our region is not new. However, the extent, frequency, duration, and severity of low-oxygen conditions have increased greatly since 1960. A National Oceanic and Atmo-

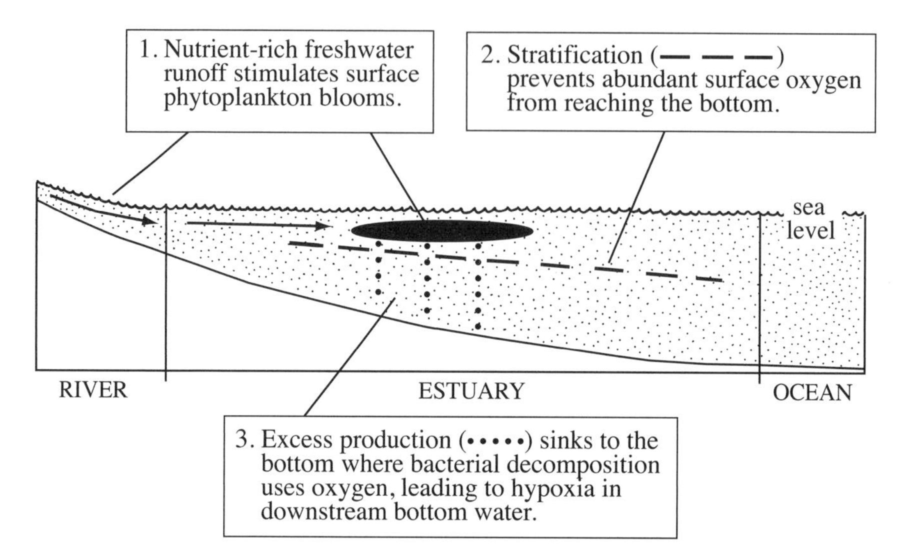

Fig. 8. Schematic model of the relationship between excess nutrients (eutrophication), water column stratification, and estuarine hypoxia. Both coastal and estuarine areas are affected by low oxygen resulting from stratification, typically in summer.

spheric Administration (NOAA) survey from 1997 indicates that 58 out of 81 major estuaries shown in Figure 9 from New England through the Gulf exhibited summer hypoxia and a large number were anoxic. An extensive 2008 survey of 562 estuarine and coastal locations along Atlantic and Gulf seaboards found 268 (48%) had documented hypoxia events during some seasons and years.

Hypoxia is not restricted to estuaries. In coastal areas, periods of calm, hot weather, combined with diluted surface water, can cause stratification in areas where tidal and other currents are not strong enough to turn over the water column. If these conditions persist for several weeks, oxygen levels drop along the bottom, a phenomenon that sometimes occurs offshore of the Hudson River and along the New Jersey coast. In most areas, coastal stratification affects small areas and is of short duration. In contrast, the "Dead Zone" in the Gulf of Mexico may cover an area the size of New Jersey in some summers.

Documented instances of hypoxia were rare before 1960. The 1970s marked a dramatic increase in hypoxia that has continued through the 1990s, where the incidence of hypoxic events increased more than 30-fold. Despite efforts at nutrient control, the incidence of hypoxia remains high in nearshore and estuarine waters, especially from the Mid-Atlantic through the Gulf of Mexico. Why are problems with hypoxia increasing? Most available evidence points to excess productivity fueled by increased nutrient runoff (eutrophication). Fertilizer, farm animal waste, and sewage effluent all contain high levels of nitrogen and phosphorus. Increasing human population puts additional stress on the system because of more agricultural output, more sewage, and more runoff from overfertilized lawns. Efforts to reduce nutrients from sewage treatment plants and agricultural runoff have resulted in

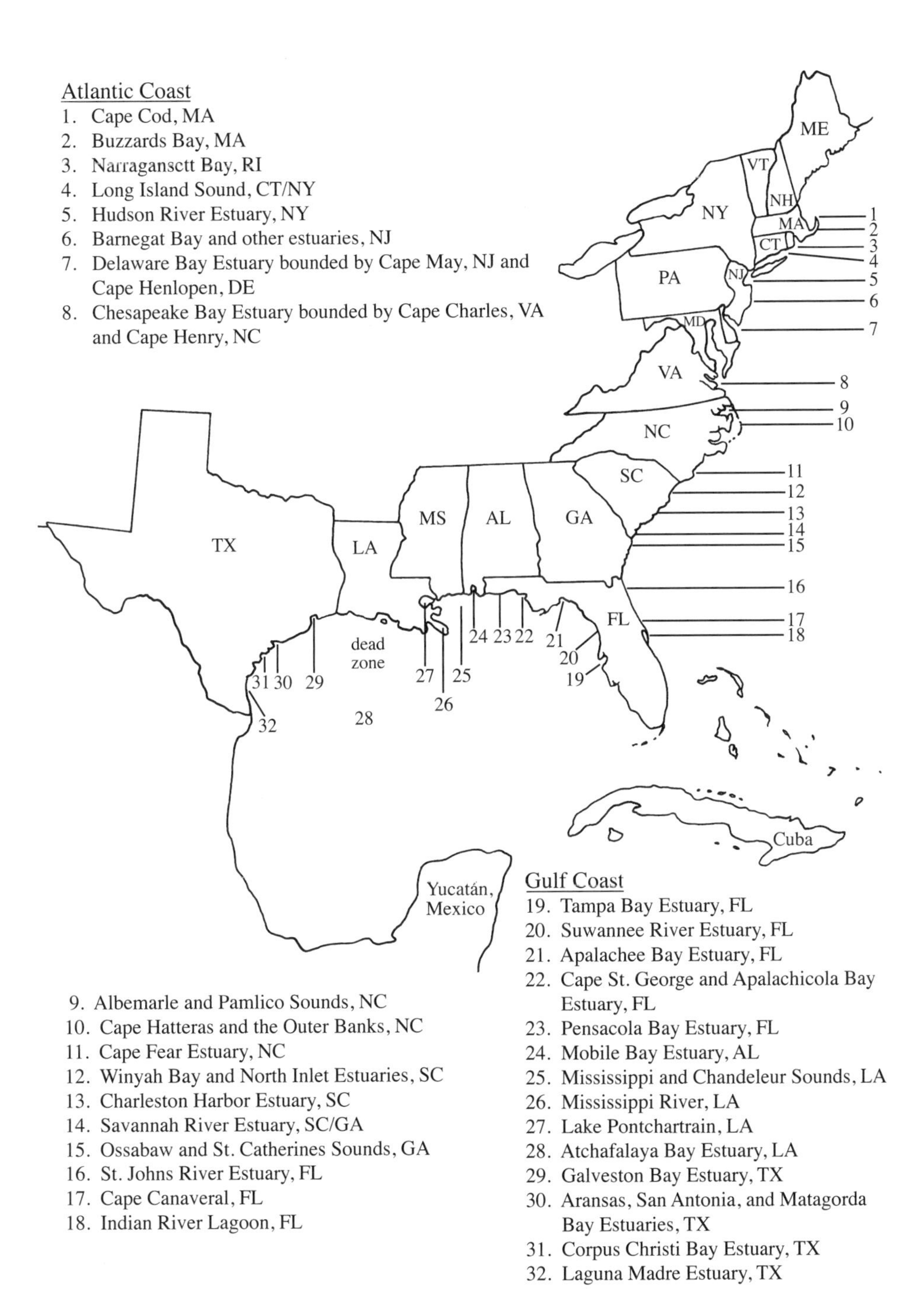

Fig. 9. Many of the coastal and estuarine areas shown experience periodic severe hypoxia or anoxia, usually in summer. The "dead zone" of low oxygen associated with the Mississippi plume in the Gulf of Mexico appears south of Louisiana in most summers but varies in location and size from year to year.

improved water and habitat quality in Tampa and Sarasota Bays and parts of the Hudson-Raritan Estuary.

The effect of low oxygen on zooplankton dynamics is a subject of active investigation. Some studies are focused on how individual species respond to low oxygen and others on how low oxygen affects planktonic food web structure. Some zooplankton avoid low-oxygen conditions by migrating to surface waters, but doing so could expose them to a different suite of predators. Meroplanktonic larvae must eventually settle to the bottom, and recruitment of benthic invertebrates in low-oxygen conditions is a major concern. The complexity and variability of zooplankton dynamics makes it difficult to predict long-term responses to reduced oxygen.

Harmful Algal Blooms

Massive blooms of phytoplankton, such as red tides and brown tides, are a widespread phenomena. Some blooms produce toxins capable of killing fish and other organisms. When these toxins are concentrated in fish or shellfish, they threaten humans who eat them. In other cases where no toxins are produced, massive blooms result in lowered oxygen levels when the dead algal cells decompose.

Red tides are perhaps the most well-known toxic, or hazardous, phytoplankton blooms and have been documented since biblical times. Red tides result from explosive blooms of highly pigmented phytoplankton where numbers often exceed 1 million cells per milliliter of water, causing a visible discoloration of the water. Often red tides cause no problems; not all reddish algae are toxic. Other harmful algal blooms (HABs) have come to our attention only recently. Among the most notable are the nontoxic **brown tides** produced by persistent and massive blooms of tiny flagellated (4–5 μm) phytoplankton in coastal systems, most impressively in Long Island Sound and Laguna Madre, Texas. The blooms around Long Island are episodic but seem to inhibit shellfish growth and may lower oxygen along the bottom. One recent Laguna Madre bloom lasted seven years (1990–96) without interruption. Mysterious fish kills in the Neuse River in North Carolina and in the Chesapeake Bay during the 1990s have been attributed to the dinoflagellate *Pfiesteria piscicida*. Recent research suggests that these strange organisms can produce a powerful neurotoxin that kills fish and causes neurological problems for humans. Concern over the recent increase in the number and the extent of HABs is widespread, and increased research into the causes of harmful algal blooms is under way.

Invasive and Introduced Species

Nonnative species may be either intentionally or accidentally introduced into marine systems. Once established, these exotic, or "alien," species may have pronounced and unpredictable effects in their newly adopted habitat. Introduced species may displace native species, bring diseases or parasites, or affect other species through predation. Ballast water from ships, which is acquired in home ports and discharged elsewhere, is a common vehicle for long-distance dispersal and introduction into new habitats. The most serious example is the introduction of our comb jelly (*Mnemiopsis leidyi*) into the Black Sea in 1982 and thence to the Sea of Azov and to the Caspian and Mediterranean Seas by 1990. Its unrestrained growth so altered the Black Sea planktonic food web that local fisheries collapsed. A separate invasion of *Mnemiopsis* resulted in its spreading to the Baltic and North Seas in 2006. Closer to home, recent outbreaks of the Australian spotted jellyfish (*Phyllorhiza punctata*) along the northern Gulf of Mexico shoreline have caused concern because these jellyfishes show a voracious appetite for plankton, especially larval fishes. Along the Atlantic Coast, the green shore crab (*Carcinus maenas*) was introduced from Europe in the early 1800s. More recently, the Asian shore crab (*Hemigrapsus sanguineus*) has spread north to Maine and south to North Carolina since its discovery on the New Jersey coast in 1988. The introduction of these two crabs has not had a major effect on either benthic or planktonic communities, but they do illustrate the establishment of introduced species along our coasts. In contrast, the large predatory rapa whelk (*Rapana venosa*), a recently discovered arrival from the Sea of Japan, threatens commercial bivalves in the Chesapeake Bay.

Last, we note the potential for introductions of alien species through aquaculture. Repeated escapes of cultivated shrimps and fishes threaten native species. With exotic species come exotic diseases. These, too, can endanger native fauna. Shrimp culture using alien species (Pacific white shrimp and Asian tiger prawns) from South Carolina to Texas and *Tilapia* fish farms in Florida are currently the largest coastal producers of nonnative species. Stocks of the native Chesapeake Bay oyster (*Crassostrea virginica*) are at record lows, prompting both Virginia and Maryland to consider introducing the Asian Suminoe oyster (*Crassostrea ariakensis*) into the Chesapeake. The potential risks and benefits of this introduction have sparked a lively controversy and revealed the complexity of issues involved. Economic issues aside, the potential for large populations of suspension feeders to remove excess sediment and phytoplankton must be weighed against the concerns typically associated with alien introductions. Major changes in the Potomac River ecosystem near Washington, DC, due to another introduced bivalve (*Corbicula*) suggest authorities should proceed cautiously before introducing alien species.

Not all "invasions" involve alien species. Many "introductions" of new species into an area result from range extensions, which are often attributed to changes in environmental conditions (some associated with climate change) that allow planktonic larvae to survive in areas that were previously unacceptable or because of physiological changes in environmental tolerances of some species. A good example is the spread of the porcelain crab (*Petrolisthes armatus*) northward hundreds of miles in just the past few decades from southern Florida to South Carolina.

Climate Change

Evidence of long-term increases in temperature and carbon dioxide in the atmosphere as well as changes in physical, chemical, and biological characteristics of estuaries and oceans is widespread and well documented, although questions remain about the extent, rate, and causes of climate change. Zooplankton are especially good sentinels of climate change because they cannot regulate their own temperature, have relatively short life cycles, and largely drift with currents and tides. Ocean temperatures are at the highest levels in about 12,000 years, and they are within 1°C of the highest level in the last 1 million years. The rate of change has increased over the past 50 years, and the average temperature of the coastal ocean on the East Coast of the United States has increased by about 1°C over that period. Although this does not appear to be a large change, it has been sufficient to affect water column mixing, nutrient availability, phytoplankton productivity, and zooplankton assemblages in many areas. Shifts in latitudinal distributions northward by hundreds to thousands of kilometers have been documented in the Northeast Atlantic, changing size, density, and abundance patterns of copepods and affecting higher trophic levels as well. Changes in zooplankton assemblages can have substantial consequences for ecosystem structure, and in extreme cases, precipitate regime shifts in which food webs change from one stable state to another.

Changing temperature affects growth rates, fecundity, and survival, especially when increases approach the physiological tolerances of the species. In temperate areas, even slight changes in temperature can affect zooplankton life cycles and the timing of biological events, known as phenology. Many reports of phenological changes, including the timing of migrations, reproduction, and peak abundances, appear in the scientific literature. Some zooplankton appear to be more responsive to changing climate than others, which results in changes in the structure of seasonal assemblages. One outcome is a mismatch between the timing of occurrence of organisms and the food on which they rely. For example, entire year classes of larval fishes may fail if their specific zooplankton prey is absent during the short window when it is required. Another consequence of changing temperature and environmental conditions is the possibility of providing introduced species with an advantage in becoming established and potentially displacing native species which may be stressed by changing conditions.

Both coastal and estuarine hypoxic events are most prevalent during the warm seasons. This is due, in part, to more rapid decomposition rates at higher temperatures. Thus, increasing water temperatures would increase the possibility of anoxia and hypoxia in deep estuarine and ocean areas. This situation would be further facilitated by earlier and more widespread thermal stratification, which would prevent mixing of oxygen-rich surface waters from reaching the bottom. Increased stratification events can also affect the timing and extent of water column turnover necessary for initiating phytoplankton blooms. In more northern areas, increased melting of ice could change water temperature and salinity patterns and alter ocean currents.

The influence of major cyclic climate events such as the El Niño Southern Oscillation (ENSO) and North Atlantic Oscillation (NAO) on zooplankton in the Atlantic, Gulf, and their estuaries is becoming better understood. In the upper reaches of our coverage area, NAO is known to change temperature, which regulates copepod and jellyfish composition and abundances. In the Southeast, El Niño events during winter and spring bring more rainfall. Increases in river discharge and nutrients can stimulate primary productivity and, subsequently, zooplankton densities. In Southeast estuaries, El Niño–related reductions in salinity have been shown to reduce densities of coastal ocean taxa and to alter the abundance and timing of larval production by resident invertebrates and fishes. Climate models predict that future warming will likely alter these global scale phenomena.

Ocean acidification resulting from increased atmospheric carbon dioxide concentrations has become an increasing concern because even slight alterations in pH inhibit biological uptake of calcium carbonate. Decreases in the pH of the ocean, where variations are typically slight, have been observed over recent decades. The threat to reef-building corals has been widely publicized, but shell-forming organisms in the ocean plankton are also at risk. Experiments show that organisms such as pteropods are vulnerable to minor decreases in pH. Foraminifera might also be particularly vulnerable, which might impact food webs over large areas of the open ocean. In estuaries, where much larger variations in pH occur over tidal and seasonal cycles, effects on zooplankton might not be as acute, but larval gastropod and bivalve molluscs, ostracods, and other groups are potentially at risk.

Changes in sea level can also affect zooplankton assemblages over longer periods in estuaries as the amount of inhabitable area decreases for intertidal populations of adults such as fiddler crabs (*Uca* spp.), killifishes (*Fundulus* spp.), and grass shrimps (*Palaemonetes* spp.). Changes in the proportions of these and other abundant larvae can affect food web dynamics of estuarine areas, especially those with large salt marsh borders. Should sea level rise cause a reduction in beaches, supplies of many other larvae could decline, including the sand and mole crabs (*Emerita* and *Albunea*).

Although concerns and predictions outlined earlier note some of the most serious issues related to rising CO_2 levels and global climate change, the overall impact and consequences of climate change on oceanic and estuarine zooplankton are difficult to predict with assurance. Time series studies that have continued for several or many decades continue to provide insights into changing zooplankton assemblages and the physical drivers of change in their environment. The only way to assess changes in biological assemblages is through long-term measurements as conditions change. Only more efforts devoted to systematic and long-term programs of integrated physical and biological collections will increase understanding and predictions.

SUGGESTED READINGS

Biology of Zooplankton: Classic Overviews

Fraser, J. 1962. *Nature Adrift: The Story of Marine Plankton*. G. T. Foulis, London. 178 pp.

Hardy, A. 1970. *The Open Sea: Its Natural History*. Pt. 1, *The World of Plankton*. Houghton Mifflin, Boston. 322 pp.

Raymont, J. E. G. 1983. *Plankton and Productivity in the Oceans*. 2nd ed. Vol. 2, *Zooplankton*. Pergamon Press, London. 824 pp.

Sieburth, J. M., Smetacek, V., Lenz, J. 1978. Pelagic ecosystem structure: Heterotrophic compartments of the plankton and their relationship to plankton size fractions. *Limnology and Oceanography* 23:1256–1263.

Adaptations of Zooplankton (Feeding, Migration, Defenses)

Bollens, S. M., Frost, B. W. 1991. Diel vertical migration in zooplankton: Rapid individual response to predation. *Journal of Plankton Research* 13:555–564.

Bullard, S. G., Hay, M. E. 2002. Palatability of marine macro-holoplankton: Nematocysts, nutritional quality, and chemistry as defense against predators. *Limnology and Oceanography* 47:1456–1467.

Bullard, S. G., Lindquist, N. L., Hay, M. E. 1999. Susceptibility of invertebrate larvae to predators: How common are post-capture larval defenses? *Marine Ecology Progress Series* 191:153–161.

Bushek, D., Allen, D. M. 2005. Motile suspension-feeders in estuarine and marine ecosystems. In: Dame, R. F., Olenin, S., eds. *The Comparative Roles of Suspension-Feeders in Ecosystems*. Springer, The Netherlands, 53–71.

Chia, F. S., Buckland-Nicks, J., Young, C. M. 1984. Locomotion of marine invertebrate larvae: A review. *Canadian Journal of Zoology* 62:1205–1222.

Cohen, J. H., Forward, R. B., Jr. 2009. Zooplankton diel vertical migration—a review of proximate control. *Oceanography and Marine Biology: An Annual Review* 47:77–110.

Hart, M. W., Strathmann, R. R. 1995. Mechanisms and rates of suspension feeding. In: McEdward, L., ed. *Ecology of Marine Invertebrate Larvae*. CRC Press, New York, 193–221.

Kerfoot, W. C., Kellogg, D. L., Jr., Strickler, J. R. 1980. Visual observations of live zooplankters: Evasion, escape, and chemical defenses. In: Kerfoot, W. C., ed. *Evolution and Ecology of Zooplankton Communities*. University Press of New England, Hanover, NH, 10–27.

Kiørboe, T. 2008. *A Mechanistic Approach to Plankton Ecology*. Princeton University Press, Princeton, NJ. 209 pp.

LaBarbera, M. L. 1984. Feeding currents and particle capture mechanisms in suspension feeding animals. *American Zoologist* 24:71–84.

Landry, M. R., Fagerness, V. L. 1988. Behavioral and morphological influences on predatory interactions among marine copepods. *Bulletin of Marine Science* 43:509–529.

McEdward, L., ed. 1995. *Ecology of Marine Invertebrate Larvae*. CRC Press, New York. 464 pp.

Ohman, M. D. 1988. Behavioral responses to zooplankton to predation. *Bulletin of Marine Science* 43:530–550.

Strathmann, R. R. 1987. Larval feeding. In: Giese, A. C., Pearse, J. S., Pearse, V. B., eds. *Reproduction in Marine Invertebrates*. Blackwell Scientific, Palo Alto, CA, 9:465–550.

Visser, A. W., Mariani, P., Pigolotti, S. 2009. Swimming in turbulence: Zooplankton fitness in terms of foraging efficiency and predation risk. *Journal of Plankton Research* 31:121–133.

Larval Ecology

Able, K. W., Fahay, M. P. 1998. *The First Year in the Life of Estuarine Fishes in the Middle Atlantic Bight*. Rutgers University Press, New Brunswick, NJ. 342 pp.

Butman, C. A. 1987. Larval settlement of soft-sediment invertebrates: The spatial scales of pattern explained by active habitat selection and the emerging role of hydrodynamical processes. *Oceanography and Marine Biology Annual Review* 25:113–165.

Chia, F.-S., Rice, M. E., eds. 1978. *Settlement and Metamorphosis of Marine Invertebrate Larvae*. Elsevier, New York. 290 pp.

Dame, R. F., Allen, D. M. 1996. Between estuaries and the sea. *Journal of Experimental Marine Biology and Ecology* 200:169–185. (Depicts seven different invertebrate and fish larval recruitment patterns involving inshore/offshore transport.)

Epifanio, C. E., Garvine, R. W. 2001. Larval transport on the Atlantic continental shelf of North America: A review. *Estuarine and Coastal Shelf Science* 52:51–77.

Garstang, W. 1985. *Larval Forms and Other Zoological Verses*. University of Chicago Press, Chicago. 98 pp. (In 1929, Garstang was among the first to suggest that some of the curious features of larval forms were specific adaptations for life in the plankton. Here, these ideas are expressed in humorous rhyme.)

Hadfield, M. G. 1998. The D. P. Wilson Lecture: Research on settlement and metamorphosis of marine invertebrate larvae; Past, present and future. *Biofouling* 12:9–29.

McEdward, L., ed. 1995. *Ecology of Marine Invertebrate Larvae*. CRC Press, New York. 464 pp. (Specific articles here deal with feeding and behavior.)

Mileikovsky, S. A. 1971. Types of larval development in marine bottom invertebrates, their distribution and ecological significance: A re-evaluation. *Marine Biology* 10:193–213.

Morse, A. N. C. 1991. How do planktonic larvae know where to settle? *American Scientist* 79:154–167.

Norcross, B. L., Shaw, R. F. 1984. Oceanic and estuarine transport of fish eggs and larvae: A review. *Transactions of the American Fisheries Society* 113:153–165.

Pawlik, J. R. 1992. Chemical ecology of settlement of benthic marine invertebrates. *Oceanography and Marine Biology: An Annual Review* 30:273–335.

Pechenik, J. A. 1999. On the advantages and disadvantages of larval stages in benthic marine invertebrate life cycles. *Marine Ecology Progress Series* 177:269–297.

Queiroga, H., J. Blanton. 2004. Interactions between behavior and physical forcing in the control of horizontal transport of decapod larvae. *Advances in Marine Biology* 47:107–214.

Rodriguez, S. R., Ojeda, F. P., Inestrosa, N. C. 1993. Settlement of marine benthic invertebrates. *Marine Ecology Progress Series* 97:193–207.

Sastry, A. N. 1983. Pelagic larval ecology and development. In: Vernberg, F. J., Vernberg, W. B., eds. *Biology of Crustacea*. Vol. 7, *Behavior and Ecology*. Academic Press, New York, 214–282.

Shanks, A. L. 2009. Pelagic larval duration and dispersal distance revisited. *Biological Bulletin* 216:373–385.

Tamburri, M. N., Finelli, C. M., Wethey, D. S., et al. 1996. Chemical induction of larval settlement behavior in flow. *Biological Bulletin* 191:367–373.

Thorson, G. 1950. Reproduction and larval ecology of marine bottom invertebrates. *Biological Reviews* 25:1–45. (A classic.)

Young, C. M., ed. Rice, M. E., Sewell, M., assoc. eds. 2001. *Atlas of Marine Invertebrate Larvae*. Academic Press, New York. 646 pp. (Comprehensive treatment of larval morphology and metamorphosis.)

Young, C. M., Chia, F.-S. 1987. Abundance and distribution of pelagic larvae as influenced by predation, behavior, and hydrographic factors. In: Giese, A. C., Pearse, J. S., Pearse, V. B., eds. *Reproduction of Marine Invertebrates*. Blackwell Scientific, Palo Alto, CA, 9:385–463.

Marine Zoogeography

Briggs, J. C. 1974. *Marine Zoogeography*. McGraw-Hill, New York. 475 pp.

Dana, J. D. 1853b. On an isothermal oceanic chart, illustrating the geographical distribution of marine animals. *American Journal of Science and Arts*, 2nd ser., 16:153–167, 314–327. (Early ideas on zoological provinces.)

Sewell, R. B. S. 1940. The extent to which the distribution of marine organisms can be explained by and is dependent on the hydrographic conditions present in the great oceans, with special reference to the plankton. Pt. 3. *Proceedings of the Linnean Society Session* 152(3), 256–286.

Physical Oceanography and Its Effects on Zooplankton

Able, K. W., Fahay, M. P. 1998. *The First Year in the Life of Estuarine Fishes in the Middle Atlantic Bight*. Rutgers University Press, New Brunswick, NJ. 400 pp. (Hydrography of the Middle Atlantic Bight.)

Bowman, M. J., Iverson, R. L. 1978. Estuarine and plume fronts. In: Bowman, M. J., Esaias, W. E., eds. *Oceanic Fronts in Oceanic Processes*. Springer-Verlag, New York, 87–105.

Garland, E. D., Zimmer, C. A., Lentz, S. J. 2002. Larval distributions in inner-shelf waters: The roles of wind-driven cross-shelf currents and diel vertical migrations. *Limnology and Oceanography* 47:803–817.

Grimes, C. B., Finucane, J. H. 1991. Spatial distribution and abundance of larval and juvenile fish, chlorophyll and macrozooplankton around the Mississippi River discharge plume, and the role of the plume in fish recruitment. *Marine Ecology Progress Series* 75:109–119.

Hare, J. A., Thorrold, S., Walsh, H., et al. 2005. Biophysical mechanisms of larval fish ingress into Chesapeake Bay. *Marine Ecology Progress Series* 303:295–310.

Møller, J. S. 1996. Water masses, stratification and circulation. In: Jorgensen, B. B., Richardson, K., eds. *Coastal and Estuarine Studies: Eutrophication in Coastal Marine Ecosystems*. American Geophysical Union, Washington, DC, 51–66.

Scotti, A., Pineda, J. 2004. Observation of very large and steep internal waves of elevation near the Massachusetts coast. *Geophysical Research Letters* 31:L22307. doi:10.1029/2004GL021052.

Shanks, A. L. 1995. Mechanisms of cross-shelf dispersal of larval invertebrates and fish. In: McEdward, L., ed. *Ecology of Marine Invertebrate Larvae*. CRC Press, New York, 323–367.

Shanks, A. L., Largier, J., Brink, L., et al. 2000. Demonstration of the onshore transport of larval invertebrates by the shoreward movement of an upwelling front. *Limnology and Oceanography* 45:230–236.

Sverdrup, H. U., Johnson, M. W., Fleming, R. H. 1942. *The Oceans: Their Physics, Chemistry, and General Biology*. Prentice Hall, Englewood Cliffs, NJ. 1100 pp. (A classic, and still one of the most comprehensive treatments on physical, chemical, and biological oceanography.)

Inshore Areas: Estuaries, Bays, Sounds, Lagoons, and Salt Marshes

Britton, J. C., Morton, B. 1989. *The Shore Ecology of the Gulf of Mexico*. University of Texas Press, Austin. 387 pp.

Bulger, A. J., Hayden, B. P., Monaco, M. E., et al. 1993. Biologically-based estuarine salinity zones derived from a multivariate analysis. *Estuaries* 16:311–322. (Presents an alternative to the Venice salinity classification.)

Buskey, E. J. 1993. Annual cycle of micro- and mesozooplankton abundance and biomass in a subtropical estuary. *Journal of Plankton Research* 15:907–924. (Describes the Nueces Estuary, Texas, a hypersaline coastal lagoon.)

Carlson, D. M. 1978. The ecological role of zooplankton in a Long Island salt marsh. *Estuaries* 1:85–92.

Dame, R., Alber, M., Allen, D., et al. 2000. Estuaries of the South Atlantic Coast of North America: Their geographical signatures. *Estuaries* 23:793–819. (Includes information on human impacts on southeastern estuaries.)

Dame, R. F., Allen, D. M. 1996. Between estuaries and the sea. *Journal of Experimental Marine Biology and Ecology* 200:169–185. (Includes a discussion of estuarine circulation and its role in larval transport.)

Epperly, S. P., Ross, S. W. 1986. Characterization of the North Carolina Pamlico–Albemarle estuarine complex. National Oceanographic and Atmospherics Administration Technical Memorandum NMFS-SEFC1-175.

Hoese, H. D., Moore, R. H. 1998. *Fishes of the Gulf of Mexico, Texas, Louisiana, and Adjacent Waters*. Texas A&M Press, College Station. 422 pp.

Houser, D. S., Allen, D. M. 1996. Zooplankton dynamics in an intertidal salt-marsh basin. *Estuaries* 19:659–673.

Kennish, M. J. 1986. *Ecology of Estuaries: I. Physical and Chemical Aspects.* CRC Press, Boca Raton, FL. 254 pp.

Lippson, A. J., Haire, M. S., Holland, A. F., et al. 1979. *Environmental Atlas of the Potomac Estuary.* Martin Marietta Corporation for Maryland Power Plant Siting Program, Annapolis, MD. 279 pp. (A thorough treatment of estuarine hydrography and the biology of phytoplankton and zooplankton in a typical mid-Atlantic estuary. Available from Maryland Department of Natural Resources, Annapolis.)

Mallin, M. A., Burkholder, J. M., Cahoon, L. B., et al. 2000. North and South Carolina coasts. *Marine Pollution Bulletin* 41(1–6):56–75.

Mann, K. H., Lazier, J. R. N. 1991. *Dynamics of Marine Ecosystems.* Blackwell, Boston. 466 pp.

Paul, R. W. 2001. Geographical signatures of middle Atlantic estuaries: Historical layers. *Estuaries* 24:151–166.

Roman, C. T., Jaworski, N., Short, F. T., et al. 2000. Estuaries of the northeastern United States: Habitat and land use signatures. *Estuaries* 23:743–764.

Roman, M. R., Holliday, D. V., Sanford, L. P. 2001. Temporal and spatial patterns of zooplankton in the Chesapeake Bay turbidity maximum. *Marine Ecology Progress Series* 213:215–227.

Turner, R. E. 2001. Of manatees, mangroves, and the Mississippi River: Is there an estuarine signature for the Gulf of Mexico? *Estuaries* 24:139–150.

Planktonic Food Webs

Allen, D. M., Johnson, W. S., Ogburn-Matthews, V. 1995. Trophic relationships and seasonal utilization of salt-marsh creeks by zooplanktivorous fishes. *Environmental Biology of Fishes* 42:37–50.

Baird, D., Ulanowicz, R. E. 1989. The seasonal dynamics of the Chesapeake Bay ecosystem. *Ecological Monographs* 59:329–364.

Calbet, A., Saiz, E. 2005. The ciliate-copepod link in marine ecosystems. *Aquatic Microbial Ecology* 38:157–167.

Coats, D. W., Revelante, N. 1999. Distributions and trophic implications of microzooplankton. In: Malone, T. C., Malej, A., Harding, J., Smodlaka, N., Turner, R. E., eds. *Ecosystems at the Land-Sea Margin.* American Geophysical Union, Washington, DC, 207–240.

Dolan, J. R., Gallegos, C. L. 1991. Trophic coupling of rotifers, microflagellates, and bacteria during the fall months in the Rhode River Estuary. *Marine Ecology Progress Series* 77:147–156.

Ducklow, H. W., Purdie, D. A., Williams, P. J. LeB. 1986. Bacterioplankton: A sink for carbon in a coastal marine plankton community. *Science* 232:865–867.

Fenchel, T. 1988. Marine plankton food chains. *Annual Review of Ecology and Systematics* 19:19–38.

Fisher, T. R., Hagy, J. D., Rochelle-Newall, E. 1998. Dissolved and particulate organic carbon in Chesapeake Bay. *Estuaries* 21:215–229.

Kiørboe, T. 1998. Population regulation and role of mesozooplankton in shaping marine pelagic food webs. *Hydrobiologia* 363:13–27.

Legendre, L., Rassoulzadegan, F. 1995. Plankton and nutrient dynamics in marine waters. *Ophelia* 41:153–172. (Includes treatment of coastal systems.)

Lonsdale, D. J., Cosper, E. M., Kim, W.-S., et al. 1996. Food web interactions in the plankton on Long Island bays, with preliminary observations on brown tide effects. *Marine Ecology Progress Series* 134:242–263. (A review of estuarine and coastal food webs, including ciliates and micrometazoa.)

Mallin, M. A., Paerl, H. W. 1994. Planktonic trophic transfer in an estuary-seasonal, diel, and community structure effects. *Ecology* 75:2168–2184.

Monaco, M. E., Ulanowicz, R. E. 1997. Comparative ecosystem trophic structure of three U.S. mid-Atlantic estuaries. *Marine Ecology Progress Series* 161:239–254. (Comparisons of Narragansett, Delaware, and Chesapeake Bays.)

Montagnes, D. J. S., Allen, J., Brown, L., et al. 2010. Role of ciliates and other microzooplankton in the Irminger Sea (NW Atlantic Ocean). *Marine Ecology Progress Series* 411:101–115.

Sanders, R. W., Wickham, S. A. 1993. Planktonic protozoa and metazoa: Predation, food quality and population control. *Marine Microbial Food Webs* 7:197–223.

Sherr, E. B., Sherr, B. F. 1994. Bactivory and herbivory: Key roles of phagotrophic protists in pelagic food webs. *Microbial Ecology* 28:223–235.

Stibor, H., Vadstein, O., Diehl, S., et al. 2004. Copepods act as a switch between alternative trophic cascades in marine pelagic food webs. *Ecological Letters* 7:321–328.

Turner, J. T., Bruno, S. F., Larson, R. J., et al. 1983. Seasonality of plankton assemblages in a temperate estuary. *P.S.Z.N.I Marine Ecology* 4:81–99. (Note links between physical factors, phytoplankton, and zooplankton cycles.)

White, J. R., Roman, M. R. 1992. Seasonal study of grazing by metazoan zooplankton in the mesohaline Chesapeake Bay. *Marine Ecology Progress Series* 86:251–261.

History of Zooplankton Research

Batten, S. D., Burkill, P. H. 2010. The continuous plankton recorder: Towards a global perspective. *Journal of Plankton Research* 32:1619–1621.

Fraser, J. H. 1968. The history of plankton sampling. In: UNESCO, ed. *Zooplankton sampling*. UNESCO, Paris, 11–18.

Gislason, A., Silva, T. 2009. Comparison between automated analysis of zooplankton using ZooImage and traditional methodology. *Journal of Plankton Research* 31:1505–1516.

Gorsky, G., Ohmn, M. D., Picheral, M., et al. 2010. Digital zooplankton image analysis using the ZooScan integrated system. *Journal of Plankton Research* 32:285–303.

Haeckel, E. 1893. *Plankton Studies: Report of the US Fisheries Commission, 1889–1891*. Government Printing Office, Washington, DC, 565–641. (Translated by G. W. Field from Professor Haeckel's "Plankton Studien" that first appeared in 1890 in *Jenaische Zeitschrift*, vol. 25.)

Herman, A. W., Beanlands, B., Phillips, E. F. 2004. The next generation of optical plankton counter: The laser-OPC. *Journal of Plankton Research* 26:1135–1145.

Kiesling, T. L., Wilkinson, E., Rabalais, J., et al. 2002. Rapid identification of adult and naupliar stages of copepods using DNA hybridization methodology. *Marine Biotechnology* 4:30–39.

Mills, E. L. 1989. *Biological Oceanography. An Early History, 1870–1960*. Cornell University Press, Ithaca, NY. 378 pp.

Reid, P. C., Colebrook, J. B. L., Matthews, J. M. 2003. The continuous plankton recorder: Concepts and history, from plankton indicator to undulating recorders. *Progress in Oceanography* 58:117–175.

Remsen, A., Hopkins, T. L., Samson, S. 2004. What you see is not what you catch: A comparison of concurrently collected net, optical plankton counter, and shadowed image particle profiling evaluation; Recorder data from the northeast Gulf of Mexico. *Deep Sea Research Part I: Oceanographic Research Papers* 51:129–151.

Young, C. M. 1990. Larval ecology of marine invertebrates: A sesquicentennial history. *Ophelia* 32:1–48.

***Collection Techniques* (also see Appendix 1)**

Doherty, K. W., Butman, C. A. 1990. A time- or event- triggered automated, serial plankton pump sampler. In: Frye, D., Stone, E., Martin, A., eds. *Advanced Engineering Laboratory Project Summaries*. Technical Report 90–20. Woods Hole Oceanographic Institution, Woods Hole, MA.

Omori, M., Ikeda, T. 1984. *Methods in Marine Zooplankton Ecology*. Wiley Interscience / Wiley & Sons, New York. 332 pp. (A comprehensive classic and still useful.)

Sameoto, D., Wiebe, P., Runge, J., et al. 2000. Collecting zooplankton. In: Harris, R., Wiebe, P., Lenz, et al., eds. *ICES Zooplankton Methodology Manual*. Academic Press, New York, 55–81.

Tranter, D. J., Fraser, J. H. 1968. *Zooplankton Sampling*. Monographs on Oceanographic Methodology 2. UNESCO Press, Paris. 146 pp. (A classic and still useful.)

Environmental Considerations

Eutrophication and Hypoxia

Bricker, S., Longstaff, B., Dennison, W., et al. 2007. *Effects of Nutrient Enrichment in the Nation's Estuaries: A Decade of Change*. National Estuarine Eutrophication Assessment Update, NOAA Coastal Ocean Program Decision Analysis Series No. 26. National Centers for Coastal Ocean Science, Silver Spring, MD. 328 pp. Available online: <http://ccma.nos.noaa.gov/publications/eutroupdate/> and <http://ccma.nos.noaa.gov/news/feature/Eutroupdate.html>. Both accessed August 21, 2011.

Clement, C., Bricker, S. B., Pirhalla, D. E. 2001. Eutrophic conditions in estuarine waters. In: *NOAA's State of the Coast Report*. National Oceanic and Atmospheric Administration, Silver Spring, MD. <http://oceanservice.noaa.gov/websites/retiredsites/supp_sotc_retired.html>. Accessed August 21, 2011. (This site provides a summary of the data in the NOAA Estuarine Eutrophication survey, including many color photographs, illustrations, and data figures.)

Cloern, J. E. 2001. Our evolving conceptual model of the coastal eutrophication problem. *Marine Ecology Progress Series* 210:223–253.

Committee on Environment and Natural Resources. 2010. *Scientific Assessment of Hypoxia in U.S. Coastal Waters*. H. Interagency Working Group on Harmful Algal Blooms, and Human Health of the Joint Subcommittee on Ocean Science and Technology, Washington, DC. 164 pp. <www.whitehouse.gov/administration/eop/ostp/nstc/oceans>. Accessed August 21, 2011.

Diaz, R. J., Rosenberg, R. 2008. Spreading dead zones and consequences for marine ecosystems. *Science* 321:926–929.

Estuaries. 2002. Vol. 25, no. 4B. (Contains 18 symposium contributions devoted to the causes, consequences, and potential solutions to coastal overenrichment.)

Jørgensen, B. B., Richardson, K. 1996. *Coastal and Estuarine Studies: Eutrophication in Coastal Marine Ecosystems*. American Geophysical Union, Washington, DC. 272 pp.

Kolesar, S. E., Breitburg, D. L., Purcell, J. E., et al. 2010. Effects of hypoxia on *Mnemiopsis leidyi*, ichthyoplankton and copepods: Clearance rates and vertical habitat overlap. *Marine Ecology Progress Series* 411:173–188.

Lewitus, A. J., Kidwell, D. M., Jewett, E. B., et al., eds. 2009. Ecological impacts of hypoxia on living resources. *Journal of Experimental Marine Biology and Ecology* 381(suppl.) 1:S1–S216.

Livingston, R. J. 2001. *Eutrophication Processes in Coastal Systems: Origin and Succession of Plankton Blooms and Effects on Secondary Production in Gulf Coast Estuaries*. CRC Press, New York. 327 pp. (Coverage is focused on the Florida Panhandle.)

Park, G. S., Marshall, H. G. 2000. Estuarine relationships between zooplankton community structure and trophic gradients. *Journal of Plankton Research* 22:121–136. (Links estuarine eutrophication to major shifts in planktonic food webs.)

Rabalais, N. N., Turner, R. E. 2001. Coastal hypoxia: Consequences for living resources and ecosystems. *Coastal and Estuarine Studies* 58. American Geophysical Union, Washington, DC. 460 pp.

Rabalais, N. N., Turner, R. E., Díaz, R. J., et al. 2009. Global change and eutrophication of coastal waters. *ICES Journal of Marine Science* 66:1528–1537.

Rabalais, N. N., Turner, R. E., Wiesman, W. J., Jr. 2002. Hypoxia in the Gulf of Mexico, a.k.a. "The Dead Zone." *Annual Review of Ecology and Systematics* 33:235–263.

Smith, D. E., Leffler, M., Makiernan, G., eds. 1992. *Oxygen Dynamics in the Chesapeake Bay: A Synthesis of Recent Research.* Maryland Sea Grant, College Park, MD. 234 pp.

Stalder, L. C., Marcus, N. H. 1997. Zooplankton responses to hypoxia: Behavioral patterns of three species of calanoid copepods. *Marine Biology* 127:599–607.

Harmful Algal Blooms (HABs)

Anderson, D. M., Glibert, P. M., Burkholder, J. M. 2002. Harmful algal blooms and eutrophication: Nutrient sources, composition, and consequences. *Estuaries* 25:704–725.

Bricelj, V. M., Lonsdale, D. J. 1997. *Aureococcus anophagefferens:* Causes and ecological consequences of brown tides in U.S. mid-Atlantic coastal waters. *Limnology and Oceanography* 42:1023–1038.

Burkholder, J. M. 1998. Implications of harmful microalgae and heterotrophic dinoflagellates in management of sustainable marine fisheries. *Ecological Applications* 8:S37–S62.

Burkholder, J. M., Glasgow, H. B., Jr., Hobbs, C. W. 1995. Fish kills linked to a toxic ambush-predator dinoflagellate: Distribution and environmental conditions. *Marine Ecology Progress Series* 124:43–61.

Buskey, E. J., Liu, H., Collumb, C., Bersano, J. E. F. 2001. The decline and recovery of a persistent Texas brown tide algal bloom in the Laguna Madre (Texas, USA). *Estuaries* 24:337–346.

Glibert, P. M., Magnien, R., Lomas, M. W., et al. 2001. Harmful algal blooms in the Chesapeake and coastal bays of Maryland, USA: Comparison of 1997, 1998, and 1999 events. *Estuaries* 24:875–883.

Hall, N. S., Litaker, W., Fensin, E., et al. 2008. Environmental factors contributing to the development and demise of a toxic dinoflagellate (*Karlodinium veneficum*) bloom in a shallow, eutrophic, lagoonal estuary. *Estuaries and Coasts* 31:402–418.

Landsberg, J. H. 2002. The effects of harmful algal blooms on aquatic organisms. *Reviews in Fisheries Science* 10:113–390.

Lassus, P., Arzul, G., Erard-Le Denn, E., et al., eds. 1995. Harmful marine algal blooms. In: *Proceedings of the Sixth International Conference on Toxic Marine Phytoplankton.* Lavoisier, Paris. 878 pp.

Lonsdale, D. J., Cosper, E. M., Kim, W. S., et al. 1996. Food web interactions in the plankton on Long Island bays, with preliminary observations on brown tide effects. *Marine Ecology Progress Series* 134:247–263.

Richardson, K. 1997. Harmful or exceptional phytoplankton blooms in the marine ecosystem. *Advances in Marine Biology* 31:301–385.

Smayda, T. J. 1997. Harmful algal blooms: Their ecophysiology and general relevance to phytoplankton blooms in the sea. *Limnology and Oceanography* 42:1137–1153.

Turner, J. T., Tester, P. A. 1997. Toxic marine phytoplankton, zooplankton grazers and pelagic food webs: A review. *Limnology and Oceanography* 42:1203–1214.

Toxic Substances and Effects of Sedimentation

Kennish, M. J. 1992. *Ecology of Estuaries: Anthropogenic Effects.* CRC Press, Boca Raton, FL. 494 pp. (Includes chapters on nutrient loading, contaminants, and dredging.)

Levinton, J. S. 2001. *Marine Biology.* 2nd ed. Oxford University Press, New York. 515 pp. (See the section "Toxic Substances," pp. 473–491, for a general introduction.)

Vernberg, F. J., Vernberg, W. B. 2001. *The Coastal Zone: Past, Present, and Future*. University of South Carolina Press, Columbia. 191 pp.

Invasive and Introduced Species

Bax, N., Carleton, J. T., Mathews-Amos, A., et al. 2001. The control of biological invasions in the world's oceans. *Conservation Biology* 15:1234–1246.

Carlton, J. T., Geller, J. B. 1992. Ecological roulette: The global transport of nonindigenous marine organism. *Science* 261:78–82.

Costello, J. H., Bayha, K. M., Mianzan, H. W., et al. 2012. The ctenophore *Mnemiopsis leidyi*—transitions from a native to an exotic species. *Hydrobiologia*. Published online: doi: 10.1007/s10750-012-1037-9.

Mills, C. E., Sommer, F. 1995. Invertebrate introductions in marine habitats: 2 species of hydromedusae (Cnidaria) native to the Black Sea, *Maeotias inexspectata* and *Blackfordia virginica,* invade San Francisco Bay. *Marine Biology* 122:279–288.

Naylor, R. L., Williams, S. L., Strong, D. R. 2001. Aquaculture—a gateway for exotic species. *Science* 294:1655–1656.

Purcell, J. E., Uye, S., Lo, W. T. 2007. Anthropogenic causes of jellyfish blooms and their direct consequences for humans: A review. *Marine Ecology Progress Series* 350:153–174.

Ruiz, G. M., Carlton, J. T., Grosholz, E. D., et al. 1997. Global invasions of marine and estuarine habitats by non-indigenous species: Mechanisms, extent, and consequences. *American Zoologist* 37:6211–6632.

Stone, R. 2002. Caspian ecology teeters on the brink. *Science* 295:430–433. (Documents the effects of the invasion by the comb jelly, *Mnemiopsis*.)

Williams, R. J., Griffiths, F. B., Van der Wal, E. J. 1988. Cargo vessel ballast water as a vector for the transport of non-indigenous marine species. *Estuarine, Coastal and Shelf Science* 26:409–420.

Climate Change

Allen, D. M., Ogburn-Matthews, V., Buck, T., et al. 2008. Mesozooplankton responses to climate change and variability in a southeastern U.S. estuary (1981–2003). Special issue, *Journal of Coastal Research* 55:95–110.

Balch, W. M., Fabry, V. J. 2008. Ocean acidification: Documenting its impact on calcifying phytoplankton at basin scales. *Marine Ecology Progress Series* 373:239–247.

Beaugrand, G., Luczak, C., Edwards, M. 2009. Rapid biogeographical plankton shifts in the North Atlantic Ocean. *Global Change Biology* 15:1790–1803.

Costello, J. H., Sullivan, B. K., Gifford, D. J. 2006. A physical-biological interaction underlying variable phenological responses to climate change by coastal zooplankton. *Journal of Plankton Research* 28:1099–1105.

Dijkstra, J. A., Westerman, E. L., Harris, L. G. 2011. The effects of climate change on species composition, succession, and phenology: A case study. *Global Change Biology* 17:2360–2369.

Doney, S. C., Fabry, J. F., Feely, R. A., et al. 2009. Ocean acidification: The other problem. *Annual Review of Marine Science* 1:169–192.

Edwards, M., Beaugrand, G., Johns, D. G., et al. 2010. *Ecological Status Report: Results from the CPR Survey 2009*. SAHFOS Technical Report, 7:1–8. Plymouth, UK.

Edwards, M., Richardson, A. J. 2004. Impact of climate change on marine pelagic phenology and trophic mismatch. *Nature (London)* 430:881–884.

Hays, G. C., Richardson, A. J., Robinson, C. 2005. Climate change and marine plankton. *Trends in Ecology and Evolution* 20:337–344.

Hoegh-Guldberg, O., Bruno, J. F. 2010. The impact of climate change on the world's marine ecosystems. *Science* 328:1523–1528. (Predicts that the size shift in phytoplankton will have cascading effects throughout the food web.)

Justić, D., Rabalais, N., Turner, R. E. 1996. Effects of climate change on hypoxia in coastal waters: A doubled CO_2 scenario for the northern Gulf of Mexico. *Limnology and Oceanography* 41:992–1003.

Kamenos, N. A. 2010. North Atlantic summers have warmed more than winters since 1353, and the response of zooplankton. *Proceedings of the National Academy of Sciences* 107:22442–22447.

Kennedy, V. S. 1990. Anticipated effects of climate change on estuarine and coastal fisheries. *Fisheries* 15(6):16–24. (An early and insightful article.)

Kimmel, D. G., Miller, W. D., Roman, M. R. 2006. Regional climate forcing of mesozooplankton dynamics in Chesapeake Bay. *Estuaries and Coasts* 29:375–387.

Kimmel, D. G., Newell, R. E. 2007. The influence of climate variation on eastern oyster (*Crassostrea virginica*) juvenile abundance in Chesapeake Bay. *Limnology and Oceanography* 52:959–965.

Kimmel, D. G., Roman, M. R. 2004. Long-term trends in mesozooplankton abundance in Chesapeake Bay, USA: Influence of freshwater input. *Marine Ecology Progress Series* 267:71–83.

Kurihara, H. 2008. Effects of CO2-driven ocean acidification on the early developmental stages of invertebrates. *Marine Ecology Progress Series* 373:275–284.

McGinty, N., Power, A. M., Johnson, M. P. 2011. Variation among northeast Atlantic regions in the responses of zooplankton to climate change: Not all areas follow the same path. *Journal of Experimental Marine Biology and Ecology* 400:120–131.

Moran, X. A. G., Lopez-Urrutia, A., Calvo-Diaz, A., et al. 2010. Increasing importance of small phytoplankton in a warmer ocean. *Global Change Biology* 16:1137–1144. (Evidence that temperature alone is causing a shift toward smaller phytoplankton in the North Atlantic.)

Piontkovski, S. A., O'Brien, T. D., Umani, S. F., et al. 2006. Zooplankton and North Atlantic Oscillation: A basin-scale analysis. *Journal of Plankton Research* 28:1039–1046.

Richardson, A. J. 2008. In hot water: Zooplankton and climate change. *ICES Journal of Marine Science* 65:279–295.

Richardson, A. J., Schoeman, D. S. 2004. Climate impact on plankton ecosystems in the Northeast Atlantic. *Science* 305:1609–1612.

Turner, J. T., Borkman, D. G., Libby, P. S. 2011. Zooplankton trends in Massachusetts Bay, USA: 1998–2008. *Journal of Plankton Research* 33:1066–1080.

Twilley, R. R., Barron, E. J., Gholz, H. L., et al. 2001. *Confronting Climate Change in the Gulf Coast Region: Prospects for Sustaining Our Ecological Heritage.* Union of Concerned Scientists, Cambridge, MA; Ecological Society of America, Washington, DC. 82 pp.

Winder, M., Jassby, A. D. 2011. Shifts in zooplankton community structure: Implications for food web processes in the upper San Francisco Estuary. *Estuaries and Coasts* 34:675–690.

Winder, M., Jassby, A. D., McNally, R. 2011. Synergies between climate anomalies and hydrological modifications facilitate estuarine biotic invasions. *Ecology Letters* 14:749–757.

IDENTIFICATION and BIOLOGY of COMMON ZOOPLANKTON

HOW TO USE THIS BOOK TO IDENTIFY ZOOPLANKTON

Quick Picks is the first step in identifying unfamiliar organisms. This section divides the taxa into distinctive groups in which members share similar size, shape, and general appearance. Within each group, find the best match for your specimen, then turn to the proper taxonomic section for identification.

A description of the key anatomical features needed for identification can be found at the end of the general introduction to the biology and ecology of most taxonomic groups, labeled **Identification Hints**. New users will find it easier to read the introductory information and identification hints *before* attempting to match the specimen with the labeled illustrations. Many anatomical terms are defined in the section that precedes the illustrations of taxa and are not in the **Glossary**.

We strongly encourage users to start the identification process with "Quick Picks" and proceed to the "Identification Hints" for each group before attempting identification. In the long run, this approach will save frustration and faulty identifications.

Use the **line drawings and bulleted identification notes** to identify individual taxa. Distinctive features that distinguish each taxon from similar forms are listed and identified in the drawings. Check the notes under **Occurrence** once you have a tentative identification to be sure that the information on range, habitat, and salinity preference matches your collection site. If not, recheck your identification. Because this book does not cover all species or developmental stages that could occur in your area, it is possible that you will arrive at an identification that is not correct. Absolute identifications can only be provided by trained experts in taxonomy, but there is a high probability that when using this book you will correctly identify the specimen or determine the name of a taxon that is closely related to the specimen in hand.

Regional variability means that some specimens will not look exactly like the drawings, although the bulleted features should be reliable over our entire range unless noted otherwise.

TERMINOLOGY

Throughout the identification section, terms are used as follows:

Geographic Terms

Our geographic range. Nearshore ocean and estuarine waters from Cape Cod, Massachusetts, south to Cape Canaveral, Florida, along the Atlantic Coast and from Fort Myers, Florida, through Texas, along the Gulf Coast.
Estuarine. From the upper extent of tidal influence to the inlet or mouth of estuaries.
Nearshore or coastal ocean. From the shore (beach) out to 10 km.
Coastal. From the upper extent of tidal influence in rivers to 10 km into the ocean.
Offshore. Beyond 10 km from shore.
Open ocean. At least 10 km from shore, usually in clear water.
New England. Southern New England, from Cape Cod to Sandy Hook, New Jersey.
Mid-Atlantic region. From Sandy Hook south to Cape Hatteras, North Carolina.
Southeast region. From Cape Hatteras, North Carolina, to Cape Canaveral, Florida.
Gulf. Gulf of Mexico from Fort Myers, Florida, to the Rio Grande, Texas.

In many cases, we indicate that a taxon occurs from the Atlantic into (or throughout) the Gulf; that statement may carry an element of uncertainty regarding extreme southern Florida. When we state that a species ranges from Massachusetts (the northern limit of our coverage) southward, that does not mean that Massachusetts is necessarily the northern limit of its range.

Salinity Terms

psu. Salinity in practical salinity units (the symbols ‰, or ppt, are equivalent to psu). Current convention favors dropping the "psu" after the numerical value if the term "salinity" occurs in the sentence.
Estuarine. Brackish areas, including bays, sounds, inlets, creeks, and rivers, from low-salinity (<5) headwaters to the coastal ocean.
Brackish. Waters ranging from 0.5–30 psu.
Oligohaline. Salt content of 0.5–5 psu.
Mesohaline. Salt content of 5–18 psu.
Polyhaline. Salt content of 18–30 psu.
Euhaline. Salt content >30 psu.

Scientific Names

Genus and species names (in italics) are used throughout. Taxonomists commonly redefine associations among taxa that require changes in existing scientific names. In the headings used in this guide, current names are sometimes followed by older (more familiar) names. These older names are preceded by the term "formerly." Scientific names following an equal sign identify cases in which alternative scientific names are in use. *The World Reg-*

ister of Marine Species <www.marinespecies.org/aphia.php?p=match> lists the currently accepted scientific names. When common names are also given, they are usually those suggested by the American Fisheries Society.

REFERENCES AND ADDITIONAL READING

Identification References, listed with each group, provide detailed treatments, including original descriptions and coverage of less common species. References with details on the occurrence and abundance of each taxonomic group appear in **Appendix 5: Regional Surveys.** These published regional studies usually report the results of general zooplankton surveys that contain information on the occurrences of many taxonomic groups. These references are not always listed under the individual identification entries, so they can provide additional information on the taxon of interest. Users will find the supplemental information of the biology of invertebrates and fishes in the books at the end of this section. **Suggested Readings**, listed after the introduction to each group, include selected books and research articles that treat the group as a whole. Additional references that treat the biology or ecology of individual taxa are listed by the author's name and date under each taxon. Full citations for references that appear in the identification section of the guide are given in the **Literature Cited** section, at the end of the book.

Additional sources of information may be found online. The website *Zooplankton of the Atlantic and Gulf Coasts* <http://zooplankton-online.net/> is an electronic companion to this book. It contains photographs of both common and rare zooplankton, including some oceanic species plus additional information on those displayed. There are sections with new information and tips and resources for teachers.

SUGGESTED READINGS

Invertebrate Zoology Texts and Larval Fish References

Able, K. W., Fahay, M. P. 1998. *The First Year in the Life of Estuarine Fishes in the Middle Atlantic Bight.* Rutgers University Press, New Brunswick, NJ. 342 pp.

Able, K. W., & Fahay, M. P. 2010. *Ecology of Estuarine Fishes: Temperate Waters of the Western North Atlantic.* Johns Hopkins University Press, Baltimore. 496 pp.

Anderson, D. T., ed. 2002. *Invertebrate Zoology.* 2nd ed. Oxford University Press, New York. 476 pp.

Brusca, R. C., Brusca, G. J. 2003. *Invertebrates.* 2nd ed. Sinauer Associates, Sunderland, MA. 936 pp.

Meglitsch, P. A., Schram, F. R. 1991. *Invertebrate Zoology.* 3rd ed. Oxford University Press, Oxford. 640 pp.

Pechenik, J. A. 2009. *The Biology of Invertebrates.* McGraw-Hill, New York. 624 pp.

Ruppert, E. E., Fox, R. S., Barnes, R. D. 2004. *Invertebrate Zoology: A Functional Evolutionary Approach.* 7th ed. Brooks Cole / Thompson, Belmont, CA. 963 pp.

Zooplankton Guides Primarily Devoted to Fauna of Other Geographic Regions

Boltovskoy, D., ed. 1999. *South Atlantic Zooplankton.* 2 vols. Backhuys, Leiden, The Netherlands. 1706 pp. (Introduction to the biology of each group. Some species listed also occur in the North Atlantic.)

Boltovskoy, D., ed. 2005. *Zooplankton of the South Atlantic Ocean.* World Biodiversity Database. DVD. ETI Bioinformatics, Amsterdam.

Gerber, R. P. 2000. *An Identification Manual to the Coastal and Estuarine Zooplankton of the Gulf of Maine Region.* Part. 1: Text and identification keys. 80 pp. Part. 2: Figures. 98 pp. Acadia Productions, Brunswick, ME. (Extensive treatment of copepods and cladocerans, many of which range much farther south.)

Newell, G. C., Newell, R. C. 1963. *Marine Plankton: A Practical Guide.* Hutchinson Educational, London. 244 pp. (This guide to British zooplankton includes some species that also occur along the East Coast of North America, including some from the Gulf Stream and continental shelf.)

Shanks, A. L. 2001. *An Identification Guide to the Larval Marine Invertebrates of the Pacific Northwest.* Oregon State University Press, Corvallis. 256 pp. (Although most of the species are Pacific, the reviews of the larval forms of individual phyla and classes are generally applicable to all coasts.)

Smith, D. L., Johnson, K. B. 1996. *A Guide to Marine Coastal Plankton and Marine Invertebrate Larvae.* 2nd ed. Kendall/Hunt, Dubuque, IA. 221 pp. (This general guide is based on Pacific zooplankton with minimal attention to Atlantic and Gulf fauna. Limited information on distribution and ecology.)

Todd, C. D., Laverack, M. S., Boxshall, G. A. 1996. *Coastal Marine Zooplankton: A Practical Manual for Students.* 2nd ed. Cambridge University Press, Cambridge. 106 pp. (This guide to British zooplankton has a number of photographs but is more selective than comprehensive. Some of the species covered also occur on the East Coast of North America.)

Wickstead, J. H. 1965. *An Introduction to the Study of Tropical Zooplankton.* Hutchinson Tropical Monographs, London. 160 pp. (The only general guide to Caribbean plankton, including some species found in the southern United States. Interesting biological notes on specific taxa.)

General Guides to Marine Life That Cover Some Planktonic Groups

Gosner, K. L. 1999. *Field Guide to the Atlantic Seashore: From the Bay of Fundy to Cape Hatteras.* Peterson Field Guides. Houghton Mifflin, Boston. 476 pp. (See the sections on jellyfishes and comb jellies.)

Miner, R. W. 1950. *Field Book of Seashore Life.* Putnam, New York. 888 pp. (A classic. Now dated but still useful for protozooplankton, the worms, peracarids, comb jellies, and jellyfishes of the Northeast and Mid-Atlantic.)

Ruppert, E. E., Fox, R. S. 1988. *Seashore Animals of the Southeast.* University of South Carolina Press, Columbia. 429 pp.

Smith, R. I., ed. 1964. *Keys to Marine Invertebrates of the Woods Hole Region.* Marine Biological Laboratory Contribution 11. Systematics-Ecology Program, Spaulding Co., Randolph, MA. 208 pp. (Primarily a guide to benthic fauna with spotty treatment of planktonic groups. Information on distribution and abundance of the taxa treated. The Woods Hole keys, including some updated sections, are now on the web.)

Suthers, I. M., Rissik, D., eds. 2009. *Plankton. A Guide to Their Ecology and Monitoring for Water Quality.* CSIRO, Collingswood, Victoria, Australia. 256 pp.

Weiss, H. M. 1995. *Marine Animals of Southern New England and New York.* State Geological and Natural History Survey of Connecticut, Department of Environmental Protection, Hartford. 344 pp. (See sections on jellyfishes, comb jellies, worms, larger crustaceans, and insects.)

QUICK PICKS

Use this section to find the group that most closely matches your specimens or interests, and then go directly to the pages indicated for species identifications. Since some groups share many of the same general characteristics, the right match may not be found on the first try. Remember that this guide does not cover many of the less widely distributed members of some of these groups, so you should not expect to find a perfect match. The anatomical figures, ID Hints, and features on the plate illustrations in the sections that most closely match your choices should help you to determine whether your specimen has the key characteristics to place it in that category.

1. Small protozoans, usually < 0.3 mm

Although often numerous, many organisms on this page pass through meshes typically used for zooplankton collections. They are at the lower limit of resolution for observation using a dissecting microscope.

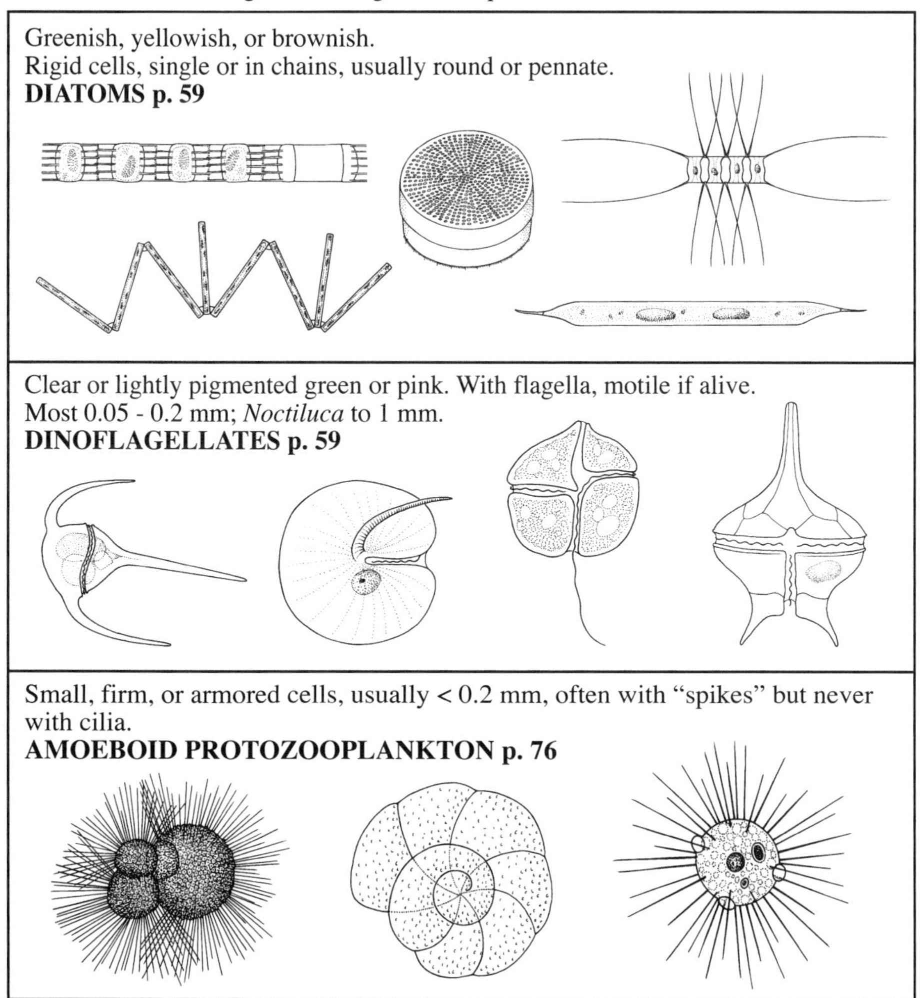

2. Ciliated protozoans and multicellular animals usually < 2 mm

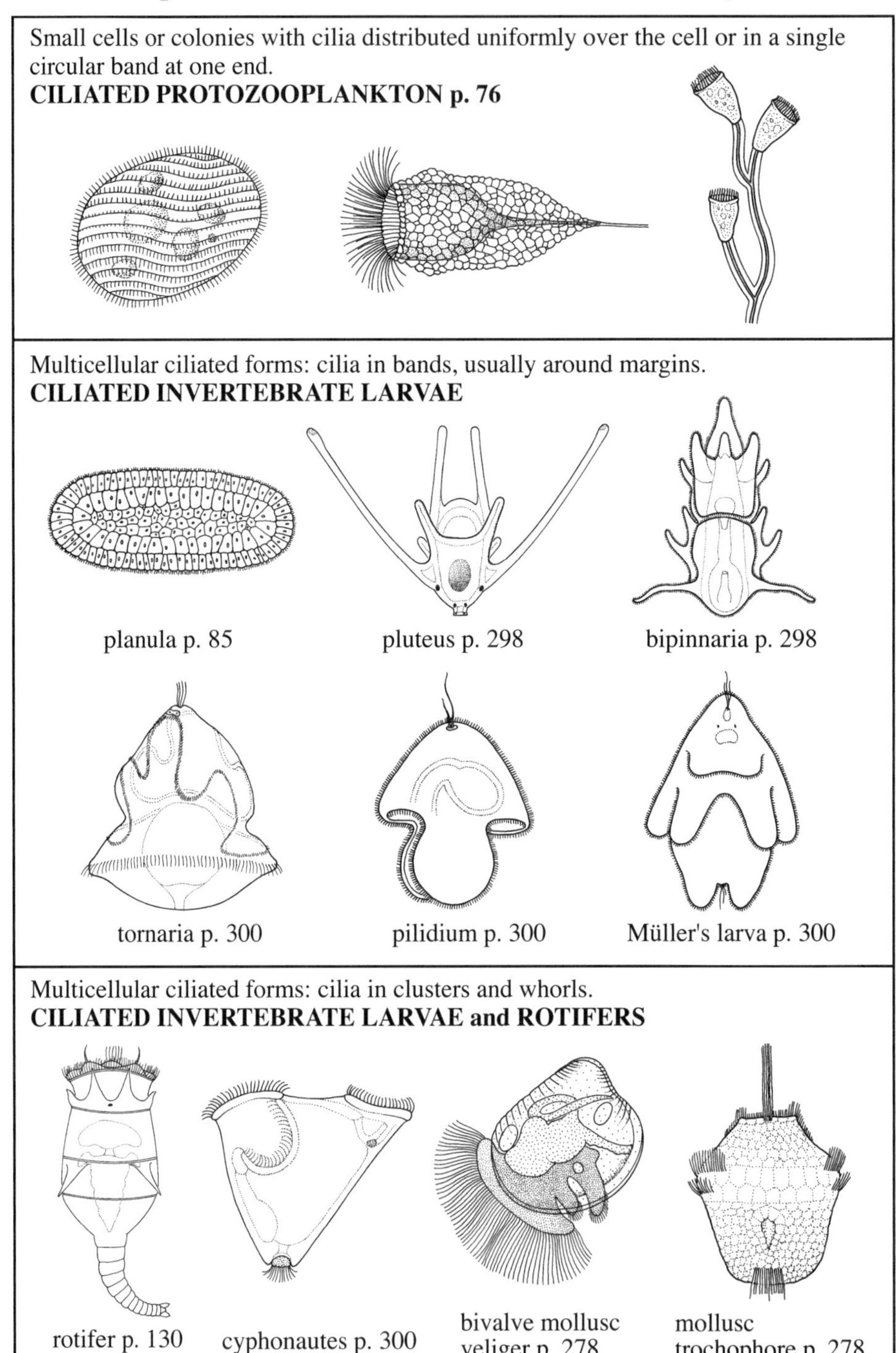

3. Gelatinous, usually transparent

< 30 mm; delicate medusae with tentacles (sometimes lost); pulsing movement.
HYDROZOANS p. 85

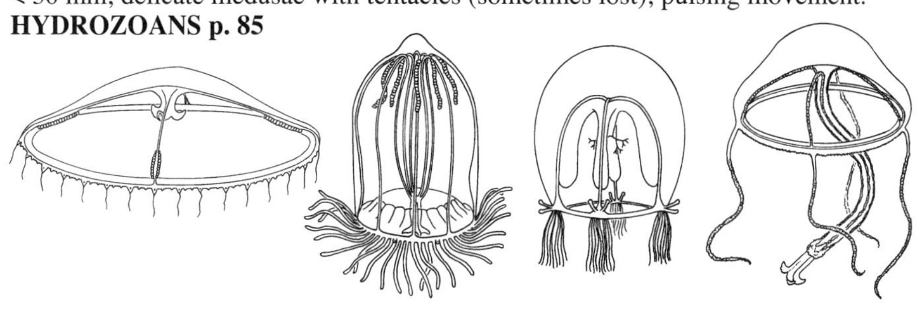

< 5 mm; flat, star shaped, without tentacles; pulsing movement.
EPHYRA STAGE OF SCYPHOZOANS p. 110

Nonmotile, spherical, or ovoid with firm covering; granular material within; with or without tadpole-shaped embryo; most < 5 mm.
FISH EGGS

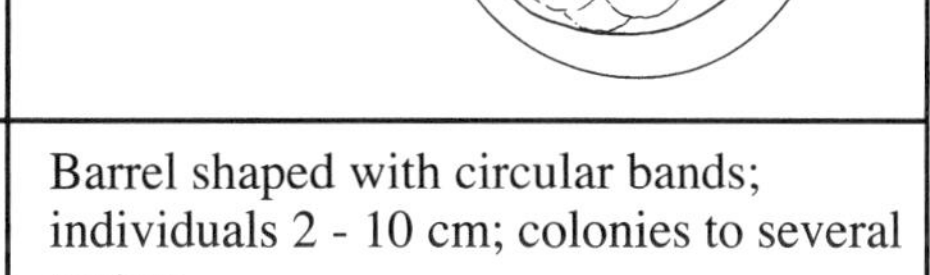

1 - 10 cm; spherical or ovoid with 8 linear rows of cilia; no circular bands; gliding movement.
CTENOPHORES (comb jellies) p. 124

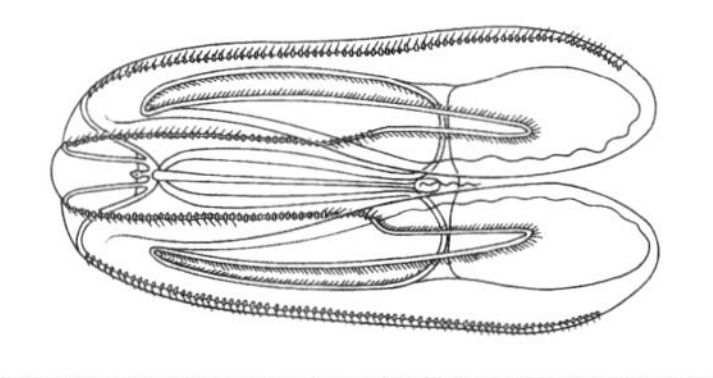

Barrel shaped with circular bands; individuals 2 - 10 cm; colonies to several meters.
SALPS and DOLIOLIDS p. 302

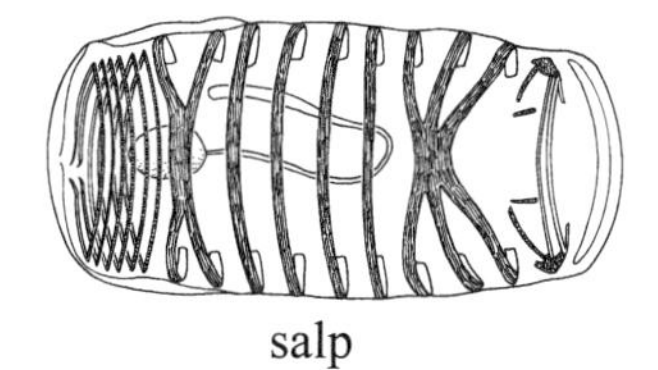

salp

Large (3 - 40 cm) umbrella or bell-shaped medusae, with or without tentacles; pulsing movement.
JELLYFISHES or SCYPHOZOANS p. 110

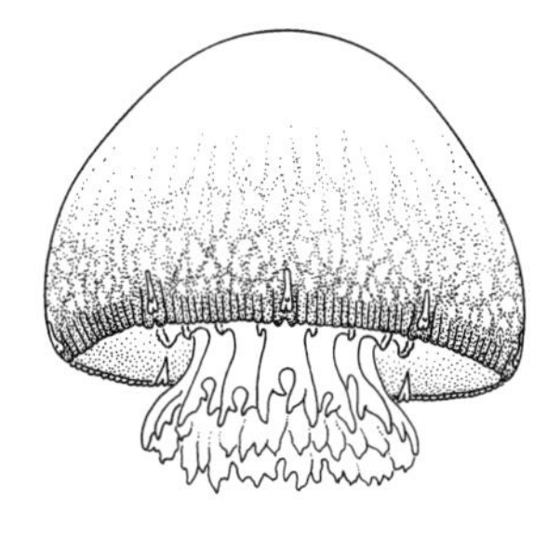

4. Small crustaceans usually < 1.5 mm, jointed appendages with bristle-like setae and no cilia

Body somewhat elongated; > 3 pairs of jointed appendages; head end usually with antennae; tail end usually forked.

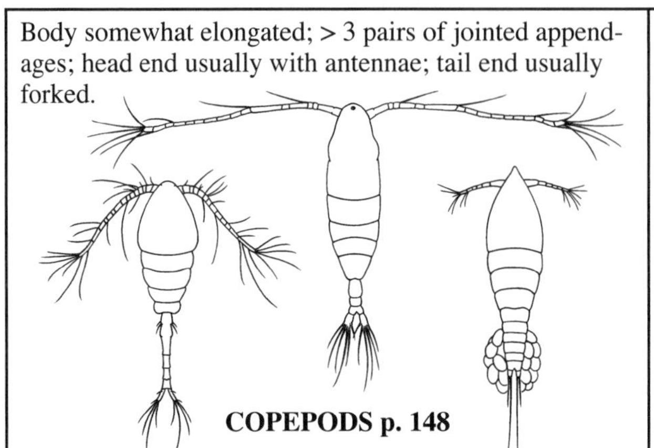

COPEPODS p. 148

Large branched antennae
CLADOCERAN p. 138.

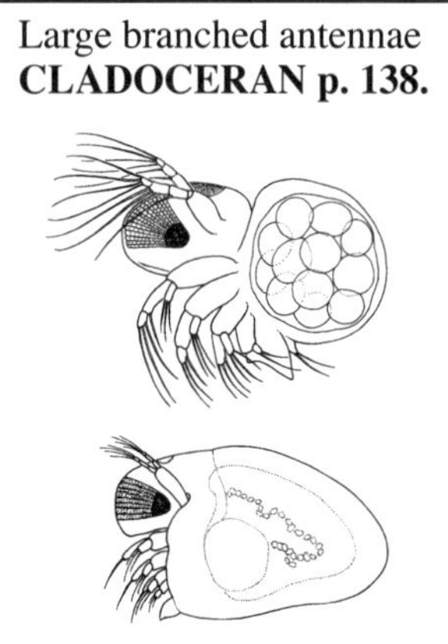

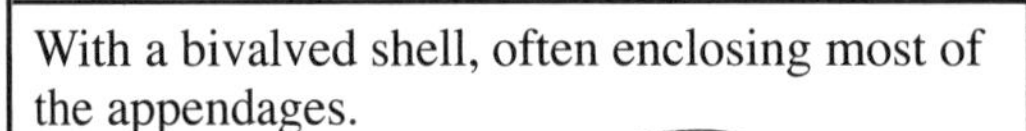

With a bivalved shell, often enclosing most of the appendages.

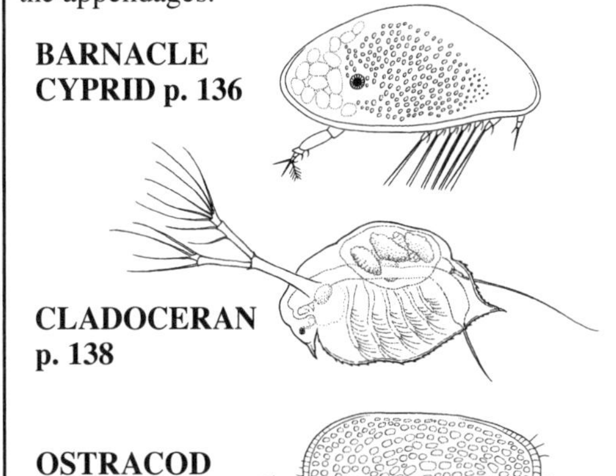

BARNACLE CYPRID p. 136

CLADOCERAN p. 138

OSTRACOD p. 138

3 pairs of jointed appendages.

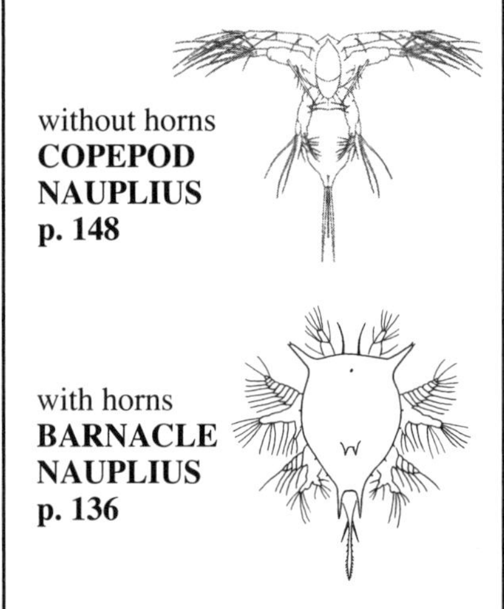

without horns
COPEPOD NAUPLIUS p. 148

with horns
BARNACLE NAUPLIUS p. 136

Firm, helmet-like, main body with eyes and one or more prominent spines on top or sides; rest of body usually curled with many small segments and prominent tail flap.
DECAPOD CRAB ZOEAE p. 218

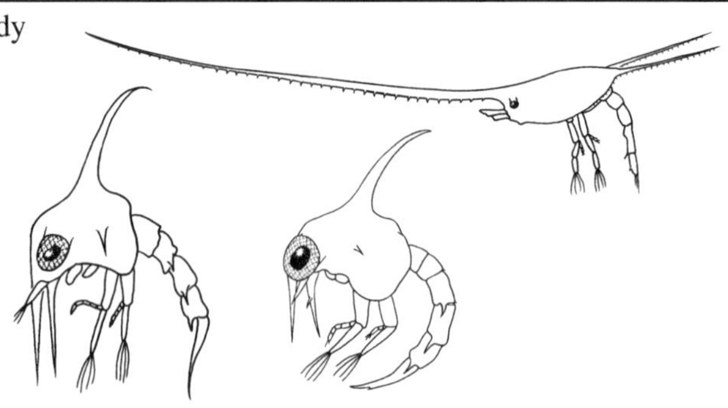

5. Larger crustaceans usually > 1.5 mm; prominent eyes and jointed appendages with bristle-like setae

Firm, elongate, barrel-like main body with or without spine between eyes; usually spines elsewhere on carapace; rest of body relatively straight with many sections and prominent tail flap and tail fin; no internal spherical organ at base of tail fins (see mysids below).

DECAPOD SHRIMP LARVAE p. 218

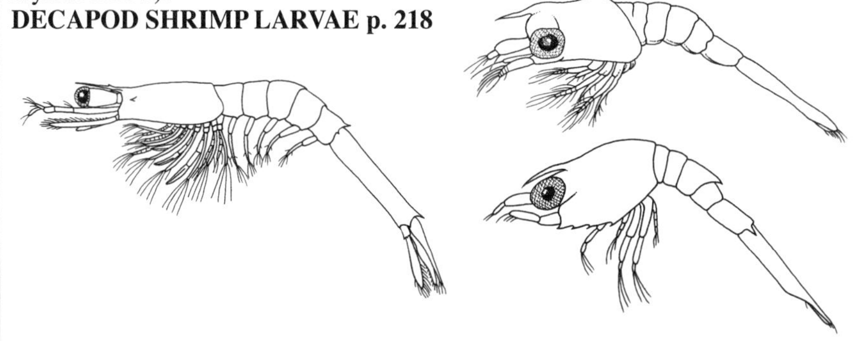

Elongate, barrel-like main body without prominent spines; rest of body with many sections; tail flap present; inner tail fins each have an internal spherical organ (inset) that is lacking in decapod shrimps.

MYSID SHRIMPS p. 184

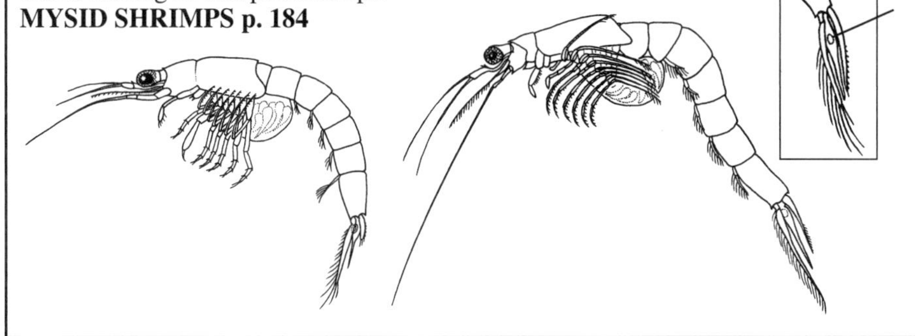

Box-like main body with eyes on corners.

CRAB MEGALOPAE p. 218

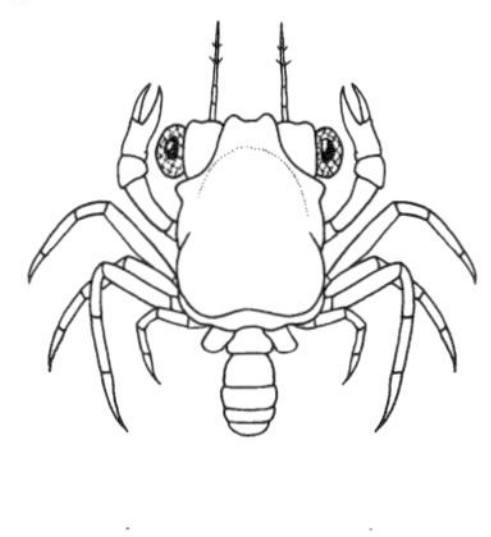

Very large (> 5 mm); flat, wide body with long, stalked eyes and long legs.

SPINY AND SLIPPERY LOBSTERS and MANTIS SHRIMP LARVAE

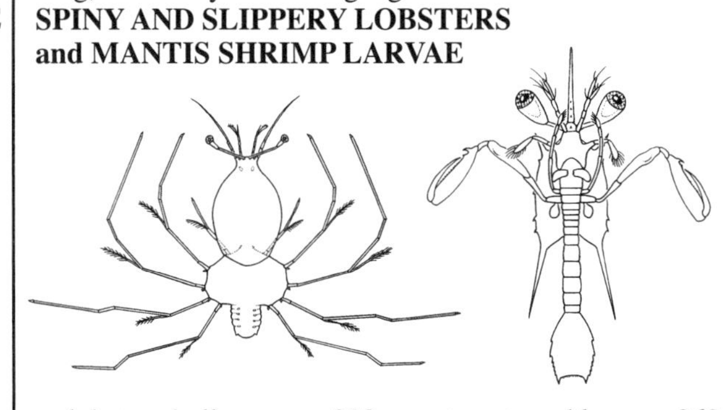

lobster phyllosoma p. 218 stomatopod larva p. 262

6. Other, often common crustaceans with firm segmented bodies and complex jointed appendages

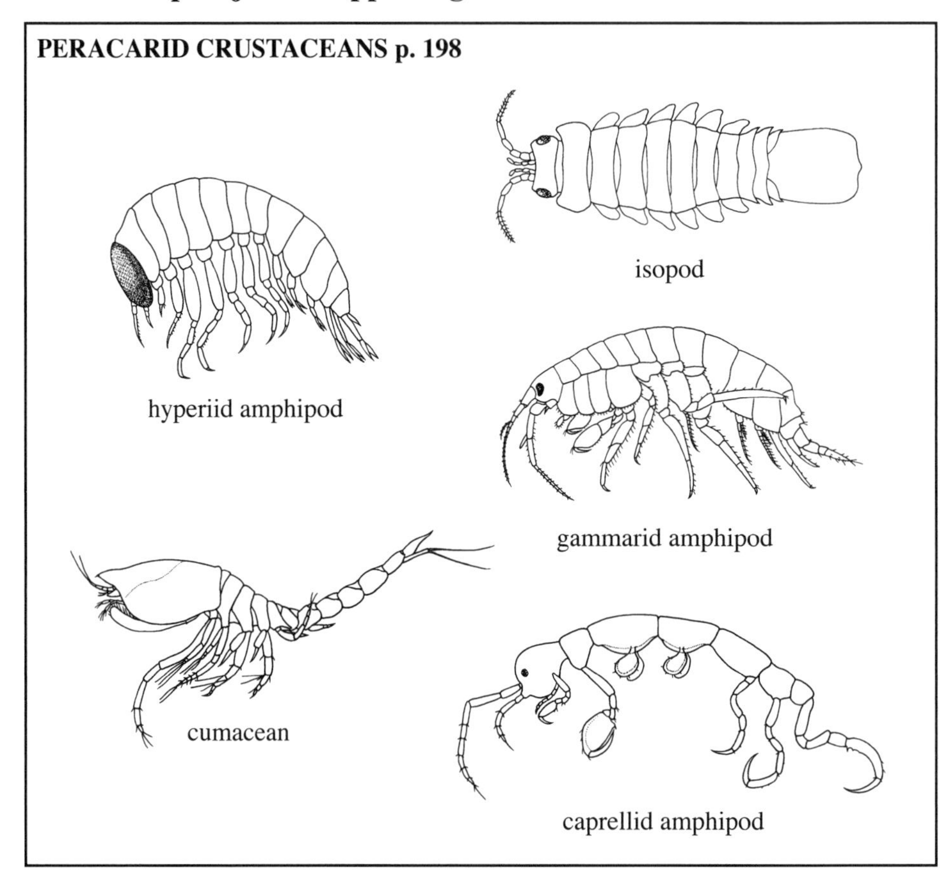

7. Other, less common, firm-bodied forms with jointed appendages but without prominent segments

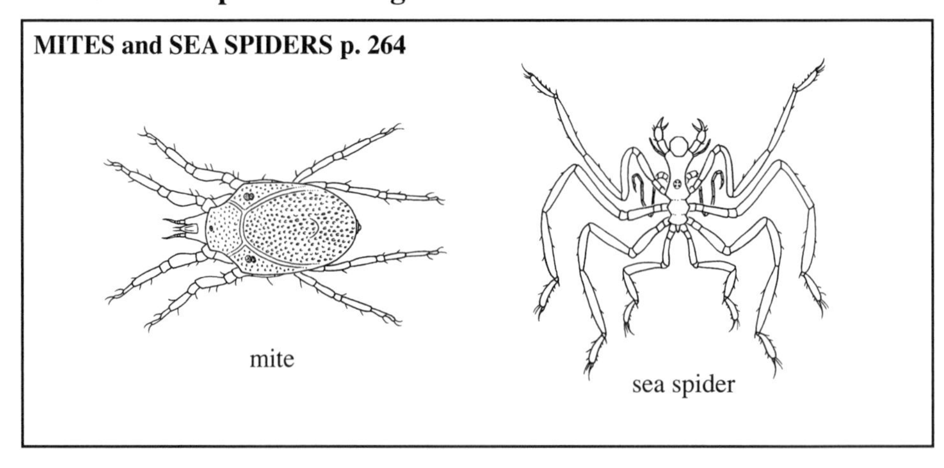

8. Elongated, worm-like

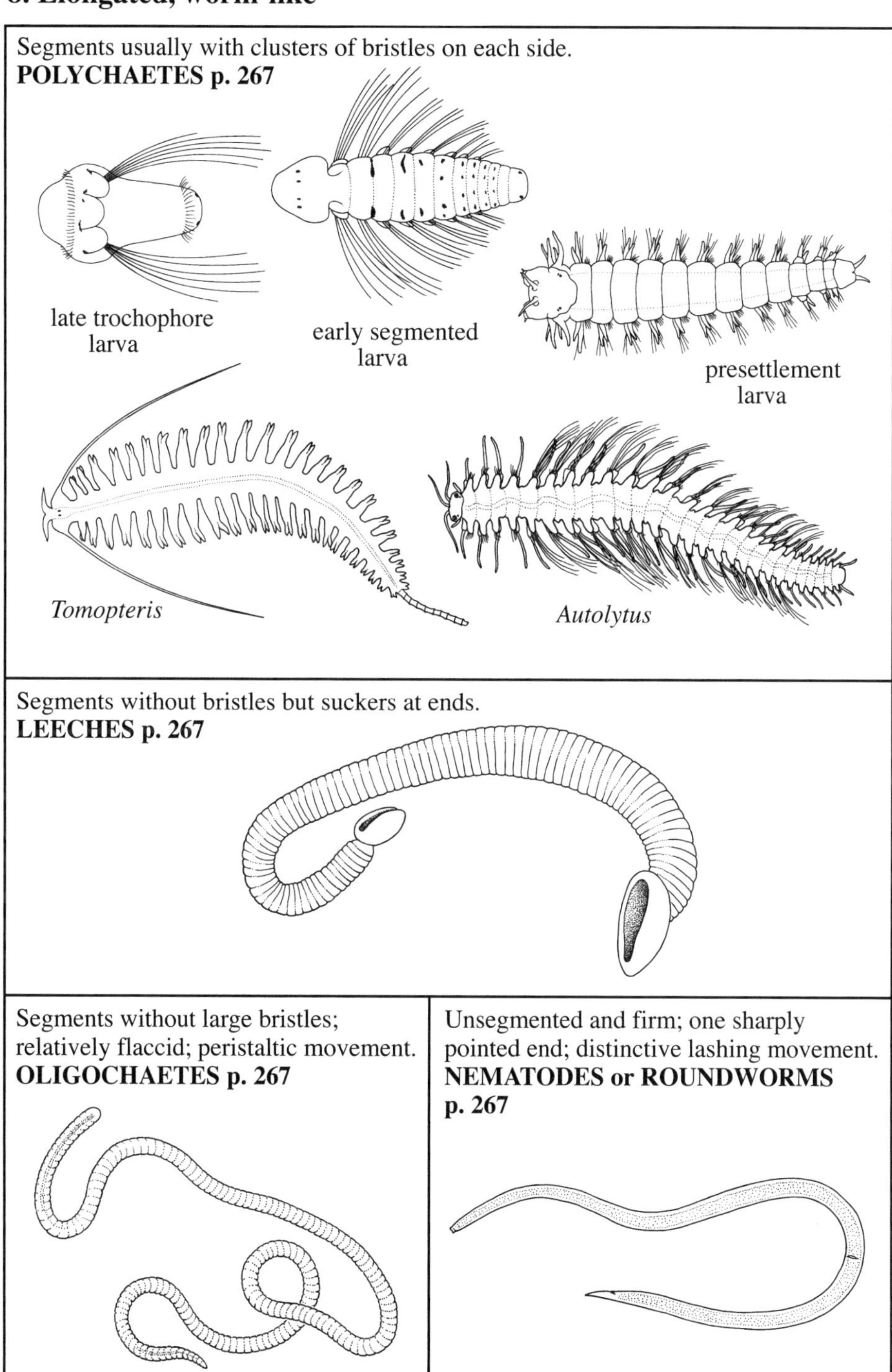

9. Fishlike with fins; usually > 2 mm

Usually lie on their sides in a sample dish; head end broader than tail end; fins not paired (in lateral view) but usually with supporting spines; typically with one eye visible, with a lens and pupil; often with pigment or pigmented spots.
FISH LARVAE p. 310

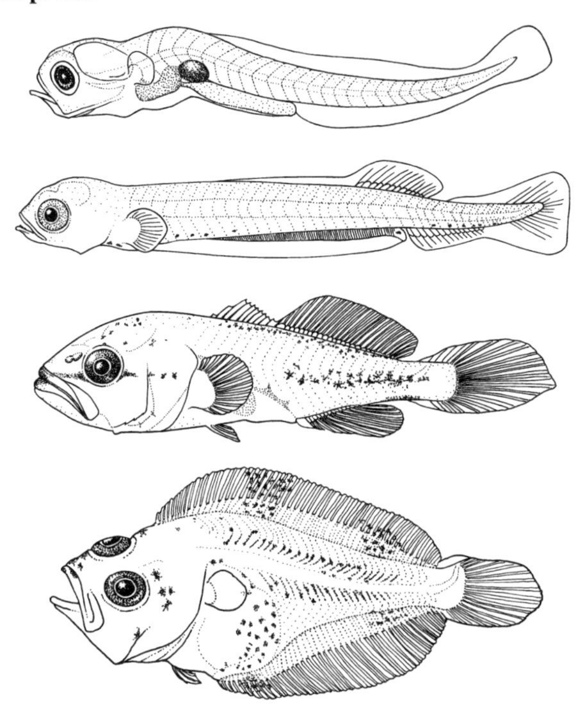

Dorsoventrally flattened; paired symmetrical fins and broad symmetrical tail fin; paired, small dashed eye spots lack lenses or pupils; typically transparent or milky.
CHAETOGNATHS or ARROW WORMS p. 290

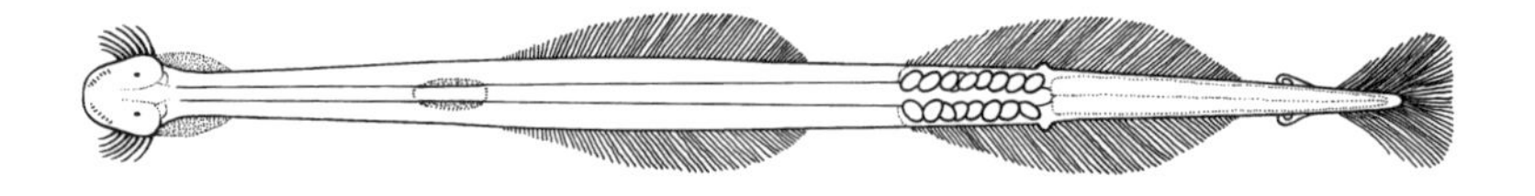

A fish like tail and swollen anterior "body"; no other fins; no eyes.
UROCHORDATE (ASCIDIAN) LARVAE p. 302

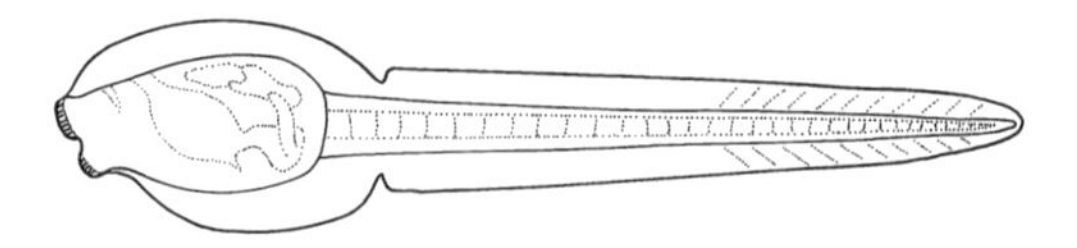

10. Other distinct forms

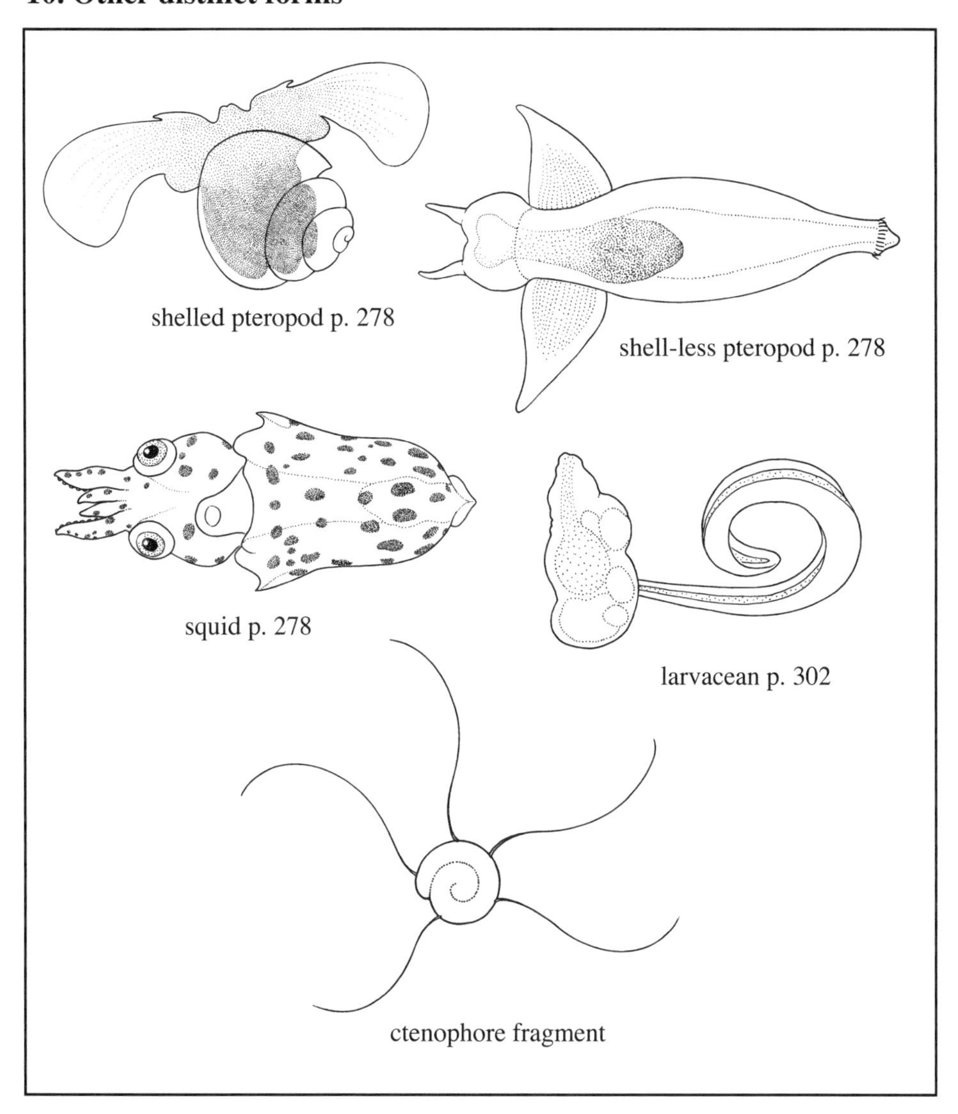

PHYTOPLANKTON

Phytoplankton is a general term for largely photosynthetic single-celled organisms found in the water column. Phytoplankton form the base for many planktonic food webs, and the type and size of the phytoplankton often dictate which zooplankton will be abundant. However, many of the common phytoplankton are too small (2–20 μm) to be caught or noticed in zooplankton collections. Indeed, production by small photosynthetic picoplankton and nanoplankton often exceeds that of the larger species. Of the many different groups of phytoplankton, the two most likely to be retained in zooplankton collections are larger species of diatoms and dinoflagellates. These two ancient groups of uncertain taxonomic affinities have many distinctive peculiarities. Both groups are widespread and abundant from the fresh headwaters of estuaries to the open sea.

Both dinoflagellates and diatoms are present all year, but their relative proportions vary. Along the Atlantic Coast and in many estuaries, there is a large spring diatom bloom and a less predictable secondary fall/winter bloom. In summer, dinoflagellates and a mixed assemblage of other flagellates usually dominate. This is a general cycle, and local exceptions are common. Phytoplankton blooms, often dominated by a single species, can produce cell densities in excess of a half-million cells per milliliter. These blooms are dynamic, appearing suddenly and lasting a few days or weeks as a changing suite of phytoplankton species assume dominance. The abundance and types of phytoplankton present often control, at least in part, the dynamics and species composition of the zooplankton grazers and thereby exert a major influence on the overall dynamics of planktonic food webs. Blooms of pigmented species produce yellow, green, red, or brown "tides," collectively referred to as harmful algal blooms (HABs). Some blooms produce toxins implicated in the deaths of fish and marine mammals and in the poisoning of humans who eat contaminated shellfish or are otherwise exposed to the toxin. Even nontoxic blooms can result in hypoxic or anoxic conditions.

Most nearshore planktonic diatoms are photosynthetic, but dinoflagellate nutrition is less readily categorized. Although many dinoflagellates are entirely autotrophic, some are heterotrophic, and others (mixotrophs) switch between autotrophy and heterotrophy, depending on their developmental stage or their nutritional state. Some dinoflagellates feed on other phytoplankton and steal and retain the still functional chloroplasts (kleptoplastidy). The prevalence of mixotrophy, especially in many species associated with blooms, has received recent attention. This nutritional flexibility may be advantageous in nutrient-poor conditions where heterotrophic feeding could supply both food and nutrients for growth. At times, grazing of heterotrophic and mixotrophic dinoflagellates on photosynthetic diatoms and flagellates may exceed that of mesozooplankton. Some dinoflagellates also feed on dissolved organic material and on bacteria and are an important link in the microbial loop. Heterotrophic dinoflagellates include the predacious *Noctiluca*, a number of parasites, and the enigmatic *Karlodinium* and *Pfiesteria* associated with fish kills in North Carolina and Maryland.

The life cycles of phytoplankton often involve both sexual and asexual stages. During blooms, diatoms reproduce asexually by binary fission to produce two diploid cells. The silica test surrounding each cell is made of two pieces (called valves or frustules) that overlap slightly, much like the two pieces of a petri dish. When cells divide, each daughter gets one of the valves and regenerates the inner half. In this way, the average cell size diminishes with time. Sexual reproduction involving gametes occurs at intervals in the life cycle. At the end of blooms or in unfavorable conditions, diatoms produce asexual resting spores that settle to the bottom. Dinoflagellate life cycles are more complex. Asexual reproduction of the haploid vegetative stages is the norm. The sexual phase, if present, is brief. As in diatoms, a resting cyst or spore may form part of the life cycle. Some species are apparently capable of multiple transformations. See the review in Coats (2002) and Litaker et al. (2002) for more details on the complexities of dinoflagellate life cycles.

Dinoflagellates and diatoms differ with respect to locomotion. Dinoflagellates typically have two flagella to propel them through the water. In contrast, diatoms lack flagella. Some diatoms employ oil droplets to assist with buoyancy or cytoplasmic projections to slow sinking. Nevertheless, diatoms ultimately rely on mixing or upwelling to keep them within the photic zone.

IDENTIFICATION HINTS

Diatoms come in two broad groups, centric and pennate. Centric diatoms (the cells, not the colonies) are typically radially symmetrical, whereas pennate diatoms are slender, often rodlike, and bilaterally symmetrical. Whereas most centric diatoms are round in cross section, this symmetry is not always evident, especially in chain-forming species. The bilaterally symmetrical pennate diatoms may have elliptical, linear, or sigmoid shapes, often with a median cleft (raphe) along their length.

In many planktonic diatoms, the cells combine to form colonies of distinctive chains joined by cytoplasmic connections. Use the shape of the cells, the length and number of threadlike projections, and the nature of the connections between cells for identification rather than the number of cells in a chain, which varies. Since diatom cell sizes decrease with successive divisions, cell size is not especially useful in identification, with a few exceptions. When visible, the number, the type, and the location of chloroplasts are distinctive for some species. Sometimes species identification requires higher magnification and special preparation to reveal differences in the intricate patterns on the siliceous skeletons.

Dinoflagellates range from a few micrometers (μm) to more than 2 mm in diameter, but most are less than 100 μm. Most dinoflagellates occur singly, but some, like *Ceratium tripos*, may join to form chains. Dinoflagellates have two dissimilar flagella. In most species, one lies in a transverse groove, the girdle, encircling the body and one protrudes from the "posterior" end. These flagella produce a distinctive spiraling motion. Flagella and other characteristics are virtually impossible to see in preserved samples. The cell wall contains a cellulose-like polysaccharide, sometimes in the form of sculptured thecal plates, or “armor.” The so-called naked (=athecate) dinoflagellates lack distinctive armor plates. The size of the cells, their shape, the position and size of the girdle, and the sculpturing of the plates on armored species are the best identification clues. Naked dinoflagellates are delicate and best identified while alive. Color is not reliable in identification. While size is useful in identification, environmental conditions can change the morphology and the size of some species. For example, the dinoflagellate *Gyrodinium spirale* reaches 150 µm in tropical ocean waters but seldom exceeds 30 µm in brackish Mid-Atlantic waters.

USEFUL IDENTIFICATION REFERENCES

Campbell, P. H. 1973. *Studies on Brackish Water Phytoplankton.* Sea Grant Publication UNC-SG-73-07. NTIS publication COM-73-10672. Sea Grant Program, School of Public Health, University of North Carolina, Chapel Hill. 406 pp. . (Covers some of the smaller phytoplankton groups not treated here.)

Horner, R. A. 2002. *A Taxonomic Guide to Some Common Marine Phytoplankton.* BioPress, Bristol, UK. 195 pp. (This guide has color photographs of specimens from around the globe.)

Marshall, H. G. 1986. *Identification Manual for Phytoplankton of the United States Atlantic Coast.* US Environmental Protection Agency Publication EPA/600/4–86/003. US Environmental Protection Agency Center for Environmental Research Information, Cincinnati, OH.

Maryland Department of Natural Resources. Chesapeake coastal bay life: Algae (Phytoplankton) <www.dnr.state.md.us/bay/cblife/algae/index.html>. Accessed July 22, 2011. (Access species by name to get photomicrographs and ecological information. Based on species found in the Chesapeake but widely applicable in other areas.)

Tomas, C. R., ed. 1997. *Identifying Marine Phytoplankton.* Academic Press, New York. 858 pp. (Comprehensive and authoritative. Contains an introduction to the biology of each group and a summary of name changes of the different taxa.)

Wood, R. D., Lutes, J. 1967. *Guide to the Phytoplankton of Narragansett Bay, Rhode Island.* University of Rhode Island Printing, Kingston. 65 pp. (A guide designed for nonspecialists.)

SUGGESTED READINGS

Capriulo, G. M., ed. 1990. *Ecology of Marine Protozoa.* Oxford University Press, New York. 357 pp.

Coats, D. W. 2002. Dinoflagellate life-cycle complexities. *Journal of Phycology* 38:417–419.

Elbrächter, M. 2003. Dinoflagellate reproduction: Progress and conflicts. *Journal of Phycology* 39:629–632.

Jeong, H. J. 1999. The ecological roles of heterotrophic dinoflagellates in marine planktonic community. *Journal of Eukaryotic Microbiology* 46:390–396.

Litaker, R. W., Tester, P. A., Duke, C. S., et al. 2002. Seasonal niche strategy of the bloom-forming dinoflagellate *Heterocapsa triqueta. Marine Ecology Progress Series* 232:45–62.

Morris, I., ed. 1980. *The Physiological Ecology of Phytoplankton.* Blackwell, Oxford. 625 pp.

Pfiester, L. A., Anderson, D. M. 1989. Dinoflagellate reproduction. *Botanical Monographs* 21:611–648.

Round, F. E., Crawford, R. M., Mann, D. G. 1990. *The Diatoms: Biology and Morphology of the Genera.* Cambridge University Press, New York. 747 pp.

Stoecker, D. M. 1999. Mixotrophy among dinoflagellates. *Journal of Eukaryotic Microbiology* 46:397–401.

Taylor, F. J. R., ed. 1987. *The Biology of Dinoflagellates.* Blackwell, Oxford. 875 pp.

Representative common phytoplankton large enough to be noticed and identified when scanning a sample for zooplankton are discussed next.

CYANOBACTERIA (FORMERLY KNOWN AS BLUE-GREEN "ALGAE")

Trichodesmium spp.

Trichodesmium is a floating cyanobacterium (blue-green alga) found worldwide in tropical and subtropical waters where it can form extensive blooms capable of discoloring the water. Coloration varies from blue green to bright pink. The filaments often clump together in small tufts that provide both food and habitat for other plankton, including the copepods *Macrosetella* and *Miracia*. *Trichodesmium* is a nitrogen fixer and an important primary producer in nutrient-poor waters. *Trichodesmium* occurs in the Gulf and is typically associated with Gulf Stream water along the Atlantic Coast. Species of *Trichodesmium* are difficult to distinguish visually. *Trichodesmium* blooms are reportedly toxic to invertebrates and humans, but no toxin has yet been isolated.

References. Capone et al. 2007; Guo and Tester 1994; Hawser et al. 1992; Janson et al. 1995; O'Neil and Roman 1994; Sellner 1997.

DIATOMS (BACILLARIOPHYCEAE)

Skeletonema costatum

Skeletonema costatum is perhaps the single most common and most abundant coastal diatom. It is typically a dominant component of the winter-spring bloom each year and may constitute more than 85% of the total diatoms in northern waters. Although it has a wide brackish distribution, its maximum abundance occurs in higher salinities. *Skeletonema potamos* is a winter-spring dominant in low salinities.

References: Anning et al. 2000; Borkman and Smayda 2009; DeManche et al. 1979; Granum et al. 2002; Smayda and Boleyn 1966.

Corethron criophilum

Corethron criophilum is widely distributed in coastal waters and bays from Maine to Florida and in moderate to high-salinity reaches of larger bays and estuaries.

Paralia sulcata

Paralia sulcata is reportedly a benthic species, but it is commonly suspended in the water column in shallow coastal waters and high-salinity reaches of bays and estuaries in the Gulf and in the Atlantic from New England south to at least South Carolina. Valves are 8–130 μm in diameter.

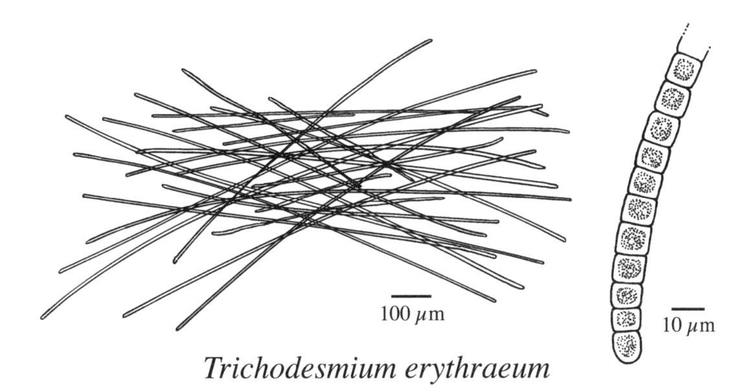

Trichodesmium erythraeum

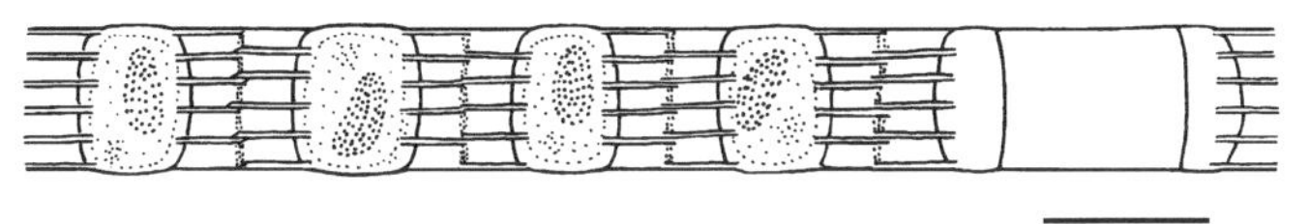

Skeletonema costatum

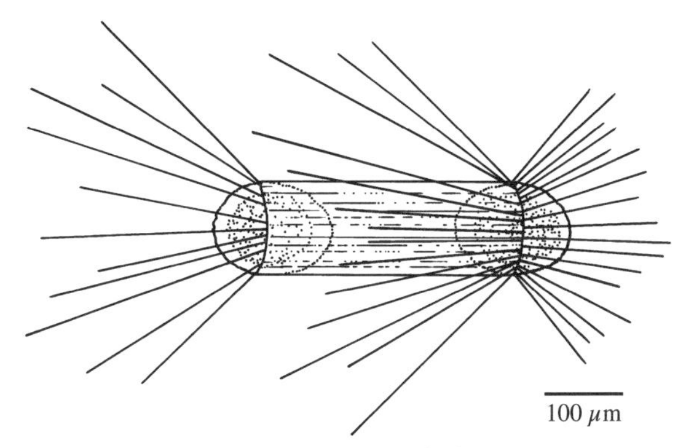

Corethron criophilum

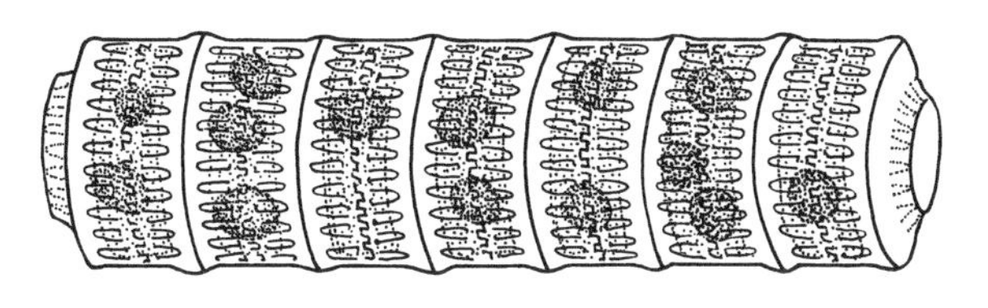

Paralia sulcata

Chaetoceros spp.

Chaetoceros spp. consist of long chains with cells joined at their outside edges. One or two pairs of long, thin projections extend from the junction between cells. Valve diameter is 7–80 μm. Of the more than two dozen species of *Chaetoceros* reported from local waters, *C. decipiens*, *C. lorenzianum*, and *C. affine* are among the most widespread and abundant in both the Atlantic and the Gulf. *Chaetoceros lorenzianum* and *C. affinis* are common in estuaries. They may be present much of the year in both coastal and estuarine habitats along the Atlantic Coast and are often part of the spring diatom bloom.

Reference. Rines and Hargraves 1988.

Coscinodiscus spp.

Coscinodiscus appear as solitary round discs (25–500 μm) without projections. *Coscinodiscus* spp. are widespread along both Atlantic and Gulf Coasts but seldom dominant. Some species are most common in nearshore ocean waters and coastal bays (*C. concinnus*), but several (*C. radiatus, C. marginatus*) are common in estuaries.

Odontella mobiliensis

Odontella may occur as single squarish cells or in short chains. Cells are 60–130 μm long. It is common along the Atlantic and Gulf Coasts, predominantly in higher salinities. Many other species of *Odontella* occur in both estuarine and nearshore waters.

Thalassiosira spp.

Thalassiosira species are usually found in chains held together by a central thread. *Thalassiosira decipiens, T. eccentrica,* and *T. nordenskioeldii* are all widespread and abundant in coastal and estuarine waters. *Thalassiosira nordenskioeldii* and *T. dicipiens* seem to favor cooler temperatures and are often dominant in both coastal and estuarine blooms in winter/spring. *Thalassiosira rotula* occurs in coastal waters of the Atlantic and Gulf and may be toxic. *Thalassiosira gravida* is reported as a dominant off of Sandy Hook, New Jersey. The much smaller *T. pseudonana* is abundant in the Atlantic and predominates in the Gulf but is too small (5 × 7 μm) to be noticed in most zooplankton collections. *Thalassiosira pseudonana* was the first diatom selected for genome sequencing.

Reference. Sugie and Kuma 2008.

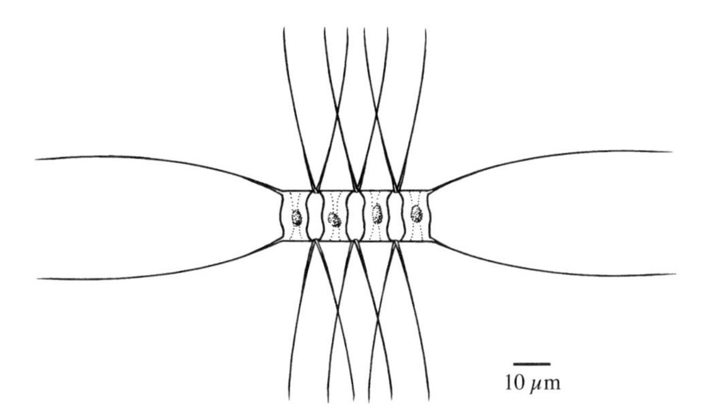

Chaetoceros lorenzianum

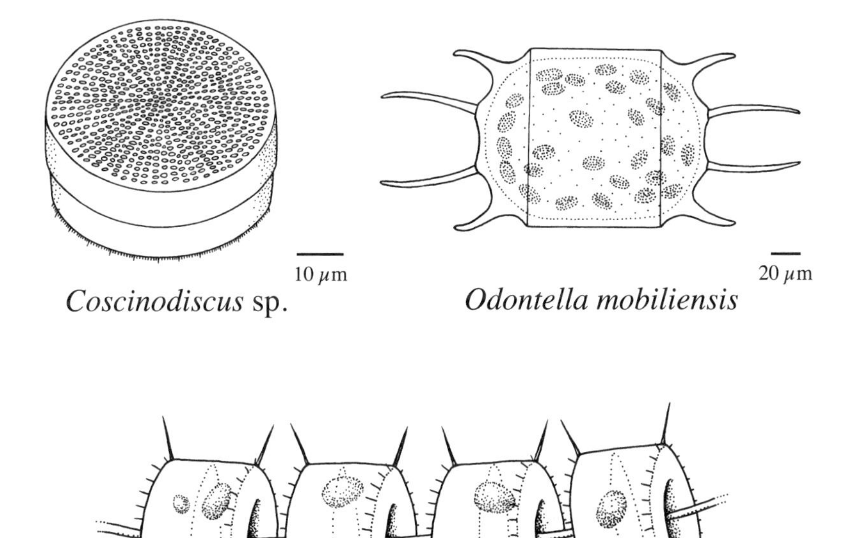

Coscinodiscus sp.

Odontella mobiliensis

100 μm

Thalassiosira decipiens

Proboscia spp.

Various *Proboscia* species are found worldwide and in nearshore and estuarine waters from New England to Florida and throughout the Gulf. They occur in salinities from 10 to hypersaline conditions and are often abundant in estuarine plumes. The several "forms" of the previous classification of *Rhizosolenia alata* are now separate species of *Proboscia*.

Pseudosolenia calcar-avis

Pseudosolenia calcar-avis is widespread along the Gulf and Atlantic Coasts and may be one of the spring bloom dominants, especially in warm or warm-temperate waters. Cell diameters range from 4.5–190 μm. The similar *Rhizosolenia setigera* is also common in warmer waters.

Pseudo-nitzschia spp.

Pseudo-nitzschia may be found singly or in chains of parallel cells attached by their overlapping ends. *Pseudo-nitzschia* species are often abundant in spring blooms along both Atlantic and Gulf Coasts. Highest abundances occur in coastal waters, but they are sometimes found in brackish areas. *Pseudo-nitzschia pungens* is one of the most common species, but *P. seriata* is also commonly reported. *Pseudo-nitzschia pseudodelicatissima* is abundant in the Mississippi River plume. *Pseudo-nitzschia* may produce domoic acid, the toxin responsible for amnesic shellfish poisoning (ASP).

References. Anderson et al. 2010; Davidovich and Bates 1999; Dortch et al. 1997; Hasle 1995; Olson and Lessarda 2010; Thessen and Stoecker 2008.

Cylindrotheca closterium, formerly *Nitzschia closterium*

Cylindrotheca closterium is most prevalent spring to fall in estuaries and bays from New England through the Mid-Atlantic. It is also reported from the Gulf. This diatom may be benthic on mudflats.

Thalassionema nitzschioides and *T. pseudonitzschioides*

Thalassionema nitzschioides is a cosmopolitan species, often present all year. *Thalassionema pseudonitzschioides* is important in diatom blooms in the Mid-Atlantic, including brackish estuaries. A scanning electron microscope may be required to distinguish these two species.

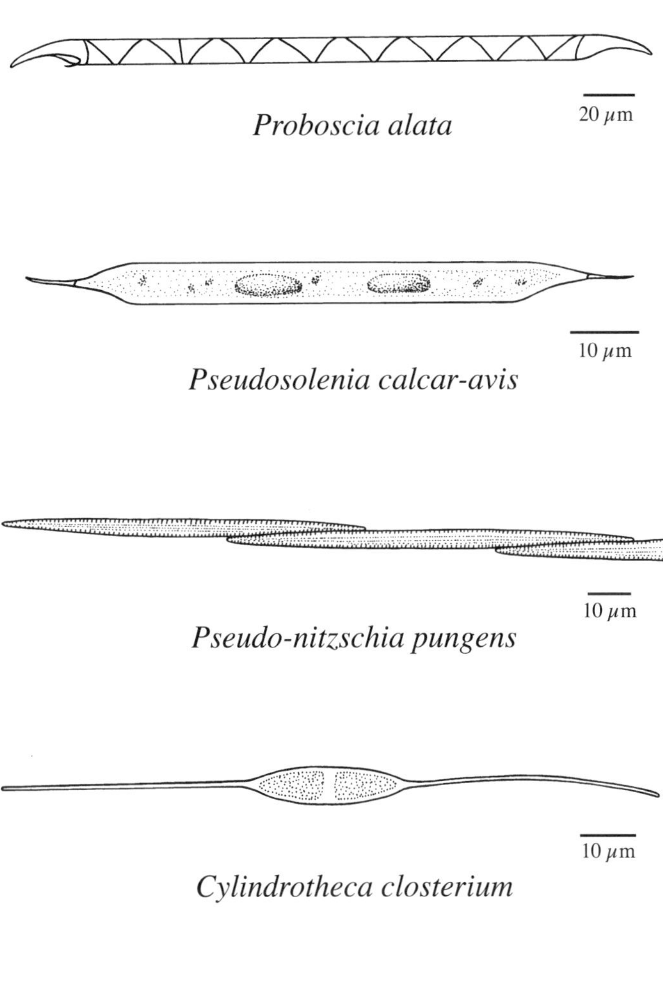

Proboscia alata

Pseudosolenia calcar-avis

Pseudo-nitzschia pungens

Cylindrotheca closterium

20 μm

Thalassionema nitzschioides

Leptocylindrus danicus

Cells are 30–60 µm long and 6–16 µm wide and form short chains. *Leptocylindrus danicus* is often a codominant (with *Skeletonema costatum*) in spring and fall diatom blooms along the Atlantic and Gulf Coasts. Although primarily coastal, it may be found in the middle reaches of estuaries. *Leptocylindrus minimus* is often a major component of nanoplankton production but too small to be noticed in most zooplankton collections.

References. Davis et al. 1980; French and Hargraves 1986.

Cerataulina pelagica and *Dactylosolen fragilissimus*

Cerataulina pelagica is widespread in coastal and estuarine waters of the Atlantic and Gulf Coasts. It is part of the spring diatom bloom along the Mid-Atlantic and a dominant in the Chesapeake Bay spring bloom. This diatom reestablishes its populations in the upper Chesapeake Bay by moving upstream in saline bottom waters. The common estuarine *Dactylosolen fragilissimus* (formerly *Rhizosolenia fragilissima*) is similar in size and shape but has a single central connection between cells instead of the two points on either side as seen in *C. pelagica.*

Ditylum brightwellii

Ditylum brightwellii is an elongate, triangularly shaped diatom with spiked ends. It is abundant from cold-temperate to tropical waters in both coastal and estuarine habitats. It occurs in all seasons and can be dominant on both Atlantic and Gulf Coasts.

Guinardia spp., formerly *Rhizosolenia*

Of the several *Guinardia* species known, *G. delicatula, G. striata,* and *G. flaccida* are relatively common. They can be present in any season and may be important in the diatom blooms along the Atlantic and Gulf Coasts. They penetrate estuaries at least to mesohaline waters. *Guinardia striata* (formerly *Rhizosolenia stolterfothii*) is one of the most abundant diatoms in the Gulf of Mexico, where it can account for >90% of the phytoplankton volume in nearshore and in brackish areas.

Reference. Bledsoe and Philips 2000.

Asterionellopsis glacialis

Cells are often united into star-shaped colonies. *Asterionellopsis glacialis* is a common and widespread Atlantic nearshore and estuarine genus from New England to Florida and through the Gulf. It is a dominant in spring diatom blooms at the mouth of the Delaware and Chesapeake Bays.

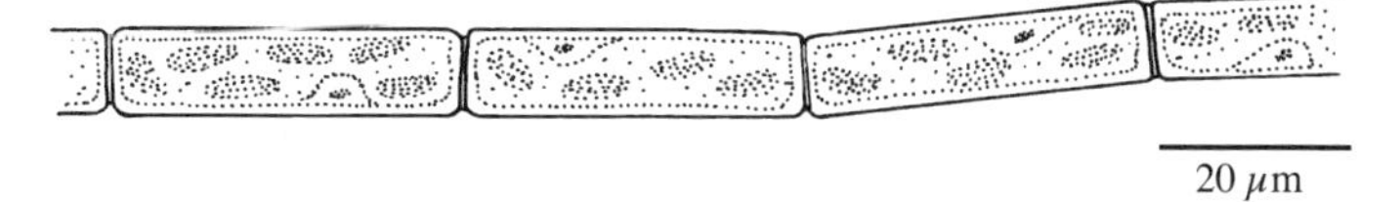

Leptocylindrus danicus

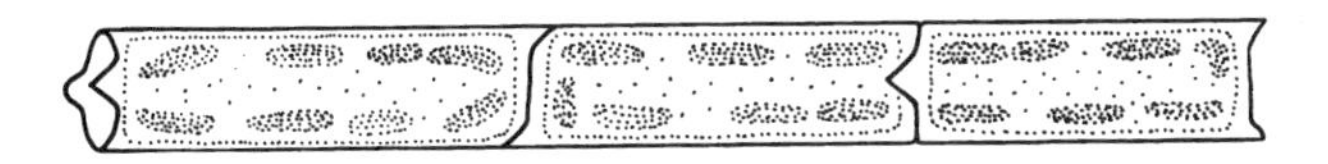

20 μm

Cerataulina pelagica

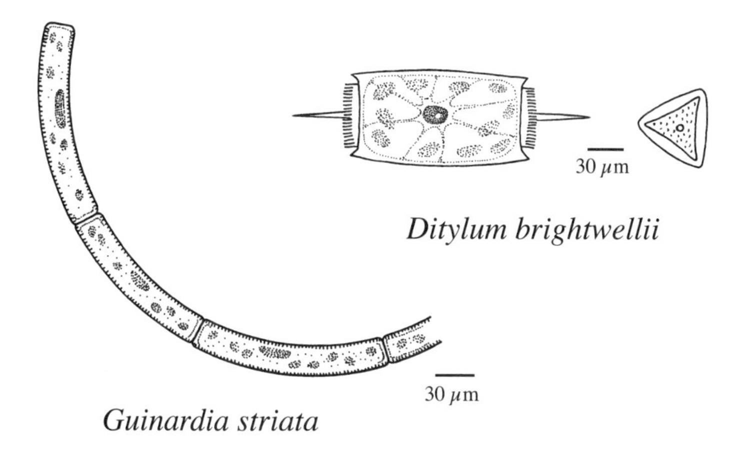

Ditylum brightwellii

Guinardia striata

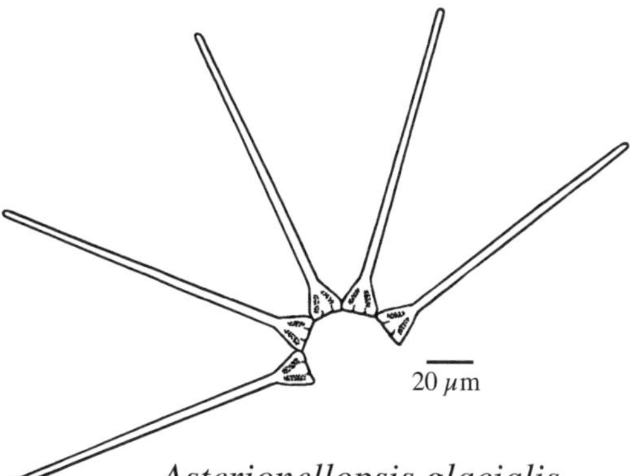

Asterionellopsis glacialis

DINOFLAGELLATES (DINOPHYCEAE)

Karenia brevis, formerly *Gymnodinium breve*

The estuarine and coastal *Karenia brevis* produces toxic red tide blooms in the Gulf of Mexico and in the southeastern Atlantic. Fish kills, deaths of manatees and dolphins, and respiratory distress in humans are attributed to "brevetoxin," a potent neurotoxin. Although *Karenia* causes a red discoloration of the water, the cells have green pigment.

References. Brand and Compton 2007; Flaherty and Landsberg 2010; Lekan and Tomas 2010; Lester et al. 2008; Leverone et al. 2007; Tester and Steidinger 1997; Vargo 2009.

Akashiwo sanguinea

Akashiwo sanguinea is a common 40–75 μm red tide dinoflagellate occurring in estuaries and along the Atlantic Coast from New England south and in the Gulf. Occasional massive blooms result in >10,000 cells ml^{-1}. Although no toxin is specifically associated with these blooms, postbloom decomposition can deplete oxygen, causing fish kills. *Akashiwo* is mixotrophic, feeding on nanociliates. The predatory dinoflagellate *Polykrikos kofoidii* and parasitic dinoflagellates feed on *Akashiwo sanguinea* in the Chesapeake Bay.

References. Bockstahler and Coats 1993a, 1993b; Hansen 1991; Johansson and Coats 2002; Park et al. 2002; Yih and Coats 2000.

Lingulodinium polyedrum, formerly *Gonyaulax polyedra*

This bioluminescent dinoflagellate occurs in warm-temperate and tropical nearshore waters. It produces red tides with yessotoxins associated with fish and shellfish poisoning. At the ends of blooms, it produces resistant cysts that sink to the bottom and initiate subsequent blooms under favorable conditions.

References: Akimoto et al. 2004; Figueroa and Bravo 2005; Kokinos and Anderson 1995; Paz et al. 2004.

Gyrodinium spp.

Gyrodinium spirale is one of the most distinctive species within this genus. It ranges widely along the Atlantic and Gulf Coasts but seems most abundant in New England. *Gyrodinium uncatenum* is a large dinoflagellate common and occasionally dominant in summer in mesohaline bays from Chesapeake Bay to New England. Blooms cause red tides. Some workers in the Chesapeake prefer the term *mahogany tide* for these reddish-brown blooms. *Gyrodinium instriatum* blooms occur in summer in the Chesapeake Bay, the Neuse River (North Carolina), and the Gulf. These mixotrophs and heterotrophs consume other phytoplankton and ciliates, including tintinnids. *Gyrodinium aureolum*, responsible for "brown tides" in the Northeast, and most other *Gyrodinium* species are too small to be easily noticed in zooplankton samples.

References. Anderson et al. 1985; Bockstahler and Coats 1993b; Hansen 1992; Johnson et al. 2003; Kim and Jeong 2004; Li et al. 2000a, 2000b; Park et al. 2002; Uchida et al. 1997; Verity 2010; Yih and Coats 2000.

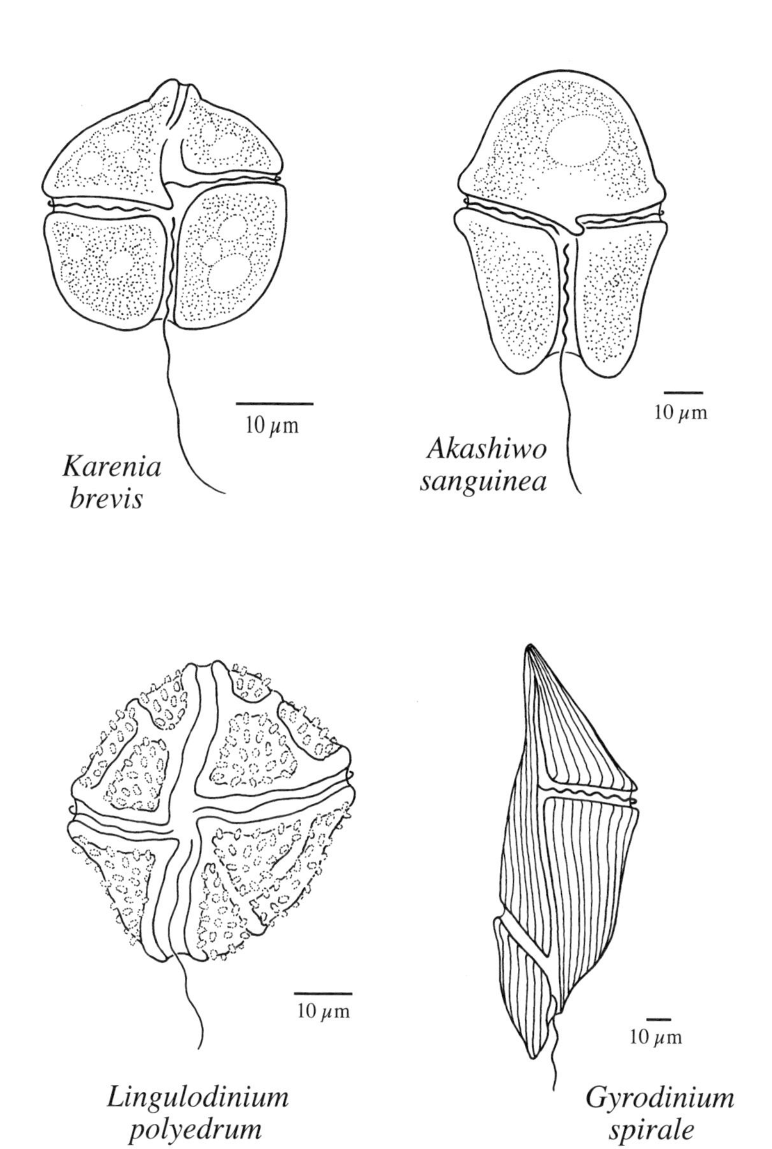
10 μm
Karenia brevis
10 μm
Akashiwo sanguinea
10 μm
Lingulodinium polyedrum
10 μm
Gyrodinium spirale

Polykrikos kofordii

Polykrikos kofordii occurs from New England south to at least South Carolina on the Atlantic Coast and in the Gulf, primarily in salinities >20. It lacks chloroplasts and engulfs dinoflagellates and, on occasion, small diatoms. It also feeds on the red-tide dinoflagellate *Gymnodinium catenatum.*

References. Jeong et al. 2001; Matsuoka et al. 2000; Matsuyama et al. 1999.

Pheopolykrikos hartmannii (=*Polykrikos hartmannii*)

Pheopolykrikos hartmannii is abundant in summer in the Mid-Atlantic bays but may range more widely. It is also widespread in the Gulf in estuaries and along the coast. This heterotroph feeds on other dinoglagellates and diatoms. It forms aggregations (pseudocolonies) of 4–16 “zooids” (individuals). Based on recent findings, taxonomists have been discussing which name (*Polykrikos* or *Pheopolykrikos*) should have priority.

Reference. Sellner and Brownlee 1990.

Dinophysis spp.

Dinophysis norvegica is a large and distinctive species occurring from the Chesapeake Bay north along the coast and in higher-salinity reaches of estuaries. It produces okadaic acid, a toxin associated with diarrhetic shellfish poisoning (DSP), especially in mussels along the New England coast where it is abundant. *Dinophysis caudata* is a temperate species common from the Mid-Atlantic south, including the Gulf. *Dinophysis accuminata* and *D. punctata* are temperate species of high to moderate salinities. Some *Dinophysis* species have chloroplasts and are mixotrophic. Others are heterotrophic predators that feed, at least in part, on ciliates. In turn, some ciliates dine on *Dinophysis.*

References. Carvalho et al. 2007; Garcia-Cuetos et al. 2010; Verity 2010.

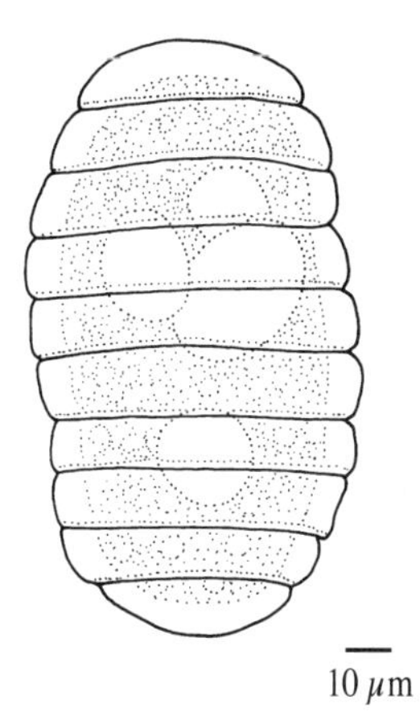

Polykrikos kofordii

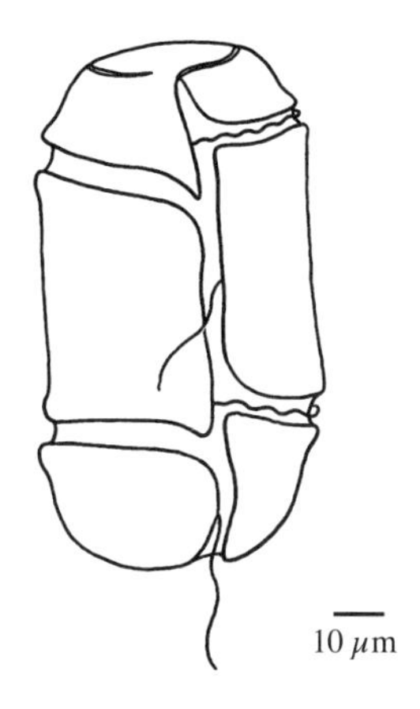

Pheopolykrikos hartmannii

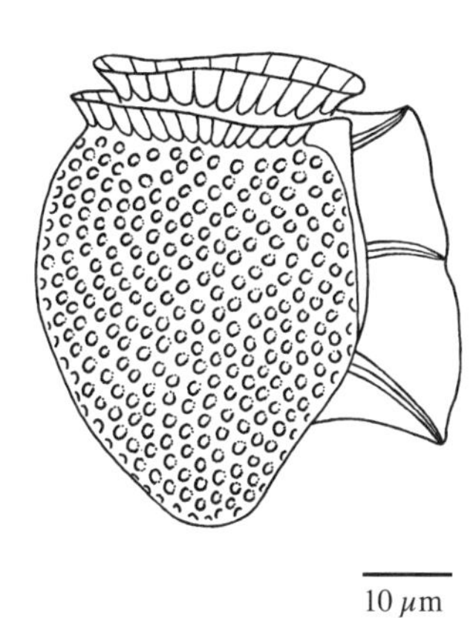

Dinophysis norvegica

Protoperidinium spp.

The various *Protoperidinium* species are widespread in coastal and estuarine waters. The distribution of individual species is problematic. *Protoperidinium depressum* and *P. pellucidum* seem common on both the Atlantic and Gulf Coasts; *P. depressum* is also found in estuaries. The genus engulfs its food and preys preferentially on diatoms. The diet may also include dinoflagellates and copepod eggs and nauplii. Many species are bioluminescent.

References. Buskey 1997; Buskey et al. 1992, 1994; Gribble et al. 2009; Jeong 1994a, 1994b; Latz and Jeong 1996; Li et al. 1996.

Ceratium spp.

Ceratium is one of the most common and abundant of the dinoflagellate genera along the Atlantic and Gulf Coasts, especially in summer. *Ceratium tripos* has distinctive "horns" and occurs singly or in chains throughout the Atlantic and Gulf. It may produce massive blooms in both coastal and estuarine areas, even in low salinities. *Ceratium lineatum* appears more abundant in high to moderate salinities. *Ceratium furca* is a common red or "mahogany" tide species in the mesohaline Chesapeake. Although it is not usually toxic, massive blooms of *C. tripos* may result in hypoxia and fish kills. Some species are mixotrophic, feeding on ciliates. *Ceratium fusus* can be bioluminescent.

References. Baeka et al. 2009; Bockstahler and Coats 1993b; Smalley and Coats 2002.

Prorocentrum spp.

Prorocentrum has an apical horn and two smaller horns at the opposite end. *Prorocentrum* species are dominant inshore and estuarine dinoflagellates in both the Atlantic and the Gulf. The larger *P. micans* is more likely to be retained and noticed in zooplankton collections. *Prorocentrum micans* prefers higher-salinity coastal and lower-estuarine areas and may be abundant along the Mid-Atlantic and New England coasts and in coastal bays. *Prorocentrum minimum* is common along the southeastern Atlantic Coast and in the Gulf. In mid- to low-salinity brackish waters of the Mid-Atlantic, its blooms often result in reddish "mahogany" tides. *Prorocentrum triestinum* (formerly *P. redfieldii*) is common from New Jersey north. *Prorocentrum* spp. are mixotrophic dinoflagellates with diets that include photosynthetic microflagellates and oligotrich ciliates. Some species may produce okadaic acid, a toxin associated with diarrhetic shellfish poisoning (DSP). Some *P. minimum* blooms are toxic to bivalve larvae.

References. Ana et al. 2010; Harding and Coats 1988; Harding et al. 1991; Johnson et al. 2003; Perry and McClelland 1981; Stoecker et al. 1997, 2008; Wikfors 2005.

Noctiluca scintillans, formerly *N. miliaris*

This very large (200–1000 µm), bioluminescent dinoflagellate is common and often abundant from the Mid-Atlantic through the Gulf, primarily in coastal but occasionally in brackish areas. Intense blooms may produce red tides. *Noctiluca* is an obligate heterotroph and an active predator on other phytoplankton, copepod eggs, and possibly fish eggs, which it engulfs whole. Named for its bioluminescence, *Noctiluca* emits a greenish light when disturbed. *Noctiluca* is essentially nonmotile but can control its buoyancy. *Noctiluca* is not toxic in our waters.

References. Buskey et al. 1992; Elbrächter and Qi 1998; Hansen et al. 2004; Kiørboe and Titelman 1998.

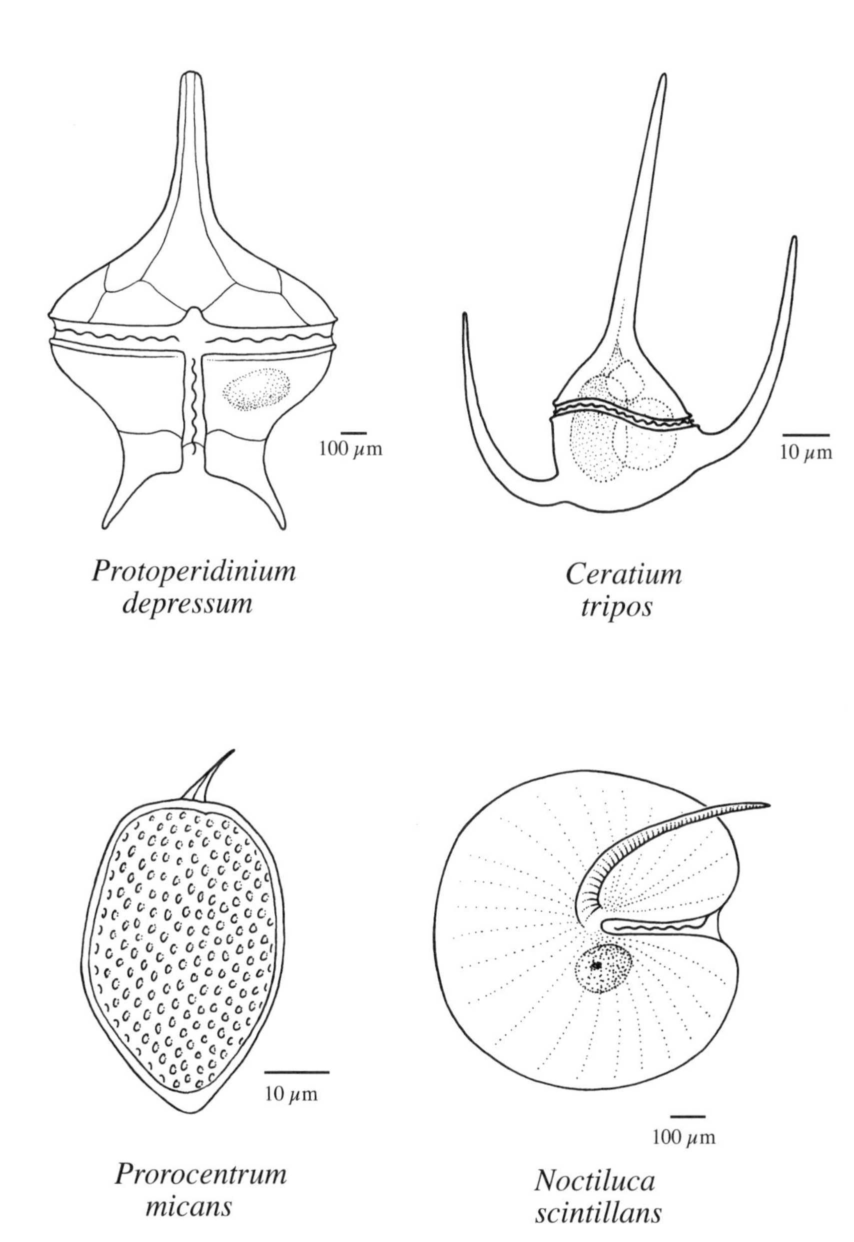

Protoperidinium depressum

Ceratium tripos

Prorocentrum micans

Noctiluca scintillans

PROTOZOOPLANKTON

The nonphotosynthetic, single-celled organisms historically referred to as **Protozoa** are important components of the marine plankton. This is a diverse assemblage encompassing many unrelated groups. Our treatment of protozooplankton is limited to the larger, single-celled heterotrophic groups likely to be retained and noticed in zooplankton collections: ciliates, foraminiferans, radiolarians, acantharians, and heliozoans. Recent studies show how abundant these heterotrophic protozoans are in estuarine and coastal areas where they form a vital connection in the microbial loop, linking bacterial and nanoplankton production to larger zooplankton and fishes. At times, protozooplankton biomass may exceed that of mesozooplankton. This fauna is relatively unstudied in our area, except in a few localities. Heterotrophic dinoflagellates appear under phytoplankton in the previous section.

AMOEBOID PROTOZOA: FORAMINIFERA, RADIOLARIA, ACANTHARIA, AND HELIOZOA

Foraminifera, Radiolaria,[1] Acantharia, and Heliozoa are all single-celled heterotrophic protozoans belonging to the Sarcodina. Like the more familiar sarcodine, *Amoeba*, they use their adhesive cellular extensions (pseudopods) to catch prey. Unlike *Amoeba*, these three protozoan groups possess a characteristic hard outer covering. Lacking cilia, flagella, or other means of locomotion, these planktonic species drift passively in the water. Although Radiolaria, Acantharia, and Heliozoa are unlikely to be abundant in nearshore waters, they are common, distinctive, and sometimes large enough to be noticed in zooplankton collections. The naked amoebae, while sometimes abundant in estuarine areas, are too small and delicate to be noticed without special collection and observation procedures.

The Foraminifera, generally called "forams," are covered by an outer shell (the test) usually composed of calcium carbonate. The shell is often divided into interconnected chambers added in distinctive patterns as the foram grows. Shells can have spines, visible pores, perforations, or some combination of these features. Some harbor photosynthetic symbionts. About 40 of the 4,000 species are planktonic and can be abundant, especially in the open ocean.

The Radiolaria, Acantharia, and Heliozoa have spherical cells with fine, stiff cytoplasmic projections known as actinopods that radiate outward, giving them a starlike appearance. The Radiolaria have a glassy skeleton made of silica through which the actinopods

1. We use the traditional name Radiolaria, but its status is disputed.

extend. These beautiful plankton arc strictly marine and prefer clear, warm oceanic waters. They are found nearshore when Gulf Stream eddies or Loop Current intrusions come ashore. Radiolarian cells sometimes aggregate into large colonies. The similar Acantharia have strontium sulfate skeletons. Acantharia are common in deep ocean waters but can occur in nearshore waters. They are most commonly reported from the Gulf of Mexico but may occur along the Atlantic Coast when intrusions of Gulf Stream waters arrive nearshore. Most heliozoans live in freshwater, a few species thrive in oligo- and mesohaline reaches of estuaries, and one species is exclusively marine. Heliozoa should be observed while alive because they are difficult to preserve.

Each of these three groups uses its own particular type of pseudopod to catch food, ranging from bacteria to algal cells to surprisingly large ciliate or copepod prey. Food items are ingested by phagocytosis, much like ingestion seen in *Amoeba*. Some warm-water radiolarians and forams retain algal cells (or their chloroplasts) and thus become functionally mixotrophic. Limited information on reproduction suggests that forams and *Actinophrys sol*, the common brackish heliozoan, reproduce asexually and sexually. Among the few radiolarians in which the life cycle is known, asexual reproduction (binary fission) is common; sexual reproduction has been identified in some colonial species. The life cycle is typically two to four weeks.

CILIATES

Often overlooked because of their small size, ciliates are ubiquitous in coastal and estuarine waters where they play a crucial role in planktonic food webs. Of the several groups of ciliates, tintinnids, with their hard outer coverings, are often conspicuous in plankton samples. Other ciliate groups that lack a protective skeleton are less likely to be collected in recognizable form with plankton nets. Despite the active reclassification of higher groupings within both ciliates and sarcodines, many genus and species names have remained relatively stable for a century or more.

Ciliates form a vital link, connecting bacteria and pico-/nanophytoplankton to larger zooplankton. Up to 50% of both bacterial and phytoplankton production may be removed by ciliates each day. In turn, ciliates are eaten by larger zooplankton, including copepods, decapod larvae, fish larvae, rotifers, and ctenophores. The trophic link provided by ciliates is especially crucial when tiny cells dominate algal productivity because few of the larger zooplankton feed on particles as small nano- or picophytoplankton. Ciliates are capable of rapid growth with cell divisions every few hours under favorable conditions. Many shallow-water ciliates reproduce both asexually, by binary fission, and sexually, by conjugation and produce resting cysts in response to unfavorable conditions.

IDENTIFICATION HINTS

Most of the protozooplankton are <0.3 mm. The tintinnids, often encrusted with sand grains, should not be confused with the greenish to golden-brown diatoms. Other soft-bodied, ciliated forms in this size range include rotifers and the smallest of the ciliated invertebrate larvae, including the planula of cnidarians and the early trochophores of polychaetes and molluscs.

USEFUL IDENTIFICATION REFERENCES

Bé, A. W. H. 1967. Foraminifera, families: Globigerinidae and Globorotaliidae. *ICES Identification Leaflets for Plankton.* Leaflets 108. Available online by number through <www.ices.dk/products/fiche/Plankton/SHEET108.PDF>. Accessed August 18, 2011.

Borror, A. C. 1973. *Marine Flora and Fauna of the Northeastern United States—Protozoa: Ciliophora.* NOAA Technical Report. NMFS Circular 378, Seattle, WA. 62 pp.

Cavalier-Smith, T. 1993. Kingdom Protozoa and its 18 phyla. *Microbiological Reviews* 57:953–994. (Extensive treatment and yet another classification scheme.)

Hemleben, C., Spindler, M., Anderson, R. O. 1989. *Modern Planktonic Foraminifera.* Springer, Heidelberg. 363 pp.

Laval-Peuto, M., Brownlee, D. C. 1986. Identification and systematics of the Tintinnina (Ciliophora): Evaluation and suggestion for improvement. *Annales de l'Institut Océanographique, Paris* 62:69–84. (European, but many genera and species occur on both sides of the Atlantic.)

Lee, J. J., Leedale, G. F., Bradbury, P., eds. 2002. *Second Illustrated Guide to the Protozoa.* Society of Protozoologists, Lawrence, KS. 368 pp.

Marshall, S. M. 1969. Protozoa, order Tintinnia. *ICES Identification Leaflets for Plankton.* Leaflets 117–127. Available online by number through <www.ices.dk/products/fiche/Plankton/SHEET108.PDF>. Accessed August 18, 2011.

Taylor, W. T., Sanders, R. W. 2010. Protozoa. In: Thorpe, J. H., Covich, A. P., eds. *Ecology and Classification of North American Freshwater Invertebrates.* 3rd ed. Academic Press, New York, 43–96.

SUGGESTED READINGS

Most readers will find good basic information in an invertebrate zoology text, such as Ruppert et al. (2004), Brusca and Brusca (2003), or Smith (2001). More detailed treatments follow.

Anderson, O. R. 1983. *Radiolaria.* Springer, New York. 355 pp.

Anderson, O. R. 1993. The trophic role of planktonic Foraminifera and Radiolaria. *Marine Microbial Food Webs* 7:31–51.

Bockstahler, K. R., Coats, D. W. 1993. Spatial and temporal aspects of mixotrophy in Chesapeake Bay dinoflagellates. *Journal of Eukaryotic Microbiology* 40:49–60.

Boltovskoy, D., ed. 1999. *South Atlantic Zooplankton.* Vol. 1. Backhuys, Leiden, The Netherlands. (See sections by Petz on Ciliophora, pp. 265–319; by Alder on Tintinnoinea, pp. 321–384; and by Bernstein et al. on Acantharia, pp. 75–147.)

Capriulo, G. M. 1982. Feeding of field collected tintinnid microzooplankton on natural food. *Marine Biology* 71:73–86.

Capriulo, G. M., ed. 1990. *Ecology of Marine Protozoa.* Oxford University Press, New York. 357 pp.

Capriulo, G. M., Carpenter, E. J. 1983. Abundance, species composition and feeding impact of tintinnid micro-zooplankton in central Long Island Sound. *Marine Ecology Progress Series* 10:277–288.

Caron, D. A., Swanberg, N. R. 1990. The ecology of planktonic sarcodines. *Review of Aquatic Science* 3:147–180.

Coats, D. W., Revelante, N. 1999. Distributions and trophic implications of microzooplankton. In:

Malone, T. C., Malej, A., Harding, J., et al., eds. *Ecosystems at the Land-Sea Margin.* American Geophysical Union, Washington, DC, 207–240.

Corliss, J. O. 1979. *The Ciliated Protozoa.* 2nd ed. Pergamon Press, Oxford. 455 pp. (Excellent introduction to ciliates; taxonomy is dated.)

Febvre-Chevalier, C. 1990. Phylum Actinopoda: Class Heliozoa. In: Margulis, L., Corliss, J. O., Melkonian, M., et al., eds. *Handbook of Protoctista.* Jones & Bartlett, Boston, 347–362.

Gifford, D. 1991. The protozoan-metazoan trophic link in pelagic ecosystems. *Journal of Protozoology* 38:81–86.

Jakobsen, H. H. 2001. Escape response of planktonic protists to fluid mechanical signals. *Marine Ecology Progress Series* 214:67–78.

Laybourn-Parry, J. 1992. *Protozoan Plankton Ecology.* Chapman & Hall, New York. 213 pp.

Lee, J. J., Anderson, O. R. eds. 1991. *Biology of Foraminifera.* Academic Press, London. 368 pp.

Lynn, D. H. 2008. *The Ciliated Protozoa: Characterization, Classification, and Guide to the Literature.* Springer, New York. 606 pp.

McManus, G. B., Zhang, H., Lin, S. 2004. Marine planktonic ciliates that prey on macroalgae and enslave their chloroplasts. *Limnology and Oceanography* 49:308–313.

Pierce, R. W., Turner, J. T. 1992. Ecology of planktonic ciliates in marine food webs. *Review of Aquatic Science* 6:139–181.

Stoecker, D. K., Johnson, M. D., Vargas, C. de, et al. 2009. Acquired phototrophy in aquatic protists. *Aquatic Microbial Ecology* 57:279–310.

Examples of the "amoeboid" protozoan groups Foraminifera, Radiolaria, and Heliozoa found in estuarine and in nearshore waters are discussed. As with ciliates, this fauna has not been systematically sampled along our coasts except in a few locations.

FORAMINIFERA

ID hint: Forams have multichambered shells, with or without fine radiating spines. Some species (e.g., *Globigerina*) appear as clusters of spheres of different sizes. Others (e.g., "other" sp.) are snail shaped but are not open at the terminal end. Most are 0.1–1 mm in size. Almost all are euhaline.

Globigerina bulloides

Globigerina bulloides is one of the more common temperate, nearshore forams, often found near the surface. The nonspinose *G. bulloides*, unlike tropical species, does not harbor symbiotic algae.

References. Lee et al. 2007; Schiebel et al. 1997.

Other foraminiferans

Many other forams occur in nearshore collections, including some with a "snail-like" shape (illustrated). In shallow areas, benthic forams are frequently swept into the water column. The specimen shown was collected in a saltmarsh creek in South Carolina and is typical of those appearing in many plankton collections.

RADIOLARIA AND ACANTHARIA

ID hint: Most Radiolaria and Acantharia have fine spines radiating from a single spherical cytoplasmic mass, but other shapes occur. Most are 30–300 μm in size. All are marine and typically occur in clear, oceanic waters.

Acanthometron spp.

Acanthometron (sometimes called *Acanthometra*) is one of the more common nearshore acantharians, but many others occur. They appear sporadically when ocean-water masses come ashore and can be numerous when present. They capture small prey, primarily tintinnids, with sticky pseudopods.

HELIOZOA

ID hint: If radiolarian-like protozoans are found in brackish waters, they are likely to be the heliozoan *Actinophrys.*

Actinophrys sol

Actinophrys sol has numerous readily visible vacuoles and a granular interior. It is widely distributed in US marine and brackish waters where it is the only common heliozoan reported thus far. It is common to abundant in Chesapeake Bay; summer to fall, in 10–24 psu (practical salinity units). *Actinophrys* can be maintained on ciliate prey in the lab but also eats small cryptophytes and dinoflagellates, including *Pfiesteria.*

References. Hausmann and Patterson 1982; Patterson 1979; Pierce and Coats 1999.

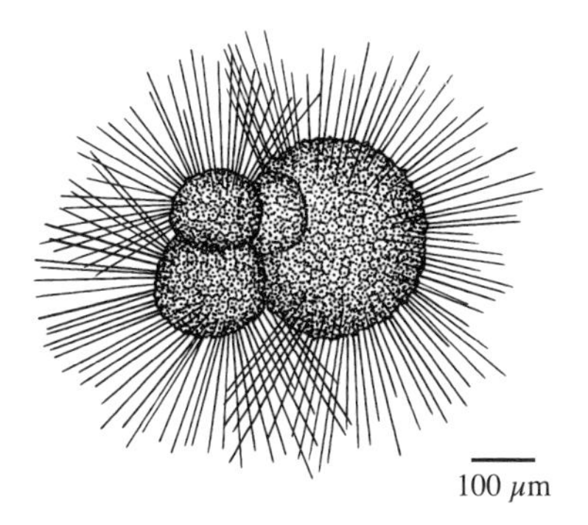

Globigerina bulloides
(foraminiferan)

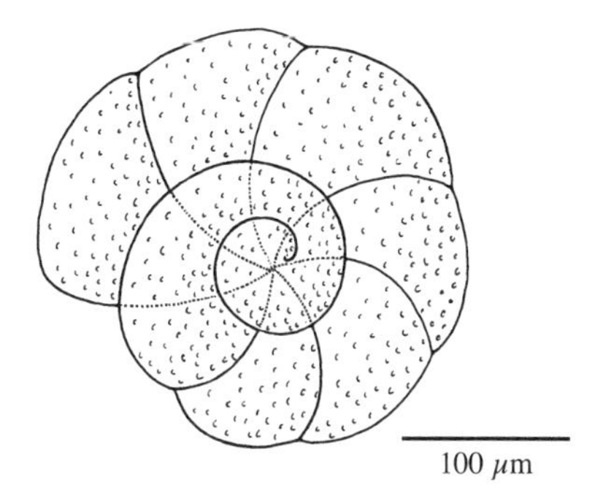

"other" foraminiferan

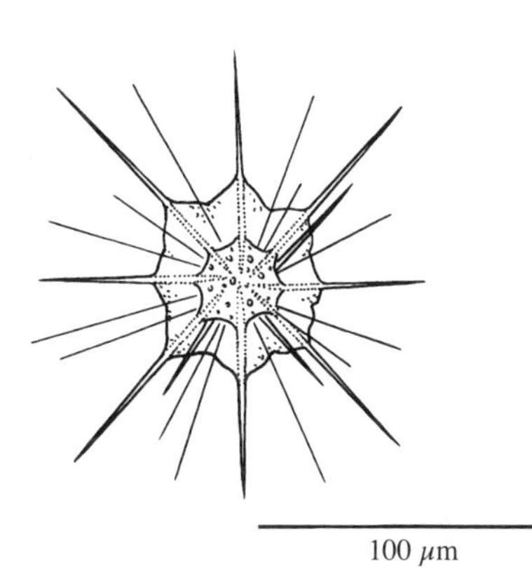

Acanthometron sp.
(acantharian)

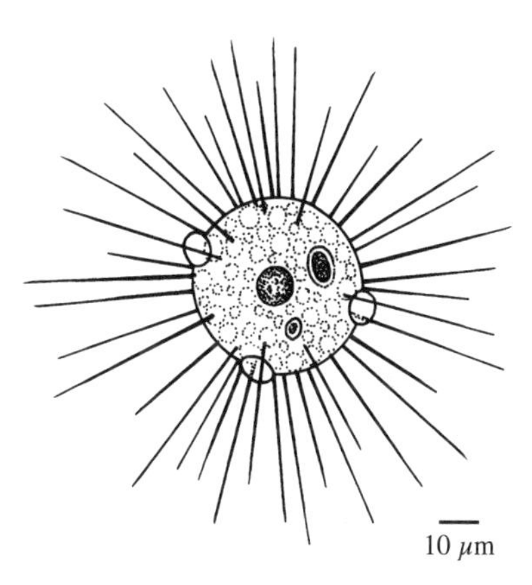

Actinophrys sol
(heliozoan)

CILIATE PROTOZOA: TINTINNIDS

ID hint: Tintinnid ciliates have a hard covering, or lorica.

Tintinnopsis, Tintinnidium, and *Favella*

These three genera are among the most common tintinnids and occur from open ocean to mesohaline reaches of estuaries. Their abundance peaks during nanophytoplankton blooms. Copepods, jellyfish, ctenophores, and fish larvae eat tintinnids.

References. Capriulo 1982; Capriulo and Carpenter 1983; Kivi and Setälä 1995; Laval-Peuto 1981; Stoecker et al. 1981, 1995; Sun et al. 2007; Verity and Villareal 1986.

NAKED CILIATES

ID hint: These almost transparent mobile cells lack a hard outer covering. Many small invertebrate larvae and rotifers are also ciliated. Refer to "Quick Picks" for comparisons.

Paranassula microstoma

Paranassula microstoma is a representative of the marine/estuarine ciliates completely covered with cilia. In freshwater, members of the closely related genus *Nassula* eat entire strands of filamentous Cyanobacteria.

Strombidium, Strobilidium, and *Laboea*

These genera are widespread along the Atlantic and Gulf Coasts and may be abundant in estuaries, especially in the summer. They feed on bacteria and on eukaryotic plankton in the 2–10 μm size range. *Laboea* and *Strombidium capitatum* feed on small phytoplankton and steal their chloroplasts for photosynthesis. Mixotrophic dinoflagellates, jellyfishes and their ephyrae, and copepods feed on these ciliates.

References. Doherty et al. 2010; Dolan 1991; Johansson and Coats 2002; Kivi and Setälä 1995; McManus et al. 2004; Montagnes 1996; Ohman and Snyder 1991; Sanders 1995; Setälä et al. 2005; Stoecker et al. 1988; Sun et al. 2007.

Myrionecta rubra (=*Mesodinium rubrum*)

Myrionecta rubra is a large ciliate (30–100 μm) divided into two lobes by a waistlike furrow. A band of cilia surround the girdle of the cell, with a cilia skirt on the posterior end and tentacle-like cilia on the anterior end. Chloroplasts visible in the interior often produce a greenish or pinkish color. *Myrionecta rubra* is common in coastal oceans to oligohaline waters, from Maine to South Carolina but are less reported in the Gulf. This obligate autotroph captures small cryptophyte phytoplankton and sequesters their chloroplasts. Dense blooms may cause red tides. Predators include ciliates, hetero- and mixotrophic dinoflagellates, and ctenophores. *Mesodinium pusilla* is common in the Gulf.

References. Bulit et al. 2004; Fenchel and Hansen 2006; Gustafson et al.; 2000; Hansen and Fenchel 2006; Johnson and Stoecker 2005; Johnson et al. 2007; Lindholm 1985; Litaker et al. 2002; Sanders 1995; Stoecker et al. 2009; Yih et al. 2004.

Zoothamnium spp.

Zoothamnium species are colonial ciliates with up to 20 or more zooids connected to a single stalk. The colony contracts as a unit when disturbed. *Zoothamnium* often attaches to planktonic crustaceans, including copepods, shrimps, and amphipods. Infestations of copepods may be so dense that they hinder swimming and feeding.

References. Herman and Mihursky 1964; Herman et al. 1971; Utz and Coats 2005, 2008; Weissman et al. 1993.

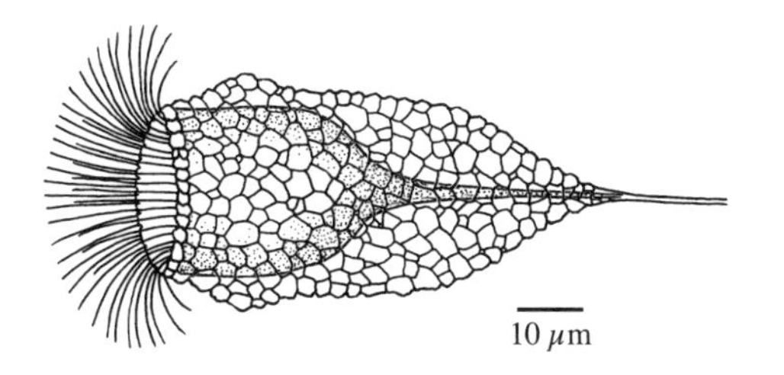

Tintinnopsis sp.

10 μm

Paranassula microstoma

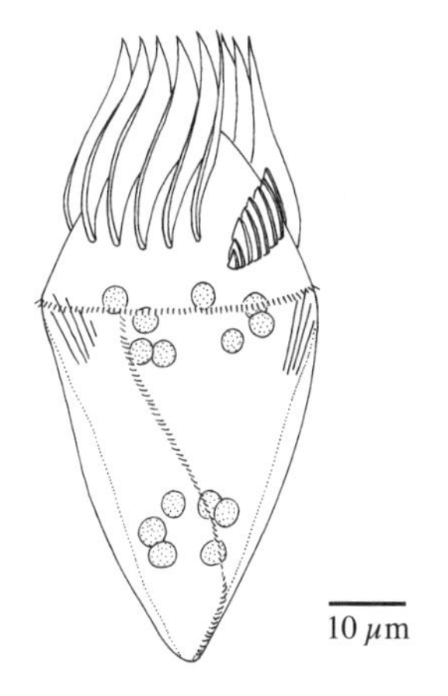

Strombidium sp.

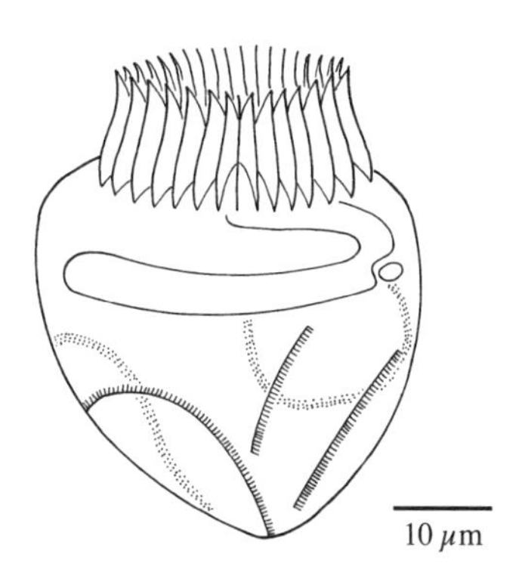

Strobilidium sp.

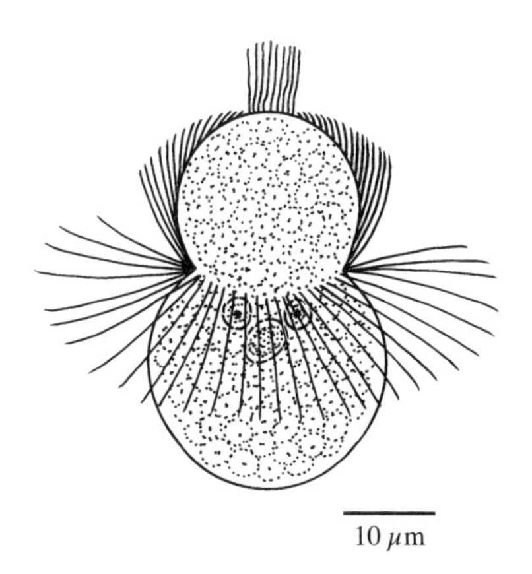

Myrionecta rubra

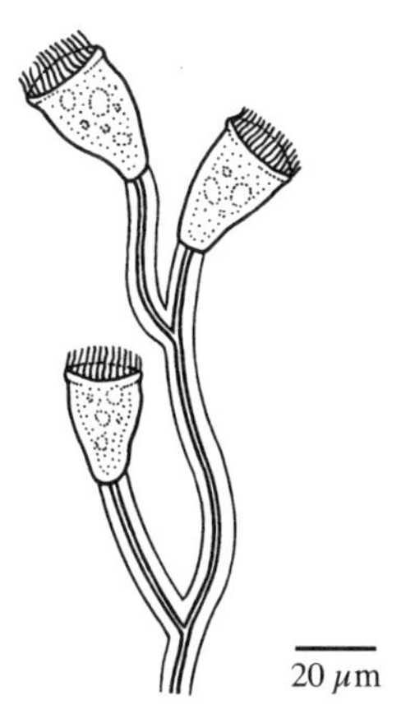

Zoothamnium sp.

CNIDARIANS

Anemones, Jellyfishes, and Related Metazoans

The Phylum Cnidaria, sometimes called Coelenterata, contains the simplest of the true metazoans. These radially symmetrical animals have a central mouth, usually surrounded by tentacles. The simple tissues, an outer epidermis and inner gastrodermis, surround a layer of gelatinous mesoglea. Cnidarians are distinguished by having unique stinging cells, cnidoblasts, which are largely responsible for the remarkable success of the phylum. Many cnidarians have two distinct body forms in their life cycle, a sessile polyp stage and swimming medusa stage. The classes Hydrozoa, Scyphozoa, and Cubozoa have planktonic medusoid stages. The Staurozoa are sessile medusoids, and the Anthozoa (e.g., anemones, corals, sea whips) are typically sessile polyps. Only the larvae of the latter two are planktonic.

DISTINGUISHING CHARACTERISTICS OF PLANKTONIC CNIDARIANS

Hydrozoans (Class Hydrozoa)

Hydromedusae have a thin, transparent bell usually <2 cm in diameter. The mouth extends into a narrow tube but typically lacks oral arms. A thin velum extends inward from the edge of the bell. A tubelike manubrium may extend from the stomach.

Porpitids (*Velella*, the by-the-wind sailor, and *Porpita*, the blue button) are colonial, open-ocean animals that consist of a flattened rigid or semirigid disc surrounded by tentacle-bearing dactylozooids extending downward from the subsurface. Porpitids are a distinctive bluish color and float on the surface, often in aggregations.

Siphonophores, comprising a distinct order of Hydrozoa, have one or more elongated swimming bells that trail a string of tentacles. An exception is the Portuguese man-of-war, which has a large float and many long tentacles.

Scyphozoans (Class Scyphozoa) are large "jellyfishes" with a diameter often exceeding 6 cm. Oral arms surround the mouth, and tentacles may or may not also be present.

Cubozoans (box jellies, Class Cubozoa) are jellyfishes with a squarish bell and with tentacles attached at the corners.

The inclusion of large and often surprisingly fast-swimming scyphozoan and cubozoan medusae among the zooplankton is often debated. Since adults are often caught in collections along our coasts, are readily observable, begin life as much smaller forms, and consume many types of smaller zooplankton, we include them in this guide. There are relatively few species of large planktonic hydrozoans but many species of small hydromedusae.

HYDROZOANS

Hydromedusae

Hydromedusae are roughly umbrella or bell shaped, transparent, and usually only a few millimeters in diameter. They are primarily marine to mesohaline; few occur in low salinities. These small, delicate creatures are often mangled in plankton samples. If alive, their pulsing movement is unmistakable. Many species occur, but most are planktonic for only a brief period each year. Only the most common are treated here.

Hydromedusae represent the sexual medusoid stage in the life cycle of most hydroids. Hydromedusae are solitary individuals that shed either sperm or eggs into the water. The resulting planula larvae settle to develop into the benthic polyp form. An exception is *Liriope* where the planulae transform directly into the medusa stage. In some groups, the planula develops into a polyp-like actinula larva (Fig. 10) that then transforms into either a medusa or a polyp, depending on the species. Release of hydromedusae seems largely controlled by temperature. Hydromedusae may vary in abundance from year to year, but they are usually present at the same times each year and for the same duration.

Contractions of the bell propel water backward and the hydromedusa forward. In nearly all species of hydromedusae, the opening across the bell may be partially constricted by a thin velum that acts as a nozzle to increase the velocity of flow and helps direct the water, resulting in turns and maneuvers. Swimming in species with a shallow bell and poorly developed velum is slower. Pigmented ocelli (light receptors) and statocysts (gravity receptors) on the margin of the bell help determine orientation. Hydromedusae use their stinging capsules (nematocysts) to capture a variety of zooplankton and can be categorized as either

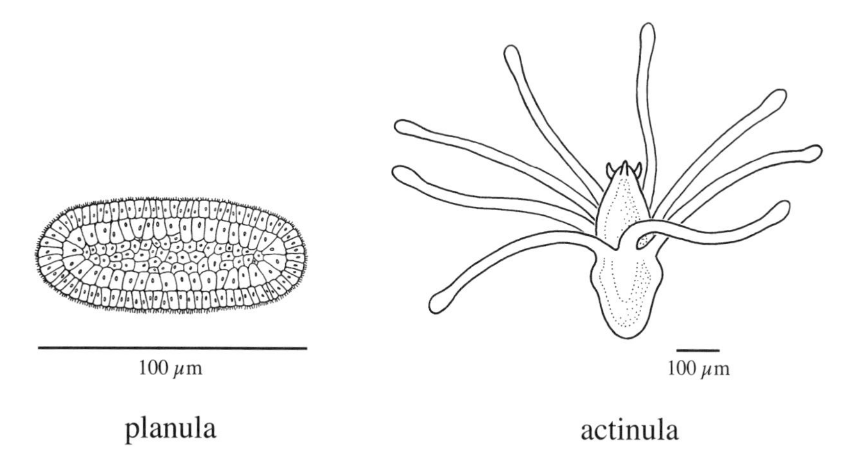

Fig. 10. The planktonic stages of hydroids include the planula larva, the actinula stage (in some species), and the hydromedusa (Fig. 11). The hydromedusae mature into separate sexes and shed their gametes into the plankton where fertilization occurs. The resulting planula larva settles to the bottom to initiate a new polyp. The actinula, when present, occurs at different stages in the life cycle for different species.

ambush or cruising predators. Ambush feeders hang motionless in the water and wait for mobile prey, especially small crustaceans and fish larvae, to swim into their long tentacles. In contrast, cruising predators swim more constantly and rely on the vortices created by swimming pulses to draw smaller and less motile prey into their short tentacles. Ciliated larvae, invertebrate and fish eggs, other jellyfishes, and larvaceans often fall prey to the cruising predators.

Siphonophores

Siphonophores in our area are entirely planktonic hydrozoans. Most are animals of the open ocean, but a few occur in nearshore waters in high salinities. These are complex, colonial animals made of individuals (zooids), including both polyp and medusoid forms. Although the zooids are variously modified for locomotion, feeding (**gastrozooids**), reproduction (**gonozooids**), prey capture and defense (**dactylozooids**), or flotation (pneumatophores) the colony functions as a single animal. The colony grows by asexual budding of individual zooids, but details vary, depending on the group. Colonies contain both male and female zooids and reproduce sexually. The swimming bells (nectophores) of most siphonophores contract to provide propulsion. In *Nanomia*, a gas float at the head of the colony can inflate or deflate to adjust vertical position. *Physalia*, the Portuguese man-of-war, is the largest and most famous animal in this group. Although *Physalia* lacks any form of active locomotion, its sail-like float is ideally suited for wind-assisted propulsion. Wind and currents may take these high-seas sailors thousands of kilometers.

Siphonophores trail tentacles arrayed with batteries of stinging nematocysts from their feeding polyps. The common nearshore siphonophores spread their tentacles and feed on small crustaceans while drifting passively. The Portuguese man-of-war (*Physalia*) trails long tentacles for feeding. When food is captured, the tentacles contract and bring food, primarily small fish, to the mouth of feeding polyps for digestion. Sea turtles, jellyfishes (scyphomedusae), and some comb jellies eat siphonophores.

Porpitids

Porpitids, once known as "chondrophores," or "capitate hydrozoans," are widespread from tropical to temperate waters where they float at the surface. Like the siphonophores, porpitids are complex colonies composed of specialized polyps, or zooids, each modified for a specific job. Some polyps form a central, gas-filled disc for floatation. Tentacle-like dactylozooids suspended from beneath the margin capture small zooplankton. Reproductive zooids or gonozooids release tiny medusae that give rise to new colonies. A large central gastrozooid carries out digestion and other functions. The two most common genera, *Porpita* (blue buttons) and *Velella* (by-the-wind sailors), typically drift in large aggregations at the surface. Beachcombers often find groups of *Velella* or *Porpita* stranded on the shore. Once included with the siphonophores, this group has proved difficult to classify.

IDENTIFICATION HINTS (HYDROMEDUSAE ONLY)

The following characteristics are consistently useful, and three to five are noted for most hydromedusae treated. Refer to Figure 11 when reviewing the list of anatomical features below. Immature specimens may not have all of the adult characteristics as shown in Figure 12 where immature specimens of *Turritopsis* do not have fully developed gonads and may not have the same number of tentacles and shape as the adults. Immature individuals are usually smaller than the sizes listed for the species.

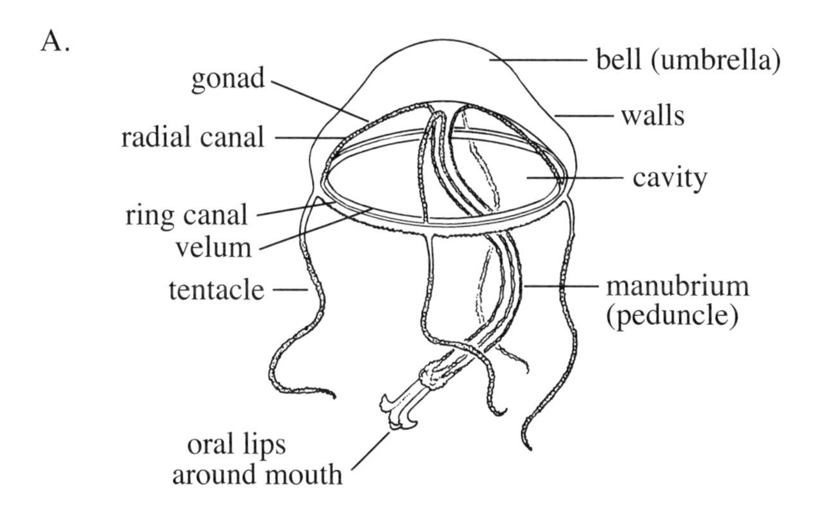

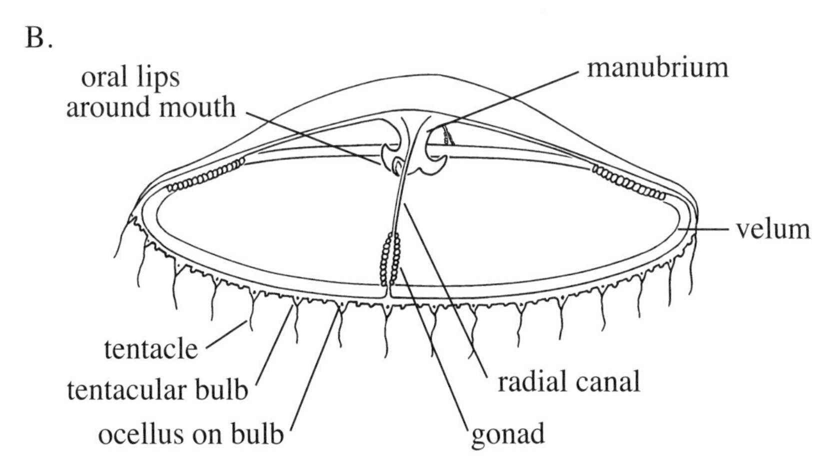

Fig. 11. Anatomical features of hydrozoan medusae. The presence and configuration of anatomical features varies considerably among hydrozoans; these two species represent major variations in several features used to identify members of this group. *A*, *Eutima mira* with a long manubrium, few complex tentacles without ocelli, and elongate gonads associated with both the radial canals and manubrium. *B*, *Clytia hemisphaerica* with a short manubrium, ocelli at the bases of numerous tentacles and oval gonads associated with the radial canals.

Bell shape and thickness are often distinctive, although medusae with thin-walled bells tend to lose their shape after death.

Manubrium thickness, shape, and/or length are reliable characteristics. Especially useful is the presence or absence of "lips" at the manubrium tip. The manubrium is an elongated extension of the connection between the mouth and the stomach. Some species have thin, branched oral tentacles on the manubrium.

The **velum** is a thin flap of tissue on the inside margin of the bell. Note its presence, width, or absence, but because of its fragility and variation among developmental stages, the velum may not be a very reliable feature.

Gonad location, shape, and size are diagnostic features on mature specimens.

Tentacle length is relatively fixed in some species and highly variable in others as tentacles extend and contract.

Tentacle numbers are predictable for the largest (mature) specimens, but since there are usually fewer in young individuals, numbers may be misleading identification features.

Marginal bulbs and ocelli. Swellings or bulbs often occur along the bell margin. They are usually at the bases of the tentacles. Note the presence or absence of these bulbs, their size and shape, and the presence or absence of pigmented ocelli within the bulbs.

The number of **radial canals** is often a reliable identification feature for hydromedusae.

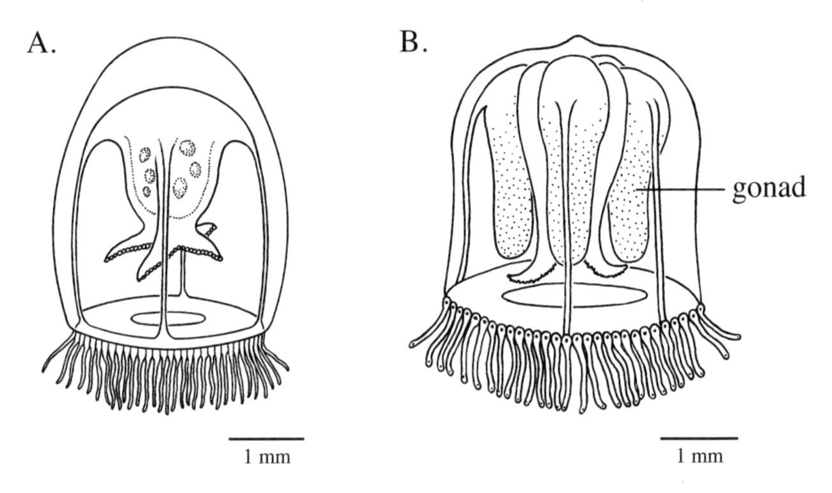

Fig 12. Anatomical changes with development in *Turritopsis nutricula*. The immature form (*A*) is only slightly smaller than the mature form (*B*). Overall shape changes from a tall, thimble-shaped form to a squarish form with a fairly flat dorsal surface. The manubrium and stomach, which are move visible in the immature form, become obscured as the gonads grow. The numbers and shapes of tentacles of *Turritopsis* do not change much during development, but dark ocelli become conspicuous and the color of this orange-yellow medusa intensifies. These and other changes with development are widespread among hydromedusae but not consistent for all species

USEFUL IDENTIFICATION REFERENCES

Allwein, J. 1967. North American hydromedusae from Beaufort, NC. *Videnskabelige Meddeleserfra Dansk Naturhistorisk Foreningi Kobenhavn* 130:117–136.

Burke, W. D. 1975. Pelagic Cnidaria of Mississippi Sound and adjacent waters. *Gulf Research Reports* 5:23–38.

Calder, D. R. 1971. Hydroids and hydromedusae of southern Chesapeake Bay. Virginia Institute of Marine Science. *Special Papers in Marine Science* 1:1–125.

Kramp, P. L. 1959. *The Hydromedusae of the Atlantic Ocean and Adjacent Waters*. Dana Report No. 46. Carlsburg Foundation, Høst and Son, Copenhagen. 283 pp.

Mayer, A. G. 1910. *Medusae of the world*. Vols. 1, 2, *The Hydromedusae*. Carnegie Institution, Washington, DC. 513 pp.

Russell, F. S. 1953. *The Medusae of the British Isles: Anthomedusae, Leptomedusae, Limnomeduse, Trachymedusae and Narcomedusae*. Cambridge University Press, London. 530 pp. (Keys, illustrations, and notes on the biology of these hydromedusae.)

Totton, A. K. 1965. *A Synopsis of the Siphonophora*. British Museum Natural History, London. 230 pp.

SUGGESTED READINGS

Arai, M. N. 1992. Active and passive factors affecting aggregations of hydromedusae: A review. *Scientia Marina* 56 (2–3): 99–108.

Bouillon, J., Gravili, C., Pagès, F., et al. 2006. *An Introduction to Hydrozoa. Mémoires du Muséum National d'Histoire Naturelle, Paris* 194. 591 pp.

Colin, S. P., Costello, J. H. 2002. Morphology, swimming performance and propulsive mode of six co-occurring hydromedusae. *Journal of Experimental Biology* 205:427–437. (This Pacific study is generally applicable to all hydromedusae.)

Costello, J. H., Colin, S. P. 2002. Prey resource use by coexistent hydromedusae from Friday Harbor, Washington. *Limnology and Oceanography* 47:934–942. (Several common Atlantic species are treated in this review of feeding in hydromedusae.)

Daly, M., Brugler, M. R., Cartwright, P., et al. 2007. The phylum Cnidaria: A review of phylogenetic patterns and diversity 300 years after Linnaeus. *Zootaxa* 1668:127–182. (This work presents a more advanced classification system of Hydrozoa than that in Bouillon et al. [2006]).

Mackie, G. O., Pugh, P. R., Purcell, J. E. 1987. Siphonophore biology. *Advances in Marine Biology* 24:97–262.

Mills, C. E., Boero, F., Migotto, A., Gili, J. M. eds. 2000. Proceedings of the Fourth Workshop of the Hydrozoan Society, Bodega Bay, California. *Trends in Hydrozoan Biology* 64(Suppl. 1). *Scientia Marina*. 284 pp.

Pugh, P. R. 1999. Siphonophorae. In: Boltovskoy, D., ed. *South Atlantic Zooplankton*. Vol. 1. Backhuys Publishers, Leiden, The Netherlands, 467–511.

Purcell, J. E., Mills, C. E. 1988. The correlation between nematocyst types and diet in pelagic Hydrozoa. In: Hessinger, S. A., Lenhoff, H. M., eds. *The Biology of Nematocysts*. Academic Press, San Diego, CA, 463–485. (This treatment has few local species, but is one of the few articles on hydromedusan feeding available.)

Schuchert, P. 2011. *World Hydrozoa Database*. <www.marinespecies.org/hydrozoa>. Accessed August 28, 2011. (This site provides a list of all currently valid species of Hydrozoa within a contemporary classification system. Also included is an bibliography on the group.)

HYDROMEDUSAE

Species are arranged by morphological similarities rather than by taxonomic groups.

> **ID hint.** Use the position of the gonads and the shape of the manubrium to help separate the umbrella-shaped species on this page.

Clytia hemisphaerica

Occurrence. *Clytia hemisphaerica* occurs throughout our geographic range in both the Atlantic and in the eastern Gulf and may be common in estuaries. *Clytia linearis* occurs in the Gulf.

Biology and Ecology. Large numbers of the similar *Clytia gracilis* polyps were noted in the plankton of Georges Bank where they were feeding on fish larvae and other animals in the water column.

References. Adamik et al. 2006; Cornelius 1982; Lindner and Migotto 2002; Madin et al. 1996.

Clytia mccradyi, formerly *Phialidium mccradyi*, and *C. noliformis*

Occurrence. This species is reported from Florida and South Carolina but may be more widespread along both the Atlantic and Gulf Coasts. The similar *Clytia noliformis* occurs at least from South Carolina southward on the Atlantic Coast.

Comments. Hydromedusae in this genus were previously classified as *Phialidium* before they were matched to the polyps of *Clytia*, and the latter name has priority.

References. Lindner and Migotto 2002.

Obelia spp.

Occurrence. *Obelia* medusae of various species occur from Massachusetts to Florida and throughout the Gulf in warmer months, typically in moderate to higher salinities.

Biology and Ecology. Perhaps as many as six species of *Obelia* occur in our geographic range. The medusae of different species of *Obelia* are virtually indistinguishable, even by experts, and identification can be made only from the hydroid stages. These species appear to be almost flat.

Clytia hemisphaerica

5 - 20 mm wide

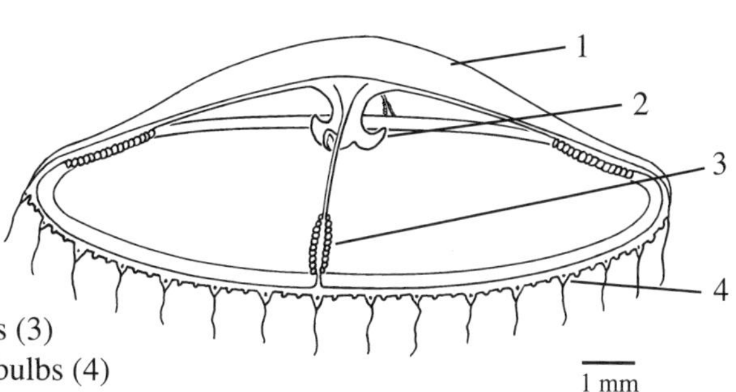

- umbrella-shaped bell with thick dome and thin walls (1)
- short manubrium with recurved lips (2)
- gonads low on radial canals (3)
- 15-58 tentacles with basal bulbs (4)
- similar species: other *Clytia* species; *Obelia* has a rectangular manubrium and sac-like gonads

---

Clytia mccradyi

8 - 20 mm wide

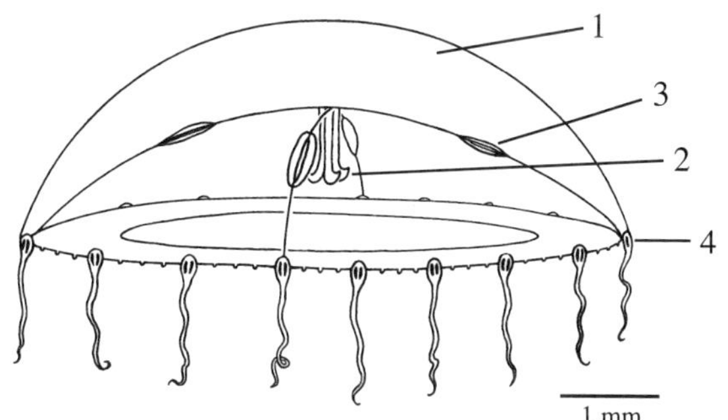

- umbrella-shaped bell with thick dome (1)
- short manubrium with recurved lips (2)
- small gonads in middle of 4 radial canals (3)
- 16-24 hollow marginal tentacles (4)
- similar species: other *Clytia* species; *Obelia* has a rectangular manubrium and globular gonads

---

Obelia sp.

2 - 6 mm wide

- umbrella almost flat with uniformly thin walls (1)
- velum very narrow
- manubrium rectangular with simple lips (2)
- sac-like gonads in middle of 4 radial canals (3)
- 16-90 stiff tentacles with swollen bases (4)
- similar species: many *Obelia* spp. are similar; *Clytia* spp. are larger and have a manubrium with recurved lips and a dome-shaped bell

Proboscidactyla ornata

Occurrence. This species is common near the northern extent of its distribution (especially in Buzzards Bay, Massachusetts, and Narragansett Bay, Rhode Island). It occurs south to Florida and throughout the Gulf.

Biology and Ecology. The hydroid stage is a remarkable two-tentacled polyp, which is commensal on tubes of sabellid polychaetes.

References. Calder 1970a.

Cunina octonaria

Occurrence. Although *Cunina octonaria* is considered oceanic, it occurs in nearshore waters from at least New Jersey to Florida and throughout the Gulf and is sometimes common in mesohaline areas.

Biology and Ecology. Holoplanktonic: The early actinula stage is parasitic on other hydromedusae, especially *Turritopsis nutricula*. It occurs in early autumn in the Chesapeake Bay.

References. Brooks 1886.

Persa incolorata

Occurrence. *Persa incolorata* is one of the more common hydromedusae along the Gulf Coast, especially in island passes, at temperatures of 10°–30°C. It occurs in the Atlantic at least as far north as South Carolina.

Proboscidactyla ornata 3 - 8 mm wide

- cup-shaped bell with thick dome (1)
- manubrium short with recurved lips (2)
- lobed stomach at base of 4 radial canals (3)
- radial canals branch toward margin (4)
- 16-20 tentacles with ocelli at bases (5)

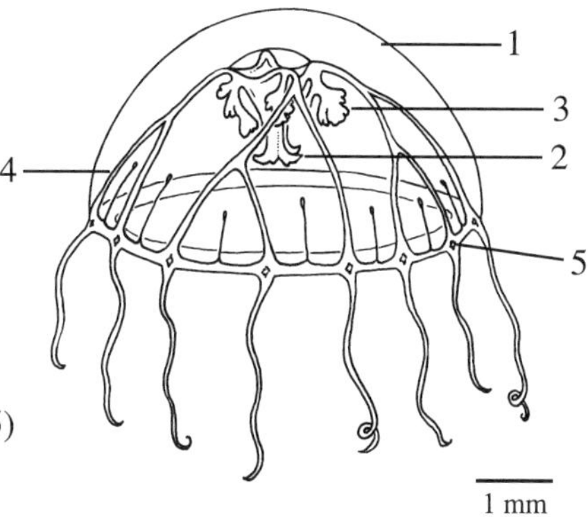

Cunina octonaria 5 - 7 mm wide

- hemispherical bell without radial canals (1)
- 8 broadly square stomach pouches (2)
- 8 tentacles project from midline just below pouches (3)
- 8 lappets with short projections (4)

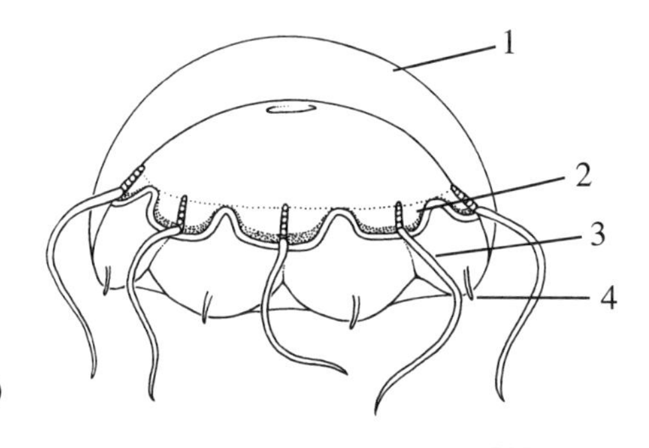

Persa incolorata 3 - 5 mm wide

- thimble-shaped bell with slight apical knob (1)
- long, tubular manubrium with rounded lips (2)
- 2 pendant gonads along opposite radial canals (3)
- up to 48 long, delicate tentacles with terminal knobs (4)

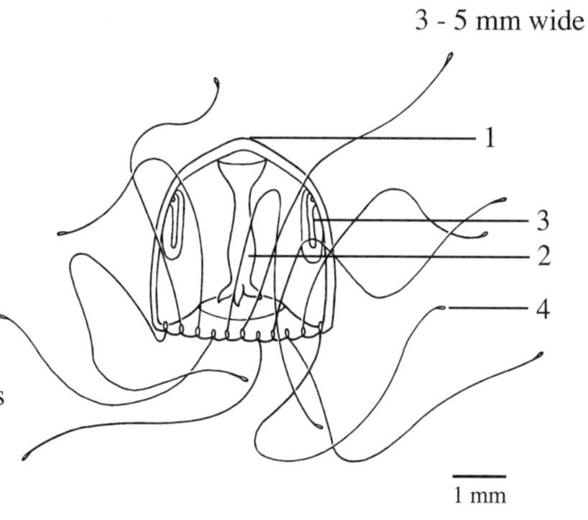

Aequorea forskalea, formerly *A. aequorea*

Occurrence. *Aequorea forskalea* is found in the Atlantic as far north as Canada and is reported in the Gulf Stream and off the Dry Tortugas, Florida. *Aequorea globosa, A. macrodactyla*, and *A. pensilis* occur in the Gulf.

Biology and Ecology. The highly bioluminescent *A. forskalea* produces light using an unusual green fluorescent protein called aequorin. Aequorin has now been incorporated in *E. coli* as a fluorescent marker widely used in molecular biology.

Maeotias marginata, formerly *M. inexspectata*

Occurrence. This native of the Black Sea was first reported from the Chesapeake Bay in 1969 where it is now well established in several tributaries in salinities of 4–10. It has spread south to at least to South Carolina.

Biology and Ecology. *Maeotias marginata* feeds on small crustaceans, including crab larvae.

References. Calder and Burrell 1969.

Margelopsis gibbesii

Occurrence. *Margelopsis gibbesii* occurs along the Atlantic Coast and occasionally in high-salinity reaches of estuaries or sounds from Maryland to South Carolina.

Biology and Ecology. *Margelopsis* is remarkable for having a planktonic polyp (actinula) stage produced from the female's manubrium via budding. This unusual actinula has two whorls of tentacles.

References. Werner 1955.

Aequorea forskalea up to 250 mm

- umbrella-shaped bell with thick center (1)
- base of manubrium about half as wide as bell (2)
- radial canals usually 60-80 (range <50 to >150) (3)
- tentacles usually less numerous than radial canals (range ½ to 2× as many) (4)

- bulbs on each tentacle small, elongate, and conical (5)

- color blue, violet, or milky white; gonads along canals sometimes colored
- similar species: wide manubrium base and numerous radial canals distinguish it from *Maeotias* and small hydrozoan medusae

---

Maeotias marginata up to 550 mm

- hemispherical bell with thick mesoglea (1)
- large manubrium with 4 long frilled palps (2)
- 4 radial canals; elongate, folded gonads attached (3)
- 300-500 similar marginal tentacles (4)
- similar species: frilled palps and abundant tentacles distinguish it from *Aequorea* and other hemispherical medusae

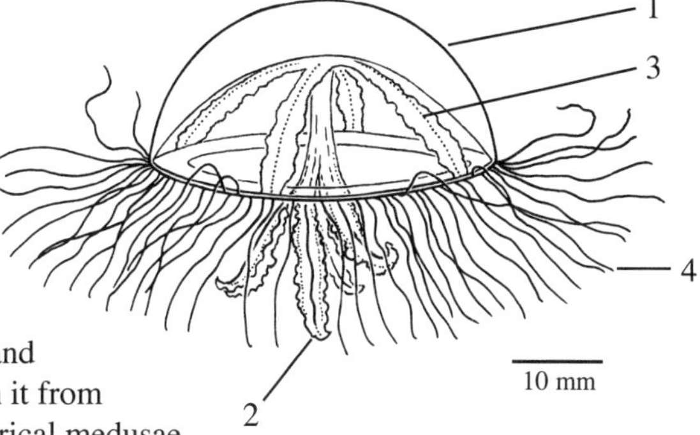

Margelopsis gibbesii (male) 2 - 3 mm wide

- high, somewhat rectangular bell (1)
- very large, unbranched manubrium without oral tentacles (2)
- in female, manubrium covered with polyp buds
- 4 radial canals with large marginal bulbs (3)
- 5 or 6 tentacles on each bulb (4)
- males greenish; females yellow to brown
- similar species: *Nemopsis* and *Bougainvillia* have small manubria

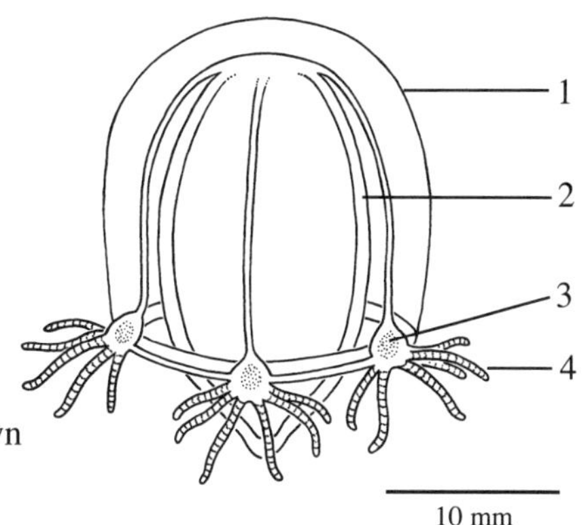

ID hint. These medusae are all tall with many closely spaced tentacles on the margin. The length of the manubrium and location of the gonads easily separate them.

Aglantha digitale (pink helmet)

Occurrence. *Aglantha digitale* is an Arctic-boreal species most common on the shelf and caught occasionally in inshore samples in winter to spring from Cape Cod, Massachusetts, to the Chesapeake Bay.

Biology and Ecology. This fast swimmer is holoplanktonic. The actinula larva transforms directly into a medusa. In Norway, *Aglantha* eats the copepods *Oithona similis* and *Temora longicornis* and the cladoceran *Evadne nordmanni*. The Atlantic mackerel eats *Aglantha* in the lab. The medusae have a life span of about one year, with a new generation beginning each spring.

References. Mackie et al. 2003; Pagès et al. 1996.

Turritopsis nutricula

Occurrence. Hydroids occur from Cape Cod through the Gulf. The medusae are reported as common in summer and in fall in mesohaline to high-salinity estuarine areas from Cape Cod to the Carolinas. They can be common in the northern Gulf in summer.

Biology and Ecology. Medusae may harbor the actinula stage of the hydromedusa *Cunina octonaria* attached to the underside of the umbrella, apparently as a parasite. *Turritopsis* medusae may settle to the bottom and transform into polyps.

References. Brooks 1886; Miglietta et al. 2007.

Blackfordia virginica

Occurrence. *Blackfordia virginica*, first recorded in Virginia and Delaware, is the most common hydromedusa in South Carolina during summer, but its distribution is uncertain in the other localities along the Atlantic Coast. It is reported in the northeastern Gulf. *Blackfordia* is euryhaline and found in salinities of 1–34.

Biology and Ecology. *Blackfordia* is a nonnative species, apparently originating from the Black Sea.

Aglantha digitale

5 - 15 mm wide

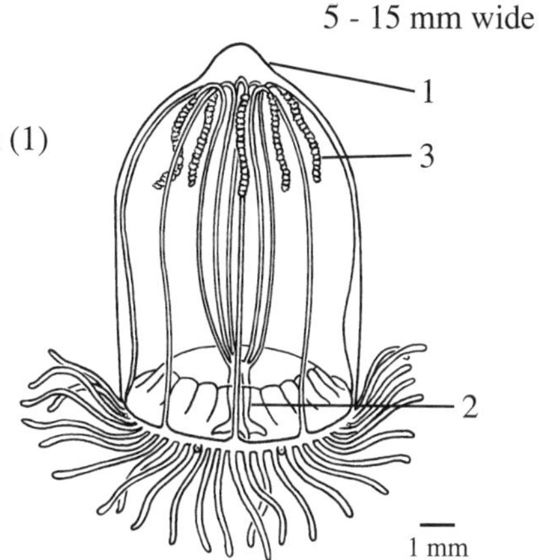

- tall, rectangular bell with apical projection (1)
- narrow manubrium almost as long as bell with simple lips (2)
- long gonads hang from top of cavity (3)
- 80 or more tentacles
- color usually pink to carmine
- similar species: *Turritopsis nutricula* has shorter, thicker manubrium and marginal bulbs with ocelli

Turritopsis nutricula

4 - 11 mm wide

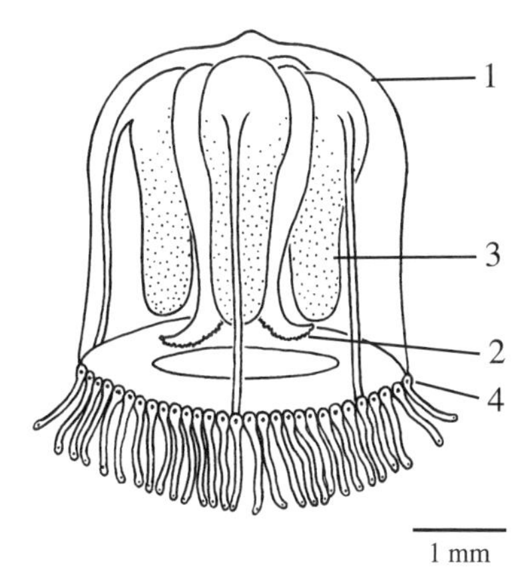

- thimble-shaped bell with thin walls (1)
- large manubrium with large lips (2)
- 4 very large gonads flank sides of manubrium (3)
- 80-120 tentacles with ocelli on bulbs (4)
- manubrium and stomach orange yellow; ocelli dark
- similar species: *Aglantha digitale* has narrow manubrium and no marginal bulbs or ocelli

Blackfordia virginica

10 - 14 mm wide

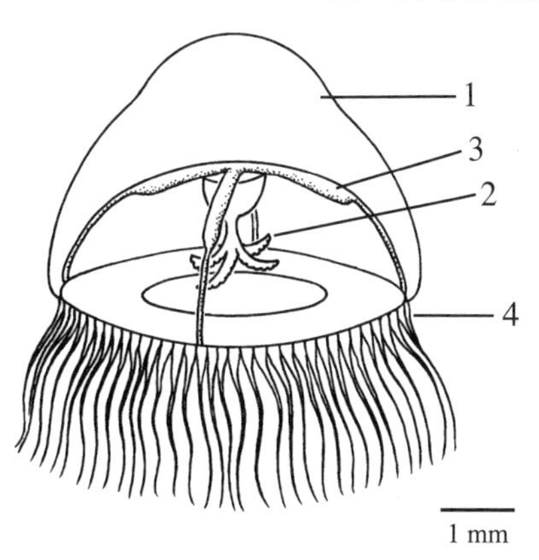

- bell shaped with rounded apex and thick walls (1)
- narrow manubrium with long recurved lips (2)
- gonads along upper half of radial canals (3)
- up to 80 tentacles with wide bases (4)

ID hint. These medusae have very long manubria and only four tentacles, but the shape and texture of both of these structures differ considerably among the species.

Eutima mira

Occurrence. *Eutima mira* occurs from Massachusetts to Florida, but its nearshore distribution is uncertain, except in the Carolinas.

Slabberia strangulata

Occurrence. Reported from Massachusetts to Florida, *Slabberia* may be abundant in estuaries in salinities as low as 18. *Slabberia strangulata* occurs spring through fall in the Chesapeake Bay. *Stauridiosarsia ophiogaster* (formerly *Dipurena ophiogaster*) may occur in the eastern Gulf.

Reference. Calder 1970b.

Sarsia tubulosa (clapper medusa)

Occurrence. This northern species occurs winter to spring from Canada to the Chesapeake Bay along the coast and in high-salinity reaches of estuaries. *Sarsia gemmifera* and *S. prolifera* are reported from the Florida Gulf Coast.

Biology and Ecology. In some areas, *Sarsia* can be brightly colored, but this coloration is variable from brown to red to blue, if present at all.

Eutima mira

15 - 30 mm wide

- triangular bell with thick dome (1)
- very long, slender manubrium with recurved lips (2)
- 8 gonads: 4 along radial canals, plus 4 along manubrium (3)
- coarse-edged margin (4)
- 4 long tentacles (5)
- similar species: *Sarsia* species and *Slabberia* do not have recurved lips or a coarse margin

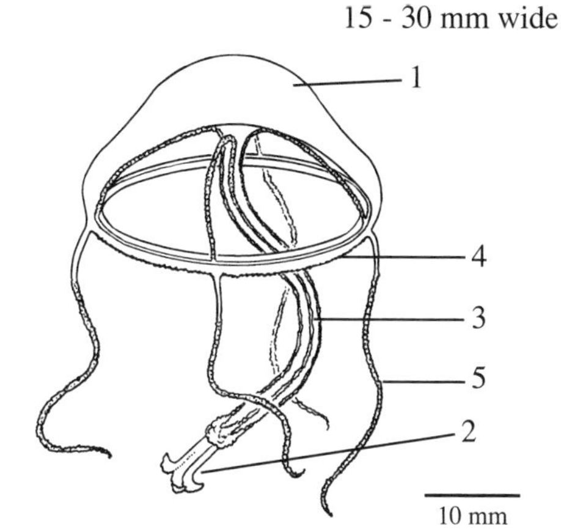

Slabberia strangulata

3 - 4 mm wide

- tall, domed bell with very thick walls (1)
- very long manubrium with surrounding gonads at middle and end (2)
- 4 stiff tentacles with terminal knobs (3)
- similar apecies: *Sarsia* species; *S. tubulosa* has a manubrium with uniform thickness and tentacles without knobs; *Eutima* has recurved lips and a coarse margin

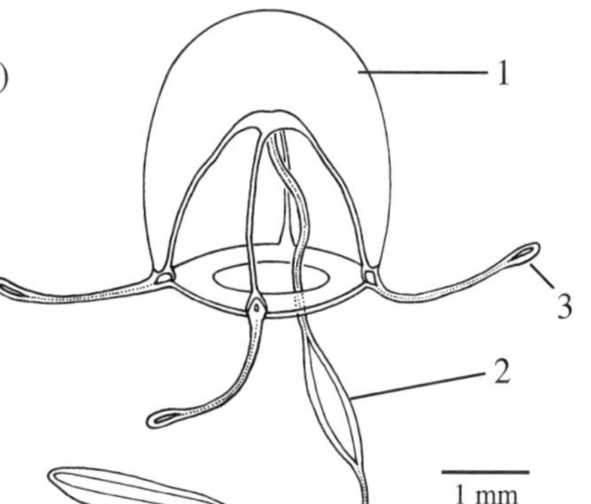

Sarsia tubulosa

8 - 12 mm wide

- high, bell-shaped medusa with moderately thick walls (1)
- very long manubrium of uniform width (2)
- distinct globular chamber at top of radial canals (3)
- 4 tentacles with clusters of large nematocysts (4)
- similar species: *Slabberia* has a manubrium with variable thickness and knobs on ends of tentacles; *Eutima* has recurved lips and a coarse margin

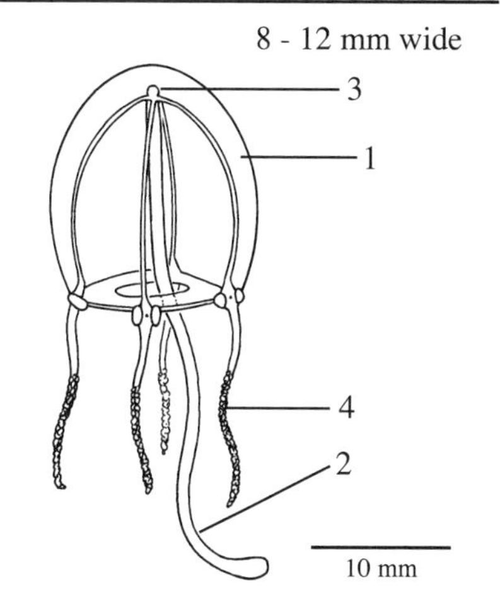

Liriope tetraphylla

Occurrence. *Liriope tetraphylla* occurs worldwide in warm waters, occasionally as far north as Cape Cod. Catches in the plankton are most commonly reported in summer and in fall from Delaware Bay south and in the Gulf where they can be locally abundant in mid to high salinities, especially in warm waters.

Biology and Ecology. This species is holoplanktonic. The actinula larva transforms directly into a medusa. Many experts say that all previous descriptions of *Liriope* represent a single species.

Moerisia lyonsi

Occurrence. *Moerisia* is common in summer in salinities <5 in estuaries from New Jersey to South Carolina. It is also reported from Lake Ponchatrain, Louisiana. Since it is small and inconspicuous, its range is likely much greater than reported.

Biology and Ecology. This presumably recent import was first reported from the Chesapeake Bay in 1967. *Moerisia lyonsi* occurs in the Middle East and may have come to the Chesapeake Bay in ship ballast water. Food consists almost entirely of copepod nauplii, copepodids, and copepod adults. It can be a pest in aquaculture operations where it feeds extensively on prawn larvae.

References. Calder and Burrell 1967; Purcell et al. 1999.

Rathkea octopunctata

Occurrence. This arctic-boreal, open-ocean species ranges south to Chesapeake Bay in winter and in spring, occasionally in nearshore waters.

Biological and Ecology. At temperatures <5°C, *Rathkea* medusae produce more medusae asexually by budding them from the sides of the manubrium. Above 5°C, sexual reproduction predominates. Newly released medusae have only four clusters of tentacles instead of the eight borne by adults.

References. Bouillon and Werner 1965; Purcell et al. 1999.

Liriope tetraphylla

10 - 30 mm wide

- hemispherical bell with thick walls (1)
- very long manubrium (peduncle) with flat lip (2)
- 4 main radial canals and 1-3 short canals between them (3)
- broad, flat gonads on main radial canals (4)
- 4 long, hollow tentacles alternate with 4 short tentacles (5)

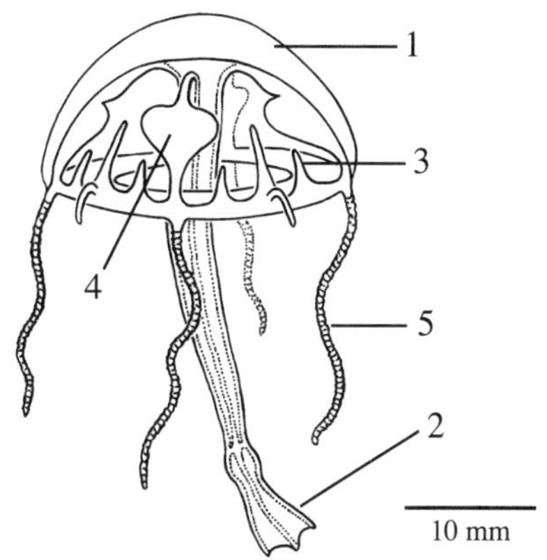

Moerisia lyonsi

2 - 8 mm wide

- globe-shaped bell with very thick walls (1)
- short cylindrical manubrium without lips (2)
- wide gonads extending two-thirds the length of radial canals (3)
- 4-24 tentacles with prominent rings of nematocysts (4)
- similar species: others with 4 single tentacles also have a very long manubrium

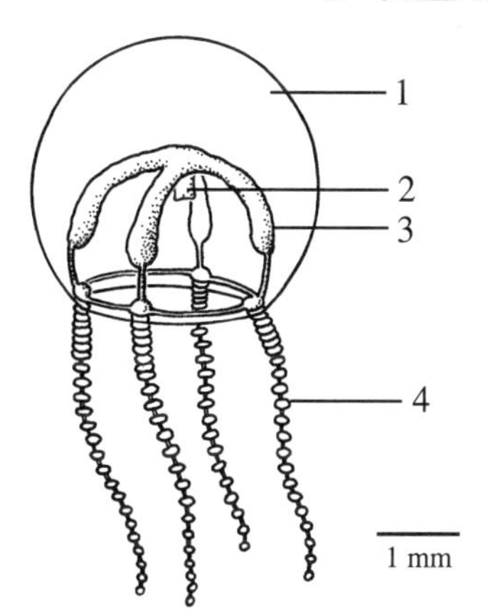

Rathkea octopunctata

3 - 5 mm wide

- wide, bell shape with thick walls (1)
- long 4-sided manubrium with forked lips (2)
- young medusae bud from manubrium (3)
- 8 groups of 3-5 solid tentacles (4)
- large marginal bulbs without ocelli (5)
- similar species: *Bougainvillia* and *Nemopsis* have branched oral tentacles instead of lips

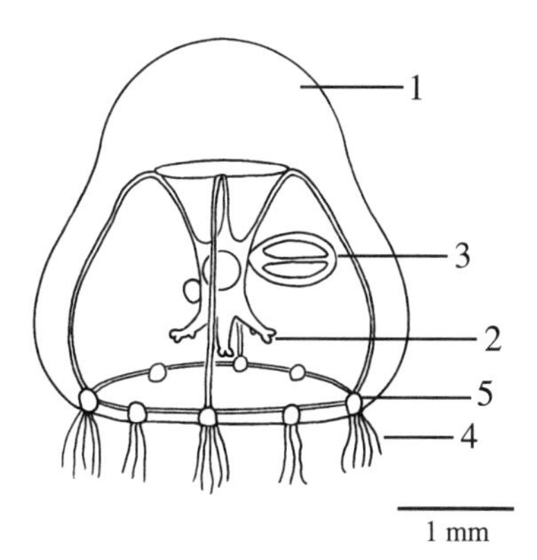

ID hint. These three similar species with thick bells all have clusters of tentacles and branched oral arms instead of a lip.

Bougainvillia carolinensis

Occurrence. *Bougainvillia carolinensis* occurs from Massachusetts to Florida where it is sporadically common in estuaries. It occurs in the Gulf all year but is most abundant in warm seasons, particularly in sounds.

Biology and Ecology. Medusa buds may arise directly from the gonads.

Bougainvillia muscus, formerly *B. ramosa* and *B. rugosa*

Occurrence. *Bougainvillia muscus* is reported from Cape Cod to South Carolina in summer to fall but probably ranges farther south in the Atlantic. It occurs in the Gulf, but reports from there are sparse. It is common in mesohaline to euhaline estuarine habitats. *Bougainvillia rugosa* occurs from the Mid-Atlantic south in summer along the coast and in high-salinity reaches of estuaries and in the Gulf.

Reference. Marshalonis and Pinckney 2008.

Nemopsis bachei

Occurrence. *Nemopsis bachei* is abundant along the Atlantic Coast from Massachusetts to Florida and in mesohaline reaches of estuaries. It occurs in summer in the northern part of its range but is more typical fall to winter in the Southeast. In the Gulf, *Nemopsis* is one of the most common hydromedusae, especially in winter to spring. It occurs in the brackish areas of the northern Gulf in salinities as low as 5 as well as in hypersaline Texas lagoons.

Biology and Ecology. Its diet includes copepods, especially *Acartia tonsa* copepodids. The medusae are flattened when first released and gradually acquire their characteristic globular shape.

References. Calder 1971; Dabiri et al. 2006; Frost et al. 2010; Marshalonis and Pinckney 2008; Purcell and Nemazie 1992.

Bougainvillia carolinensis

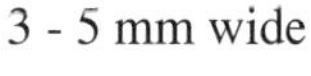
3 - 5 mm wide

- high, oval bell with very thick walls (1)
- oblong manubrium with branched oral tentacles (2)
- 4 clusters of 7-9 stiff tentacles (3)
- small marginal bulbs with dark ocelli (4)
- similar species: other *Bougainvillia*; *B. muscus* has 3-7 tentacles per cluster; *Nemopsis bachei* is larger with forks at margins; *Rathkea* has 8 clusters of tentacles and forked lips

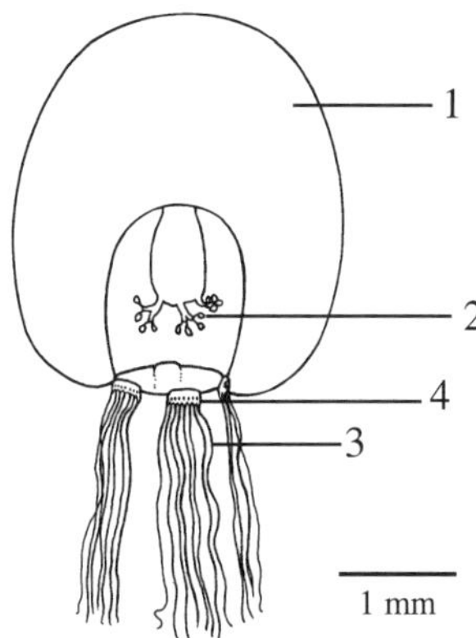

Bougainvillia muscus

2 - 4 mm wide

- cup shaped with very thick walls (1)
- short manubrium with oral tentacles branched 1-3 times (2)
- 4 clusters of 3-7 tentacles (3)
- large marginal bulbs with large black ocelli (4)
- similar species: other *Bougainvillia*; *B. carolinensis* has 7-9 tentacles per cluster; *Nemopsis bachei* is larger with forks at margins; *Rathkea* has 8 clusters of tentacles and forked lips

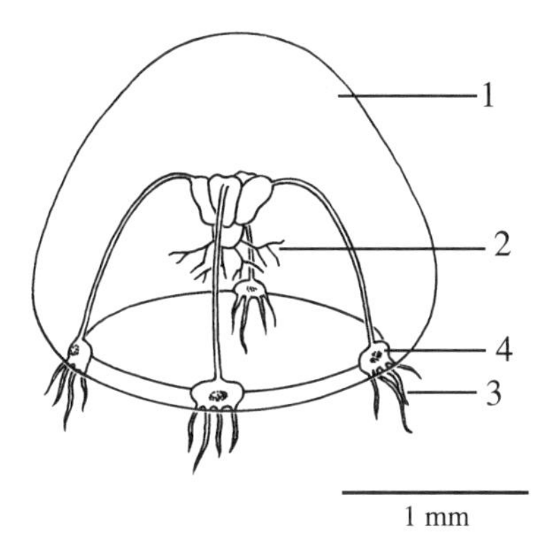

Nemopsis bachei

4 - 11 mm wide

- thimble-shaped bell with thick walls (1)
- short manubrium with oral tentacles branched 3-7 times (2)
- wide, folded gonads extend two-thirds of the length of the radial canals (3)
- 4 clusters of 14-18 tentacles (4)
- marginal bulbs with paired fork-like lateral processes (5)
- similar species: *Bougainvillia carolinensis* has 7-9 tentacles and *B. muscus* has 3-7 tentacles per cluster; *Rathkea* has 8 clusters of tentacles and forked lips

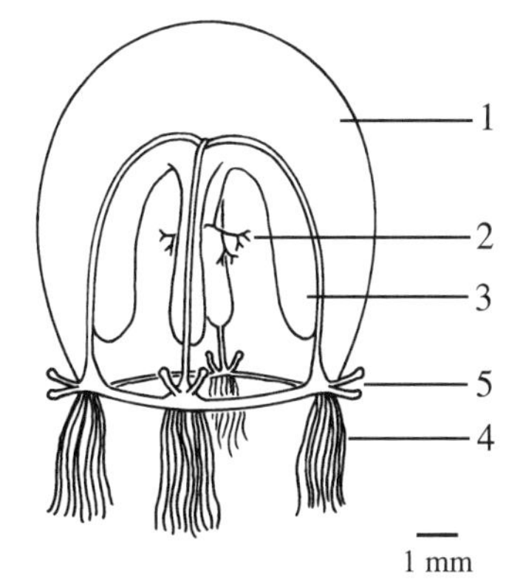

Hybocodon prolifer

Occurrence. This boreal species is found in our geographic range from Cape Cod to the lower Chesapeake Bay in winter to spring. In New England, it occurs in tide pools as well as in the ocean.

Biology and Ecology. Medusae may bud from the base of the single tentacle.

References. Calder 1971.

COLONIAL HYDROZOANS: SIPHONOPHORES

Note. Handle with care. Even the tiny and delicate siphonophores have surprisingly powerful stinging cells that can produce painful welts. Because the tentacles almost invariably become detached in plankton tows, use care when handling nets if siphonophores are present.

Physalia physalis (Portuguese man-of-war)

Occurrence. These typically oceanic, floating hydrozoans are often associated with tropical waters or the Gulf Stream. They are sometimes blown onshore by storms as far north as Canada. They also occur in the northern Gulf.

Biology and Ecology. Although *Physalia* may eat fishes up to 10 cm long, the bulk of the diet is fish larvae supplemented by arrow worms, small squid, and crustaceans. Loggerhead and other sea turtles eat *Physalia* as does the blue glaucus sea slug *Glaucus atlanticus* and the purple bubble snail *Janthina janthina*. The latter two seem immune to the stinging cells. The banded man-of-war fish, *Nomeus gronovii*, has a permanent symbiotic association with *Physalia*. Other fishes may form temporary associations, especially as juveniles. Gas glands at the bottom of the float secrete carbon dioxide. Tentacles shown are contracted, but they may extend up to 10–30 times longer than shown, sometimes reaching 10–15 m below the float.

Caution. The stinging cells can cause severe injury.

References. Bayer 1963; Jenkins 1983; Purcell 1984.

Hybocodon prolifer

3 - 5 mm wide

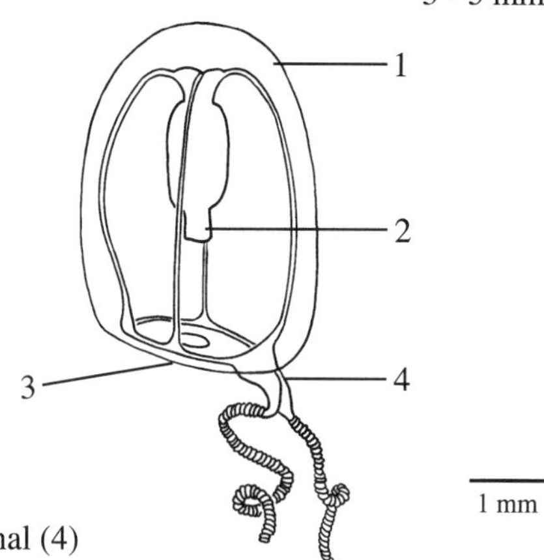

- tall, oval bell with moderately thick walls (1)
- long, broad manubrium without lips (2)
- margin oblique to vertical axis (3)
- one or more tentacles on single large bulb at end of longest radial canal (4)

Physalia physalis

float up to 350 mm wide

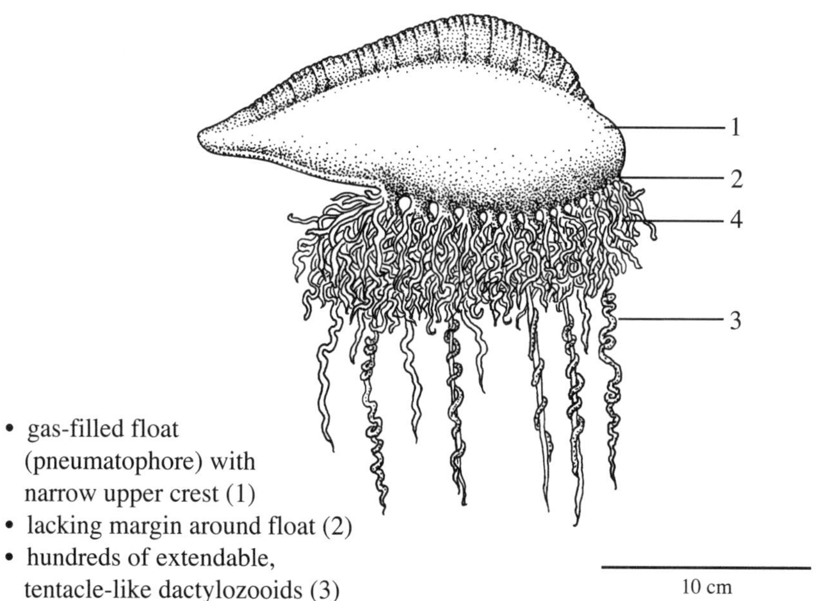

- gas-filled float (pneumatophore) with narrow upper crest (1)
- lacking margin around float (2)
- hundreds of extendable, tentacle-like dactylozooids (3)
- short-feeding polyps (gastrozooids) at centers of clusters of tentacles (4)
- other specialized zooids also attached to bottom of float
- float almost transparent with bluish base and pink-purple crest; gastrozooids often brilliant blue, but greens and reds sometimes evident
- similar species: *Porpita* and *Velella* are large and often bluish floating colonies, but only *Physalia* has a large float

Muggiaea kochi and *Diphyes dispar*

Occurrence. *Muggiaea kochi* and *Diphyes dispar* occur sporadically along the Atlantic Coast in summer and in fall, often associated with intrusions of Gulf Stream water. *Muggiaea* occurs in brackish embayments in summer, but *Diphyes* is less tolerant of reduced salinities. They also occur in the Gulf where they are more common in cooler months.

Biology and Ecology. These siphonophores feed on small prey, primarily copepods. Hyperiid amphipods feed on both *Muggiaea* and *Diphyes* while the phyllosome larvae of spiny lobsters hitch rides on these siphonophores. The fragile tentacles are usually lost during capture, leaving only the swimming bells.

References. Carre and Carre 1991; Gasca et al. 2009; Purcell 1982.

Nanomia cara

Occurrence. *Nanomia cara* is a cold-water species occurring predominantly from the Chesapeake Bay north. *Nanomia bijuga* occurs in the Gulf.

Biology and Ecology. A float at the head of the colony can inflate or deflate to adjust vertical position and perhaps facilitate vertical migration. The fragile tentacles are usually lost during capture, leaving only the swimming bells.

References. Rogers et al. 1978.

Muggiaea kochi

bell up to 7 mm

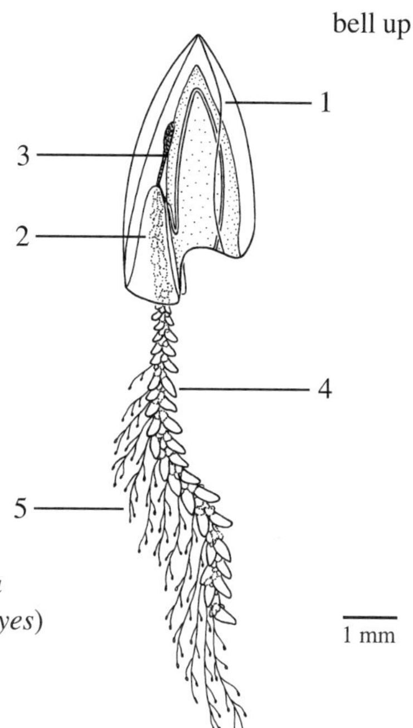

- triangular-shaped swimming bell (nectophore) with 5 longitudinal ridges bending at end (1)
- shallow cavity (2) with extension of gastrovascular system (stomatocyst) extending halfway up cavity (3)
- shield-like bracts along stem (4); each with cluster of zooids
- similar species: many other *Muggiaea* species and genera (e.g., *Lensia*, *Diphyes*) occur in deeper ocean areas

Nanomia cara

bells up to 8 mm

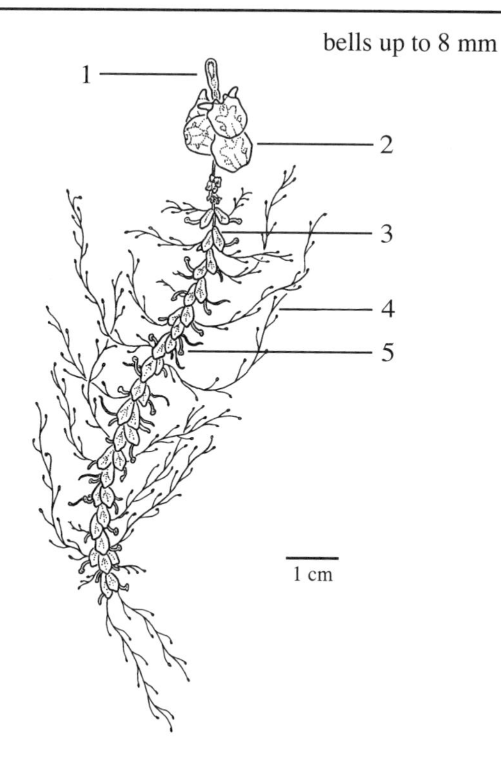

- small, single terminal float (pneumatophore) (1)
- multiple small, ovoid swimming bells (nectophores) with Y-shaped cavities (2)
- leaf-like bracts along stem (3); each with a group of zooids
- long, branched tentacle-like zooids (dactylozooids) (4)
- short and wider gastrozooids (5)
- bioluminescent nectophores and bracts

COLONIAL HYDROZOANS: PORPITIDS

Velella velella (by-the-wind sailor)

Occurrence. These typically oceanic colonial hydrozoans are often associated with tropical waters or the Gulf Stream but are sometimes blown onshore by storms as far north as New England. They are also occasional in the northern Gulf.

Biology and Ecology. All of the zooids of this unusual colonial hydroid are modified polyps. A large central feeding polyp is suspended from the float (pneumatophore) and surrounded by feeding tentacle-like dactylozooids. While free medusae are produced, they are rarely collected in the plankton, and the complete life cycle awaits clarification. Juvenile pilot fish may form temporary associations with *Velella*. Sea turtles and ocean sunfish eat *Velella*. The stinging cells are harmless to humans.

References. Bayer 1963; Brinkman 1964; Calder 1988; Francis 1991.

Porpita porpita (blue button)

Occurrence. Widely distributed at the surface of tropical and subtropical Atlantic waters and through the Gulf. They can be transported farther north in the Atlantic via the Gulf Stream. *Porpita* occur in large aggregations, which occasionally drift ashore.

Biology and Ecology. *Porpita* resembles *Velella* in lifestyle. The gas-filled central disc keeps *Porpita* afloat while the dactylozoids suspended below feed on other zooplankton. Predators include the planktonic gastropods *Glaucus atlantica* and *Janthina janthina*. The sting is not dangerous but may cause irritation.

References. Bieri 1970; Calder 1988.

Velella velella

50 - 100 mm wide

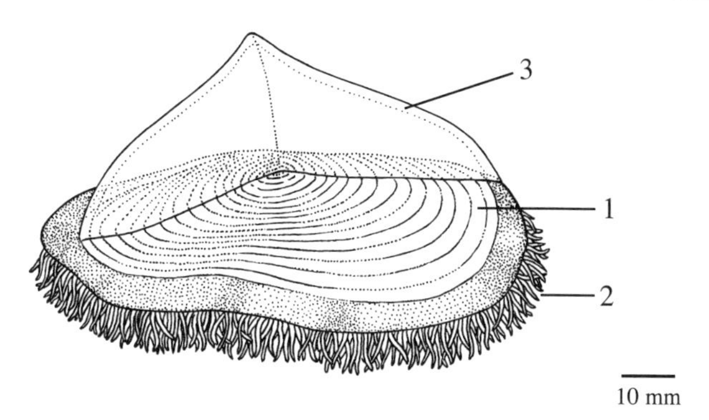

- rigid, flat, oval float with concentric internal chambers (1)
- soft margin with hundreds of tentacle-like zooids (2)
- large siphon-like gastrozooid under float (not visible here)
- rigid, triangular sail perpendicular to float (3)
- sail usually transparent-blue with purple fringe and the float is blue
- similar species: other large and often bluish floating colonies include *Physalia*, which has a large sail, and *Porpita*, which has neither a conspicuous float nor sail

---

Porpita porpita

up to 400 mm wide

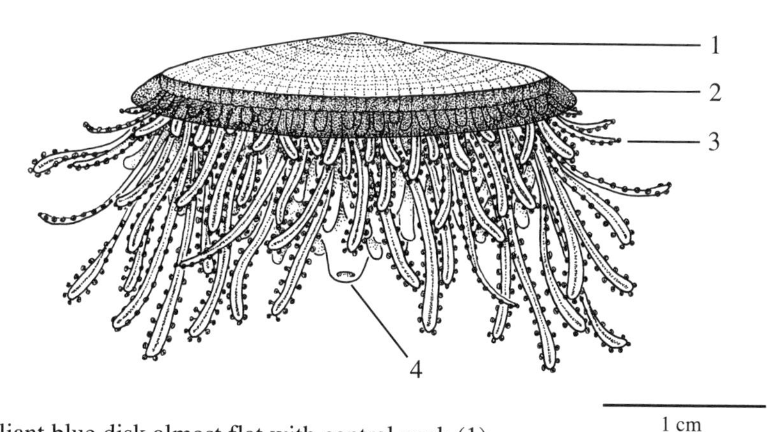

- brilliant blue disk almost flat with central peak (1)
- wide marginal veil covers bases of zooids (2)
- tentacle-like dactylozooids throughout bottom of disk (3)
- single, centrally located, and bulb-shaped gastrozooid (4)
- tentacles and margin are brilliant blue, turquoise, or yellow
- similar species: absence of either a sail or conspicuous float distinguishes *Porpita* from both *Velella* and *Physalia*

SCYPHOZOANS AND CUBOZOANS

SCYPHOZOANS (JELLYFISHES)

Scyphozoan medusae are the familiar large jellyfishes more often seen from the surface than caught in plankton tows. Many of the roughly 200 currently recognized species reportedly have broad or even global distributions. The largest Atlantic jellyfish, *Cyanea capillata*, may reach a diameter of 2 m with tentacles exceeding 60 m in Arctic or boreal waters. Jellyfishes south of Cape Cod are usually much smaller but are important predators on estuarine and coastal zooplankton and exert a considerable influence on planktonic food webs when abundant.

The typical scyphozoan life cycle includes both polyp and medusoid stages. Small, inconspicuous polyps (scyphistomae) live attached to the bottom and divide asexually by budding off a succession of 1 to 16 or more swimming ephyra stages (Fig. 13) that soon transform into young planktonic medusae. The medusae (i.e., scyphomedusae) represent the sexual stage of the life cycle and are either male or female, with gonads located in pockets in the mesoglea.

Males shed their sperm into the water, but females of some local species retain the ripe eggs in the oral region where fertilization and early development occur. The resulting tiny, ciliated planula larvae are in the plankton only briefly before they settle to the bottom and begin the next generation of polyps. Polyps of many species can produce cysts resistant to unfavorable conditions and are important for overwintering (or oversummering in the case of *Cyanea* in the south). A few species, such as the open-water *Pelagia noctiluca*, lack a benthic polyp stage altogether.

Jellyfishes move through the water using a pulsing movement of the bell as they trail their tentacles behind. The mesoglea's stiffness returns the bell to its original position for the next stroke. Most local jellyfishes are feeble swimmers, but they can control their position in the water column, and some undertake regular vertical migrations. Statocysts arranged around the margin on the bell provide crucial information on which way is up. Often statocysts are found in association with primitive photosensors (ocelli).

Temperate jellyfishes are predators that use their nematocysts to subdue a variety of primarily zooplankton prey. Several quite different feeding mechanisms are used. Species with long tentacles (e.g., sea nettles and lion's mane jellies of the Order Semaeostomeae) capture prey with their tentacles and oral arms, which have powerful nematocysts. Often they swim up toward the surface and slowly sink down while feeding. Unwary zooplank-

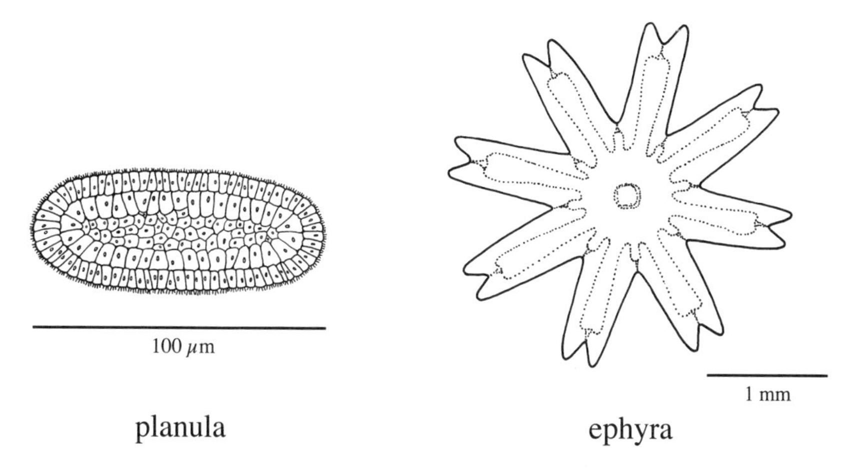

Fig. 13. The life cycle of scyphozoan jellyfishes typically includes planktonic planula larva and the ephyra stages. The larger and distinctive ephyrae are more likely to be noticed in zooplankton samples than the much smaller planulae.

ton and small fishes are caught by the passively drifting tentacles in this "ambush" style, subdued, and then transferred to the mouth and gastric cavity by the oral arms. While swimming, vortices created during bell pulsation serve both to propel the medusae forward and to carry smaller prey into the long, trailing tentacles and oral arms. The moon jelly *Aurelia aurita* employs its relatively small (<3 cm) tentacles as a filter. When the bell contracts, the curtain of fine tentacles strains zooplankton from the current. In addition to its tentacles, *A. aurita* uses the sticky mucus on both the upper and lower surfaces of the bell to catch small prey and detritus. Cilia in the underlying epithelium move food caught in the mucus to pockets on the underside of the bell where the oral arms collect and carry food to the mouth.

Another group (Order Rhizostomeae), exemplified by the cannonball jelly (*Stomolophus*), has no visible tentacles at all. These jellyfishes rely exclusively on suspension feeding using modified oral arms rather than tentacles to capture prey. The highly branched oral arms bear small microvilli-like "digitata" armed with nematocysts. As the bell refills, water currents are channeled through this sievelike plankton trap. Cilia transfer the victims through small pores, or "mouths," to canals leading to the gastric cavity. This mechanism is surprisingly effective in capturing small prey, including bivalve, copepod, and fish larvae. Larson (1991) estimated that a medium-sized *Stomolophus* (5.2 cm) catches 20,000 prey per day.

In 2000, there was an unprecedented outbreak of the large Australian spotted jellyfish (*Phyllorhiza*) in the Gulf of Mexico, and scientists were concerned about the effect of these invasive jellyfishes on larval fishes and shrimps. Another large, alien jellyfish (*Drymonema larsoni*), native to the Caribbean, has also appeared recently in the Gulf. Reasons for the sudden outbreaks and subsequent declines of these alien species are unclear.

Spadefish, harvestfish, butterfish, ocean sunfish (*Mola mola*), loggerhead turtles, and humans eat jellyfishes. In Asia, some jellyfishes are considered delicacies and eaten both fresh and salted. Along the Georgia and Florida coasts, a new commercial fishery exploits cannonball jellies (*Stomolophus*) for this market. Scyphomedusae are involved in a variety of specific symbiotic associations. Some juvenile fishes swim among the tentacles for protection, and commensal or parasitic invertebrates attach themselves to the bells (Table 6). If scyphomedusae are collected carefully with dip nets, these crabs and other associates may be captured.

CUBOZOANS (BOX JELLIES OR SEA WASPS)

The squarish cubozoan jellyfishes are similar to scyphomedusae, but in box jellies, the bell has four flattened sides with tentacles only at the corners. They also have a veil-like structure (a velum) inside the margin of the bell that partially closes the opening.

Box jellies are the most venomous jellyfishes known. Representative species that occur along the Atlantic and Gulf Coasts of the United States are far less dangerous than the Australian species, but they can produce painful welts and lesions. Each benthic polyp produces a single medusa by complete metamorphosis, in contrast to the budding of multiple planktonic ephyrae seen in scyphomedusae. Box jellies are tropical and subtropical and occur locally in the Gulf and along the Atlantic Coast, occasionally as far north as New England. They may be attracted to lights at night.

IDENTIFICATION HINTS

There are generally few species of scyphomedusae or sea wasps present at a time in a given location. Size, general shape, and the relative number and lengths of the tentacles and oral arms distinguish these. Figure 14 illustrates the features noted below:

Table 6. Symbiotic associations involving scyphomedusae in our region

Symbiotic associates		
Group	Species	Scyphomedusae involved
Fishes	Atlantic bumper Butterfish Harvestfish	*Aurelia aurita* *Cyanea capillata* *Chrysaora quinquecirrha*
Crustacea (Decapoda)	Spider crab *Libinia dubia*	*Stomolophus meleagris* *Rhopilema verrilli*
Crustacea (Hyperiid amphipods)	*Hyperia galba* And other hyperiids	*Cyanea* spp. *Aurelia aurita* *Chrysaora quinquecirrha*
Sea anemones	*Peachia parasitica*	*Cyanea capillata*
Dinoflagellates (Zooxanthellae)	*Symbiodinium*	*Cassiopea* spp.

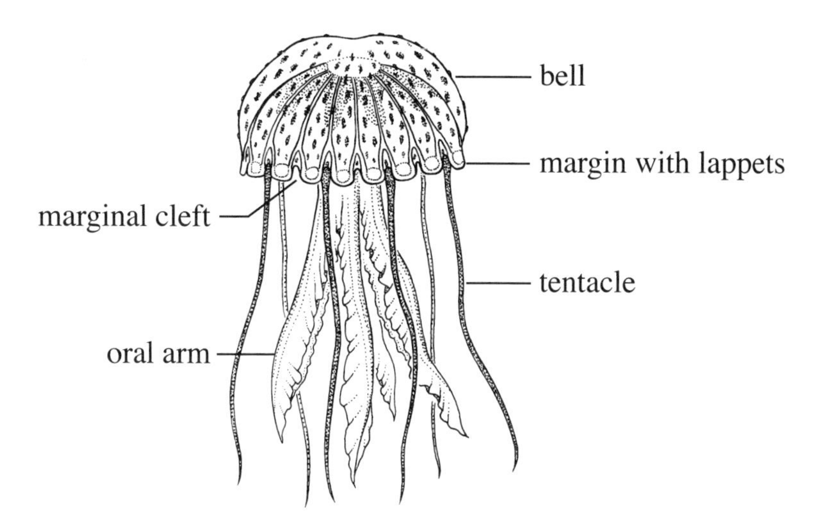

Fig. 14. *Pelagia noctiluca* displays many of the anatomical features seen in the large jellyfishes. Ruffled oral arms and thin tentacles are common in one group but some others (e.g., *Rhopilema verrilli*) have robust and often fused oral arms and no tentacles.

The **bell** may be flattened, umbrella-like, or bell shaped.

The **margin of the bell** is usually scalloped with indentations (called clefts, if large) at regular intervals leaving flaplike **lappets** between these indentations.

Tentacles, if present, arise from the margin of the bell or, in some species, from the underside of the bell (e.g., *Cyanea capillata, Drymonema larsoni*). Note the number, position, and clustering of tentacles.

Four to eight **oral arms** extend from the oral region in the center of the bell and should not be confused with the much less robust tentacles attached at or near the margin.

USEFUL IDENTIFICATION REFERENCES

Calder, D. R. 1983. Nematocysts of stages in the life cycle of *Stomolophus meleagris*, with keys to scyphistomae and ephyrae of some western Atlantic Scyphozoa. *Canadian Journal of Zoology* 61:1185–1192. (One of few guides to the ephyrae.)

Calder, D. R. 2008. An illustrated key to the Scyphozoa and Cubozoa of the South Atlantic Bight (from Calder 2009) and a glossary of terms. Available through SERTC at <www.dnr.sc.gov/marine/sertc/info.htm>. Accessed July 19, 2011.

Calder, D. R. 2009. Cubozoan and scyphozoan jellyfishes of the Carolinian biogeographic province, southeastern, USA. *Royal Ontario Museum Contributions in Marine Science* 3:1–58.

Hargitt, C. W. 1905. The medusae of the Woods Hole region. *US Bureau of Fisheries* 24:21–79.

Kramp, P. L. 1961. Synopsis of medusae of the world. *Journal of the Marine Biological Association of the United Kingdom* 40:1–469.

Larson, R. J. 1976b. *Marine Flora and Fauna of the Northeastern United States: Cnidaria, Scyphozoa*. NOAA Technical Reports NMFS Circ. 397, 1–17. 294 pp.

Mayer, A. G. 1910. Medusae of the world. Vol. 3, The Scyphomedusae. *Carnegie Institution of Washington Publications* 109:499–735.

Russell, F. S. 1970. *The Medusae of the British Isles*. Vol. 2, *Pelagic Scyphozoa, with Supplement to the First Volume on Hydromedusae*. Cambridge University Press, Cambridge. 284 pp.

SUGGESTED READINGS

Scyphozoa and Cubozoa

Arai, M. N. 1997. *A Functional Biology of Scyphozoa*. Chapman & Hall, New York. 316 pp.

Bayha, K. M., Dawson, M. N. 2010. New family of allomorphic jellyfishes, Drymonematidae (Scyphozoa, Discomedusae), emphasizes evolution in the functional morphology and trophic ecology of gelatinous zooplankton. *Biological Bulletin* 219:249–267.

Bentlage, B., Cartwright, P., Yanagihara, A. A., et al. 2010. Evolution of box jellyfish (Cnidaria: Cubozoa), a group of highly toxic invertebrates. *Proceedings of the Royal Society, B, Biological Sciences* 277:493–501.

Costello, J. H., Colin, S. P. 1995. Flow and feeding by swimming scyphomedusae. *Marine Biology* 124:399–406.

Hamner, W. M. 1995. Sensory ecology of scyphomedusae. *Marine and Freshwater Behavior and Physiology* 26:101–118.

Larson, R. J. 1976a. Cubomedusae: Functional morphology, behavior and phylogenetic position. In: Mackie, G. O., ed. *Coelenterate Ecology and Behavior*. Plenum Press, New York, 237–245.

Mackie, G. O. 1995. Defensive strategies in planktonic coelenterates. *Marine and Freshwater Behavior and Physiology* 26:119–130. (A review.)

Mianzan, H. W., Cornelius, P. F. S. 1999. Cubomedusae and scyphomedusae. In: Boltovskoy, D., ed. *South Atlantic Zooplankton*. Vol. 1. Backhuys, Leiden, The Netherlands, 513–559.

Purcell, J. E., Graham, W. M., Dumont, H. J., eds. 2001. *Jellyfish Blooms: Ecological and Societal Importance*. Kluwer Academic, Boston. 356 pp.

Purcell, J. E., Uye, S.-I., Lo, W. T. 2007. Anthropogenic causes of jellyfish blooms and their direct consequences for humans: A review. *Marine Ecology Progress Series* 350:153–174.

SCYPHOZOANS

Stomolophus meleagris (cannonball jellyfish)

Occurrence. Cannonball jellyfishes are common in summer and fall along the Atlantic Coast, south of Cape Hatteras. Although rare north of Chesapeake Bay, they are reported as far north as Massachusetts, chiefly in the ocean rather than in estuaries. In the Gulf, *Stomolophus* may be present all year, sometimes in <10 psu.

Biology and Ecology. This relatively fast-swimming suspension feeder preys on bivalve veligers, copepods (eggs, larvae, and adults), gastropod veligers, tintinnids, larvaceans, and fish eggs. Symbionts include juvenile spider crabs (*Libinia dubia*) that probably settle on medusae as larvae. One to several dozen small fishes (e.g., juvenile butterfishes and jacks) can often be observed swimming with these jellyfish. Cannonballs are a favored food of spadefish and sea turtles. At times, cannonball jellyfishes are so numerous that they can fill shrimp and fish trawls. They have been harvested in Georgia and Florida for sale to predominantly Asian markets. Other common names are jellyball and cabbagehead jellyfish. Its sting is mild.

References. Calder 1982, 1983; Larson 1991; Shanks and Graham 1987.

Phyllorhiza punctata (Australian spotted jellyfish)

Occurrence. This alien invasive species has been in the Gulf for some years and was typically uncommon to rare. Dense aggregations in the Gulf of Mexico from Alabama to Texas appeared in the summer of 2000, but this bloom has not recurred through 2009. *Phyllorhiza* occurred on the Atlantic Coast of Florida in the vicinity of the St. Johns River and Cape Canaveral in the summer of 2001. In summer 2007, several were collected in Bogue Sound, North Carolina.

Biology and Ecology. An individual can filter up to 50 m^3 (10,000 gallons) of water and remove more than 1,000 fish eggs daily. When they are present in large numbers in coastal inlets, they may pose a threat to the larval fishes and invertebrates migrating into or out of the bays and sounds. Juvenile butterfish are sometimes associated with them. Its sting is mild.

References. D'Ambra et al. 2001; Graham et al. 2003; Perry and Graham 2000.

Rhopilema verrilli (mushroom cap jelly)

Occurrence. This jellyfish occurs in the Atlantic from New England to Georgia in summer and fall but is most common south of Cape Hatteras. It also occurs in the Gulf. This is a coastal species often found in sounds and lagoons in salinities >25.

Biology and Ecology. Spider crabs and blue crabs may be attached to the oral surface. It is nonvenomous to humans.

References. Calder 1973; Cargo 1971; Harper and Runnels 1990.

Stomolophus meleagris

to 25 cm wide

- bell margin with 8 deep clefts (1)
- 128 forked lappets of two lengths (2)
- 16 short, forked oral arms fused along entire length (3)
- milky blue or yellow bell with pale, speckled red-brown margin
- similar species: *Rhopilema* and *Phyllorhiza* both have finger-like projections on the oral arms

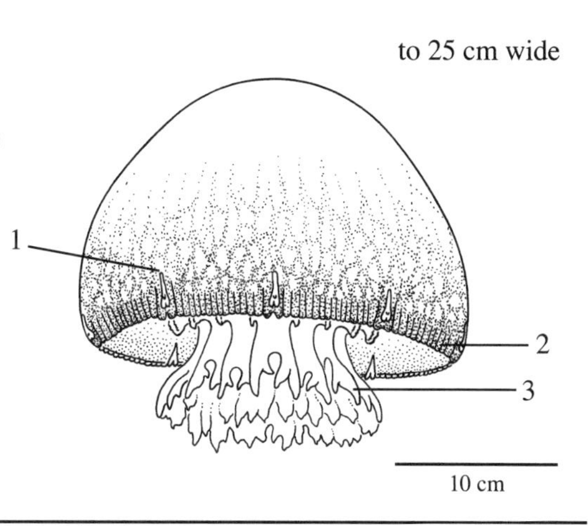

Phyllorhiza punctata

to 40 cm wide

- bell margin without deep clefts
- about 100 rectangular lappets on margin (1)
- 8 long oral arms with terminal finger-like projections (2)
- paired bundles of stinging cells on arms (3)
- bell milky, sometimes bluish brown with many evenly distributed white spots
- similar species: *Stomolophus* does not have finger-like projections; *Rhopilema* does not have distinctive white spots

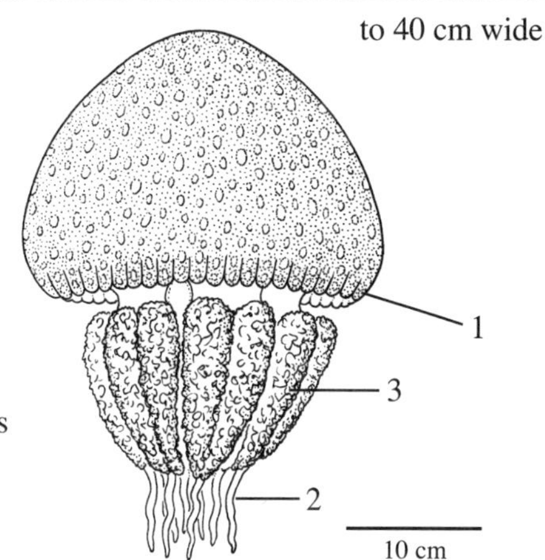

Rhopilema verrilli

to 35 cm wide

- bell margin without deep clefts
- 64 rounded lappets of two lengths (1)
- 8 long, oral arms fused near base (2)
- finger-like projections extend from ends of oral arms (3)
- milky bell with colorless margin; arms with yellow-brown marks
- similar species: *Stomolophus* is smaller with more lappets and strong marginal color; *Phyllorhiza* has prominent white spots

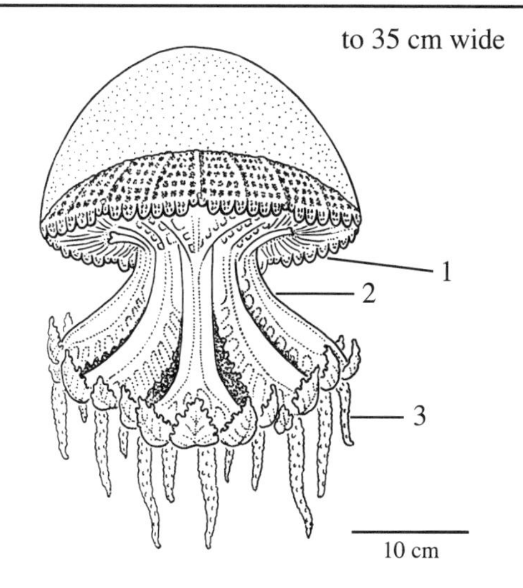

Chrysaora quinquecirrha (sea nettle)

Occurrence. The sea nettle occurs along the coast from Cape Cod to Florida and throughout the Gulf. *Chrysaora* is particularly abundant in Chesapeake Bay and other large estuaries in 5–30 psu but is also common along the coasts to at least 10 km offshore.

Biology and Ecology. Polyps prefer upper reaches of deep tidal creeks of 5–20 psu. Ephyrae are released in spring and summer at 2–3 mm diameter. Often sea nettles swim in circles while feeding on copepods, comb jellies, and fish eggs and larvae (but rarely bivalve larvae). When numerous, their feeding may control estuarine plankton dynamics. Ephyrae eat protozoans and rotifers. Small harvestfish (*Peprilus alepidotus*) use sea nettles as a refuge and swim among the tentacles, which they eat. Butterfish and loggerhead turtles also eat sea nettles, but potential predators are restricted to high salinities and pose little threat to estuarine populations. The hyperiid amphipod *Hyperia galba* may be a symbiont. In high salinities, sea nettles sometimes have beautiful red markings not seen in more brackish settings. **Note:** Powerful nematocysts readily penetrate human skin, causing painful, though seldom dangerous, reactions. High densities may prevent swimming, clog fishing (and plankton) nets, and block water intakes. Specimens washed ashore can still sting and should be handled with care.

References. Calder 1972; Decker et al. 2007; Larson 1986; Mansueti 1963; Matanoski and Hood 2006; Matanoski et al. 2001; Olesen et al. 1996; Purcell 1992; Schuyler and Sullivan 1977.

Cyanea capillata (lion's mane)

Occurrence. *Cyanea capillata* reaches its maximum size and abundance in Arctic and boreal waters. It occurs regularly in winter and in spring south to Florida. It may be common in the Gulf in late winter and spring. It prefers salinities >20 but sometimes occurs in low salinities in winter.

Biology and Ecology. *Cyanea* is the largest Atlantic jellyfish with a bell reaching 2 m in diameter and tentacles exceeding 60 m in boreal or Arctic regions. Planulae are brooded on the oral arms. When females deteriorate, planulae are deposited on the bottom as the oral arms sink. Ephyrae appear in late winter in the north and during autumn or early winter farther south. *Cyanea* swims to the surface and then drifts down while feeding on zooplankton, including *Aurelia*, fish eggs and larvae, and lobster larvae. Juvenile butterfish and whiting are sometimes commensal with *Cyanea*. Leatherback turtles and at least one fish, the Atlantic bumper, eat *Cyanea*. The anemone *Peachia parasitica* is a symbiont. From New England north, contact can result in serious stings. *Cyanea* is commonly called the "winter jellyfish" from the Mid–Atlantic Coast southward. The related and similar, large, pink jellyfish, *Drymonema larsoni*, a Caribbean species, appeared suddenly in the northeastern Gulf in 2000. It is sporadically abundant in the Caribbean and may be carried northward in the Gulf Stream off the southeastern United States.

Comments. *Cyanea* was previously split into at least four species, all now assigned to *C. capillata*. However, taxonomy of the genus is still unsettled. The isolated Gulf population may be a distinct subspecies.

References. Båmstedt et al. 1994; Brewer 1989; Colin and Costello 2007; Colin and Kremer 2002; Hansson 1997, 2006; McDermott et al. 1982; Shoji et al. 2008; Suchman and Sullivan 2000.

Chrysaora quinquecirrha

to 19 cm wide

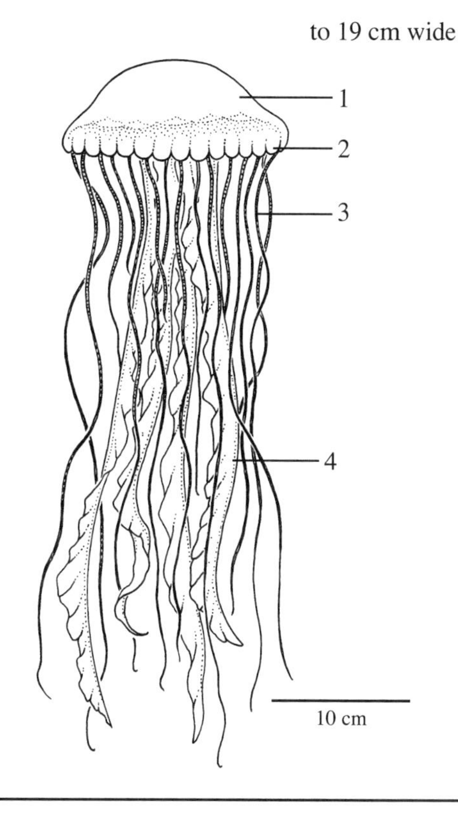

- bell with fairly thick dome (1)
- 24-28 scalloped lappets on margin (2)
- equal number of long tentacles from clefts between lappets (3)
- 4 long, ruffled oral arms (4)
- brackish form is uniformly milky; marine form is strongly colored with 16 radiating reddish, speckled bars and yellow tentacles
- similar species: reddish color (when present) leads to confusion with *Cyanea,* which has clusters of tentacles originating from under the bell; *Pelagia* has 8 or fewer tentacles

Cyanea capillata

to greater than 1 m
but usually to 20 cm wide

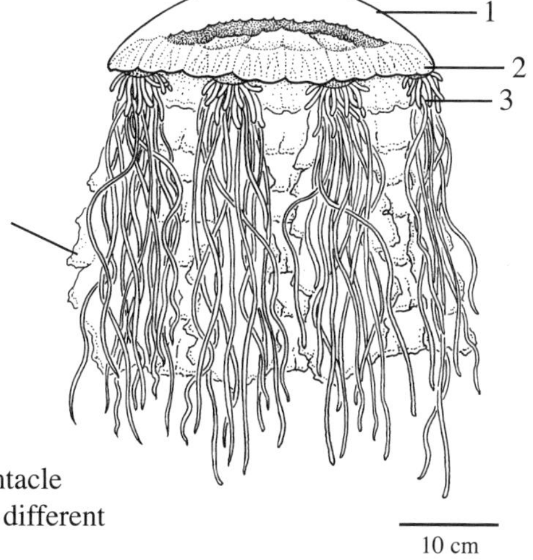

- bell fairly flat with smooth surface (1)
- about 32 wide, low-profile lappets on margin (2)
- 8 clusters of 3-60 tentacles; originate from under bell (3)
- folded curtain-like oral arms hang from bell inside tentacles (4)
- color variable from pink-yellow in young to red-brown (especially oral arms) in adults
- similar species: reddish color leads to confusion with *Chrysaora* but tentacle and oral arm arrangements are very different

Aurelia aurita (moon jelly)

Occurrence. *Aurelia aurita* is common throughout our geographic range. It is most common in summer and in early fall along the coast and in higher-salinity reaches of estuaries in New England and the Mid-Atlantic. It is also present in winter in the Gulf.

Biology and Ecology. Males release sperm into the water to fertilize eggs retained on females' oral arms where planulae are brooded before being released. *Aurelia* is a general zooplanktivore, consuming quantities of microzooplankton (copepod nauplii and tintinnids), rotifers, copepods, bivalve veligers, fish eggs, and small fish larvae. Ephyrae feed on ciliates and rotifers. Small medusae <12 mm in diameter feed on hydrozoan medusae, rotifers, and other small prey. *Aurelia* adults feed with tentacles and with cilia and mucus as described in the introduction to "Scyphozoa." The lion's mane *Cyanea capillata* preys heavily on moon jellies. *Aurelia* often swims close to the surface, sometimes in dense aggregations. This species has a mild sting but can irritate delicate skin or cuts.

Comments. Calder (2009) recently referred the population of *Aurelia* in southern states to a different species, *A. marginata*.

References. Costello and Colin 1994; Graham and Kroutil 2001; Hansson 1997; Lucas 2001; Stoecker et al. 1987; Sullivan et al. 1997.

Pelagia noctiluca (purple jellyfish)

Pelagia noctiluca is an oceanic form that occurs inshore erratically on both Atlantic and Gulf Coasts when brought inshore by specific wind or current conditions. It has a low tolerance of brackish waters.

Biology and Ecology. Fertilized eggs develop directly into ephyrae. Thus, there is no free polyp stage. *Pelagia* feeds on other cnidarians, ctenophores, crustaceans, chaetognaths, and fish eggs. At night, this bioluminescent jellyfish looks like a glowing white ball, hence the specific name *noctiluca*. Its sting is painful to humans.

References. Larson 1987; Morand et al. 1987; Rottini Sandrini and Avian 1989.

Aurelia aurita

to greater than 50 cm
but usually to 20 cm wide

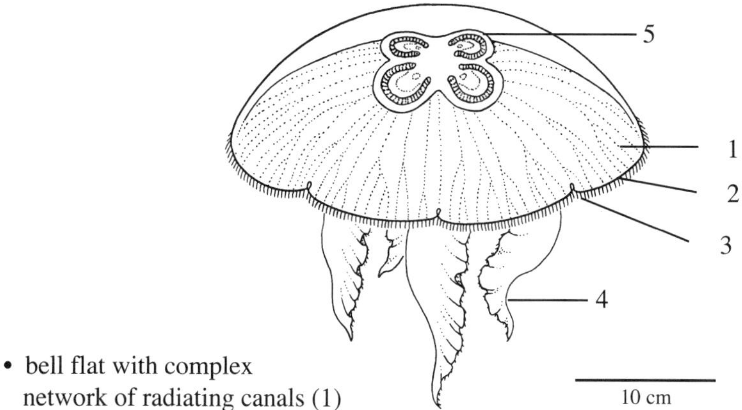

- bell flat with complex network of radiating canals (1)
- 8 shallow marginal lobes (2)
- hundreds of short marginal tentacles (3)
- 4 ruffled oral arms (4)
- 4-lobed stomach surrounded by horseshoe-shaped gonads (5)
- transparent to milky with strongly, but variably, colored gonads

---

Pelagia noctiluca

to 10 cm wide

- bell hemispherical with warty surface (1)
- 16 rectangular lappets (2)
- 8 or fewer tentacles emerge from clefts between lappets (3)
- 4 long, ruffled oral arms (4)
- rose-, purple-, or yellow-tinted bell with purple streaks and external bars; oral arms often pink
- similar species: *Chrysaora* has 24-48 tentacles

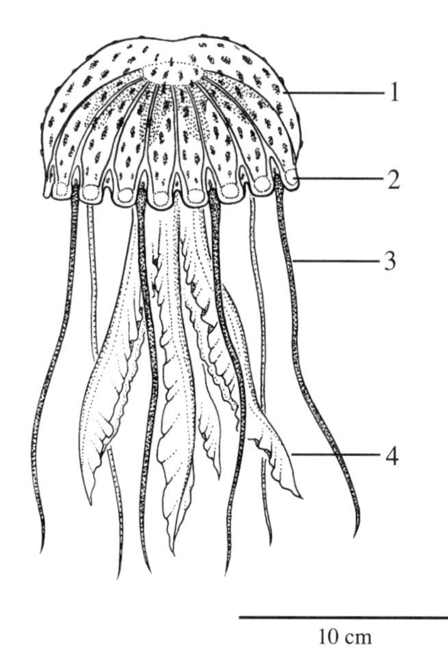

CUBOZOA (BOX JELLIES OR SEA WASPS)

Tamoya haplonema

Occurrence. This tropical to warm-temperate jellyfish occurs sporadically north as far as Connecticut in summer along the coast but seldom in estuaries. It is uncommon to rare in Gulf nearshore waters.

Biology and Ecology. *Tamoya* is a fast swimmer, often found near the bottom. Its sting produces severe discomfort but is not lethal.

Comments. *Tamoya* along the East Coast of the United States may represent an as-yet-undescribed species. For now we retain the name *T. haplonema*.

Chiropsalmus quadrumanus

Occurrence. *Chiropsalmus* occurs along the northern Gulf Coast in inshore waters in late summer and fall when it may be abundant. It prefers deeper, more saline waters along barrier islands. It occurs erratically along the Atlantic Coast in summer at least as far north as the Carolinas and may be abundant along the coastline and in bays and creeks to salinities as low as 23.

Biology and Ecology. *Chiropsalmus* feeds on crustacean zooplankton, including *Lucifer*, crab and stomatopod larvae, and amphipods. Reports of feeding on small fishes are unconfirmed. **Warning: This is a dangerous jellyfish whose severe sting may be life threatening.**

References. Gershwin 2006; Kraeuter and Setzler 1975; Phillips and Burke 1970.

Tamoya haplonema to 9 cm wide

- rectangularly shaped bell (1)
- blade-like lobes extend from each corner (2)
- one long, beaded tentacle extends from each of the 4 lobes (3)

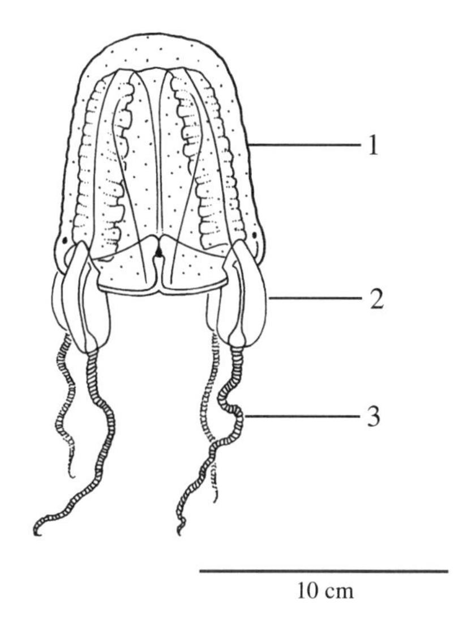

Chiropsalmus quadrumanus to 15 cm wide

- squarely shaped bell with internal septae or canals (1)
- hand-like extensions from each corner (2)
- a single, long tentacle extends from each of the multiple fingers on each "hand" (3)

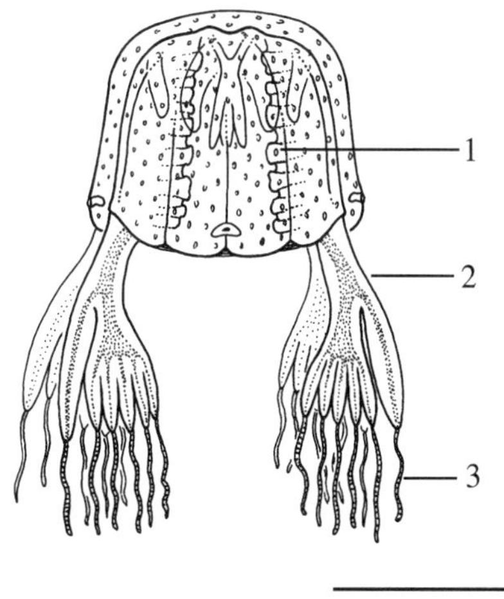

CTENOPHORES

Comb Jellies and Sea Walnuts

Ctenophores are among the most common macrozooplankton in estuarine and coastal waters in summer. The marine Phylum Ctenophora (pronounced with a silent "c," as "teen-o-four-ah" or "ten-o-four-a") has more than 150 species, but only several species are common in nearshore waters in our area. Ctenophores share many features with the Cnidaria, including radial (or biradial) symmetry and a gelatinous body with simple tissues. Unlike the cnidarians, these relatively large (up to 10 cm), wholly planktonic predators lack stinging cells and are harmless. Highly iridescent in sunlight and sometimes bioluminescent at night if disturbed, ctenophores are beautiful animals.

Nearshore ctenophores in our geographical range have a planktonic life cycle. Unlike most cnidarian jellyfishes, there is no sessile polyp phase. Fertilized eggs develop into small ciliated larvae and then into cydippid "larvae" that resemble the adult in basic body plan. The most common ctenophores in nearshore waters belong to two classes. Class Tentaculata has tentacles for at least part of the life cycle (Fig. 15). *Pleurobrachia* and *Mnemiopsis* belong to this group. Class Nuda, which includes *Beröe*, completely lacks tentacles.

Eight linear arrays of cilia (ctene, or comb rows) propel ctenophores through the water with a gliding motion quite unlike the muscular pulsing of cnidarian medusae. The extra long cilia (the longest known) beat in a synchronized fashion. Ctenophores may be the largest animals to rely on cilia for locomotion. Contact with predatory jellyfishes prompts surprisingly effective escape responses. While feeding, *Mnemiopsis* and *Pleurobrachia* hover or swim slowly. In contrast, the predaceous *Beröe* is a rapid and highly maneuverable swimmer.

The three ctenophores treated here employ different modes of feeding. *Pleurobrachia* and larval *Mnemiopsis* feed with a pair of long, trailing tentacles (Fig. 15). The tentacles have sticky colloblast cells to entangle actively swimming prey. Larger *Mnemiopsis* lack tentacles. As they swim, they catch copepods, fish eggs, and mollusc larvae on the sticky mucus lining of the preoral lobes or on the numerous small tentacles that line the inner surface of the lobes as water is pumped past. The lobes close rapidly to catch prey and then reopen. *Mnemiopsis* often occurs in high densities and can have a major impact on meszooplankton abundance, sometimes affecting entire food webs. *Beröe ovata* feeds exclusively on other ctenophores, especially *Mnemiopsis*. When these voracious predators sense prey, they increase speed in a search behavior that often involves swimming in circles or helices. Smaller prey is simply engulfed whole. *Beröe* uses its long macrocilia on the edges of the

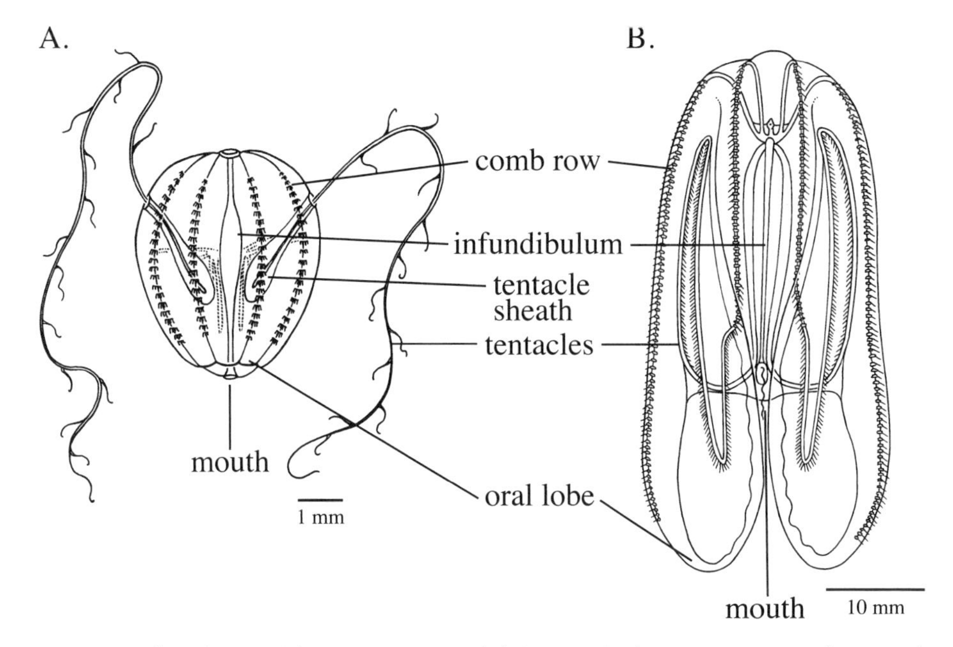

Fig. 15. *A*, Cydippid stage (about 9 mm). *B*, Adult (5–6 cm) of *Mnemiopsis* sp., showing the comb rows of cilia and feeding structures. The cydippid feeds using its two long tentacles. When the tentacles are retracted into their sheaths, prey are brought into contact with the mouth for ingestion. The adult (*B*) lacks the long tentacles. It cruises through the water with its oral lobes spread wide. Prey are caught on the sticky surface of the inside of the lobes or on the short tentacles on the interior.

mouth to slice through the soft tissue of larger prey from which it removes large bites. Cydippid stage ctenophores eat copepod nauplii.

Adult ctenophores fall prey to sea nettles *Chrysaora quinquecirrha* in brackish waters and to spadefish, butterfish, and harvestfish along the coast. Several parasites or symbionts regularly occur on local ctenophores, including the hyperiid amphipod *Hyperoche medusarum* and the larvae of the burrowing anemone *Edwardsia leidyi*.

The North American ctenophores *Mnemiopsis leidyi* and *Beröe* have been introduced to European waters. *Mnemiopsis* was first noted in the Black Sea in 1982 where it multiplied to become a dominant planktivore that disrupted the local food web and threatened a fishery's collapse until a subsequent introduction of the predaceous *Beröe ovata* began to control its numbers. *Mnemiopsis* then spread to the Sea of Azov and reached the Caspian and Mediterranean Seas by 1990. In 2006, *Mnemiopsis* appeared in the Baltic and North Seas. DNA analyses suggest that the Black Sea population originated from the Gulf of Mexico and that the northern European populations came from the northwestern Atlantic. Ctenophores inadvertently included in the ballast water routinely carried in transport vessels may have facilitated these invasions.

IDENTIFICATION HINTS

The ciliated bands separate ctenophores from other phyla. Each of the nearshore ctenophores has a distinctive shape and consistency, most obvious in adult, intact specimens. The delicate *Mnemiopsis* often disintegrates when captured, leaving pieces of comb rows with their still-beating cilia to confuse novice planktologists. If abundant, these comb jellies clog plankton nets and reduce samples to a gooey blob. Promptly remove ctenophores from the plankton sample and study them separately.

USEFUL IDENTIFICATION REFERENCES

Mayer, A. G. 1912. Ctenophores of the Atlantic Coast of North America. *Publications of the Carnegie Institution of Washington* 162:1–58.

SUGGESTED READINGS

Burrell, V. G., Jr., Van Engel, W. A. 1976. Predation by and distribution of a ctenophore *Mnemiopsis leidyi* A. Agassiz, in the York River estuary. *Estuarine and Coastal Marine Science* 4:235–242.

Cahoon, L. B., Tronzo, C. R., Howe, J. C. 1986. Notes on the occurrence of "*Hyperoche medusarum*" (Kroyer) (Amphipoda, Hyperiidae) with Ctenophora off North Carolina, U.S.A. *Crustaceana* 51:95–96.

Costello, J. H., Coverdale, R. 1998. Planktonic feeding and evolutionary significance of the lobate body plan within the Ctenophora. *Biological Bulletin* 195:247–248.

Martindale, M. Q. 2002. Phylum Ctenophora. In: Young, C. M., ed. *Atlas of Marine Invertebrate Larvae*. Academic Press, New York, 109–122.

Matsumoto, G. I., Harbison, G. R. 1993. In situ observations of foraging, feeding, and escape behavior in three orders of oceanic ctenophores: Lobata, Cestida, and Beroida. *Marine Biology* 117:279–287.

Mianzan, H. W. 1999. Ctenophora. In: Boltovskoy, D., ed. *South Atlantic Zooplankton*. Vol. 1. Backhuys, Leiden, The Netherlands, 561–573.

Reeve, M. R., Walter, M. A. 1978. Nutritional ecology of ctenophores: A review of recent research. *Advances in Marine Biology* 15:249–287. (A nice introduction to the group.)

Tamm, S. L. 1980. Cilia and ctenophores. *Oceanus* 23:50–59. (A nontechnical look.)

Mnemiopsis leidyi (sea walnut)

Occurrence. *Mnemiopsis* is the most abundant nearshore ctenophore, with densities sometimes exceeding 100 m^{-3} in summer. *Mnemiopsis leidyi* occurs from Massachusetts to Florida and throughout the Gulf. It seems restricted to higher salinities in colder months but occurs in salinities as low as 5 in summer and fall. The similar northern ctenophore *Bolinopsis* has a projecting blunt apex and occurs in Long Island Sound in winter.

Biology and Ecology. Larval (cydippid) *Mnemiopsis* feed on copepod nauplii and ciliates. The adults eat copepods of all stages and the eggs and larvae of fishes, barnacles, and bivalves. In turn, *Mnemiopsis* is eaten by the sea nettle *Chrysaora* in estuarine areas and by harvestfish and butterfish, by the ctenophore *Beröe*, and by the lion's mane jelly along the coast. *Mnemiopsis* may contain pink, immature stages of the sea anemone *Edwardsia* as a symbiont. This ctenophore is iridescent in the sun and beautifully bioluminescent at night when disturbed. Recent evidence indicates that the ctenophore formerly classified as *M. mccradyi* is a form of *M. leidyi.*

References. Burrell et al. 1976; Colin et al. 2010; Condon and Steinberg 2008; Costello et al. 1999, 2006, 2012; Kolesar et al. 2010; Kreps et al. 1997; Larson 1988; Rapoza et al. 2005; Sullivan and Gifford 2007.

Beröe ovata

Occurrence. The typically pinkish and somewhat flattened *Beröe ovata* occurs along the Atlantic Coast south of Cape Cod and throughout the Gulf. It is more common along the coast and in high-salinity bays than in estuaries. *Beröe cucumis* is a northern species that rarely strays south of Cape Cod.

Biology and Ecology. *Beröe ovata* preys on other comb jellies. It consumes smaller prey whole and takes bites of larger comp jellies using its cutting oral cilia. *Beröe* may reduce or eliminate *Mnemiopsis* where they co-occur. This species is bioluminescent and can harbor hyperiid amphipods.

References. Cahoon et al. 1986; Nelson 1925; Swanberg 1974; Tamm and Tamm 1991.

Pleurobrachia pileus (sea gooseberry)

Occurrence. *Pleurobrachia pileus* is an Arctic species that is occasional in New England in winter and spring and, more rarely, as far south as North Carolina. It occurs in the coastal ocean but may also be seen in high-salinity portions of estuaries. *Pleurobrachia brunnea* occurs further offshore.

Biology and Ecology. *Pleurobrachia* swims slowly while trailing its two, long tentacles and often drifts passively while fishing. These tentacles are invariably lost during collection unless the animals are carefully dipped from the water. Copepods are the primary prey, but other motile crustaceans, including cladocerans, amphipods, and decapod larvae, are consumed. Spiny dogfish and pollock eat *Pleurobrachia.*

References. Båmstedt 1998; Costello and Coverdale 1998; Esser et al. 2004; Frank 1986; Greene et al. 1986; Møller et al. 2010; Nelson 1925.

Mnemiopsis leidyi

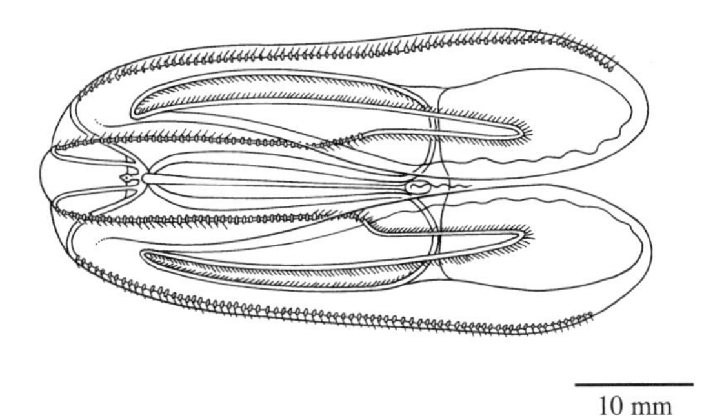

10 mm

Beröe ovata

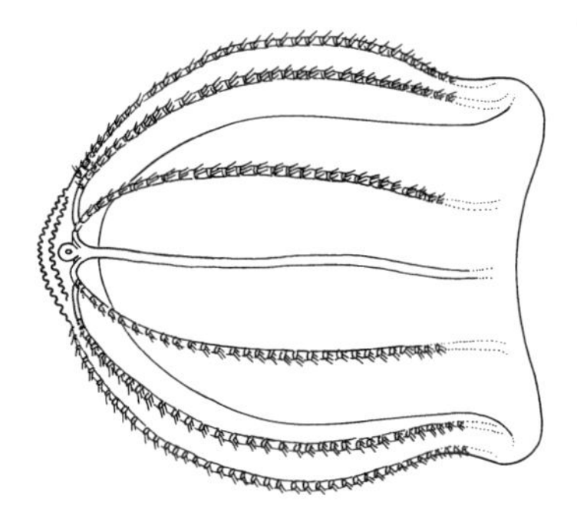

10 mm

Pleurobrachia pileus

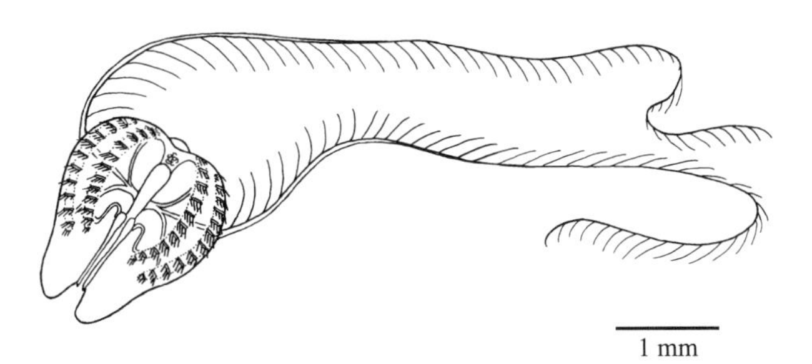

1 mm

ROTIFERS

Rotifers are among the smallest of the true metazoans; most are less than 1 mm in length. The Phylum Rotifera (or Rotatoria) includes about 2,000 described species. Most rotifer species occur in freshwater, a few occur in seawater, and some are abundant in brackish waters, especially at river-estuarine transition regions. Most rotifers have a circular corona of cilia at the anterior end that resembles a rotating wheel, which is used for both locomotion and feeding. The biology of marine and estuarine rotifers is understudied, especially in the United States. This is probably due to their small size and to difficulties in preserving them for later identification rather than to an absence of rotifers.

The common rotifers covered here have a complex and variable life cycle similar to the cyclic **parthenogenesis** seen in *Daphnia* and other cladocerans (see "Cladocerans and Ostracods" section for details). Females dominate most populations and usually reproduce asexually by parthenogenesis. Sexual reproduction, with a brief appearance by males, occurs intermittently, seemingly brought on by specific environmental cues. Sexual reproduction results in fertilized "resting" eggs (actually zygotes) that remain dormant until conditions signal the initiation of hatching and a new cycle. (See Wallace and Snell [2001] or Smith [2001] for an overview.)

Rotifers have rapid life cycles. Females can reach reproductive maturity in one day and produce two to eight eggs daily during their one to two-week life span. Thus, an individual could give rise to more than 100,000 offspring in one week under ideal conditions. Many rotifers reproducing parthenogenically release their eggs directly into the water, but a few brood their eggs.

The coronal cilia propel rotifers through the water. Some rotifers are entirely planktonic, while others cement themselves to the substrate using adhesive produced in special glands in the foot. Most rotifers are omnivorous and will consume whatever particles or prey of the appropriate size that they encounter. Some (e.g., *Brachionus*) feed primarily on small particles in suspension: phytoplankton, detritus, or small protozoans brought to the mouth by the whirling anterior cilia. The raptorial predators project their jaws from the mouth to grab their prey, including smaller rotifers. *Synchaeta* can use either method, depending on the type of food available.

Rotifer diversity and abundance decreases with increasing salinity, but some occur in both coastal and oceanic waters. Information on rotifer ecology along our coasts is limited, with the exception of ongoing research in the Chesapeake Bay where huge concentrations of rotifers can appear and disappear within a few days. When abundant, rotifers play a crucial role in the food web, especially in the spring when other microphagous (small particle) grazers may be virtually absent. Thus, rotifers can form an important link between the microbial loop and higher trophic levels. They are particularly important, perhaps essential, as food for the many larval fishes that are abundant at the freshwater-estuarine interface in spring.

IDENTIFICATION HINTS

Living rotifers are best for identification, but they need to be relaxed and immobile. When preserved, rotifers tend to contort into unrecognizable blobs unless special precautions are taken to relax them. Appendix 3 gives detailed suggestions for narcotizing and preserving rotifers. Because the species most common in the plankton are <400 μm in length, a compound microscope is recommended. Dark-field illumination may produce stunning images. Rotifers often occur in different "morphs," so your specimen may not be an exact match for any of those depicted. Definitive identification may require examination of the hard internal feeding structures (trophi).

The rotifer body generally contains three regions (Fig. 16):

The **head** contains the corona (wheel organ) at the anterior end made of two ciliated rings. Muscles may contract the corona, especially if the rotifer is disturbed. Check for rudimentary eyespots: one, two, three, or none. Hard jaws (trophi) in the pharynx are always present and are often used by specialists. Unfortunately, the jaws are not readily seen without a compound microscope after first dissolving the tissue.

The **trunk** of many rotifer species, including *Keratella*, is encased in a shell called a **lorica** adorned with hooks, spikes, and plates. Other species have a thin, flexible body

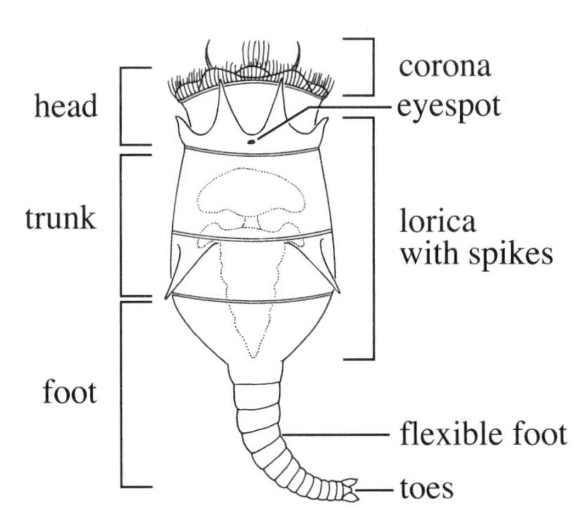

Fig. 16. Anatomy of a rotifer (*Branchionus* sp.). The presence of a corona is shared by most rotifers. The general shape of the body, presence of a lorica, projections on the lorica, eyespot(s), and a foot vary considerably between species. In some rotifers without a lorica, internal organs may be visible. Some rotifers carry eggs externally.

wall. Some species have a flexible **foot**, often with toes that attach temporarily to the substrate (e.g., *Brachionus*). One or more large eggs may be attached where the body meets the foot. Some entirely planktonic species lack feet; presence or absence of a foot is a good diagnostic feature, but be sure to check several specimens because the foot can be retracted.

USEFUL IDENTIFICATION REFERENCES

No guides specifically address the brackish or marine fauna from the western Atlantic or the Gulf of Mexico. However, most species will be found in the following references.

Jersabek, C. D., Segers, H., Dingmann, B., et al. 2003. *The Frank J. Myers Rotifera Collection: An Illustrated Catalog*. Philadelphia, Academy of Natural Sciences of Philadelphia. CD-ROM.

Koste, W. 1978. Rotatoria. In: *Die Rädertiere Mitteleuropas. Bestimmungswerk begründet von Max Voigt*. 2 vols. Borntraeger, Berlin, Stuttgart. 673 pp. (A guide to European species but still useful since many rotifers occur on both sides of the Atlantic.)

Nogrady, T., Segers, H., eds. 2002. *Guides to the Identification of the Microinvertebrates of the World*. Vol. 18, *Rotifera 6*: *Asplanchnidae, Gastropodidae, Lindiidae, Microcodidae, Synchaetidae, Trophosphaeridae and Filinia*. Backhuys, Leiden, The Netherlands. 304 pp.

Smith, D. G. 2001. *Pennak's Freshwater Invertebrates of the United States: Porifera to Crustacea*. 4th ed. Wiley & Sons, New York. 638 pp.

Wallace, R. L., Snell, T. W. 2010. Rotifera. In: Thorp, J. H., Covich, A. P., eds. *Ecology and Classifications of North American Freshwater Invertebrates*. 3rd ed. Academic Press, New York, 173–235.

SUGGESTED READINGS

Arndt, H. 1993. Rotifers as predators on components of the microbial food web (bacteria, heterotrophic flagellates, ciliates): A review. *Hydrobiologia* 255/256:231–246.

Edmondson, W. T., ed. 1959. *Fresh-water Biology*. Wiley & Sons, New York. 1248 pp.

Nogrady, R., Wallace R. L., Snell, T. W. 1993. *Guides to the Identification of the Microinvertebrates of the Continental Waters of the World*. Pt. 4, *Rotifera*. Vol. 1, *Biology, Ecology and Systematics*. SBP Academic, The Hague. 142 pp. (Provides an extensive introduction to the group.)

Park, G. S., Marshall, H. G. 2000. The trophic contribution of rotifers in tidal and estuarine habitats. *Estuarine, Coastal and Shelf Science* 51:729–742.

Pourriot, R., Snell, T. W. 1983. Resting eggs in rotifers. *Hydrobiologia* 104:213–224.

Smith, D. W. 2001. *Pennak's Freshwater Invertebrates of the United States*. 4th ed. Ronald Press, New York. 803 pp. (See chapter on rotifers.)

Wallace, R. L., Snell, T. W., Ricci, C., et al. 2006. *Rotifera*. Vol. 1, *Biology, Ecology and Systematics*. 2nd ed. Vol. 23, *Guides to the Identification of the Microinvertebrates of the Continental Waters of the World*, Segers, H., ed. Backhuys, Leiden. 299 pp. (Provides an extensive introduction to the group.)

Brachionus spp.

Occurrence. Many brackish water *Brachionus* species occur in our area. *Brachionus calyciflorus* can be dominant in early spring blooms in oligohaline regions of Mid-Atlantic estuaries, primarily near the surface. It is less abundant in mesohaline areas. *Brachionus plicatilis* may also occur in higher salinities and was found in the hypersaline Laguna Madre, Texas.

Biology and Ecology. *Brachionus plicatilis* is widely used in aquaculture as a food for both larval fishes and shrimps. It is also used extensively in ecotoxicology assays. This species has a life cycle with both sexual and asexual phases and resting eggs. *Brachionus plicatilis* is an especially variable species and can differ considerably in both size and morphology from place to place and seasonally.

References. Gilbert 2007; Hagiwara et al. 1989; Hlawa and Heerkloss 1995; Rico-Martinez and Snell 1995; Snell et al. 1983.

Keratella spp.

Occurrence. *Keratella cochlearis* may be present all year and can be abundant in early spring blooms at the freshwater/oligohaline interface of Atlantic estuaries, especially in creeks and tributaries.

Biology and Ecology. The diet includes phytoplankton. Predatory rotifers eat *Keratella* when they co-occur.

Synchaeta spp.

Occurrence. *Synchaeta baltica* is abundant in colder months from Woods Hole, Massachusetts, south to at least the Chesapeake Bay, where it can be a winter dominant in salinities of 5–22. *Synchaeta curvata, S. stylata, S. cecilia,* and *S. fennica* (among others) are reported from this region, with *S. fennica* more common in oligohaline areas. *Synchaeta stylata* and *S. cecilia* are typically the more common summer rotifers and can be locally abundant.

Biology and Ecology. *Synchaeta*, a major phytoplankton grazer, is found mainly in periods of high phytoplankton production. *Synchaeta* feeds on bacteria, nanophytoplankton, and heterotrophic microflagellates using both raptorial and filter feeding. It is eaten by some copepods, including *Acartia tonsa.*

References. Badylak and Phlips 2008; Dolan and Gallegos 1991, 1992; Egloff 1988; Heinbokel et al. 1988; Sellner and Brownlee 1990.

Brachionus calyciflorus

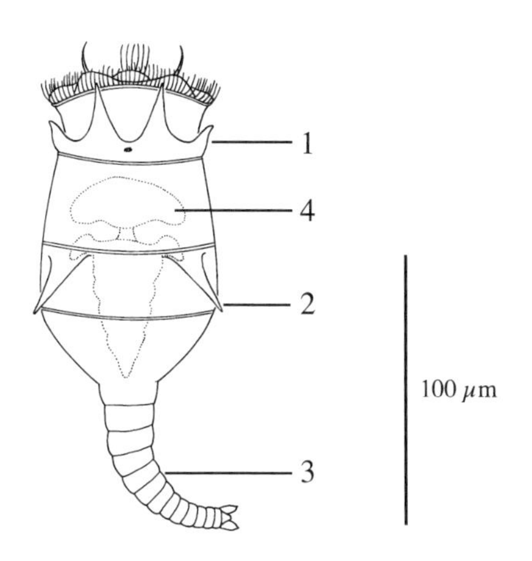

- occipital spines on dorsal margin below corona (1)
- spines near centerline of lorica not always present (2)
- flexible foot with toes (3)
- internal organs visible (4)

Keratella cochlearis

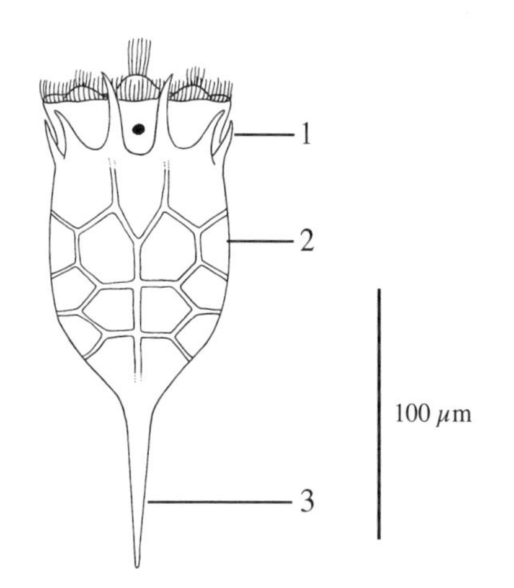

- occipital spines on dorsal margin below corona (1)
- rigid body covering (lorica) has a patterned surface (2)
- pointed rigid foot (3)
- internal organs not visible

Synchaeta sp.

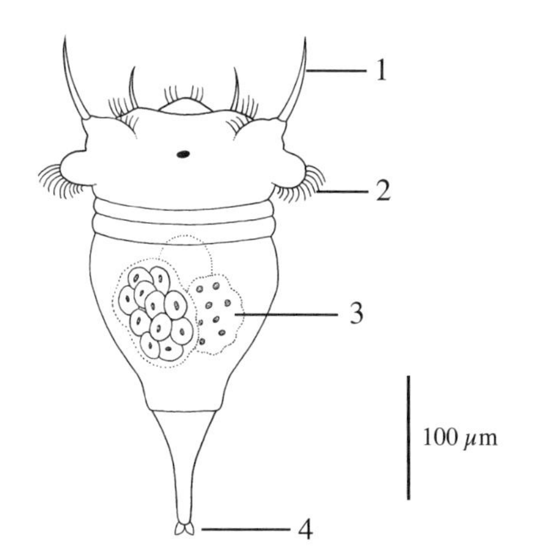

- sensory setae above the corona (1)
- lateral palps with cilia for swimming (2)
- internal organs visible (3)
- fairly rigid foot with toes (4)

CIRRIPEDES

Barnacle Larvae

Barnacle larvae are common and, occasionally, abundant in marine and estuarine plankton samples. The larvae pass through six similar naupliar stages before metamorphosing into the cyprid stage. Most nauplii use their anterior appendages for both swimming and feeding. Nauplii typically feed on phytoplankton, but some species are omnivorous. The cyprid (or cypris) larva is a nonfeeding stage specifically adapted for selecting the proper substrate for settlement. The cyprid swims using six pairs of posterior (thoracic) appendages. When testing the substrate, the cyprid "walks" around using its two antennules and senses the surface contour with the setae at the rear. Chemical cues are also used; sites with previous barnacle settlements are especially favored. At settlement, waterproof glue from the cement gland anchors the incipient barnacle in place. A dramatic metamorphosis into the adult form follows immediately.

Barnacle Nauplius Larva

Two horns on the head and their penchant for swimming upside down distinguish barnacle nauplii from the nauplii of penaeid shrimps and copepods. A pigmented eyespot is usually visible. Nauplii in the *Semibalanus* and *Amphibalanus* (=*Balanus*) group have cephalic shields distinctly different from those of *Chthamalus*. Nauplii occur at all seasons but especially in spring. The most common coastal barnacles, including *Chthamalus fragilis, Amphibalanus venustus, A. amphitrite,* and *A. improvisus,* are widespread. Barnacle nauplii are common to abundant in estuaries. *Amphibalanus eburneus* and *A. subalbidus* are the most common low-salinity barnacles of both Atlantic and Gulf estuaries and sometimes occur near freshwater. *Semibalanus balanoides* is a high-salinity northern species common in New England, ranging south to North Carolina. The first stage nauplii are about 0.2 mm. Because of their small size, retention of early nauplii in nets is often low. Later stages may reach 0.9 mm before the transition to the cyprid stage.

References. Qiu et al. 1997.

Barnacle Cyprid (or Cypris) Larva

Cyprids (usually <0.5 mm long) have a smooth, flexible, and translucent bivalved covering that may or may not enclose the jointed appendages. Color is pale yellow to light brown with a somewhat reddish eyespot near the center of the body. The seedlike shape resembles ostracods and bivalves, but those groups have hard shells with surface sculpturing. Distinguishing separate species of cyprids is seldom feasible. Cyprids are relatively common, especially near the bottom, but are seldom as abundant as nauplii. Geographic distributions follow those of nauplii above.

References. Berntsson et al. 2000; Dineen and Hines 1994; Elbourne and Clare 2010; Faimali et al. 2004; Lagersson and Hoeg 2002; Mullineaux and Butman 1991; Pechenik et al. 1993; Thiyagarajan 2010; Tremblay et al. 2007.

USEFUL IDENTIFICATION REFERENCES

Lang, W. H. 1979. Larval development of shallow water barnacles of the Carolinas (Cirripedia: Thoracica) with keys to naupliar stages. NOAA Technical Report NMFS Circular 421. 39 pp.

Chthamalus fragilis nauplius stage IV (of VI)

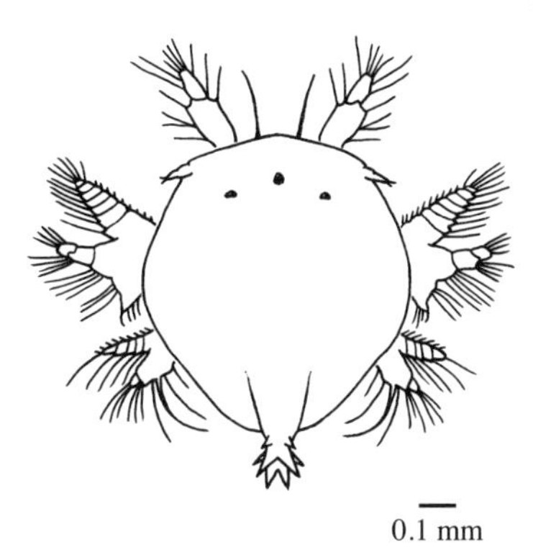

Amphibalanus venustus nauplius stage IV of (VI)

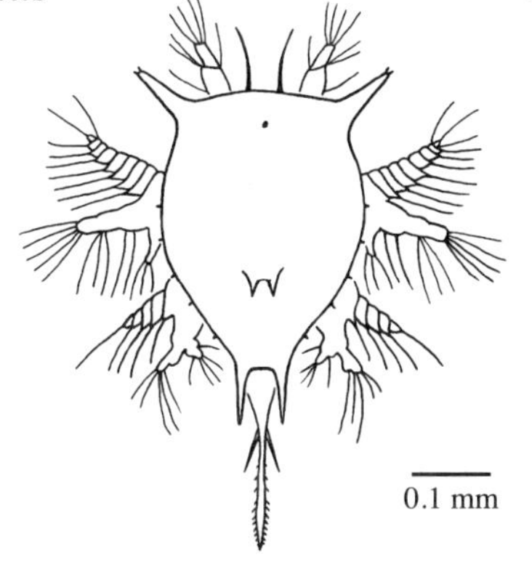

Balanus sp. cyprid

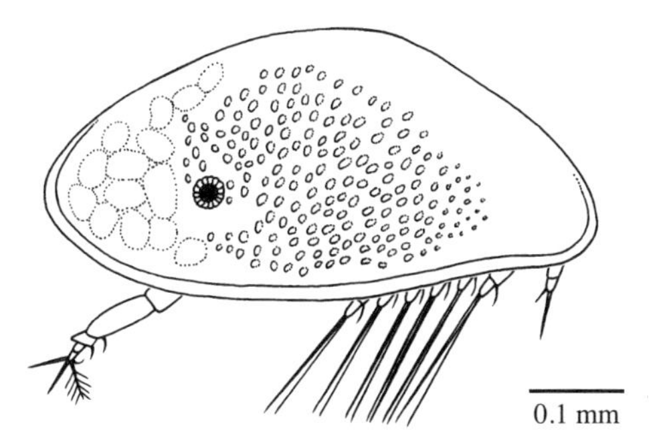

CLADOCERANS AND OSTRACODS

While water fleas like *Daphnia* are among the most common and conspicuous of freshwater zooplankton, only a few of the 600 species of cladocerans occur in brackish or marine waters. Nevertheless, marine cladocerans can be abundant and, at times, dominate the mesozooplankton. The traditional Order Cladocera has been replaced by four new orders. The orders with marine species represented here are Ctenopoda (*Penilia*), Anomopoda (*Bosmina*), Cladocera (*Leptodora*), and Onychopoda (*Evadne, Pseudevadne, Pleopsis, Podon*). Cladocerans tend to be small (<2 mm), semitransparent, constantly swimming members of the holoplankton. Large, branched second antennae propel cladocerans through the water, while ventral appendages are used in feeding. *Penilia* and *Bosmina* dine on prey in the 1–5 μm size range using suspension feeding. The other cladocerans covered here use raptorial feeding to capture diatoms and dinoflagellates (especially *Ceratium)* in the 20–170 μm size range. These cladocerans may also consume ciliates and copepod larvae. Since field data are lacking, this summary of feeding habits should be considered as preliminary.

Cladocerans brood their young within their carapace and have no free larval forms. Under favorable conditions, females produce more females asexually by parthenogenesis (without fertilization). At the end of the growing season or when conditions deteriorate, males appear and sexual reproduction ensues. This unusual mode of reproduction is called cyclic parthenogenesis, as explained in Figure 17. The fertilized resting “eggs” (actually zygotes or early embryos) produced by sexual reproduction remain dormant in the sediments until favorable conditions initiate hatching. Rapid maturation and quick succession of broods can produce rapid population increases.

OSTRACODS

Ostracods are small bivalved crustaceans, usually less than 2 mm in length. Like many cladocerans, they have a calcified bivalved shell, but the shell is usually pigmented and conceals most of the appendages. Although the majority of species are benthic, ostracods are common in nearshore plankton. Some open ocean species are holoplanktonic, but nearshore samples often contain benthic or demersal species noted for their frequent forays into the water column. Ostracods swim with their bivalved carapace slightly agape and with their antennae extended for propulsion. Most cruise at <5 cm/s, but some reach 20 cm/s in sprint mode. Ostracods are assumed to be opportunistic omnivores.

The few local marine species studied reproduce sexually, although some may be partially or entirely parthenogenic. Many planktonic ostracods can produce a bright biolumi-

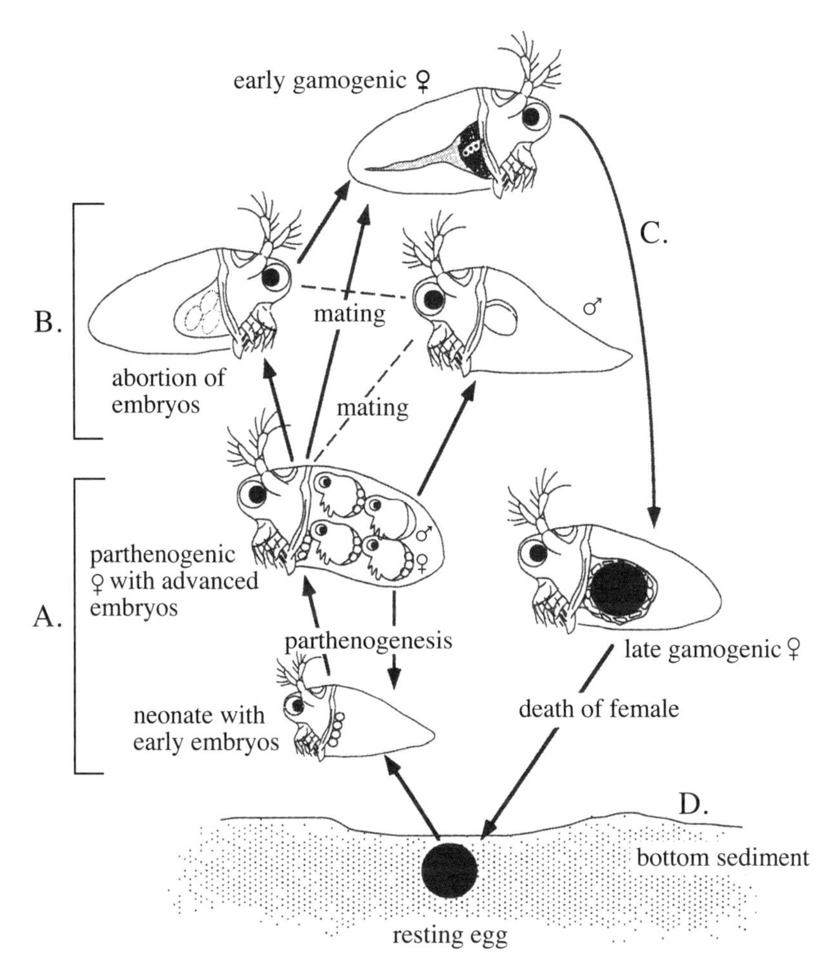

Fig. 17. Cladocerans exhibit cyclical parthenogenesis as illustrated by the life cycle of *Evadne nordmanni*. *A*, During most of the year, females reproduce asexually via parthenogenesis, resulting in successive female generations with no males in the population. At the onset of sexual reproduction, females produce male and female offspring via parthenogenesis. *B*, Females also switch from producing diploid (parthenogenic) eggs to producing haploid eggs. *C*, These "gamogenic" (mictic) females mate with the newly produced males. *D*, The resulting fertilized eggs, encased in a protective coating, sink to the bottom where they remain in a state of arrested development until favorable conditions signal hatching. The resting "eggs" give rise to parthenogenic females. A unique feature of the marine Onychopoda (*Evadne*, *Pseudevadne*, *Pleopsis*, *Podon*) is the occurrence of early embryos in the brood pouches of late stage embryos; they are born pregnant (Modified after Egloff et al. 1997).

nescence when disturbed. Considering their prevalence, the nearshore planktonic ostracods deserve more scientific attention.

IDENTIFICATION HINTS

Bivalve mollusc veligers, barnacle cyprids, cladocerans, and ostracods all have a bivalved shell or carapace. The presence of jointed appendages projecting from the shell indicates that the specimen is a crustacean rather than a mollusc. The clamlike carapace may partially or wholly enclose these appendages. In mollusc veligers and ostracods, valves are hard and calcified, though they are often brittle. Bivalve mollusc veligers often have transparent valves with smooth concentric ridges and soft external tissues. Nearshore ostracods often have valves with a rough texture or shallow pits or both.

Cladocerans and barnacle cyprids tend to have soft or at least flexible valves. Cladocerans have anterior eyes and conspicuous, external antennae. Barnacle cyprids have eyes that are set well posterior of the anterior margin and are under the carapace. Although cladocerans and ostracods brood eggs that are usually visible through the carapace, egg shapes, sizes, or positions are not reliable in identification. The shape of their valves and characteristics of their appendages usually identify ostracods.

USEFUL IDENTIFICATION REFERENCES

Cladoceran Identification

Della Croce, N. 1974. *Cladocera: ICES Zooplankton Identification Leaflets No. 143.* <www.ices.dk/products/fiche/Plankton/START.PDF>. Accessed August 18, 2011. (A guide to European species.)

De Melo, R., Hebert, P. D. N. 1994. Taxonomic reevaluation of North American Bosminidae. *Canadian Journal of Zoology* 72:1808–1825.

Rivier, I. K. 1998. The predatory Cladocera (Onychopoda: Podonidae, Polyphemidae, Cercopagidae) and Leptodorida of the world. In: Dumont, H. J., ed. *Guides to the Identification of the Microinvertebrates of the Continental Waters of the World* (p. 214). Vol. 13. Leiden: Backhuys, Leiden, The Netherlands.

Thorpe, J. H., Covich, A. P., ed. 2010. *Ecology and Classification of North American Freshwater Invertebrates.* Academic Press, New York. 1021 pp.

Ostracod Identification

The planktonic ostracod fauna in our geographical range is so poorly known that it cannot be treated in detail, so further identification is problematic; recent European guides may help.

Angel, M. V. 2000. *Marine Planktonic Ostracods.* World Biodiversity Database. CD-ROM Series. Macintosh/Windows Version. Springer, New York. (European, but likely much overlap with U.S. fauna.)

Athersuch, J., Horne D., Whittaker, J. E. 1989. *Marine and Brackish Water Ostracods.* E. J. Brill, New York. 343 pp.

Cohen, A. C. Tabular Key to the Subclasses of Ostracoda and Families of Myodocopa. <http://home.comcast.net/~fireflea2/OstracodeKeyindex.html.> Accessed August 3, 2011.

Darby, D. G. 1965. *Ecology and Taxonomy of Ostracoda in the Vicinity of Sapelo Island, Georgia.* Project GB-26. Report No. 2. National Science Foundation, Arlington, VA, 1–76.

Poulsen, E. M. 1969. *Ostracoda: ICES Zooplankton Identification Leaflets*. Nos. 115–116. <www.ices.dk/products/fiche/Plankton/START.PDF>. Accessed August 18, 2011. (An older guide to European species.)

SUGGESTED READINGS

Cladoceran Biology

Egloff, D. A., Fofonoff, P. W., Onbé, T. 1997. Reproductive biology of marine cladocerans. *Advances in Marine Biology* 31:79–167. (Includes information on distribution, including salinity range.)

Fryer, G. 1987. Morphology and the classification of the so-called Cladocera. *Hydrobiologia* 145:19–28. (Calls for splitting the Cladocera into four separate orders.)

Gieskes, W. W. W. 1971. Ecology of the Cladocera of the North Atlantic and the North Sea, 1960–1967. *Netherlands Journal of Sea Research* 5:342–376.

Kim, S. W., Yoon, Y. H., Onbé, T. 1993. Note on the prey items of marine cladocerans. *Journal of the Oceanological Society of Korea* 28:69–71.

Onbé, T. 1999. Ctenopoda and Onychopoda (=Cladocera). In: Boltovskoy, D., ed. *South Atlantic Zooplankton*. Vol. 1. Backhuys, Leiden, The Netherlands, 797–813.

Richter, S., Braband, A., Aladin, N., et al. 2001. The phylogenetic relationships of "predatory water fleas" (Cladocera: Onychopoda, Haplopoda) inferred from 12 SrDNA. *Molecular Phylogenetics and Evolution* 19:105–113.

Rivier, I. K. 1998. *The Predatory Cladocera (Onychopoda: Podonidae, Polyphemidae, Cercopagidae) and Leptodorida of the World*. Backhuys, Leiden, The Netherlands. 214 pp. (All aspects of the biology of *Podon, Pleopsis, Evadne,* and *Pseudevadne*.)

Ostracod Biology

Angel, M. V. 1999. Ostracoda. In: Boltovskoy, D., ed. *South Atlantic Zooplankton* (pp. 815–868). Vol. 1. Backhuys, Leiden, The Netherlands.

Angel, M. V. 2000. *Marine Planktonic Ostracods*. World Biodiversity Database. CD-ROM Series. Macintosh/Windows Version. Springer, New York.

Horne, D. J., Martens, K. 2000. Evolutionary biology and ecology of Ostracoda. Special issue, *Hydrobiologia* 419:1–197.

Schram, F. R. 1986. Ostracoda. In: *Crustacea* (pp. 399–422). Oxford University Press, New York.

Pleopsis polyphemoides, formerly *Podon polyphemoides* and *Podon* spp.

Occurrence. *Pleopsis polyphemoides* is the most common cladoceran in coastal embayments and estuaries along the Gulf and Atlantic Coasts. It is most common in moderate salinities of 10–20 but has been reported from nearshore ocean waters. Maximum seasonal abundance occurs in spring and summer. *Podon intermedius* is a temperate coastal species occurring from the Chesapeake Bay north to the Gulf of Maine. *Podon leuckartii* is a cold-water coastal (>25 psu) species that occurs south to Virginia during winter.

Biological Notes. In the Potomac River, Virginia, *P. polyphemoides* populations overwinter in tributaries and creeks with a salinity of about 8. During spring and summer, they disperse throughout the river, where they reach maximum densities near the mouth at about 15 psu and then enter the Chesapeake Bay main stem, where densities may exceed 50,000 m^{-3}. The diet of these omnivorous water fleas includes phytoplankton and perhaps larvae of the estuarine copepod *Eurytemora.*

References. Bosch and Taylor 1973a, 1973b; Egloff et al. 1997; Jagger et al. 1988; Katechakis and Stibor 2004; Lippson et al. 1979.

Penilia avirostris

Occurrence. *Penilia* occurs only in high salinities along the Atlantic and Gulf Coasts in warm seasons. On the Atlantic Coast, it ranges northward to Woods Hole, Massachusetts, but it is more common from Delaware Bay south. It can occur at densities equal to or greater than those of the most common copepods.

Biological Notes. This suspension feeder removes small particles 2–20 µm in diameter, particularly those <5 µm, as well as some larger foods, including diatoms and heterotrophic flagellates. The brood pouch typically contains seven to nine young. Resting eggs allow survival during colder months.

References. Atienza et al. 2008; Egloff et al. 1997; Lochhead 1954; Mullin and Onbé 1992; Paffenhöfer and Orcutt 1986; Turner et al. 1988.

Bosmina spp.

Occurrence. These widespread, predominantly freshwater cladocerans are common in Mid-Atlantic and southern estuaries in winter and spring, primarily in tidal fresh to oligohaline areas. *Daphnia* spp. also occur in oligohaline areas, especially when spring rains wash them downstream.

Biological Notes. *Bosmina* feeds on phytoplankton and small protozoans the 1–3 mm size range. It is an important food for larval white perch and striped bass in tidal freshwater and oligohaline areas. The taxonomy of this genus is unresolved in our geographic range.

References. De Melo and Hebert 1994; Kankaala 1983; Vaque et al. 1992.

Pleopsis polyphemoides

♀ 0.6 - 0.7 mm, ♂ 0.5 - 0.6 mm

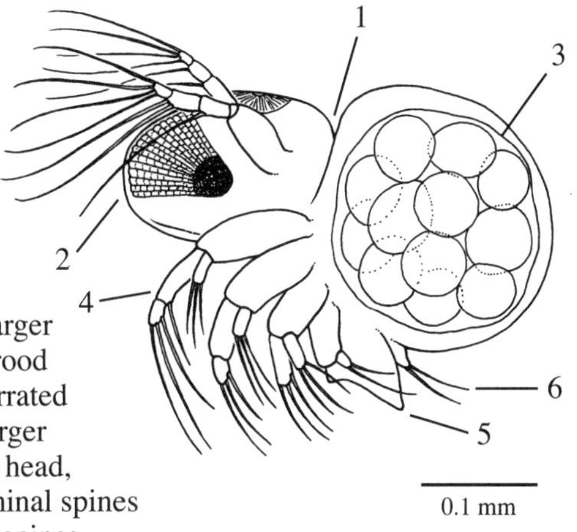

- head and brood sac (carapace) separated by dorsal groove (1)
- large compound eye (2)
- prominent round or oval brood sac (with eggs) (3)
- 4 pairs of thoracic legs (4)
- short, triangular carapace spines (5)
- short postabdominal spines (6)
- similar species: *Podon lueckartii* is larger (♀ 0.8 - 1.0 mm) and has a smaller brood sac and carapace spines with long, serrated points; *Podon intermedius* is much larger (♀ 1.0 - 1.2 mm) and has an elongate head, low-profile brood sac, and postabdominal spines that extend well beyond the carapace spines.

Penilia avirostris

♀ 0.4 - 1.2 mm, ♂ 0.7 - 0.9 mm

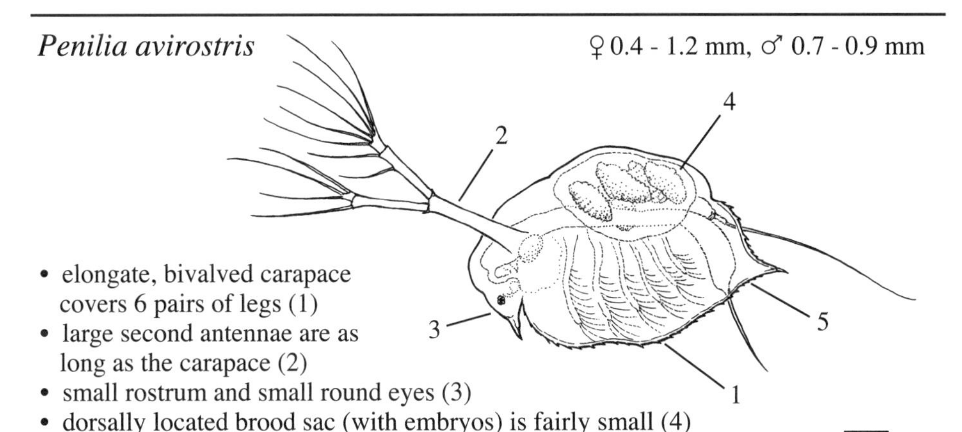

- elongate, bivalved carapace covers 6 pairs of legs (1)
- large second antennae are as long as the carapace (2)
- small rostrum and small round eyes (3)
- dorsally located brood sac (with embryos) is fairly small (4)
- posterior of the carapace is pointed, and posterior and ventral margins have many small spines (5)

Bosmina sp.

♀ 0.3 - 0.5 mm, ♂ 0.25 - 0.4 mm

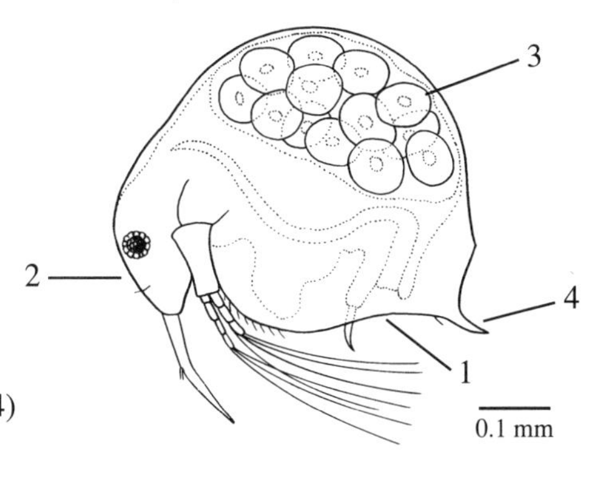

- round body with bivalved carapace covering legs (1)
- head with pair of round eyes and strong rostral spine (2)
- brood sac (with eggs) occupies dorsal region (3)
- posterior of carapace terminates in slender points; no marginal spines (4)

Pseudevadne tergestina

Occurrence. At least two species of *Evadne* and one species of *Pseudevadne* occur within our range; they are morphologically similar. *Pseudevadne tergestina* is a southern coastal species that ranges from Delaware Bay to the Gulf. It prefers waters above 20°C and salinities >30.

References. Bryan 1979; Mullin and Onbé 1992; Wong et al. 2008.

Evadne nordmanni

Occurrence. *Evadne nordmanni* is the most widespread and common *Evadne* species in both estuarine and coastal waters from Chesapeake Bay north and occurs in salinities as low as 20. It prefers temperatures <16°C. *Evadne nordmanni* is more common in neritic waters, especially in the surface layer, although some may be in deeper waters. *Evadne spinifera* is more typical of offshore waters from Delaware Bay to Florida. It prefers waters above 20°C and salinities >30.

Biology and Ecology. *Evadne* releases its brood at night or early morning.

References. Bainbridge 1958; Katechakis and Stibor 2004.

Leptodora kindtii (giant water flea)

Occurrence. *Leptodora kindtii* is widespread in Northern Hemisphere lakes and reservoirs but is rarely reported from brackish waters. However, it is widespread and frequently abundant during spring in lower-salinity (1–5 rarely to 10) reaches of the Chesapeake Bay and its tributaries. Its range in brackish waters beyond the Chesapeake is not yet documented.

Biology and Ecology. The virtually invisible *Leptodora* is an ambush predator that swims with its thoracic appendages spread wide to form a "feeding basket." When prey contacts *Leptodora*, the abdomen flexes to pin the prey against the body, where it is grasped by the thoracic appendages. *Leptodora* then bites its prey and sucks out the contents. Other cladocerans and rotifers are the primary prey in freshwaters, but feeding in brackish waters is unstudied. *Leptodora* is a strong vertical migrator, found near bottom during the day. Resting eggs, resulting from sexual reproduction, hatch into the plankton as nauplius larvae. Otherwise, free nauplii larvae are unknown in cladocerans.

References. Branstrator 2005; Browman et al. 1989; Olesen et al. 2003.

Pseudevadne tergestina

♀ 0.3 - 1.3 mm, ♂ 0.5 - 0.8 mm

- body laterally flattened; head continuous with carapace (1)
- head with large eye and no rostrum (2)
- carapace triangular but rounded at end (3)
- linear pattern evident on carapace (4)
- carapace spine is ventral to legs and directed slightly forward (5)
- females with one to a few eggs (6)
- similar species: the posterior end of the carapace (brood sac) of *Evadne nordmanni* is always pointed; *E. spinifera* has a needle-like projection from the pointed posterior margin

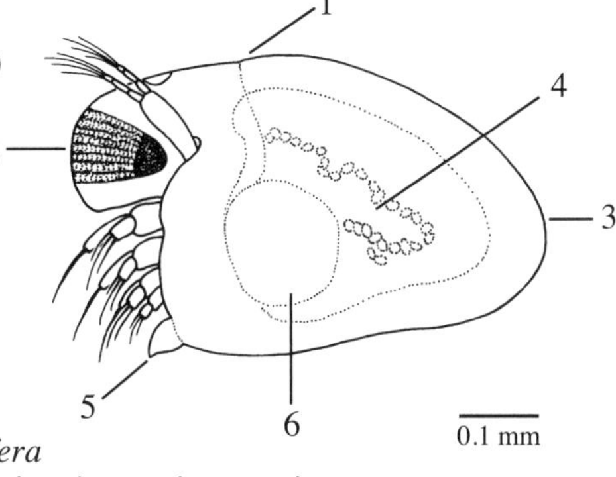

Evadne nordmanni (female)

♀ 0.5 - 1.5 mm, ♂ up to 1.0 mm

- body laterally flattened; head continuous with carapace (1)
- head with moderately large eye and no rostrum (2)
- carapace sharply pointed at posterior end (3)
- carapace spine ventral to legs is almost perpendicular to axis of body (4)
- female with one to a few eggs (5)
- similar species: *E. spinifera* has a needle-like projection from the pointed posterior margin; *Pseudevadne tergestina* has a rounded posterior

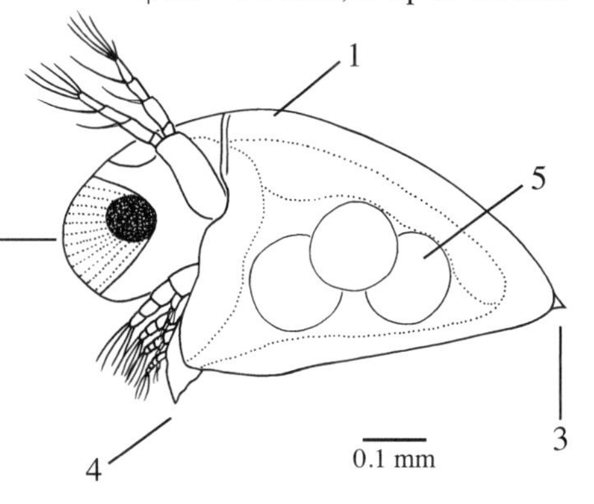

Leptodora kindtii (female)

♀ to > 12 mm

- elongate body with paired prominent antennae well posterior to the eye (1)
- single black eye reveals presence of the almost transparent body (2)
- rudimentary carapace (3)
- dorsal brood sac with eggs or embryos in females (4)
- distinctly forked posterior end (5)
- similar species: *Leptodora* is very unique and is not likely to be confused with another estuarine planktonic crustacean

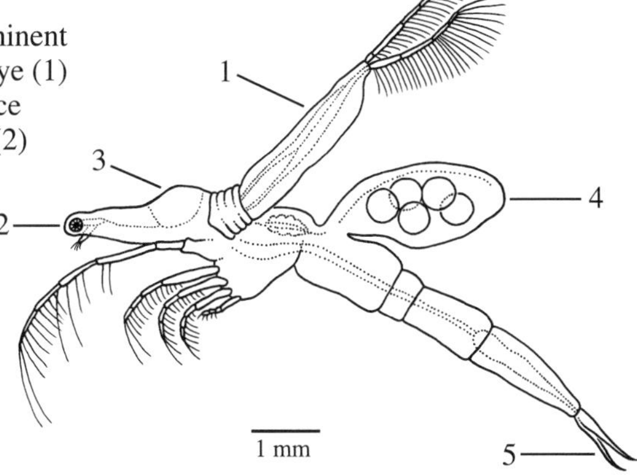

OSTRACODS

ID hint: The calcified bivalve shells typical of nearshore ostracods often have a characteristic texture of ridges or shallow pits quite unlike the smooth surface seen in the bivalved shells of cladocerans and barnacle cyprids. The shells of ostracods are pigmented unlike the semitransparent "shells" of mollusc veligers, cladocerans, and barnacle cyprids. The presence of jointed appendages when visible makes ostracods easy to distinguish from mollusc veligers, the only other one in the group that has a brittle rather than a flexible shell. Offshore holoplanktonic ostracods have a thinner, smoother carapace.

Euconchoecia chierchiae

Occurrence. This ostracod is reported from Rhode Island to Texas in coastal waters and in estuaries at salinities generally above 15 and more typically above 25. In ocean waters, it is caught within a few meters of the surface both day and night.

Biology and Ecology. *Euconchoecia* apparently feed on diatoms, detritus, and possibly fragments of gelatinous zooplankton. They typically swim slowly but are capable of rapid acceleration when disturbed. The related *Conchoecia* species can control their buoyancy by a still unknown mechanism. The sexes are separate, and the brood develops within the carapace.

References. Darby 1965.

Cytheromorpha sp. (podocopid ostracod)

Occurrence. Various species of *Cytheromorpha* are reported from the Atlantic Coast from at least as far north as the Chesapeake Bay and south to Georgia. *Cytheromorpha paracastanea* occurs in Southeast estuaries and throughout the Gulf. *Cytheromorpha curta* and *C. warneri* are widely distributed in Georgia estuaries at salinities ranging from 2 to 29, and adult *C. curta* occurred throughout the year. Other species of this genus and other genera of podocopids occur in both Atlantic and Gulf waters. Although podocopid ostracods are associated with the bottom, occurrences in the water column are not uncommon in tidal systems.

Biology and Ecology. The biology and ecology of estuarine members of podocopid ostracods is little studied despite their abundance. *Cytheromorpha* is thought to live as long as one year in Georgia estuaries.

Euconchoecia cheirchiae (female) ♀ 1.1 - 1.3 mm, ♂ 1.0 - 1.1 mm

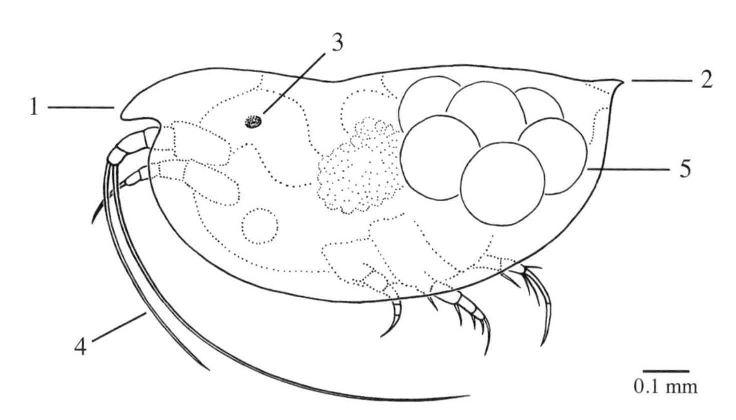

- generally D shaped, translucent carapace with anterior rostrum (1)
- dorsal posterior end with distinct point (2)
- indistinct eye located well posterior of rostrum (3)
- antennae visible, but legs are often not (4)
- 2-8 eggs usually visible in females through posterior half of carapace (5)
- similar species: cladocerans have 4-6 pairs of appendages rather than 2 or fewer; many other genera of ostracods occur in estuaries and the coastal ocean, and identification is difficult; many other ostracods have solid (more opaque) shells

---

Cytheromorpha sp. less than 2 mm

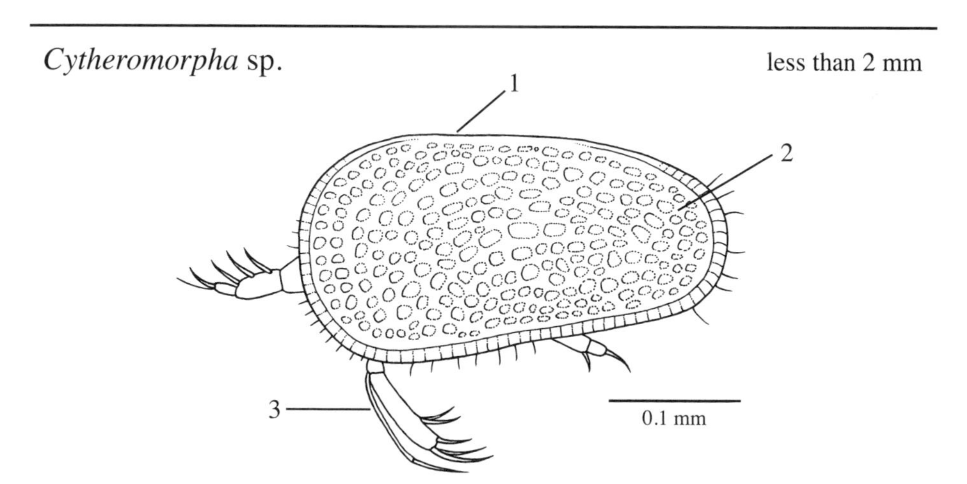

- rigid bivalve carapace hinged along dorsal margin (1)
- pitted and pigmented surface (2)
- distal portion of jointed appendages usually extend from the anterior and ventral margins (3)

COPEPODS

Copepods are the most abundant animals in most mesozooplankton collections and often outnumber all other animals combined. This diverse group of small crustaceans (adults mostly 0.5–5.0 mm) contains more than 200 families and more than 10,000 marine species. Because of their abundance, copepods are important links in virtually all marine and estuarine food webs. Of the 10 orders of copepods, **calanoids**, **cyclopoids**, **poecilostomatoids**, **monstrilloids**, and **harpacticoids** are most frequently encountered in estuarine and nearshore zooplankton samples, but parasitic **siphonostomatoids** appear occasionally.

Temperate species show seasonal cycles (Fig. 18) and have a variety of adaptations for surviving unfavorable periods. Most nearshore calanoids produce resting or diapause eggs that sink to the bottom and remain viable for months until changing temperatures induce hatching and initiate the next generation. Harpacticoids may overwinter as quiescent females that respond to increasing day length in spring to release their nauplii. In yet another variation, cyclopoids and poecilostomatoids often overwinter in copepodid stages. Species that favor cooler seasons may produce resting eggs in response to rising temperatures.

Copepods reproduce sexually, with rare exceptions. Ritualized behaviors and physical or chemical cues assist pairing. Males often grasp the female using specially modified antennae or legs while cementing a sperm packet to her genital region. A single copulation often fertilizes multiple clutches. Eggs may be shed singly into the water or enclosed in egg clusters attached to the female.

The eggs hatch as nauplius larvae, often only about 100 µm long and with only three pairs of appendages used for both swimming and feeding. Larvae go through six naupliar stages (molts) before becoming copepodids (Fig. 19). The five copepodid stages increasingly resemble the adults, adding body segments (somites), pairs of legs, or both in each molt. Most free-living copepods have a maximum life span of six months to a little more than a year. Generation times from egg to egg are temperature dependent—usually two to six weeks. Under favorable conditions, some copepods can release successive clutches daily.

Each copepod group has its own mode of swimming. As suspension feeders, larger calanoids cruise slowly using the second antennae and oral appendages. When threatened, they execute rapid evasive maneuvers, or jumps, using their thoracic "swimming" legs, sometimes in combination with flapping the abdomen. Smaller calanoids are more likely to swim in an irregular, zigzag motion. As they swim, the first antennae are usually folded against the body to reduce drag. When copepods move slowly or are at rest, the antennae are extended and act as hydrodynamic sensors to detect vibrations produced by prey

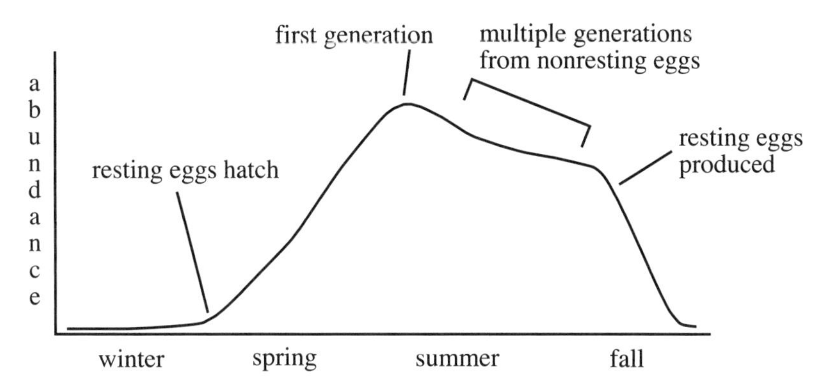

Fig. 18. Generalized "life history" and reproductive pattern for a temperate copepod that reproduces rapidly in warm seasons. Resting eggs hatch in the spring.

or predators. The extended antennae also act as underwater parachutes to retard sinking. Cyclopoids typically exhibit a "hop and sink" motion in which periods of slow sinking alternate with periodic bouts of upward swimming. Recent studies suggest that copepods can cover their hydrodynamic tracks to conceal their presence or, alternatively, create distinctive hydromechanical signals that attract mates.

Most nearshore calanoids are both omnivorous particle grazers and opportunistic predators on microzooplankton. Suspension feeding seems to be the primary feeding mechanism in smaller calanoids, but many species also employ raptorial feeding to catch individual prey, including ciliates and copepod nauplii. Planktonic cyclopoids and poecilostomatoids are primarily raptorial carnivores as adults and often feed on larger prey, including fish larvae. Planktonic harpacticoids are enigmatic. The benthic forms are grazers, or browsers, on benthic microflora and fauna. At least some feed on microalgae in the water column and detritus and its associated microflora, but the feeding mechanism is unknown.

The most studied feeding mechanism is **suspension feeding**, widely, but incorrectly, termed *filter feeding*. Sixty years after the classic treatments of feeding in *Calanus*, the mechanism of suspension feeding remains an area of active inquiry and debate. Anterior appendages beat to produce currents that pull small suspended particles, primarily phytoplankton cells, ciliates, and other small "microzooplankton," toward a ventral feeding chamber. Contrary to the old assumption that the feeding appendages simply strain particles from this feeding current in a continuous motion, high-speed motion pictures show that at least some copepods detect and trap individual particles with directed movements of the feeding appendages. Some copepods feed effectively on a surprisingly broad range of food sizes (3 μm to more than 300 μm in *Calanus*) and can adjust to the size spectrum of particles in the water. New appreciation for forces affecting movement of small particles in a viscous water medium may revise our understanding of suspension feeding.

Raptorial feeding is used to capture motile prey. Predatory copepods use special sensors on their first antennae to detect water disturbances produced by prey. Some species remain motionless while waiting to ambush prey. The large, fast calanoid *Labidocera* cruises

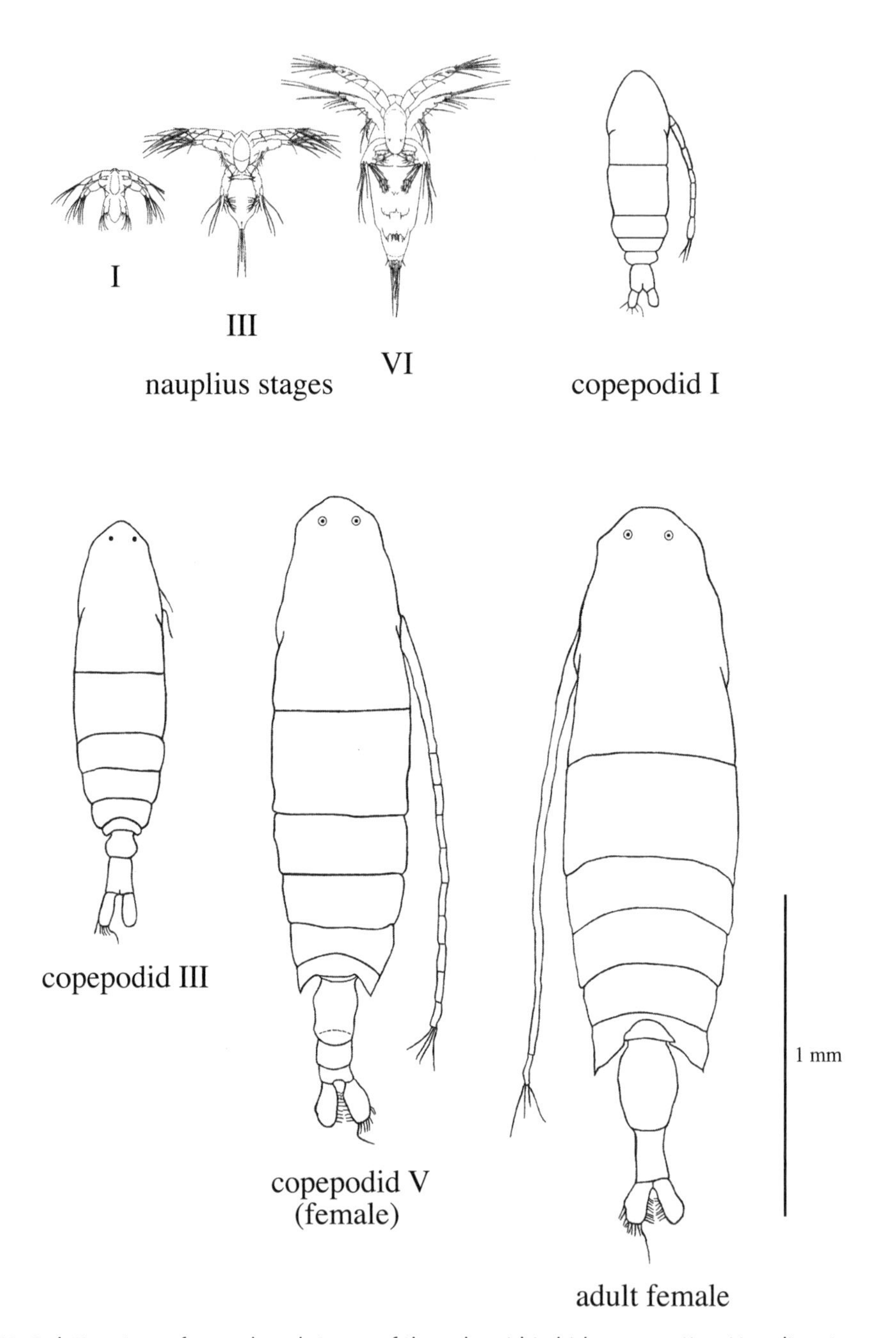

Fig. 19. Relative sizes of some larval stages of the calanoid *Labidocera aestiva*. Nauplius stages I, III, and VI (ventral view) show growth and the developing rudiments of thoracic legs before molting into the very different copepodid stages (dorsal view). The adult is actually the last copepodid stage. Note that only one antenna and some posterior setae are shown in the copepodid and adult stage illustrations.

constantly while hunting. When prey are detected, these copepods accelerate to attack and then grasp their victims using saber-shaped mandibles.

Recurring patterns of dominant species result in relatively consistent assemblages of copepods. *Acartia tonsa* is perhaps the most ubiquitous and regularly occurring coastal and estuarine copepod. Other species have more restricted distributions associated with salinity, temperature, and distance from shore (Table 7). Relatively few species are common in lower salinities, although they may be extremely abundant. Higher-salinity estuarine areas and the nearby coastal areas harbor a significantly higher diversity, with temperature preferences affecting both latitudinal and seasonal occurrences. Oceanic species constitute a separate fauna typical of clear ocean water. Cool-water species are common over the continental shelf, while the Gulf Stream or the Florida Loop Current in the Gulf of Mexico may carry warm-water species northward. Oceanic species appear occasionally in nearshore waters when storms or currents bring offshore water into these regions. In several cases, different species in the same genus may prefer specific temperatures (e.g., *Acartia*) or either coastal or offshore waters (e.g., *Centropages*).

As the most abundant mesozooplankton in most marine and brackish areas, copepods play a central role in marine food chains. They are often the dominant grazers on phytoplankton and thus a primary link between primary productivity and higher trophic levels. Copepods also feed extensively on the microheterotrophs of the microbial loop and thus form a link to bacterial productivity. The larger predaceous copepods are more important in oceanic waters, where they feed on a variety of mesozooplankton, including other co-

Table 7. An overview of basic distributional patterns of selected copepods

	Oligohaline (usually 0.1–5 psu)	Estuarine (5–20 psu)	Coastal and estuarine (>20 psu)	Oceanic (often >10 km offshore)
Cool water	*Coullana canadensis*	*Acartia hudsonica*	*Acartia hudsonica*	*Calanus finmarchicus*
	Eurytemora affinis	*Eurytemora affinis*	*Temora longicornis* *Pseudocalanus newmani, moultoni* *Tortanus discaudatus* *Centropages hamatus*	*Pseudocalanus newmani, moultoni* *Tortanus discaudatus* *Centropages typicus*
Warm water	*Acartia tonsa* *Halicyclops fosteri*	*Acartia tonsa* *Oithona colcarva*	*Acartia tonsa* *Oithona colcarva* *Labidocera aestiva* *Pseudodiaptomus pelagicus* *Parvocalanus crassirostris* *Temora turbinata* *Euterpina acutifrons*	*Paracalanus* *Oithona similis* *Anomalocera* *Pontella* *Oncaea venusta* *Microsetella* *Macrosetella* *Centropages furcatus* *Subeucalanus pileatus*

Note: Salinities and temperatures where copepods are most abundant are indicated, although they appear in lower numbers over a wider range. The designation of coastal includes high-salinity (>20 practical salinity unit, or psu) reaches of estuaries and the coastal ocean out to 5–10 km from shore.

pepods. In turn, copepods become food for benthic suspension feeders and water- column predators, including ctenophores, medusae, arrow worms, and planktivorous fishes.

IDENTIFICATION HINTS

Characteristics Useful in Distinguishing Individual Species

Adults of most common copepods are not difficult to identify based on key morphological characteristics; information on regional or seasonal distribution is helpful in the process. Females receive more detailed treatment here because they usually greatly outnumber males and tend to be more distinctive. Males are usually similar enough to females to be matched with them in any given sample. Thus, we typically figure only the females of most species and include males when they differ substantially from the females.

Identifications here are based on adult morphology. **It can be difficult to distinguish adults from the later copepodid stages.** Although copepodids strongly resemble the adult, they are always smaller. Always look for the largest individuals in the sample. If in doubt, check to see whether the specimen reaches the minimum size listed for the species. In other respects beyond size, copepodids resemble adults except that the copepodids will not look "finished" and will not have the final complement of legs and abdominal segments. In particular, the urosome and its setae do not appear fully developed in copepodid stages. These differences become more readily discernible with practice. You can usually match copepodids with their respective adults in your sample. Mesh sizes of 153 μm or less may retain a predominance of copepodids. Smaller mesh sizes retain nauplii, which are especially difficult to distinguish even at the genus level.

Below are the major characteristics used to identify adult copepods. Figure 20 identifies these and other features described in the bulleted illustrations that follow.

The Shape and Relative Proportions of the Body Regions

A major body articulation divides the body into two parts. The term **prosome**, as used by Gerber (2000) and Mauchline (1998), is the broad anterior division of the body. It consists of the head and those larger thoracic segments (somites) fused to it.
On the **prosome** note:

The length of the **antennae** (also called "first antennae") relative to the body may be useful. In males, one or both antennae may be modified for grasping females and appear "bent," or articulated (geniculate). The number of individual antennal segments is *not* reliable, but the appearance of swollen or unusually shaped antennal segments in certain positions could be diagnostic (*Centropages*). Some copepods have notably more and/or longer setae than others (e.g., *Oithona*).

Eye size, color, and location of both small median eyes and lateral eyes (often with lenses) can be diagnostic. Eyes are not conspicuous on all copepods, and the color and appearance of eyes on many can change after preservation.

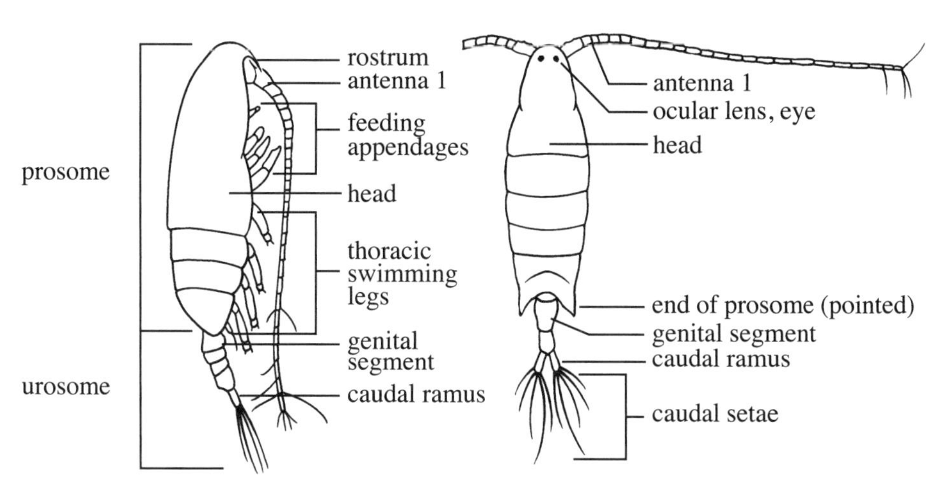

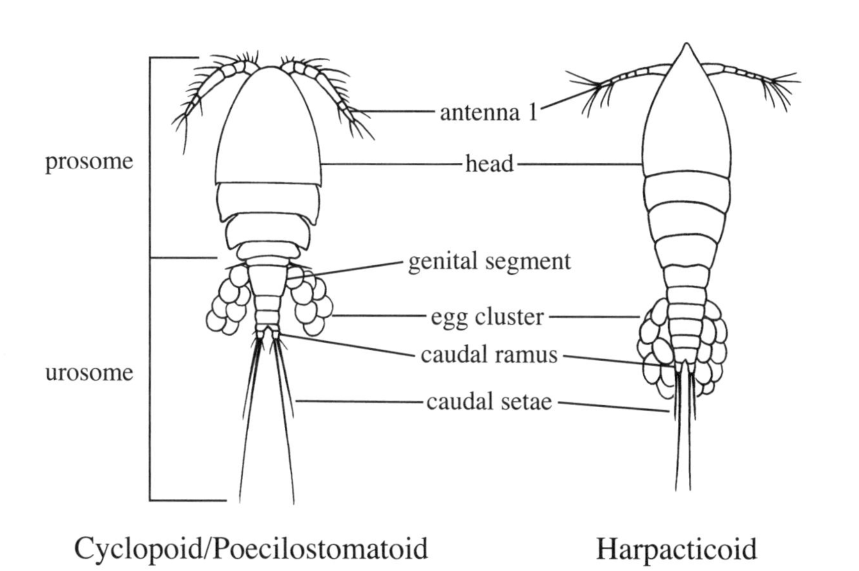

Fig. 20. Body regions and basic anatomical features of calanoid, cyclopoid (and poecilostomatoid), and harpacticoid copepods are useful in identification.

The **anterior margin of the head** may be distinctive with respect to shape or width.

The presence and shape of the **rostrum** may be very distinct for some species, but the specimen needs to be viewed from the side to see this ventrally oriented projection.

Segments behind the head often have characteristic shapes or proportions, but their number may vary among taxa.

The **posterior of the prosome** may end in spines or winglike projections.

The **urosome** is the narrow posterior body region. It may be asymmetrical. The urosome lacks jointed appendages. The size and shape of the genital segment is often a useful feature for identifying species and gender. Males may have a slightly longer urosome with more segments than females. If present, egg clusters on females usually facilitate identification of the species. The urosome ends in two projecting appendages called the **caudal rami**. Note the shape of these rami and the number and relative lengths of **caudal spines** or **caudal setae** on them. Spines or setae are easily broken, so check several specimens if in doubt.

Color is especially useful in fresh specimens. Location, shape, and color of pigmentation spots or bands can be useful.

Relative size readily separates many taxa. Sizes given are for total length from the "forehead" (extreme anterior of the body) to the end of the caudal rami (excluding spines or setae). Since most shallow-water collections are likely to include adult *Acartia,* which is between 1.0 and 1.5 mm in length, *Acartia* can serve as a good standard for comparing the sizes of many common species (Fig. 21).

Distinguishing Characteristics of Copepod Groups and Branchiura

Calanoids are common in both estuarine and coastal ocean areas, where their streamlined shape and long antennae are usually diagnostic. Most calanoids are holoplanktonic, but some shallow-water species are semibenthic. Only a few species retain attached egg clusters.

Cyclopoids and **poecilostomatoids** usually have a shorter and rounder body shape than calanoids and are generally among the smaller of the adult copepods in routine collections. Females typically have paired, elongate egg clusters.

Harpacticoids are streamlined and often recognized by their short antennae. Most harpacticoids are benthic, but a few species are common in the nearshore plankton. Egg sacs, when present, are not elongated.

Siphonostomatoids are predominantly symbiotic but are occasionally caught in the plankton. A subgroup, the **caligoids**, are ectoparasites on fishes. They are flattened and have conspicuous ventral suckers and often trail long, paired strings of eggs. They are superficially similar to the branchiurans below, and both groups are known as "fish lice" or "sea lice."

Monstrilloids are parasitic on polychaetes or molluscs as juveniles, but adults enter the plankton to reproduce. The fifth pair of swimming legs is absent or rudimentary. Females have long posterior spines.

Branchiura are not copepods but are treated here because they resemble the similarly flattened caligoids. Branchiura are distinguished by their movable compound eyes and by a carapace that covers the thorax. Females also lack the paired egg clusters seen in caligoids. Branchiura are also called fish lice or sea lice.

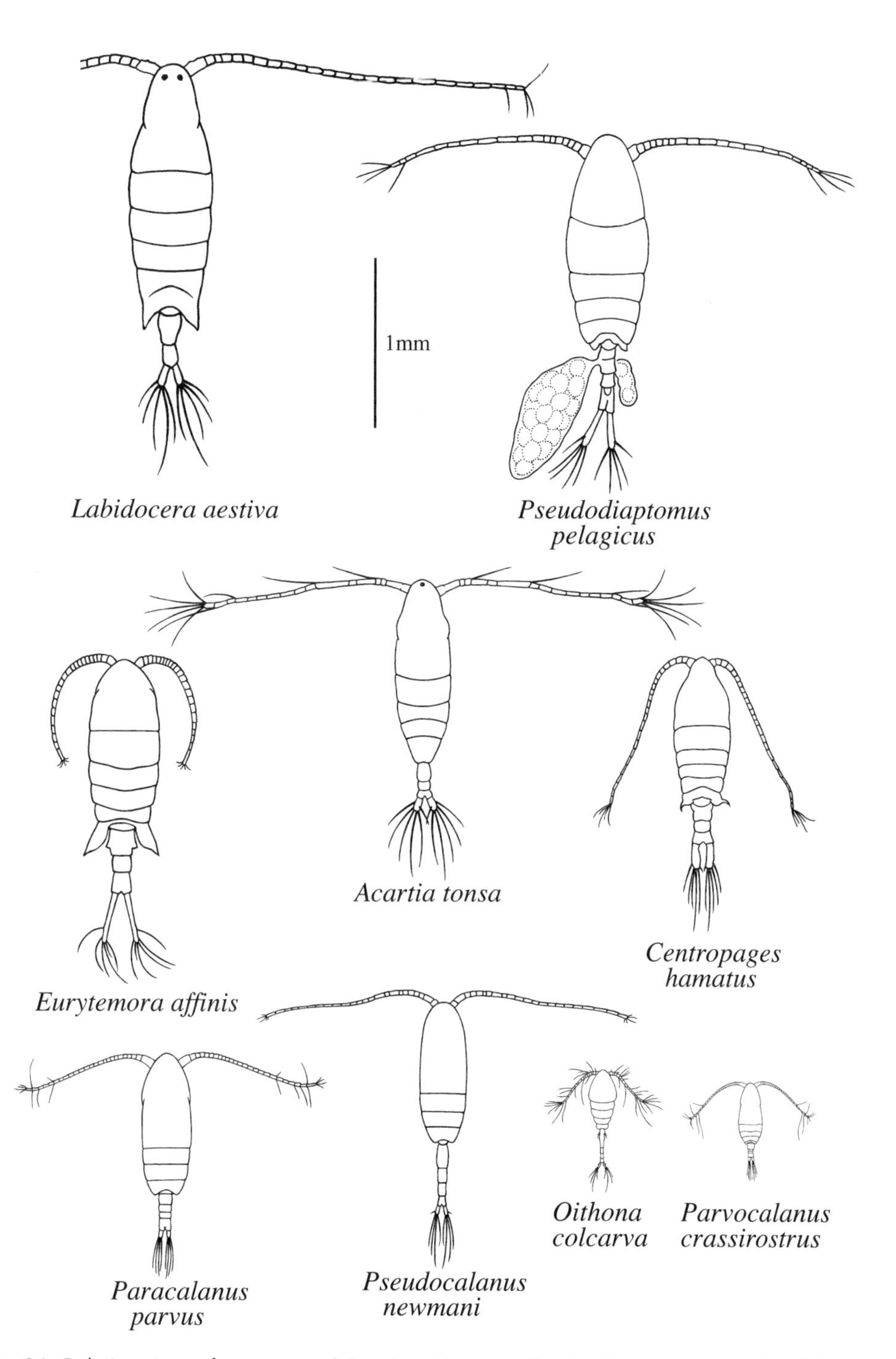

Fig. 21. Relative sizes of common adult calanoid copepods plus the common cyclopoid *Oithona*.

USEFUL IDENTIFICATION REFERENCES

Coull, B. C. 1977. *Marine Flora and Fauna of the Northeastern United States. Copepoda: Harpacticoida.* NOAA Technical Report NMFS Circular 399. 49 pp.

Cressey, R. F. 1978. *Marine Flora and Fauna of the Northeastern United States. Crustacea: Branchiura.* NOAA Technical Report NMFS Circular 413. 12 pp.

Faber, D. J. 1966. Free-swimming copepod nauplii of Narragansett Bay with a key to their identification. *Journal of the Fisheries Research Board of Canada* 23:189–205. (One of few sources for identification of nauplii.)

Gerber, R. P. 2000. *An Identification Manual to the Coastal and Estuarine Zooplankton of the Gulf of Maine Region from Passamaquoddy Bay to Long Island Sound.* Part I, Text and identification keys (80 pp.). Part II, Figures (98 pp.). Acadia Productions, Brunswick, ME.

González, J. G., Bowman, T. E. 1965. Planktonic copepods from Bahía Fosforescente, Puerto Rico, and adjacent waters. *Proceedings of the U.S. National Museum* 117:241–303.

Owre, H. B., Foyo, M. 1967. *Copepods of the Florida Current with Illustrated Keys to Genera and Species.* Fauna Caribaea. No. 1, Crustacea. Part 1, Copepoda. University of Miami Press, Miami. 137 pp. (Primarily covers southern and tropical species but includes a general introduction to copepod identification and anatomical nomenclature.)

Reid, J. W., Williamson, C. E. 2010. Copepoda. In: Thorpe J. H., Covich, A. P., eds. *Ecology and Classification of North American Freshwater Invertebrates.* Academic Press, New York, 829–899. (Contains a key that includes some brackish species.)

Wilson, C. B. 1932a. The copepod crustaceans of Chesapeake Bay. *Proceedings of the U.S. National Museum* 80:1–54. (Technical. Some names have changed.)

Wilson, C. B. 1932b. The copepods of the Woods Hole region, Massachusetts. *U.S. National Museum Bulletin* 158:1–635. (Technical. Some names have changed.)

SUGGESTED READINGS

Boxshall, G. A., Halsey, S. H. 2004. *An Introduction to Copepod Diversity.* Ray Society, London. 966 pp.

Boxshall, G. A., Schminke, H. K. 1988. Biology of copepods. *Hydrobiologia* 167/168:1–639.

Bundy, M. H., Vanderploeg, H. A. 2002. Detection and capture of inert particles by calanoid copepods: The role of the feeding current. *Journal of Plankton Research* 24:215–223.

Buskey, E. J. 1998. Components of mating behavior in planktonic copepods. *Journal of Marine Systems* 15:13–22.

Dahms, H.-U. 1995. Dormancy in the Copepoda: An overview. *Hydrobiologia* 306:199–211.

Greene, C. H. 1988. Foraging tactics and prey-selection patterns of omnivorous and carnivorous calanoid copepods. *Hydrobiologia* 167/168:295–302. (Includes descriptions of swimming patterns.)

Hartline, D. K., Lenz P. H., Herren, C. M. 1996. Physiological and behavioral studies of escape responses in calanoid copepods. *Marine and Freshwater Behavior and Physiology* 27:199–212.

Kiørboe, T. 2011. What makes pelagic copepods so successful? *Journal of Plankton Research* 33:677–685.

Koehl, M. A. R., Strickler, J. R. 1981. Copepod feeding currents: Food capture at low Reynolds number. *Limnology and Oceanography* 26:1062–1073.

Lonsdale, D. J., Cosper, E. M., Doall, M. 1996. Effects of zooplankton grazing on phytoplankton size-structure and biomass in the lower Hudson River estuary. *Estuaries* 19:874–889. (Reviews copepod feeding.)

Lowndes, A. G. 1935. The swimming and feeding of certain calanoid copepods. *Proceedings of the Zoological Society of London* 105:687–715. (An early, but important contribution.)

Marcus, N. H. 1996. Ecological and evolutionary significance of resting eggs in marine copepods: Past, present, and future. *Hydrobiologia* 320:141–152.

Marcus, N. H., Scheef, L. P. 2009. Photoperiodism in copepods. In: Nelson, R. J., Denlinger, D. L., Somers, D. E., eds. *Photoperiodism: The Biological Calendar* (pp.193–217). Oxford University Press, Oxford, 193–217.

Marshall, S. M., Orr, A. P. 1955. *The Biology of a Marine Copepod Calanus finmarchicus (Gunnerus).* Oliver & Boyd, Edinburgh. 188 pp. (One of the most comprehensive treatments of a single species of copepod ever attempted.)

Mauchline, J. 1998. The biology of calanoid copepods. *Advances in Marine Biology* 33:1–710. (A comprehensive review of all aspects of calanoid biology.)

Paffenhöfer, G.-A., Strickler, J. R., Lewis, K. D., et al. 1996. Motion behavior of nauplii and early copepodid stages of marine planktonic copepods. *Journal of Plankton Research* 18:1699–1715.

Smithsonian Institution, U.S. National Museum of Natural History. *The World of Copepods.* <http://invertebrates.si.edu/copepod/>. Accessed July 21, 2011.

Strickler, J. R . *Strickler Central: The Functional Ecology of Calanoid Copepods.* <www.uwm.edu/~jrs/COPEPODS%20CENTRAL.htm >. Accessed July 21, 2011.

Tiselius, P., Jonsson, P. R. 1990. Foraging behavior of six calanoid copepods: Observations and hydrodynamic analysis. *Marine Ecology Progress Series* 66:23–34.

Titelman, J., Kiørboe, T. 2003. Predator avoidance by nauplii. *Marine Ecology Progress Series* 247:137–149.

van Duren, L. A., Videler, J. J. 2003. Escape from viscosity: The kinematics and hydrodynamics of copepod foraging and escape swimming. *Journal of Experimental Biology* 206:269–279.

Waggett, R., Buskey, E. 2006. Copepod sensitivity to flow fields: Detection by copepods of predatory ctenophores. *Marine Ecology Progress Series* 323:205–211.

Yen, J., Strickler J. R. 1996. Advertisement and concealment in the plankton: What makes a copepod hydrodynamically conspicuous? *Invertebrate Biology* 115:191–205.

CALANOID COPEPODS

Acartia tonsa

Occurrence. *Acartia tonsa* is a dominant coastal and estuarine species in the Atlantic from Massachusetts to Florida and throughout the Gulf, sometimes exceeding 100,000 m^{-3}. *Acartia tonsa* occurs primarily in summer and fall in the north and all year from New Jersey south. A true euryhaline species, *A. tonsa* is most abundant in bays and estuaries in 5–30 psu, but it occurs from nearly freshwater to hypersaline lagoons.

Biology and Ecology. *Acartia tonsa* may be restricted to productive nearshore waters because of its high food requirements. Adults switch between suspension feeding on immobile particles and ambush feeding on motile microzooplankton. Nauplii feed on both phytoplankton and ciliates. *Acartia* is a vertical migrator, often occurring closer to the bottom by day, even in tidal systems. Swimming typically consists of a "hop-and-sink" motion. *Acartia* is food for anchovies, larval fishes, jellyfishes, and ctenophores. Stalked ciliates (*Zoothamnium*) may be attached to the cuticle.

References. Bagøien and Kiørboe 2005; Bollens et al. 1994; Buskey et al. 2002; Jakobsen et al. 2005; Kiørboe et al. 1999; Paffenhöfer and Stearns 1988; Richmond et al. 2006; Roman 1984; Sabatini 1990; Saiz and Kiørboe 1995; Stoecker and Egloff 1987; Sullivan et al. 2007; Sullivan and McManus 1986; Weissman et al. 1993; Wiadnyana and Rassoulzadegan 1989.

Acartia hudsonica

Occurrence. *Acartia hudsonica* occurs from Cape Cod to North Carolina and is especially abundant in Narragansett Bay and Long Island Sound in the spring. In the Mid-Atlantic, it is abundant in winter and spring along the coast and in higher-salinity reaches of estuaries but disappears in June when *A. tonsa* increases. Other *Acartia* species occur north and south of our coverage. *Acartia lilljeborgii, A. negligens,* and *A spinata* occur in shallow water in the Gulf, while *A. longiremis* inhabits New England waters.

Biology and Ecology. Females lay dormant eggs as temperatures warm. These hatch in spring to initiate new generations during cooler conditions. *Acartia hudsonica* feeds on phytoplankton and protozooplankton, with a preference for larger ciliates and heterotrophic dinoflagellates, especially when larger phytoplankton are scarce.

Comment. *Acartia hudsonica* on our coast appears as *A. clausi* in earlier literature.

References. Avery 2005; Bochdansky and Bollens 2004; Sullivan et al. 2007; Weissman et al. 1993.

Calanus finmarchicus

Occurrence. *Calanus finmarchicus* is one of the most abundant oceanic and coastal copepods in the entire North Atlantic, but it has a more restricted distribution in nearshore waters, where it occurs regularly south to Long Island Sound and irregularly in winter and in spring as far south as the mouth of the Delaware Bay.

Biology and Ecology. This phytoplankton grazer also eats heterotrophic protozoans. *Calanus*'s large size and abundance probably contributes to its role as a major source of food for fishes and seabirds.

References. Castellani et al. 2008; Head et al. 2000; Jiang et al. 2007; Lenz et al. 2004; Marshall and Orr 1955; Meyer-Harms et al. 1999; Ohman and Runge 1994; Turner et al. 1993.

Acartia tonsa

♀ 1.3 - 1.5 mm, ♂ 1.0 - 1.3 mm

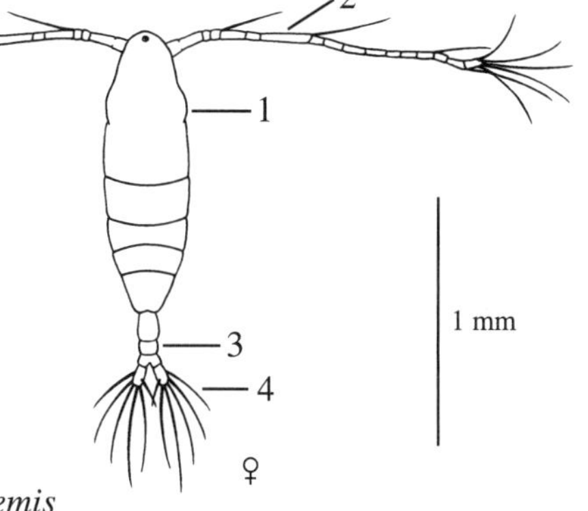

- prosome elongate and tapered; head > 50% of length
- head with lateral bulges near middle (1); prominent eye
- antennae delicate with long segments and long setae (2)
- middle segment of 3 in urosome about 30% length of first (3)
- long setae on caudal rami widely flared (4)
- males with 5 urosomal segments; right antenna with some swollen segments
- similar species: *A. hudsonica* and *A. longiremis* are much smaller, without lateral bulges, and with longer urosomes; *A. lilljeborgii* has a very long, slightly pointed head

Acartia hudsonica

♀ 1.0 - 1.3 mm, ♂ 1.0 - 1.1 mm

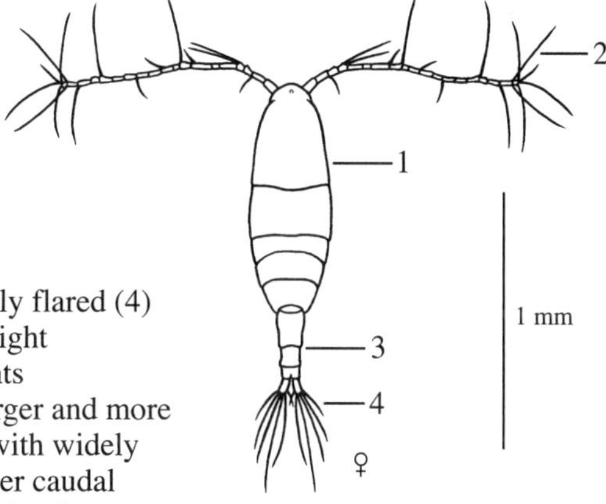

- prosome elongate and tapered; head < 50% of length
- head without lateral bulges; widest posterior to middle (1)
- antennae delicate with long segments and long setae (2)
- middle segment of 3 in urosome about 50% length of first (3)
- long setae on caudal rami not widely flared (4)
- males with 5 urosomal segments; right antenna with some swollen segments
- similar species: *Acartia tonsa* is larger and more robust, and has a shorter urosome with widely flared setae; *A. longiremis* has longer caudal rami; *A. lilljeborgii* has a very long head

Calanus finmarchicus

♀ 2.4 - 4.2 mm, ♂ 2.6 - 4.0 mm

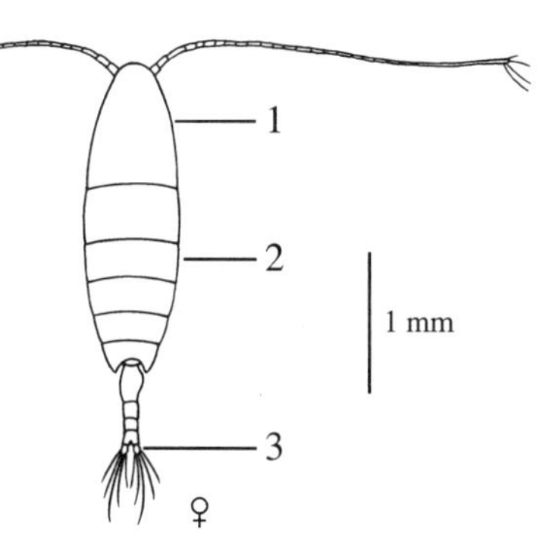

- prosome elongate and oval; cyclindrical in profile
- head < 40% of total prosome (1)
- 5 similar segments behind head (2)
- urosome < 25% of total body length; very short caudal rami (3)
- red oil sac(s) often within prosome (not shown)
- males slightly smaller and slender, without twisted antennae
- similar species: *Acartia* spp. are much smaller with larger heads

ID hint: These three similar calanoids all have a rostrum. Information on distribution may assist identification. Size may also be diagnostic (Fig. 21).

Pseudocalanus spp.

Occurrence. *Pseudocalanus newmani* and *P. moultoni* are common nearshore and in estuarine waters from Massachusetts north, especially in winter and spring. Most pre-1990 surveys referred only to "*P. minutus*," which was divided into three species (*P. minutus, P. moultoni,* and *P. newmani*) in 1989. The distribution of *P. newmani* and *P. moultoni* south of Cape Cod awaits clarification. Based on the earlier reports referring to "*P. minutus*," *P. moultoni*, and *P. newmani* are probably uncommon nearshore south of Long Island Sound. At least some *Pseudocalanus* species extend south to the Carolinas, but preliminary data suggest they occur at least several kilometers offshore. *Pseudocalanus minutus* is found offshore and north of Cape Cod.

Biology and Ecology. *Pseudocalanus* is primarily herbivorous. Chaetognaths and larval and postlarval fishes, notably Atlantic mackerel and sand lances, eat *Pseudocalanus*. Identification of *Pseudocalanus* to species is difficult.

References. Bucklin et al. 2001; Corkett and McLaren 1978; Frost 1989; Frost and Bollens 1992; Halsband-Lenk et al. 2005; McGillicuddy and Bucklin 2002; Poulet 1976.

Parvocalanus crassirostris

Occurrence. *Parvocalanus crassirostris* is widely distributed from Cape Cod to Florida and in the Gulf of Mexico. This small, high-salinity species is usually associated with warm coastal waters but is often abundant all year in high-salinity estuaries in the Southeast and Gulf.

Biology and Ecology. *Parvocalanus crassirostris* is mainly herbivorous and prefers phytoplankton >5 μm, but its diet is not well studied. At times, *P. crassirostris* occurs deeper during the day and closer to the surface at night. Highly productive estuarine areas seem to favor *Acartia tonsa* over the much smaller *P. crassirostris*. In North Carolina, *P. crassirostris* is an important food for larval fishes.

Comment. As late as the mid-1980s, *Parvocalanus crassirostris* was referred to as *Paracalanus crassirostris*, causing uncertainty with respect to reported distributions.

References. Lawson and Grice 1973; Mallin 1991.

Paracalanus parvus and other *Paracalanus* spp.

Occurrence. *Paracalanus parvus* is distributed from Cape Cod to Florida and throughout the Gulf of Mexico. *Paracalanus* is most abundant over the continental shelf along the Atlantic Coast, where it is sometimes found close to shore and in high-salinity estuaries and bays. *Paracalanus aculeatus* and *P. denudatus* are more typical of deeper and more saline coastal and continental shelf collections from the Mid-Atlantic to southern Florida and in the Mississippi River Plume. *Paracalanus quasimodo* appears occasionally in the southeastern Atlantic and eastern Gulf.

Biology and Ecology. Limited information indicates that *Paracalanus* is chiefly herbivorous. *Paracalanus* swims slowly and steadily without the jerks and hops seen in *Acartia*. This suspension feeder can remove particles as small as 2 μm but is most efficient on particles >12 μm.

Note. *Paracalanus parvus* was sometimes erroneously listed as *Parvocalanus parvus*.

References. Bartram 1981; Gomes et al. 2004; Waggett and Buskey 2007.

Pseudocalanus newmani

♀ 0.9 - 1.5 mm, ♂ 0.8 - 1.2 mm

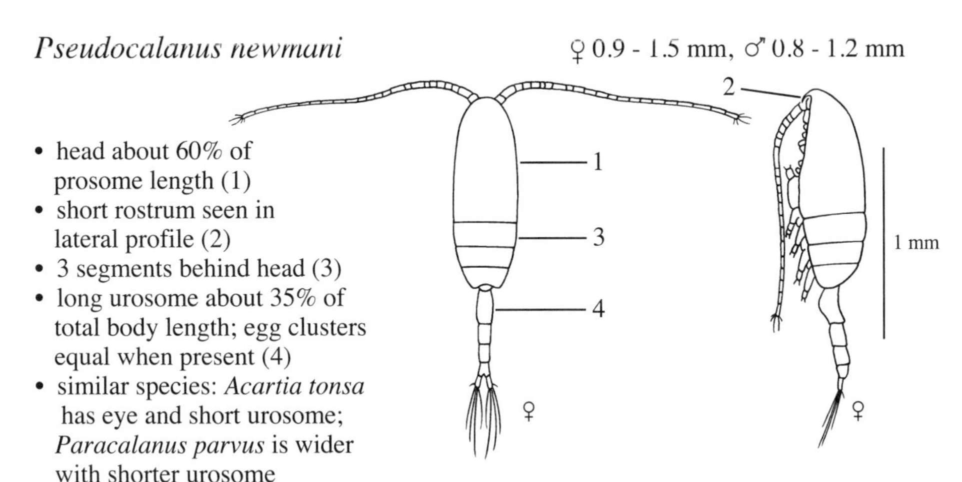

- head about 60% of prosome length (1)
- short rostrum seen in lateral profile (2)
- 3 segments behind head (3)
- long urosome about 35% of total body length; egg clusters equal when present (4)
- similar species: *Acartia tonsa* has eye and short urosome; *Paracalanus parvus* is wider with shorter urosome

Parvocalanus crassirostris

♀ 0.5 - 0.6 mm, ♂ 0.3 - 0.4 mm

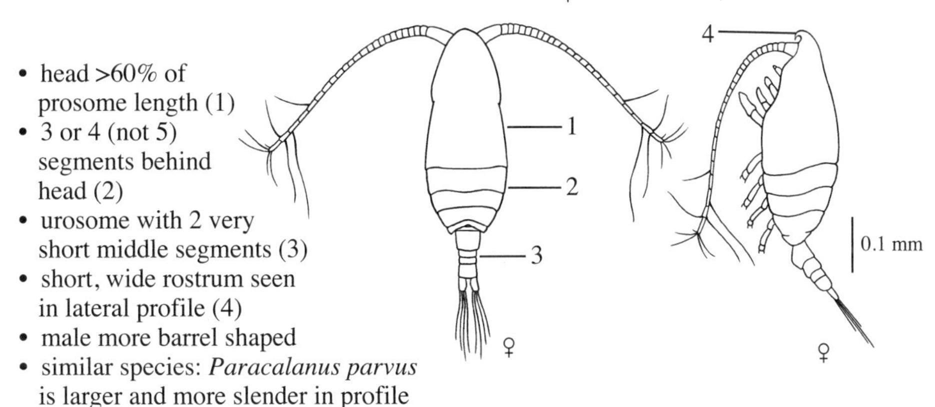

- head >60% of prosome length (1)
- 3 or 4 (not 5) segments behind head (2)
- urosome with 2 very short middle segments (3)
- short, wide rostrum seen in lateral profile (4)
- male more barrel shaped
- similar species: *Paracalanus parvus* is larger and more slender in profile

Paracalanus parvus

♀ 0.8 - 1.0 mm, ♂ 0.8 - 1.0 mm

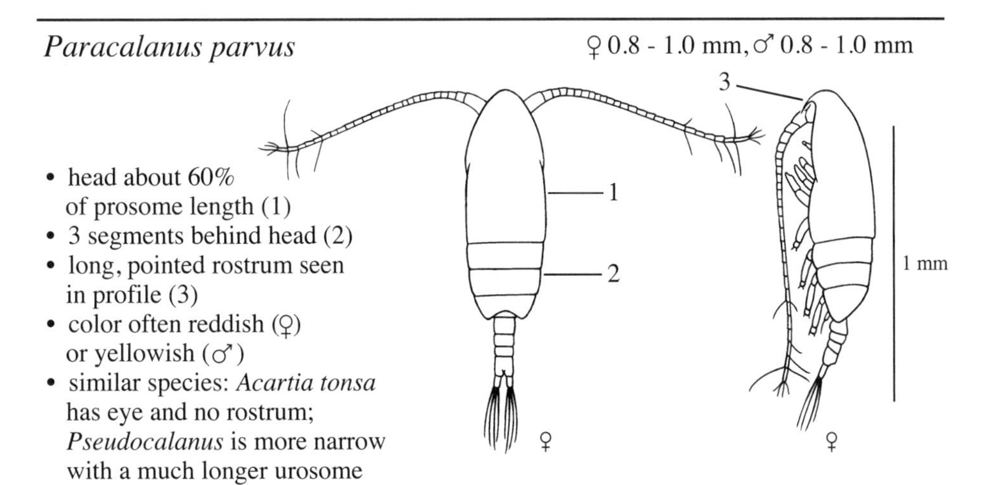

- head about 60% of prosome length (1)
- 3 segments behind head (2)
- long, pointed rostrum seen in profile (3)
- color often reddish (♀) or yellowish (♂)
- similar species: *Acartia tonsa* has eye and no rostrum; *Pseudocalanus* is more narrow with a much longer urosome

Labidocera spp.

Occurrence. *Labidocera* prefer ocean waters and high-salinity regions of estuaries. *Labidocera aestiva* is the only common *Labidocera* in nearshore Atlantic waters and is often the largest copepod in coastal collections. Its abundance peaks during summer, especially toward the north, but it occurs all year along the Florida Atlantic Coast. *Labidocera aestiva* is also the most abundant coastal *Labidocera* in the northern Gulf from the Florida Panhandle to north Texas. *Labidocera scotti* is uncommon in the northern Gulf but gradually replaces *L. aestiva* along the south Texas and south Florida coasts, especially when temperatures exceed 26°C. *Labidocera mirabilis* is also a Gulf of Mexico species and appears to be a deeper coastal ocean form.

Biology and Ecology. This raptorial feeder is omnivorous. As a "cruising" predator, it swims faster than most nearshore copepod larvae on which it feeds. *Labidocera* prey include *Acartia tonsa*, but larger diatoms and dinoflagellates are also eaten. *Labidocera* is frequently more common in the water column by day and nearer bottom at night.

References. Blades and Youngbluth 1979; Chen and Marcus 1997; Conley and Turner 1985; Gibson and Grice 1977a; Landry and Fagerness 1988.

Anomalocera spp.

Occurrence. *Anomalocera ornata* is a common component of the continental shelf neuston, where it accumulates in surface windrows. This often brilliant blue copepod has been reported in nearshore waters of the Gulf and along the Atlantic Coast north to Beaufort, North Carolina. *Anomalocera patersoni* resides in Atlantic shelf waters and comes ashore sporadically, especially north of North Carolina. Juveniles migrate to the surface at night, but adults exhibit a "reverse" vertical migration to surface waters during the day.

Biology and Ecology. These opportunistic omnivores can feed on diatoms, but they are also predators on zooplankton, including copepods and their larvae. They can kill and consume early larval stages of fishes (e.g., menhaden and spot).

References. Ianora and Santella 1991; Tester et al. 2004; Turner 1978, 1985; Turner et al. 1985.

Pontella spp.

Occurrence. At least five species of *Pontella* occur along the southeastern Atlantic and Gulf Coasts. Most are more typical of offshore waters, but several are common or infrequent in nearshore waters. *Pontella meadii* (or *meadi*) is among the most widespread, reported from Woods Hole, Massachusetts, through the Gulf.

Biology and Ecology. *Pontella* species typically occur within the top meter of the water column. These predators of microcrustaceans also eat larger diatoms.

References. Gibson and Grice 1977b; Grice and Gibson 1977; Turner 1978, 1985.

Labidocera aestiva

♀ 1.8 - 2.0 mm, ♂ 1.8 - 2.2 mm

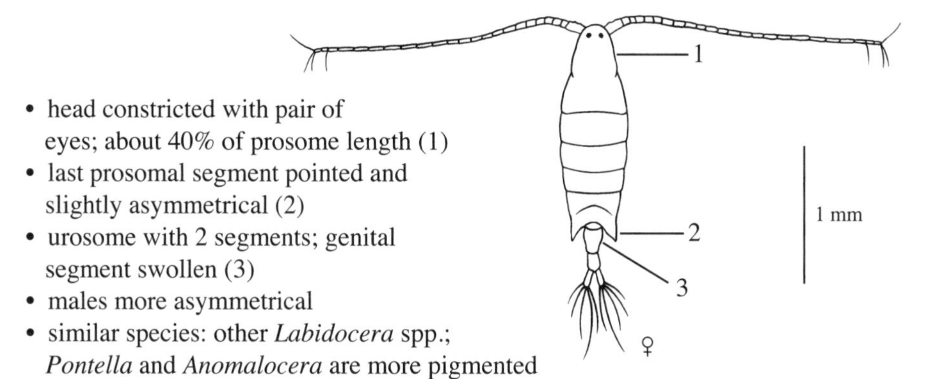

- head constricted with pair of eyes; about 40% of prosome length (1)
- last prosomal segment pointed and slightly asymmetrical (2)
- urosome with 2 segments; genital segment swollen (3)
- males more asymmetrical
- similar species: other *Labidocera* spp.; *Pontella* and *Anomalocera* are more pigmented with pointed heads; *Centropages* spp. are smaller and have single eyes

Anomalocera patersoni

♀ 3.0 - 3.3 mm, ♂ 2.5 - 3.0 mm

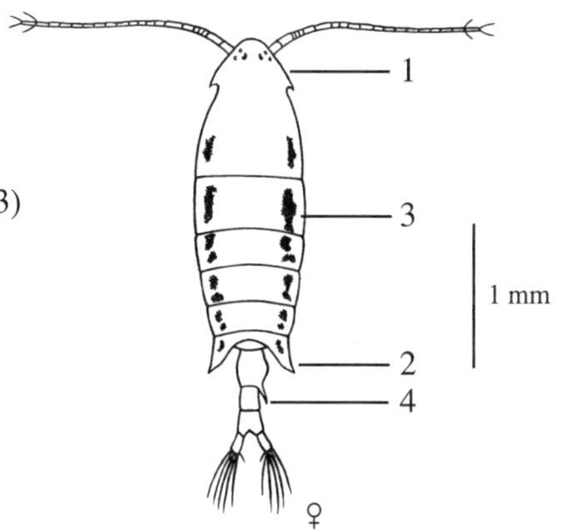

- head arrow shaped with strong lateral hooks; 2 or 3 pairs of eyes (1)
- last prosomal segment pointed (2)
- large, mostly lateral, dark pigment spots (3)
- genital segment with large process on right (♀ only) (4)
- similar species: other *Anomalocera*; *Pontella* spp. have a very asymmetrical urosome and pigment along midline

Pontella meadii

♀ 2.0 - 2.7 mm, ♂ 2.8 - 3.0 mm

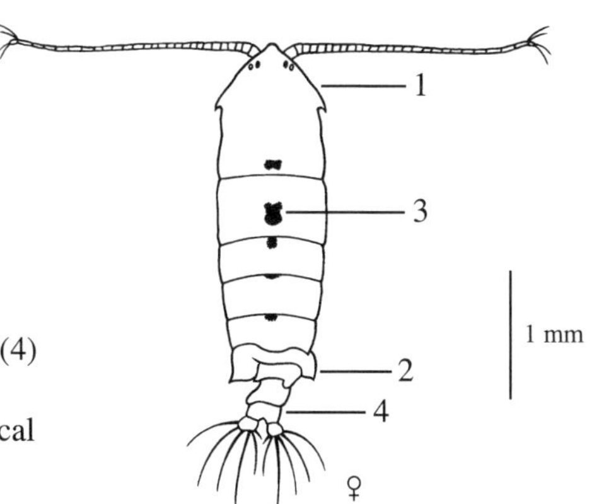

- head arrow shaped with strong lateral hooks; 2 pairs of eyes (1)
- last prosomal segment with very asymmetrical points (2)
- single, dark pigment patches along midline (3)
- urosome short and all segments asymmetrical; very short caudal rami (4)
- similar species: other *Pontella* spp.; *Anomalocera* spp. are more symmetrical with lateral pigment patches

Calanopia americana

Occurrence. *Calanopia americana* occurs throughout the Gulf and ranges north along the Atlantic Coast at least as far as North Carolina. It inhabits both nearshore and offshore waters.

Biology and Ecology. *Calanopia* is a diel vertical migrator. It lives close to the bottom during the day and then ascends to the surface at night.

References. Cohen and Forward 2005a, 2005b.

Candacia spp.

Occurrence. *Candacia* species are oceanic copepods but may be common in nearshore collections. *Candacia armata* occurs most frequently in nearshore waters from Cape Cod through the Mid-Atlantic and occasionally farther south. *Candacia curta*, *C. pachydactyla*, and several other *Candacia* species occur in southern waters and in the Gulf but are less common nearshore. While several additional *Candacia* species occur in the Gulf, they are seldom reported from nearshore U.S. waters.

Biology and Ecology. *Candacia* is strikingly pigmented and bears formidable mouthparts. Pacific studies found that *Candacia* is a raptorial predator whose diet includes larvaceans and arrow worms.

Reference. López-Urrutia et al. 2004; Wickstead 1959.

Centropages furcatus, formerly *C. velificatus*

Occurrence. *Centropages furcatus* is a warm-water species abundant offshore and occasionally nearshore in the Gulf and common along the Southeast coast north to Cape Hatteras in both nearshore and offshore waters. It appears occasionally as far north as New York.

Biology and Ecology. Although its diet includes both dinoflagellates and diatoms, both laboratory and field data suggest that *Centropages furcatus* is primarily carnivorous. It swims with a hop-and-sink motion and may remain motionless to ambush microzooplankton, including copepod nauplii.

References. Alcaraz et al. 2007; Bundy et al. 1993; Checkley et al. 1992; Grant 1988; Paffenhöfer and Knowles 1980; Turner 1987; Turner and Tester 1989.

Calanopia americana

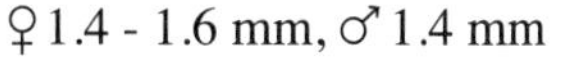
♀1.4 - 1.6 mm, ♂1.4 mm

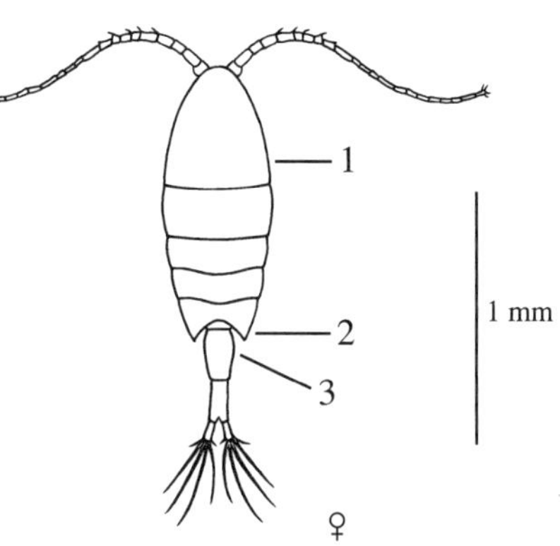

- broadly rounded head about 45% of total prosome length; no eye (1)
- 4 segments behind head; the last one with broad triangular points (2)
- urosome about 35% of total body length; genital segment long with bulges, and about the same length as the next posterior urosomal segment (3)
- similar species: *Pseudodiaptomus* and other copepods with round heads do not have a widely pointed posterior prosomal segment

---

Candacia armata

♀ 2.5 - 2.8 mm, ♂2.3 - 2.6 mm

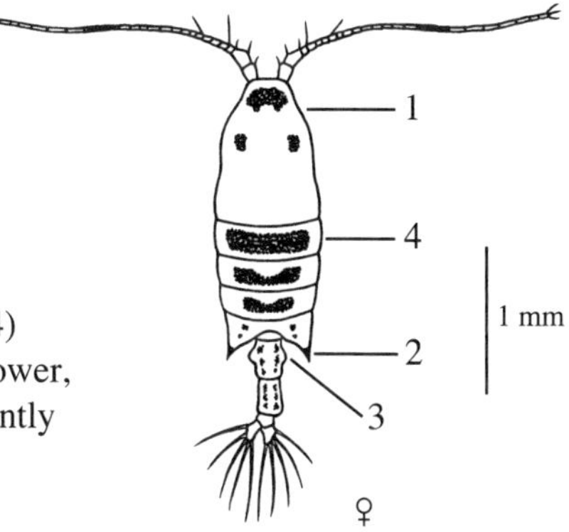

- head with flat end and slight shoulders; almost 50% length of prosome (1)
- 4 segments behind head; last segment broadly pointed (2)
- urosome asymmetrical; genital segment with lateral bulges (3)
- thoracic appendages tipped in black; dark pigment pattern on dorsal surface variable (4)
- similar species: *Centropages* spp. have narrower, asymmetrical points on last segment and bluntly pointed heads; *Labidocera*, *Pontella*, and *Anomalocera* have eyes

---

Centropages furcatus

♀ 1.6 - 1.7 mm, ♂1.2 - 1.3 mm

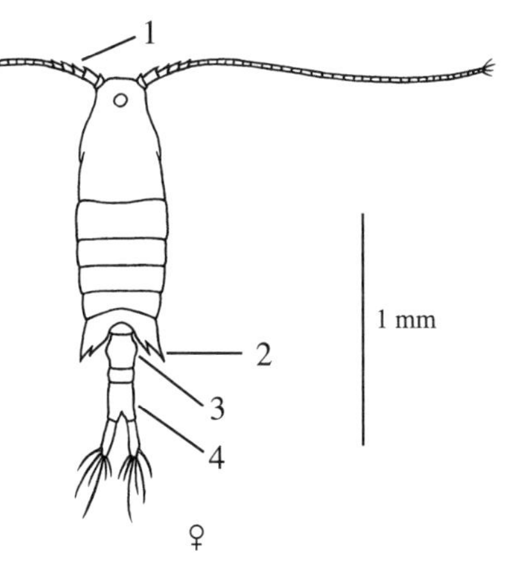

- head flat with ventral eye; sections of antennae closest to bases form jagged edge (1)
- last prosomal segment symmetrical with long external points, each with a shorter internal spine (2)
- genital segment with central bulges (3)
- anal segment twice as long as preceding segment (4)
- males with small anterior teeth on swollen antenna sections 15 and 16
- similar species: *Centropages typicus* and *C. hamatus* have bluntly rounded heads and no secondary spines on last prosomal segment; *Candacia* is heavily pigmented

Centropages hamatus

Occurrence. *Centropages hamatus* is a wide-ranging, cool-water species along both the Atlantic and Gulf Coasts. It is more common inshore than the other *Centropages* species and may be found in sheltered bays and estuaries, especially where salinities exceed 20. It is more common in winter south of Cape Hatteras and in the Gulf.

Biology and Ecology. Reportedly "broadly omnivorous," *Centropages* feeds on larger phytoplankton, ciliates, and on larval copepods and molluscs. *Centropages* prefers copepod nauplii to phytoplankton. *Centropages hamatus* is an important food for larval fishes. All of the species of *Centropages* covered here perform extensive diel vertical migrations, preferring deeper waters during the day.

Note. In addition to *C. typicus and C. furcatus*, several other *Centropages* species may be encountered. *Centropages violaceus* occurs from the Carolinas to Florida and may be common in the northeastern Gulf of Mexico. Along the Atlantic, it is usually associated with intrusions of Gulf Stream or shelf waters.

References. Alcaraz et al. 2007; Chen and Marcus 1997; Costello et al. 1990; Durbin and Kane 2007; Grant 1988; Hwang and Strickler 2001; Hwang et al. 1993; Lawrence et al. 2004; Saage et al. 2009.

Centropages typicus

Occurrence. *Centropages typicus* is one of the most abundant copepods of the continental shelf of the Mid-Atlantic and may be common in nearshore coastal waters from Cape Cod to Florida. It prefers temperatures between 13° and 20°C. In New England waters, *C. typicus* may be a summer dominant, but it is uncommon except in winter at the southern end of its range. It may enter nearshore waters sporadically along the Atlantic Coast. It is also reported from the northeastern Gulf of Mexico.

Biology and Ecology. Feeding is likely similar to *C. hamatus*, but *C. typicus* also attacks and consumes yolk-sac larvae of menhaden and spot. *Centropages typicus* is a major food of right whales on the Stellwagen Bank off Massachusetts. Egg production is variable but can exceed 200 eggs per female per day.

References. Alcaraz et al. 2007; Bagøien and Kiørboe 2005; Blades 1977; Calbet et al. 2007; Caparroy et al. 1998; Durbin and Kane 2007; Grant 1988; Turner et al. 1985; Wiadnyana and Rassoulzadegan 1989.

Centropages hamatus

♀1.0 - 1.4 mm, ♂0.9 - 1.2 mm

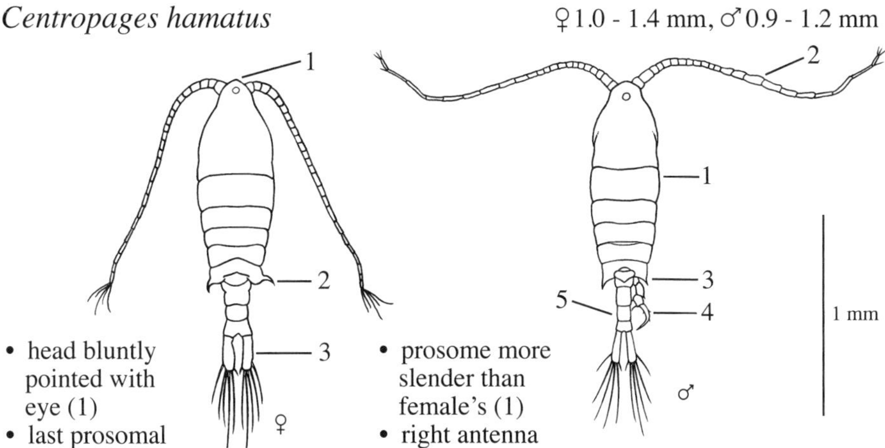

- head bluntly pointed with eye (1)
- last prosomal segment sharply pointed with right point directed outward (see *C. typicus*) (2)
- symmetrical urosome with long, wide caudal rami (3)
- similar species: *C. typicus* is larger and more asymmetrical; *Labidocera* spp. are larger and have blunt prosomal points

- prosome more slender than female's (1)
- right antenna with slightly swollen segments near middle (2)
- point on right side of last prosomal segment is shorter than left; both directed posteriorly (3)
- fifth thoracic leg modified with claw-like structure (4)
- last urosomal segment shorter than preceding 2 segments; total 4 (5)
- similar species: *C. typicus* male is larger and has a more swollen antenna; modified fifth leg is unusual in both

---

Centropages typicus

♀1.3 - 1.8 mm, ♂1.3 - 1.6 mm

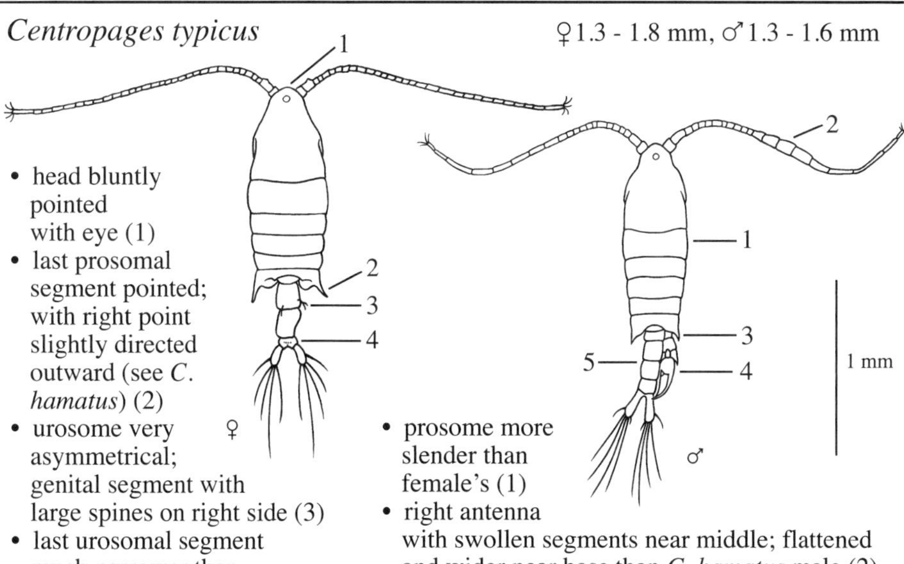

- head bluntly pointed with eye (1)
- last prosomal segment pointed; with right point slightly directed outward (see *C. hamatus*) (2)
- urosome very asymmetrical; genital segment with large spines on right side (3)
- last urosomal segment much narrower than preceding segment (4)
- similar species: *C. hamatus* is smaller and more symmetrical; *Labidocera* spp. are larger and have blunt prosomal points

- prosome more slender than female's (1)
- right antenna with swollen segments near middle; flattened and wider near base than *C. hamatus* male (2)
- point of right side of last prosomal segment shorter than left; both pointed posteriorly (3)
- fifth thoracic leg modified with claw-like structure (4)
- all 4 urosomal segments about same length (5)

Eurytemora affinis

Occurrence. *Eurytemora affinis* is found in estuaries along the entire Atlantic Coast from Cape Cod south and in the Gulf. Although broadly euryhaline, it reaches its maximum abundance in low salinities and may be the winter and spring dominant in 1–10 psu from Chesapeake Bay north. At its spring peak, it may comprise 90% of copepod collections.

Biology and Ecology. *Eurytemora* spp. are associated with spring diatom blooms but can also eat cyanobacteria. *Eurytemora* are important as a food for larval white perch, striped bass, shad, and herring when they co-occur in low-salinity areas in spring. These copepods are also potential prey of the mysid *Neomysis americana*.

Comments. *Eurytemora hirundoides* is now considered a subspecies of *E. affinis*. Alekseev and Souissi (2011) report that putative *E. affinis* from the Chesapeake Bay differ appreciably from the European *E. affinis* and have described them under the name *Eurytemora carolleeae*. Thus, the new *E. carolleeae* could replace, or partially replace, *E. affinis* along the Atlantic and Gulf Coasts of the United States.

References. Devrecker et al. 2004; Hough and Naylor 1991; Jeffries 1962; Katona 1971; Kimmel et al. 2006; Lawrence et al. 2004; Merrell and Stoecker 1998; Michalec et al. 2010; Roman et al. 2001; Seuront 2006; Tackx et al. 2003.

Eurytemora herdmani

Occurrence. *Eurytemora herdmani* is a northern species that ranges south to Sandy Hook, New Jersey, and prefers 20–30 psu.

Biology and Ecology. See comments under *E. affinis* above.

References. Grice 1971; Jeffries 1962; Katona 1971.

Eurytemora americana

Occurrence. *Eurytemora americana* is a northern brackish water species that occasionally ranges south to New Jersey and perhaps to Chesapeake Bay. Its abundance alternates with *E. affinis* in Raritan Bay, New Jersey, corresponding to seasonal temperature and salinity changes. Where the two species co-occur, *E. americana* is more likely in cooler temperatures and higher salinities.

Biology and Ecology. *Eurytemora americana* is native to the northeastern United States but is an invasive species now found in Europe, South America, and Alaska.

References. Grice 1971; Jeffries 1962.

Eurytemora affinis

♀ 1.1 - 1.5 mm, ♂ 1.1 - 1.4 mm

- short, often curved antennae (1)
- last prosomal segment with large angular projections ("wings") reaching beyond genital segment (♀ only) (2)
- very long caudal rami (3)
- males very slender without wings; strongly twisted right antenna (4)
- similar species: other *Eurytemora*; males very similar among species; *Centropages* and *Tortanus* have long antennae and no "wings"; *Corycaeus* has eyes

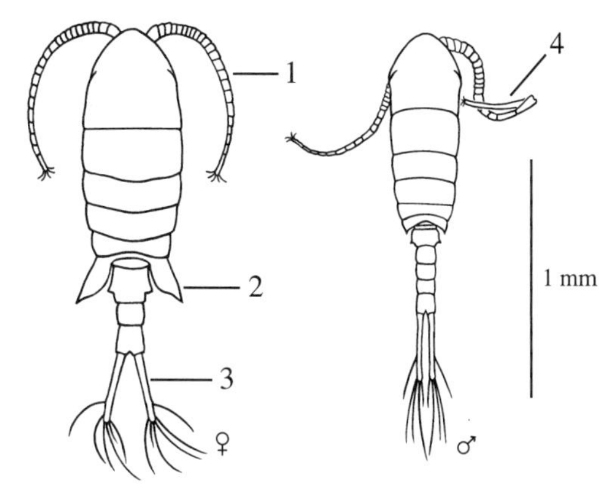

Eurytemora herdmani

♀ 1.5 - 2.6 mm, ♂ 1.2 - 1.5 mm

- very large wing-like ends to prosome; left one longer and narrower (1)
- genital segment expanded into wide, flat, rounded processes (2)
- similar species: see comments under *E. affinis*

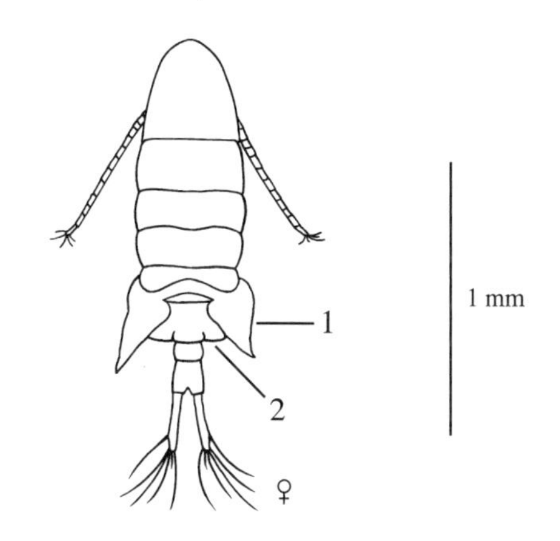

Eurytemora americana

♀ 1.4 - 1.6 mm, ♂ 1.2 - 1.4 mm

- short, wing-like ends to prosome, slightly asymmetrical (1)
- genital segment asymmetrical with slight bulges (2)
- similar species: see comments under *E. affinis*

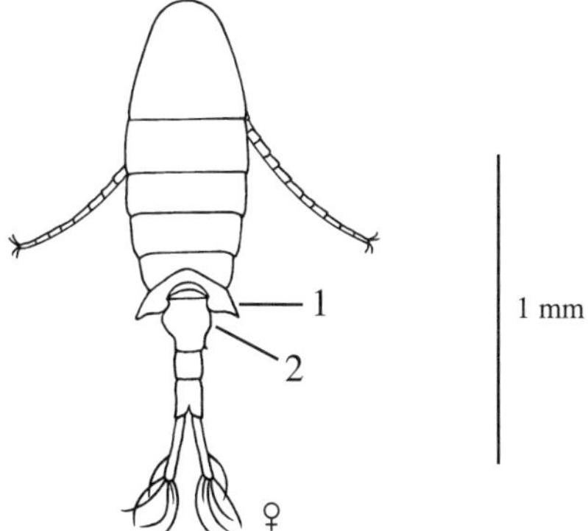

Temora longicornis

Occurrence. *Temora longicornis* is typically more common along the coast than in estuaries. It occurs from Cape Cod to Florida. It may be a dominant calanoid in the Mid-Atlantic region in winter and spring and in Long Island Sound in summer and fall. *Temora stylifera* occurs in coastal and continental shelf waters from the Chesapeake Bay to Florida and in the Gulf, including hypersaline lagoons.

Biology and Ecology. *Temora longicornis* is an important phytoplankton grazer in Long Island Sound where its biomass may be two to five times that of *Acartia hudsonica*, the numerical dominant in the spring. *Temora* is the major prey of the sand lance. A strong vertical migrator, *Temora* is usually found near bottom by day and near the surface at night. Its eggs are buoyant, with successive larval stages found progressively deeper in the water column. Diatoms are a primary food of this broadly omnivorous copepod. *Temora* feeds poorly on phytoplankton <20 μm.

References. Dam and Peterson 1991, 1993; Doall et al. 1998; Gentsch et al. 2009; Jakobsen et al. 2005; van Duren and Videler 2003; Weissburg et al. 1998; Yule and Crisp 1983.

Temora turbinata

Occurrence. *Temora turbinata* occurs throughout the Atlantic and Gulf Coasts but is most abundant in shallow ocean and estuarine waters from Chesapeake Bay south, where it is typically more abundant than *T. longicornis*. *Temora turbinata* also occurs in the brackish areas of the Mississippi Plume.

Biology and Ecology. Presumably similar to *T. longicornis* above.

References. Bird and Kitting 1982; Jakobsen et al. 2005; Turner 1984a; Waggett and Buskey 2006, 2007; Wu et al. 2010.

Subeucalanus pileatus

Occurrence. *Subeucalanus pileatus* occurs from Cape Cod to Florida and is common in the nearshore waters of the Gulf. It is frequently a codominant along with *Acartia tonsa* in Mississippi Sound. In the Atlantic, it is typically oceanic, but it appears sporadically in coastal collections.

Biology and Ecology. *Subeucalanus* is primarily herbivorous.

References. Koehl and Strickler 1981; Turner 1984b.

Temora longicornis

♀ 1.0 - 1.5 mm, ♂ 1.0 - 1.8 mm

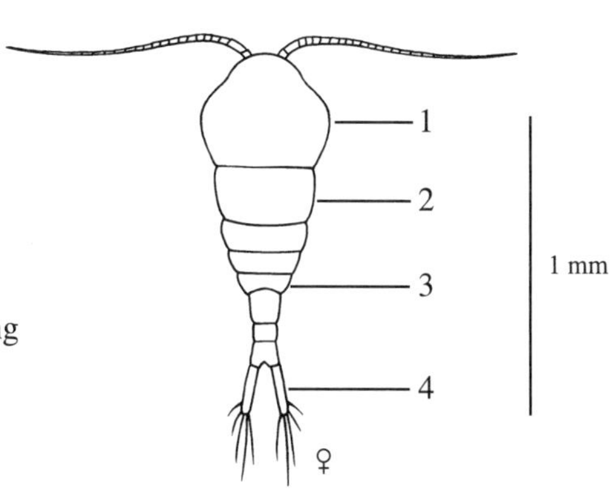

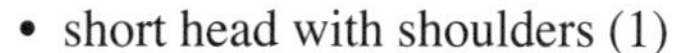

- short head with shoulders (1)
- 4 segments behind head; each with bulge in center (2)
- last prosomal segment tapers to urosome (3)
- caudal rami straight and long (4)
- prosome often dark and legs are long
- males more slender
- similar species: other *Temora* spp.; *Oithona* spp. have more pointed heads and are much smaller

---

Temora turbinata

♀ 1.4 - 1.6 mm, ♂ 1.3 - 1.5 mm

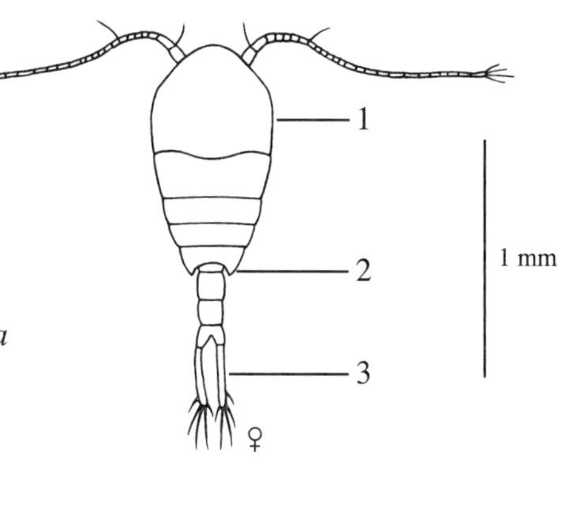

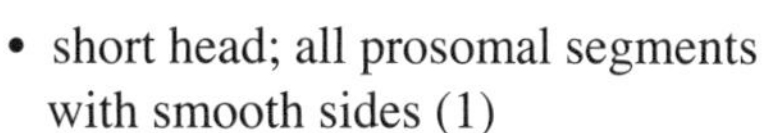

- short head; all prosomal segments with smooth sides (1)
- last prosomal segment pointed (2)
- long, slightly curved caudal rami (3)
- males more slender
- similar species: the prosome of *Temora longicornis* has central lateral bulges that make the entire body appear less uniformly tapered than *T. turbinata*. *Oithona* spp. are much smaller.

Subeucalanus pileatus

♀ 5 - 9 mm♂ 3 - 6 mm

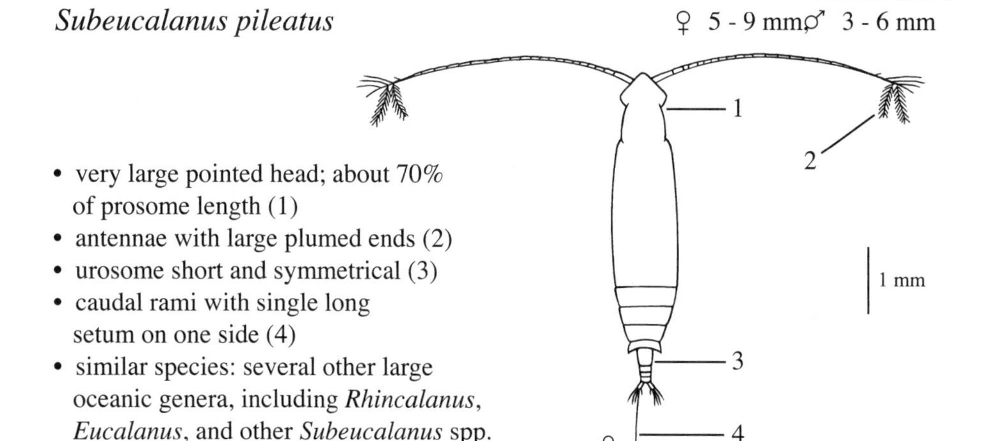

- very large pointed head; about 70% of prosome length (1)
- antennae with large plumed ends (2)
- urosome short and symmetrical (3)
- caudal rami with single long setum on one side (4)
- similar species: several other large oceanic genera, including *Rhincalanus*, *Eucalanus*, and other *Subeucalanus* spp.

Tortanus discaudatus

Occurrence. *Tortanus discaudatus* occurs from Cape Cod to central Florida and in the Gulf, primarily east of the Mississippi. Along much of the Atlantic Coast, it is an uncommon coastal and estuarine copepod occurring in winter and spring. In the northern Gulf, *T. discaudatus* is often found in estuaries or lagoons along with *Acartia*, *Parvocalanus*, and *Pseudodiaptomus*.

Biology and Ecology. *Tortanus discaudatus* is a nonvisual predator that responds actively to motile zooplankton. In the laboratory, it darts randomly and catches copepod nauplii or copepodids on contact. It is one of few predatory copepods in our geographical range that also feeds on adult copepods, including *Acartia*.

References. Ambler and Frost 1974; Landry and Fagerness 1988; Mullin 1979.

Tortanus setacaudatus

Occurrence. *Tortanus setacaudatus* occurs from Massachusetts south but is more common from the southeast Atlantic to Florida and through the Gulf, often in estuarine waters.

Biology and Ecology. This *Tortanus* species is relatively unstudied, but its biology is presumably similar to *T. discaudatus*.

References. Koehl and Strickler 1981.

Pseudodiaptomus pelagicus, formerly *P. coronatus*

Occurrence. *Pseudodiaptomus pelagicus* ranges from Cape Cod to Florida and occurs in the Gulf. It is most abundant along the coast and in warm embayments in salinities of 15–30. *Pseudodiaptomus pelagicus* only occurs during summer in the north but is year-round and especially abundant in summer in the Mid-Atlantic. It may disappear from the Gulf in the warmest months.

Biology and Ecology. *Pseudodiaptomus* appears to live close to the bottom, at least during the day where it is often found attached to the substrate. Standard plankton collections may therefore underestimate actual abundance. When mating, the male grabs the female with his geniculate right first antenna. He then uses his modified fifth leg to hold her urosome. The pair may remain together for hours or days.

References. Anraku 1964; Grice 1969; Jacobs 1961.

Tortanus discaudatus

♀ 1.4 - 2.3 mm, ♂ 1.4 - 2.0 mm

- last prosomal segment with laterally directed points (1)
- long caudal rami with one side broad and spined (2)
- egg clusters often present
- similar species: other *Tortanus* spp.; *Centropages* and *Candacia* with asymmetrical or posteriorly directed points

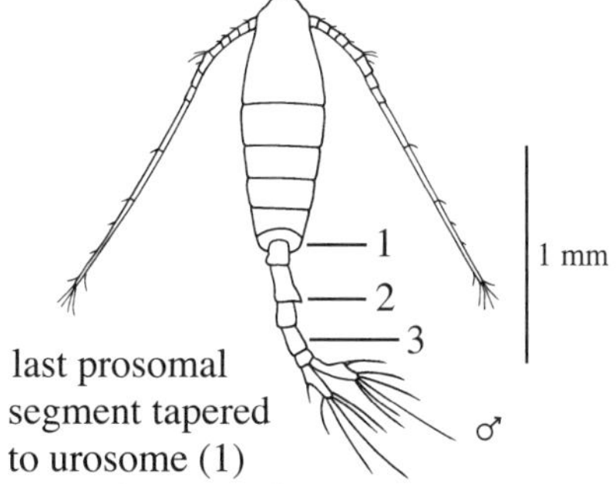

- last prosomal segment tapered to urosome (1)
- second urosomal segment with process on right side (2)
- long urosome is about 40% of total body length (3)
- similar species: males of many other copepods have twisted antennae

---

Tortanus setacaudatus

♀ 1.2 - 1.4 mm
♂ 0.8 - 1.0 mm

- prosome broadly rounded; widest at segment behind head (1)
- last prosomal segment not pointed and much wider than urosome (2)
- long, symmetrical caudal rami (3)
- male shorter and more slender
- similar species: other *Tortanus*; *Pseudodiaptomus* has a larger second segment, shorter caudal setae, and pigment bands; *Temora* spp. have wider heads

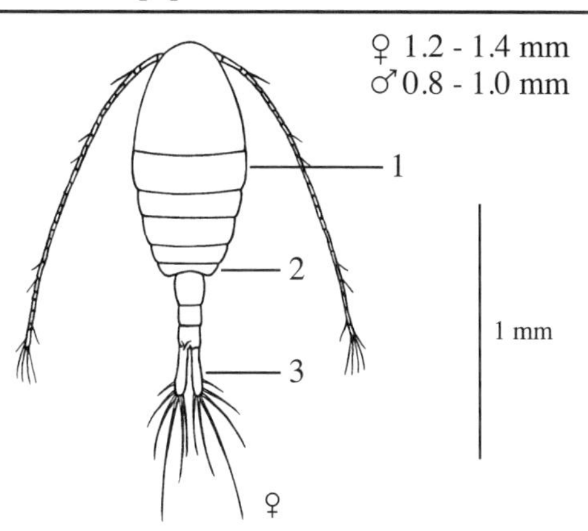

Pseudodiaptomus pelagicus

♀ 1.3 - 1.5 mm, ♂ 1.0 - 1.3 mm

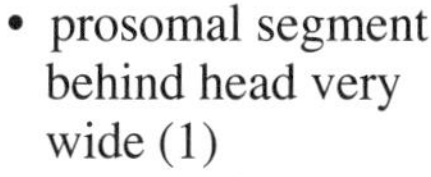

- prosomal segment behind head very wide (1)
- no eye; short head round from above, squarish in profile (2)
- urosome about 35% of total body length; very long caudal rami (3)
- often two egg clusters with right one smaller (4)
- opaque with rusty bands in mid-section
- similar species: *Tortanus setacaudatus* has a more symmetrical urosome, some very long caudal setae, and no color bands

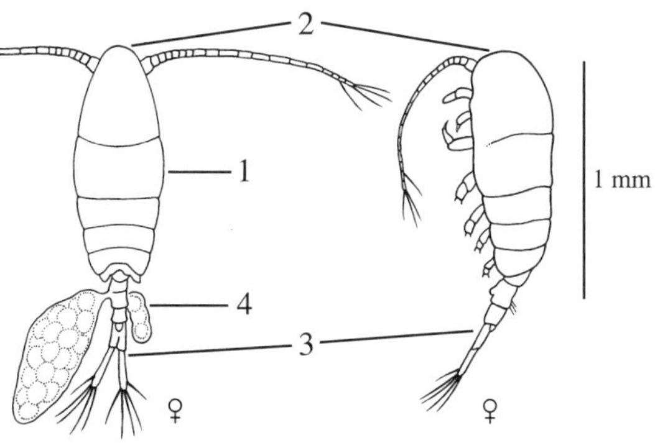

POECILOSTOMATOID COPEPODS

Hemicyclops spp. (*Saphirella* stage)

Occurrence. *Hemicyclops* is reported from mesohaline waters of the Mid-Atlantic and New England, but it is likely more widespread.

Biology and Ecology. The planktonic genus formerly identified as "*Saphirella*" is actually the first copepodid stage of *Hemicyclops*. The adults of all *Hemicyclops* studied are commensal or parasitic on benthic invertebrates, including polychaetes and molluscs. Thus, the adults do not occur in the plankton.

Notes. Itoh and Nishida (1995) and Kim and Ho (1992) reared two different species of Asian *Hemicyclops* in the laboratory to confirm long-held suspicions that the copepod formerly known as *Saphirella* is, in fact, a juvenile stage of *Hemicyclops*.

References. Gooding 1988; Itoh and Nishida 1995; Kim and Ho 1992; Lee 1978.

ID hint: The next two poecilostomatoids have highly distinctive pigmentation.

Oncaea venusta

Occurrence. This widely distributed oceanic and coastal copepod occurs from spring through fall along the entire Atlantic Coast and throughout the Gulf, occasionally in high-salinity estuarine waters. Additional *Oncaea* species occur offshore.

Biology and Ecology. *Oncaea* is omnivorous, feeding on phytoplankton and, perhaps preferentially, on a variety of zooplankton, including crustaceans. It may also feed on the surface of gelatinous zooplankton, especially larvacean houses, where it is often found. In turn, larval spot and croaker in the Gulf of Mexico eat *Oncaea*.

References. Go et al. 1998; Turner 1986.

Corycaeus amazonicus

Occurrence. *Corycaeus amazonicus* occurs from Rhode Island south to Florida and throughout the Gulf in summer and fall but is infrequent nearshore.

Biology and Ecology. This largely carnivorous copepod feeds on zooplankton, including small crustaceans, but it also feeds on phytoplankton when they are abundant.

References. Turner 1986.

Hemicyclops sp.

juvenile 0.5 - 0.6 mm

- head triangulate and accounts for most of prosome (1)
- long caudal setum on each ramus (2)
- similar species: *Euterpina* has a smaller head and segments form long, smoothly tapered sides

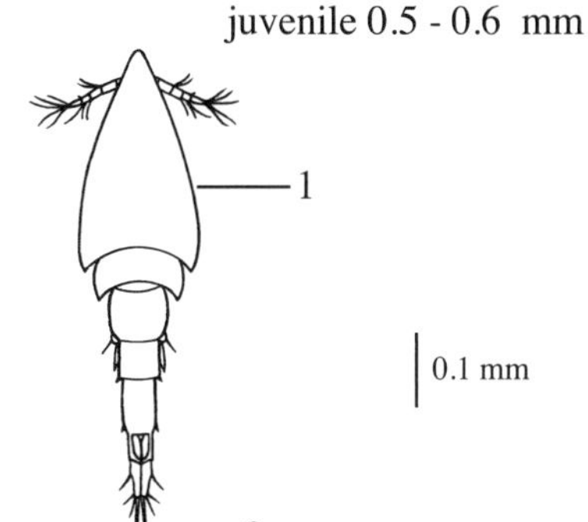

Oncaea venusta

♀ 1.1 - 1.3 mm, ♂ 0.8 - 1.0 mm

- head very wide at end (1)
- forehead flat and antennae shorter than head (2)
- very long genital segment with swollen anterior end (3)
- brightly colored in life; yellow-orange, sometimes red
- similar species: other *Oncaea* spp.; *Temora* spp. have longer antennae and shorter genital segments

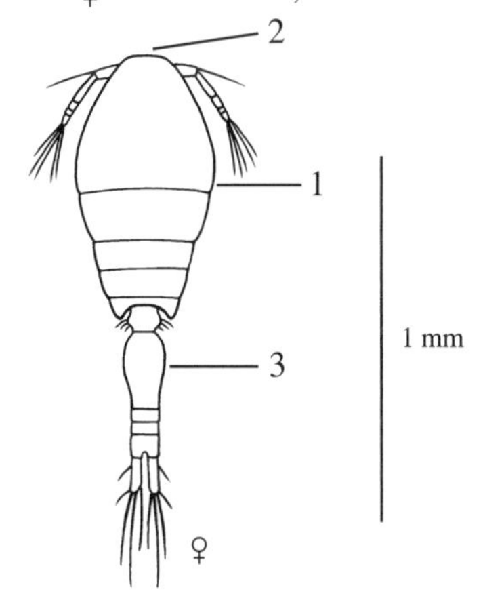

Corycaeus amazonicus

♀ 0.8 - 1.0 mm, ♂ 0.7 - 0.8 mm

- very short antennae (1)
- very large crystalline eyes, almost touching in males (2)
- last prosomal segment with symmetrical points (3)
- genital segment with strong spine on each side (4)
- distinctive blue to violet color (not uniform)
- similar species: other *Corycaeus* spp.; *Eurytemora* spp. have smaller heads, no eyes, and longer antennae

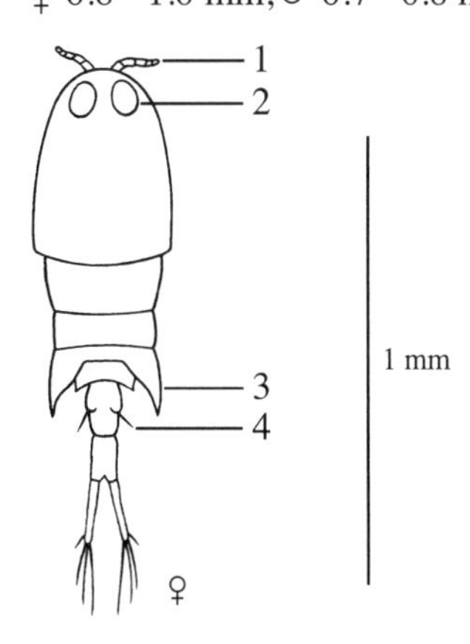

CYCLOPOID COPEPODS

Oithona colcarva and other *Oithona* spp.

Occurrence. *Oithona colcarva* occurs all year along the Atlantic, from Cape Cod to Florida and in the Gulf of Mexico. It may be the dominant cyclopoid in estuarine (down to 5 psu) and nearshore collections. *Oithona plumifera* and *O. robusta* are tropical oceanic species sometimes found in northern Gulf nearshore waters and distributed sporadically along the Atlantic Coast by the Gulf Stream. *Oithona nana* is common nearshore and in estuaries along the Florida west coast. *Oithona atlantica* is predominately a northern species reported offshore from the mouths of the Chesapeake and Delaware Bays in cool months. Even more *Oithona* species occur offshore.

Biology and Ecology. *Oithona* is among the smallest copepods. It is a relatively weak swimmer and apt to remain motionless while awaiting motile prey to ambush. Juveniles feed on dinoflagellates, while adults are more carnivorous and feed largely on calanoid nauplii and heterotrophic protozoans. *Oithona* is an important food for larval fishes. Because of its small size, many *Oithona* may pass through 153 μm mesh nets.

Notes. Until 1975, *Oithona colcarva* was frequently identified as *O. brevicornis*.

References. Castellani et al. 2005; Nakamura and Turner 1997; Nielsen and Sabatini 1996; Svensen and Kiørboe 2000; Turner 1986; Uchima and Hirano 1988.

Oithona similis

Occurrence. The euryhaline *Oithona similis* ranges from Canada to the Chesapeake Bay and probably farther south since it also occurs in the northeastern Gulf. In the southern parts of its range, it is most common in colder seasons.

Biology and Ecology. The ambush feeder *Oithona similis* alternates short jumps with periods of remaining motionless. While quiescent, *O. similis* detects its prey at a distance, perhaps by using hydromechanical signals. It feeds on microzooplankton, especially ciliates. Additional prey include flagellates and young copepod nauplii.

References. Castellani et al. 2005; Maar et al. 2006; Nakamura and Turner 1997; Shuvayev 1979; Svensen and Kiørboe 2000.

Halicyclops fosteri

Occurrence. *Halicyclops fosteri* is a common copepod in the oligohaline and low-salinity mesohaline reaches of estuaries from the Mid-Atlantic through the Gulf. It seems abundant in Louisiana and Texas but may be underreported in other areas.

References. Aurand and Daiber 1979; Gillespie 1971; Wilson 1958.

Oithona colcarva

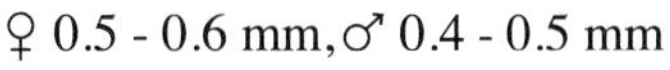

- head roundly pointed (1)
- rostrum present (not visible here); thick, pointed, and curved
- 4 segments behind head form irregular outer edges (2)
- antennae with many long setae (3)
- urosome about 40% of total body length; short setae on first urosomal segment (4)
- male stockier and both antennae bent and twisted; rostrum thick and short
- similar species: *O. similis* has a more blunt head; *O. plumifera* has 4-pointed head; *O. nana* and *O. simplex* females without rostrum

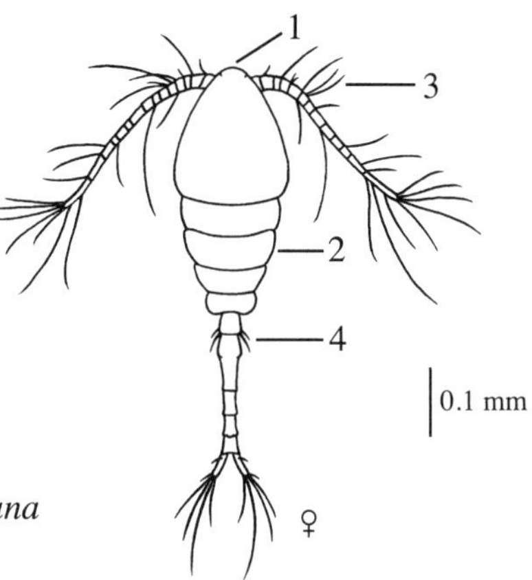

Oithona similis

♀ 0.65 - 0.85 mm, ♂ 0.50 - 0.65 mm

- head slightly round to blunt (1)
- rostrum present (not visible here); thin, pointed, and curved
- 4 segments behind head form smooth outer edges of prosome (2)
- antennae with many long setae (3)
- urosome almost as long as prosome; long setae on first urosomal segment (4)
- male stockier and both antennae bent and twisted; rostrum thick and short
- similar species: *O. colcarva* has a wide pointed head; *O. atlantica* (north) and *O. plumifera* (south) are much larger; combination of plumose antennae and long urosomes of *Oithona* species is distinctive

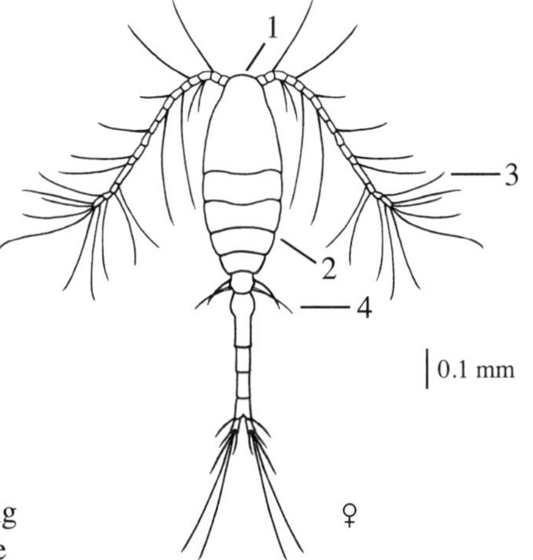

Halicyclops fosteri

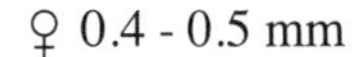

- rounded head is about 60% of prosome length (1)
- segments following head do not form smooth sides (2)
- antennae very stout (3)
- very short caudal rami with long setae (4)
- males with strongly bent and twisted right antenna
- similar species: other *Halicyclops*; *Oncaea* has smooth sides and a longer genital segment

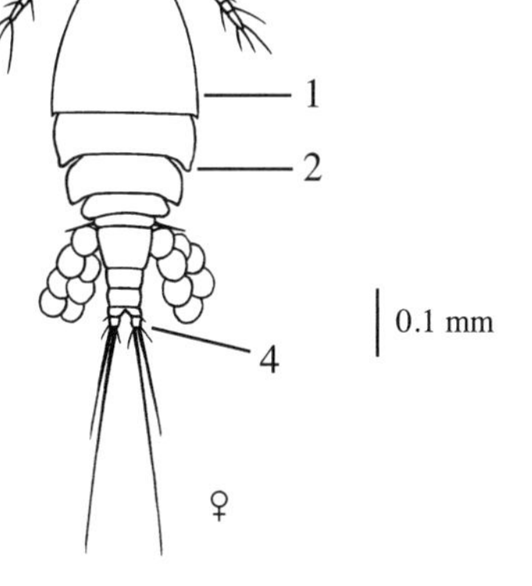

Mesocyclops edax

Occurrence. This freshwater species also occurs in oligohaline reaches of estuaries. Most reports from estuaries are from tributaries of the Chesapeake Bay and the Gulf, but its geographical distribution within our area is unclear. A number of other freshwater cyclopoids are flushed into low-salinity brackish areas. *Mesocyclops* is one of the most widespread and regular of these.

Biology and Ecology. *Mesocyclops edax* is a voracious predator that may actually pursue its victims before "pouncing" on them. Prey include copepod nauplii, cladocerans, rotifers, and protozoans. In oligohaline areas, *Mesocyclops* sometimes preys on fish eggs and larvae, including those of striped bass and white perch and may cause extensive damage in fish hatcheries. *Mesocyclops* detects the hydromechanical disturbances of active prey but is often eluded by slower, smooth-swimming rotifers and protozoans, which are only detected by direct contact. Most studies of *Mesocyclops* have been conducted in lakes; its biology in estuaries deserves more attention.

References. Brandl and Fernando 1978; Janicki and DeCosta 1990; Williamson 1986.

MONSTRILLOID COPEPODS

Monstrilla helgolandica

Occurrence. Monstrilloid copepods are widespread but are relatively rare in most collections. *Monstrilla helgolandica* seems more common in cooler months from Maine south. Other species may range farther south on the Atlantic Coast, but there are few reports from the Gulf.

Biology and Ecology. The monstrilloids are unique among copepods. The planktonic nauplius stage becomes endoparasitic on benthic polychaete or mollusc hosts where development occurs. The planktonic, nonfeeding adults are entirely reproductive.

References. Davis 1984.

HARPACTICOID COPEPODS

ID hint: Harpacticoids appear more tapered in dorsal profile than most copepods.

Longipedia americana

Occurrence. *Longipedia americana* is reported from Massachusetts to the Caribbean and in the Gulf, typically in coastal waters.

Biology and Ecology. These semibenthic copepods make periodic excursions into the plankton, sometimes in large aggregations. Feeding is poorly known in *L. americana*, but other *Longipedia* species use their mouthparts to produce a feeding current to feed on diatoms. One mating is sufficient to allow females to produce multiple batches of eggs.

References. Nicholls 1935; Onbé 1984.

Mesocyclops edax

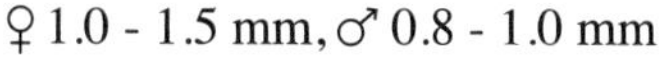

- very round head is about 70% of prosome length (1)
- urosome almost as wide as last prosomal segment (2)
- genital segment long; often with paired egg clusters (3)
- similar species: *Oithona* spp. have plumose antennae and narrow urosomes; *Oncaea* spp. have flat foreheads

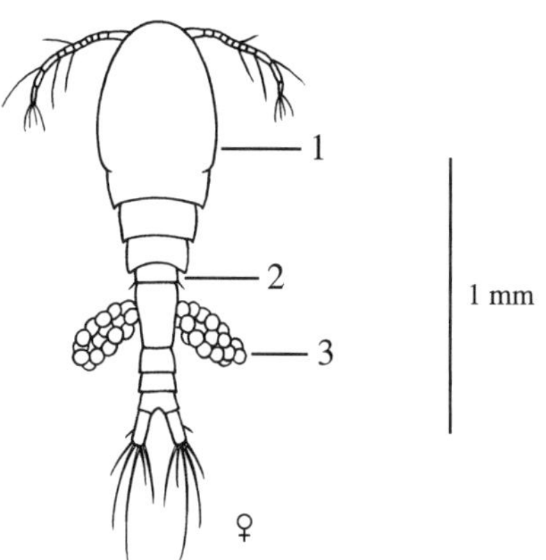

Monstrilla helgolandica

♀ 1.4 - 5.3 mm
♂ 1.1 - 2.0 mm

- head about 60% of length of elongate prosome (1)
- head transparent except for pigmented central gut (2)
- stout antennae with setae at joints (3)
- posterior prosomal segments wide and pigmented (4)
- fifth pair of thoracic legs rudimentary (5)
- similar species: other elongate copepods have mouthparts and five pairs of swimming legs; many other species of *Monstrilla* are similar

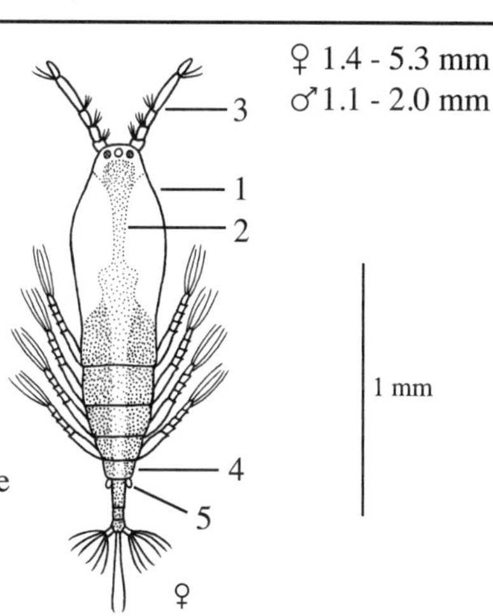

Longipedia americana

♀ 0.8 - 1.0 mm
♂ 0.6 - 0.7 mm

- wide, blunt head with prominent anterior knob (1)
- 3 segments behind head; last prosomal segment with sharp points (2)
- urosome tapers toward end with segments 3 and 4 being especially long (3)
- long, narrow, and sharply pointed tip on last urosomal segment (4)
- similar species: difficult to separate from *L. helgolandica* and other *Longipedia* species in the region; knob on head and pointed posterior segment distinguish it from other copepods

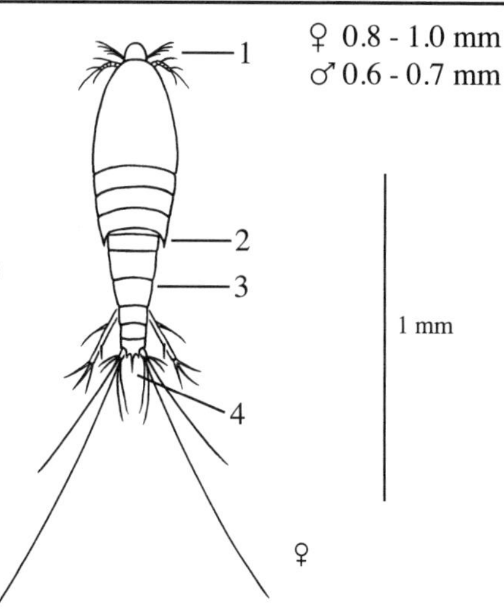

Euterpina acutifrons

Occurrence. *Euterpina acutifrons* is a widespread tropical and subtropical species and is the only common nearshore harpacticoid in the water column in high-salinity (>19) waters of our geographical range. It is reported as far north as Delaware Bay in the Atlantic and from the Gulf, where it is most abundant from spring to fall. It may reach densities of 500 m^{-3}.

Biology and Ecology. Although well studied in South American, European, and Asian waters, little recent information is available for *E. acutifrons* along our coasts.

References. Lonsdale and Levinton 1985; Nassogne 1970.

Coullana canadensis

Occurrence. *Coullana* may be abundant in winter to early summer in shallow oligohaline (0–15 psu) areas from Cape Cod south to Florida and in the Gulf but is most abundant in 5–10 psu. Densities of adults may reach 300,000 m^{-3}.

Biology and Ecology. These semibenthic harpacticoids are frequently in the water column. Males and females pair for two to three days while mating, and males exhibit mate-guarding behavior. These harpacticoids swim using a gliding motion.

References. Buffan-Dubau and Carman 2000; Gupta et al. 1994; Harris 1977; Lonsdale et al. 1998; Pace and Carman 1996.

Macrosetella gracilis

Occurrence. *Macrosetella gracilis* occurs in both the Atlantic and the Gulf and is more common in warm offshore waters than in nearshore waters. *Macrosetella gracilis* is associated with blooms of the large filamentous cyanobacterium *Trichodesmium*, which it uses for both habitat and food.

References. O'Neil 1998; O'Neil and Roman 1994; Roman 1978.

Microsetella norvegica

Occurrence. *Microsetella* is usually associated with clear warm water, but it is reportedly the most common planktonic harpacticoid in the Neuse River, North Carolina, where it reaches densities of >20,000 m^{-3}.

Biology and Ecology. *Microsetella* sometimes attaches to larvacean houses.

References. Maar et al. 2006; Mallin 1991; Ohtsuka et al. 1993.

Euterpina acutifrons

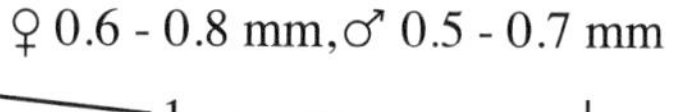
♀ 0.6 - 0.8 mm, ♂ 0.5 - 0.7 mm

- head strongly pointed (1)
- prosome widest behind head; smooth taper to urosome (2)
- urosomal segments similar in size; egg clusters often present (3)
- short caudal rami with long setae (4)
- body with distinct hunched appearance in side view
- similar species: *Hemicyclops* has a larger head and segments of different widths that do not produce evenly tapered sides

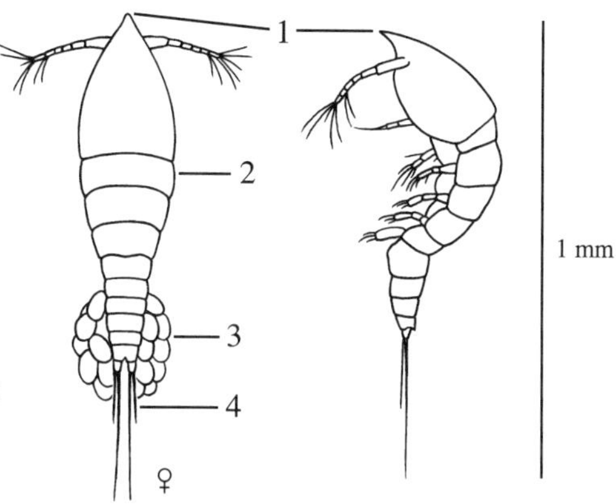

Coullana canadensis

♀ 1.2 - 1.4 ♂ 1.1 - 1.3 mm

- head with narrow tongue-like anterior end (1)
- body nearly cylindrical in cross-section
- prosome and urosome segments with similar width; slight taper to posterior end (2)
- caudal rami long, wide, and directed laterally (3)

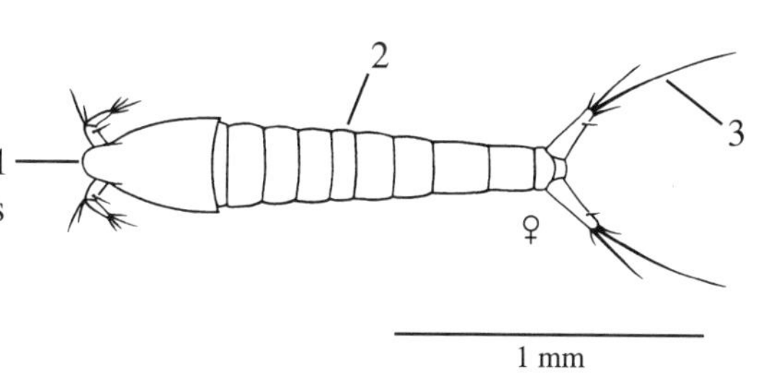

Macrosetella gracilis

♀ 1.4 - 1.5 mm, ♂ 1.2 - 1.3 mm

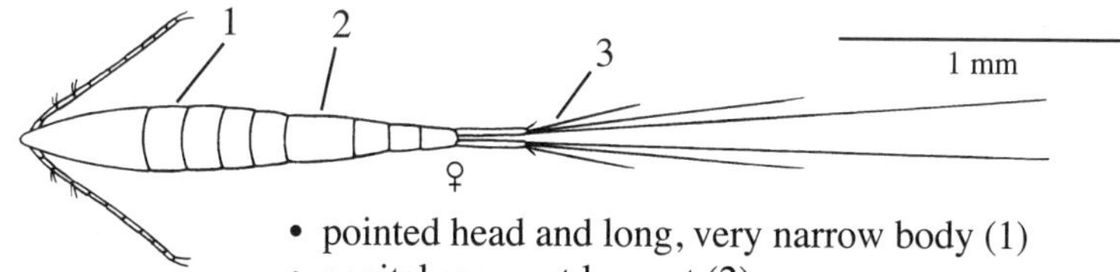

- pointed head and long, very narrow body (1)
- genital segment longest (2)
- long caudal rami with setae longer than body (3)
- violet or pink in color with yellow oil drops

Microsetella norvegica

(♀ 0.4 - 0.6 mm, ♂ 0.3 - 0.5 mm)

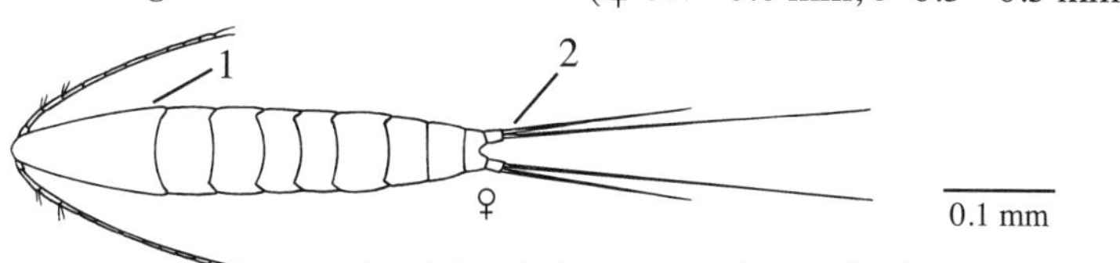

- pointed head about as wide as all other segments (1)
- short caudal rami with setae about as long as body (2)
- yellowish to reddish in color

PARASITIC COPEPODS AND BRANCHIURANS

Ergasilus spp. (poecilostomatoid copepod)

Occurrence. Many species of *Ergasilus* occur in U.S. waters, but only a few are common in salt or brackish waters. *Ergasilus manicatus* occurs in brackish to high-salinity areas, whereas *E. versicolor* is common in oligohaline and mesohaline regions of the Atlantic and Gulf Coasts. Both are more common in creeks and bays.

Biology and Ecology. Females of the genus *Ergasilus* are parasitic on the gills of fishes. *Ergasilus versicolor* is especially common on various catfishes, while *E. manicatus* occurs on killifishes and mullet.

References. Barse 1998; Bere 1936; Harris and Vogelbein 2006.

Caligus spp. (caligoid copepod, fish lice)

Occurrence. Close to two dozen species of *Caligus* occur in our geographical range. Adults are parasitic on fishes but are frequently caught in the plankton, especially males. Many *Caligus* species are common in coastal areas and in high-salinity embayments of both the Gulf and Atlantic Coasts. Migrating fish may carry them into brackish water.

Biology and Ecology. Adult caligoids are common ectoparasites on the oral or gill cavities, operculum, or body surface of fishes. Most species are associated with a certain group of fishes. *Caligus schistonyx* is especially common on menhaden. The complex life cycle includes several larval stages seldom found in the plankton. The parasite typically attaches during the copepodid stage. Mating may take place in the plankton before final attachment to the host. *Caligus* infestations plague many aquaculture operations.

References. Piasecki 1996; Piasecki and MacKinnon 1995; Yamaguti 1963.

Argulus spp. (branchiuran, fish lice)

Occurrence. *Argulus* is the only marine genus in the Subclass Branchiura. Both larvae and adults are frequently encountered in the plankton.

Biology and Ecology. Branchiuran fish lice are flattened crustaceans specifically adapted as fish ectoparasites. The newly hatched juveniles swim as they search for a host. Initial attachment is by the clawlike first antennae. Two large ventral suckers provide long-term attachment to the fish's outer surface, mouth, or gill chamber. Larval *Argulus* feed on mucus or body fluids. The adults are also accomplished swimmers and may become dislodged or leave hosts voluntarily to search for mates. Males attach to and mate with females while they are fastened to their host. Females produce multiple clutches, each containing several hundred eggs. Eggs are not carried in external egg clusters but are often fastened to the bottom. There is little host specificity, and a diverse array of fishes serve as hosts.

Note. Once included with the parasitic copepods, fish lice now constitute the Subclass Branchiura. There are at least 30 species of *Argulus*.

References. Mikheev et al. 2000; Møller et al. 2007; Pasternak et al. 2000; Rushton-Mellor and Boxshall 1994; Taylor et al. 2009; Walker et al. 2004; Yamaguti 1963.

Ergasilus manicatus

♀ 0.65 - 0.85 mm without eggs

- head fused with first segment, longer than wide (1)
- first two segments of second antennae rounded (2)
- terminal antennal segments are stout claws (3)
- genital segment barrel shaped (4)
- milky white body and eggs (not shown) and red eye
- similar species: other short, wide crustaceans (e.g., *Argulus*, *Caligus*) do not have wide antennal segments; *E. versicolor* has a more slender shape, much larger claws, violet spots, and a red gut

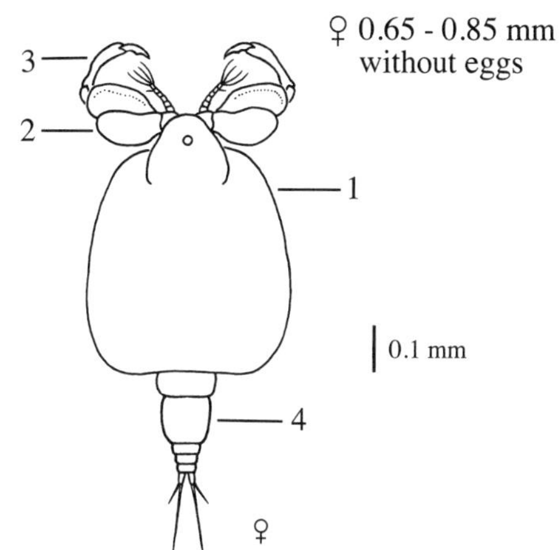

Caligus sp.

2 - 8 mm

- body very flat
- head fused to thoracic segments forming carapace ≥ 50% total length of body (1)
- very short antennae (2)
- prominent paired eyes on anterior margin (3)
- crescent-shaped suckers under anterior margin (not shown)
- one pair of legs adjacent to genital segment (seen in dorsal view) (4)

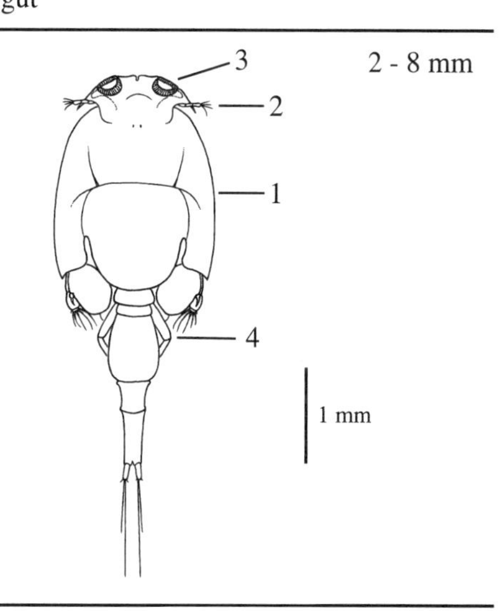

Argulus sp.

1 - 6 mm

- body very flat and wide
- carapace forms bilobed dorsal shield (1)
- paired eyes not on anterior margin (2)
- antennae and antennules hooked; not visible in dorsal view (3)
- pair of round, extended suckers prominent (4)
- abdomen not segmented (5)

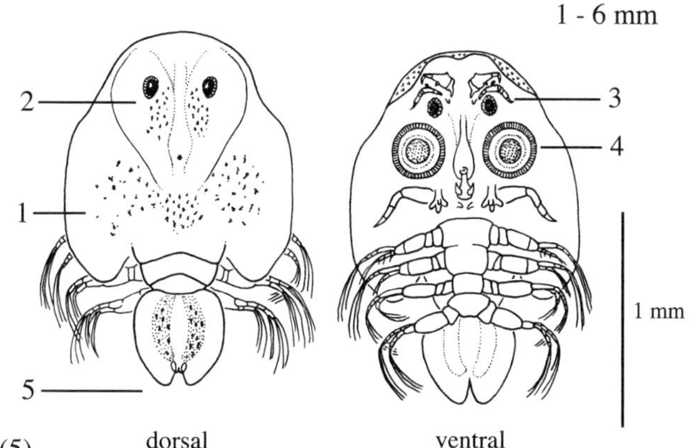

MYSIDS

Opossum Shrimps

Mysids are small, shrimplike crustaceans that, like the amphipods, isopods, and cumaceans, brood their young in a ventral brood pouch. Most of the nearly 1,000 described mysids are coastal and estuarine species <15 mm in total length. Most nearshore species are epibenthic and are found near the bottom during the day, where they are readily collected using sleds fitted with plankton nets. Epibenthic mysids typically rise higher in the water column at night, often in swarms. Other shallow-water mysid species spend most of their time swimming higher in the water column and a few species are able to burrow in sediments regardless of light conditions.

After mating, clutches of fertilized eggs are retained and brooded by females. Newly released mysids are almost completely developed and resemble adults in almost every regard. Females can be identified by the presence of developing brood plates (oöstegites) or complete brood pouches (Fig. 22). Temperate zone mysids usually have a longer-lived overwintering generation of large individuals, which releases young in spring. One to several shorter "summer" generations follow. Reproduction ceases with the onset of cold temperatures. Many species in warm, temperate areas probably reproduce all year.

These agile swimmers seem to glide through the water but can change direction quickly and use sudden abdominal flexions to escape danger. They can also hover or remain stationary on the bottom for long periods. Some mysids undergo tidal and seasonal migrations between shallow and deep waters.

Most shallow-water mysids are omnivorous and employ two primary types of feeding; individuals can switch between removing small particles (e.g., algae, detritus) from the water and raptorial feeding in which they seize smaller living zooplankton (e.g., copepods and rotifers). They also scavenge dead carcasses, including other mysids. Mysids are important and often primary sources of food for a variety of estuarine and coastal fishes. Because they are small, motile, and often live near the bottom of deeper waterways, the importance of mysids in coastal systems is often overlooked and largely underestimated.

IDENTIFICATION HINTS: DISTINGUISHING MYSIDS FROM DECAPOD SHRIMPS

Mysids resemble small decapod shrimps but differ from them in several basic ways. The following features are typical of shallow-water mysids:

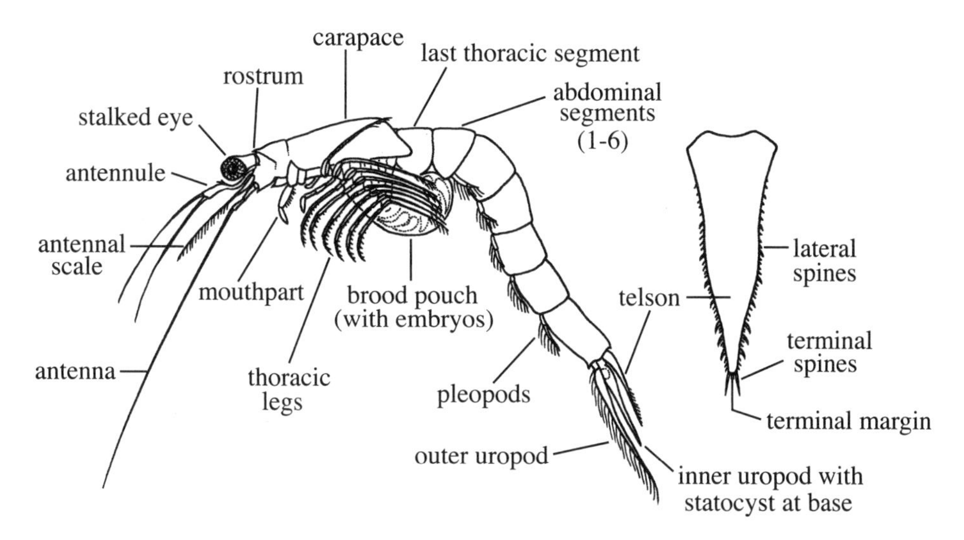

Fig. 22. Anatomical features of a typical mature female mysid, *Neomysis americana*.

Spherically shaped organs called **statocysts** at the bases of the inner uropods help mysids maintain balance (Fig. 22). Statocysts are definitive features for the group, but they are sometimes difficult to see and should not be used as the only factor to determine whether a specimen is a mysid or a decapod shrimp. See detail in Section 5 of "Quick Picks."

The **carapace** is attached to no more than the first four thoracic segments, leaving the posterior portion of the carapace flared and free of the body.

No **gills** are present under the carapace, whereas decapod shrimps (*Lucifer* is an exception) have structures that generally resemble the gills of fishes. Lift up the lower edge of the carapace to make this determination.

Mature female mysids have a conspicuous ventral **brood pouch** made of paired transparent plates that retain the eggs and young (see Fig. 22).

In many species, mature male mysids have one especially long pair of **pleopods** that is used in mating.

IDENTIFICATION HINTS: SEPARATING MYSID SPECIES

The **telson** (see Fig. 22) is often the primary feature for separating mysid species. Its margins usually have multiple projections technically known as spiniform setae. In the following descriptions, we refer to these rigid and often pointed extensions as spines. The numbers, lengths, and arrangements of spines on the telson and the presence or absence of a median cleft are usually diagnostic. Telson characteristics usually apply regardless of the size or sex of the mysid. Note that decapods also have telsons with unique arrangements of spines or setae.

Additional features below are useful for separating closely related species.

The shape and length of the **uropods** and the patterns of spines are sometimes unique.
Antennal scales of some mysids are unusually long or short, and their shapes are often diagnostic.
Pigmentation, especially overall color (e.g., translucent, brown, pink), is useful for distinguishing some species but beware that subtle patterns and small spots vary between individuals and with the condition of the specimens.

USEFUL IDENTIFICATION REFERENCES

Heard, R. W., Price, W. W., Knott, D. M., et al. 2006. *A Taxonomic Guide to the Mysids of the South Atlantic Bight* (pp. 1–45). NOAA Professional Paper NMFS 4. U.S. Department of Commerce, Washington, DC. <www.dnr.sc.gov/marine/sertc/Mysid%20key.pdf>. Accessed August 20, 2011.

Price, W. W. 1982. Key to the shallow water Mysidacea of the Texas coast with notes on their ecology. *Hydrobiologia* 93:9–21.

Stuck, K. C., Perry, H. M., Heard, R. W. 1979. An annotated key to the Mysidacea of the north central Gulf of Mexico. *Gulf Research Reports* 6:225–238.

Tattersall, W. M. 1951. A review of the Mysidacea of the United States National Museum. *US National Museum Bulletin* 201:1–292.

Whittmann, K. J. 2009. Revalidation of *Chlamydopleon aculeatum* Ortmann, 1893, and its consequences for the taxonomy of Gastrosaccinae (Crustacea: Mysida: Mysidae) endemic to coastal waters of America. *Zootaxa* 2115:21–33.

SUGGESTED READINGS

Fulton, R. S., III. 1983. Interactive effects of temperature and predation on an estuarine zooplankton community. *Journal of Experimental Marine Biology and Ecology* 72:67–81.

Johnson, W. S., Stephens, M., Watling, L. 2000. Reproduction and development of marine peracaridans. *Advances in Marine Biology* 39:107–260.

Mauchline, J. 1980. The biology of mysids and euphausids. Pt. 1, The biology of mysids. *Advances in Marine Biology* 18:1–369.

Wittmann, K. J. 1984. Ecophysiology of marsupial development and reproduction in Mysidacea (Crustacea). *Oceanography and Marine Biology: An Annual Review* 22:393–428.

Neomysis americana

Occurrence. *Neomysis* is often referred to as the most abundant shallow-water mysid from Cape Cod to northern Florida; peak abundance occurs in the Middle Atlantic states, where it inhabits almost all major waterways. It occurs from the upper ends of estuaries (almost freshwater) to 36 psu in the ocean but appears most abundant in the deeper portions of brackish and high-salinity estuaries. Although generally present all year, peak numbers occur in the summer from the Chesapeake Bay north. In the southern part of its distribution, mysids can be most abundant in estuaries during winter and spring.

Biology and Ecology. *Neomysis* is a strong swimmer capable of directed movements even in strong tidal systems. Recent studies have found that most individuals remain near the bottom during day and night; however, at night, there is a tendency for the population distribution to expand vertically. The extent of the partial vertical redistribution varies according to stage of tide, depth, light penetration, and season. *Neomysis* has a tendency to aggregate and may occur in large shoals with densities of hundreds to thousands per cubic meter. Reproduction occurs in all but the coldest months, with large overwintering adults reproducing and dying off in the spring, giving rise to a first summer generation of mysids composed of smaller mature individuals. Females typically produce 10–40 young, depending on adult size. Juvenile *Neomysis* are more commonly encountered than adults in collections high in the water column. *Neomysis* may be the most important source of food for many young estuarine fishes, including weakfish, summer flounder, striped bass, and American shad.

References. Herman 1963; Hulburt 1957a; Sato and Jumars 2008; Williams et al. 1974; Winkler et al. 2007; Zagursky and Feller 1985.

Mysis stenolepis and *M. mixta*

Occurrence. *Mysis stenolepis* occurs from Massachusetts to New Jersey, where it occupies intertidal and shallow subtidal areas, including seagrass beds. *Mysis mixta* is a less common open-ocean relative that occurs as far south as New York.

Biology and Ecology. Unlike most cold-water crustaceans, both species reproduce during the winter. Large females may brood and release more than 150 young mysids in early spring before dying. Some older publications (1902–15) place these species in the genus *Michtheimysis*. A comparison of the predation and feeding ecology of co-occurring *Mysis stenolepis* and *Neomysis americana* showed that the diets of these omnivorous species showed a flexibility in foraging behavior that minimized direct competition for prey.

References. Amaratunga and Corey 1975, 1979; Richoux et al. 2004; Winkler et al. 2007.

Neomysis americana

♀ to 16 mm, ♂ to 12 mm

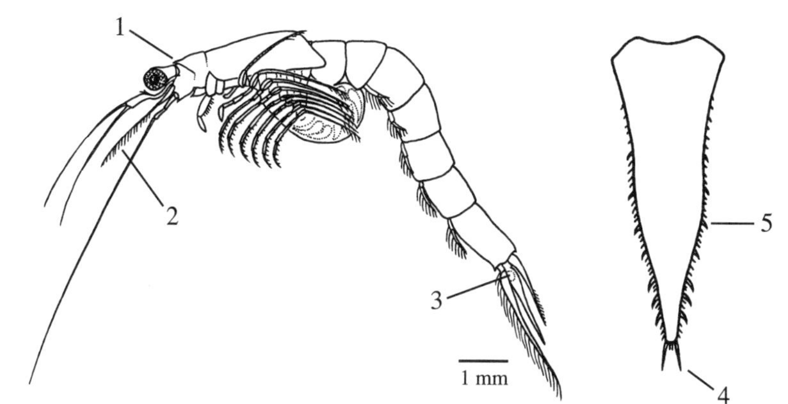

- carapace without distinct rostrum (1)
- antennal scale long; about 10 times longer than wide (2)
- statocyst large and prominent (3)
- telson:
  - terminal margin has two pairs of spines with the outer pair much longer than the central pair (4)
  - 35-40 lateral spines on each side arranged in clusters of short and long ones (5)

Mysis stenolepis

♀ to 30 mm, ♂ to 25 mm

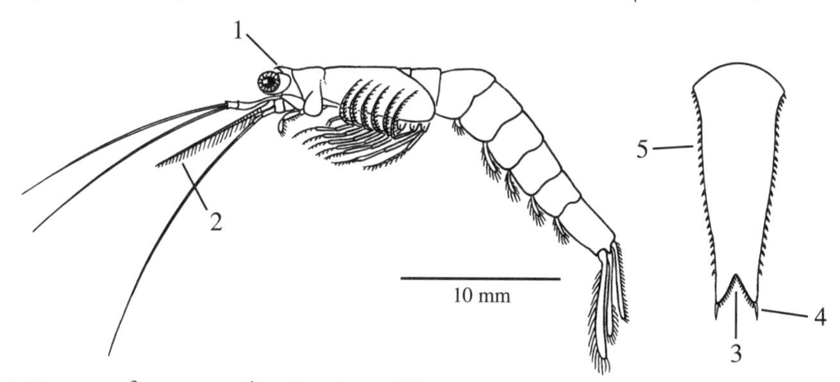

- carapace forms prominent rostrum (1)
- antennal scale long; about 12 times longer than wide (2)
- telson:
  - deep cleft lined with small, closely spaced spines (3)
  - strong corner spines (4)
  - about 25 similar lateral spines on each side, but not along posterior ends (5)
- body is heavily pigmented
- similar species: telson of *Mysis mixta* has lateral spines along the entire length and the body is opaque with few dark spots

Americamysis bigelowi, formerly *Mysidopsis bigelowi*

Occurrence. This commonly collected mysid occurs from Cape Cod to northern Florida. Many earlier records indicated the occurrence of *Americamysis bigelowi* in the Gulf of Mexico; however, it is now defined as an Atlantic Coast species, with two recently designated species (*A. alleni* and *A. stucki,* described next) that account for those previous records in the Gulf. *Americamysis bigelowi* prefers sandy bottoms and tends to occur in high-salinity estuarine and coastal ocean areas.

Biology and Ecology. *Americamysis bigelowi* shares the same life history pattern with *Neomysis americana*, reproducing from spring through fall in much of its range. In New Jersey, females produced from 5 to 40 young, with larger clutches associated with the overwintering generation. Females produce multiple broods during the warmest months. *Americamysis bigelowi* often co-occurs with *N. americana* where their ranges overlap. It is a common prey item for fishes. Less is known about the life history and habits of the Gulf species of *Americamysis.*

References. Allen 1984; Price et al. 1994; Williams 1972.

Americamysis alleni

Occurrence. *Americamysis alleni* occurs in the Gulf from northern Florida to Texas from the shoreline to about 15 m. It is found in medium to high-salinity bays and other open coastal habitats.

References. Price et al. 1994.

Americamysis stucki

Occurrence. *Americamysis stucki* occurs in the Gulf from southwestern Florida to northern Texas from the shoreline to 48 m but appears to be more common in shallow water. It is found in high-salinity, open coastal areas and bays.

Americamysis bigelowi

♀ and ♂ to 8.5 mm

1

2

5

4

3

1 mm

- carapace forms short pointed rostrum (1)
- second thoracic limb large; more so in females than males but unlike any other mysid on the Atlantic Coast (2)
- telson:
  - terminal margin with 3 pairs of strong spines; central pair slightly longer than next outer pair (3)
  - corner pair about one half the length of central pair (4)
  - 9-12 lateral spines on each side (5)

Americamysis alleni

♀ and ♂ to 7.0 mm

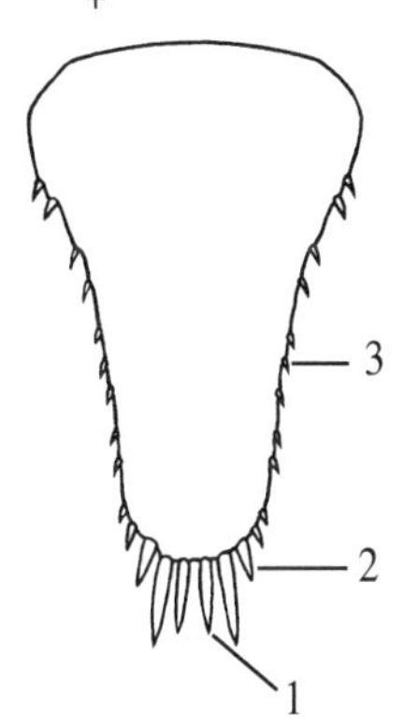

- telson:
  - terminal margin with 3 pairs of strong spines
  - central pair shorter than or equal to next outer pair (1)
  - corner pair shorter than the other two pairs of terminal spines, but longer than the lateral margin spines (2)
  - 10-15 lateral spines on each side (3)
- very large second thoracic limb as in *A. bigelowi* and *A. stucki*

Americamysis stucki

♀ to 6.0 mm, ♂ to 6.4 mm

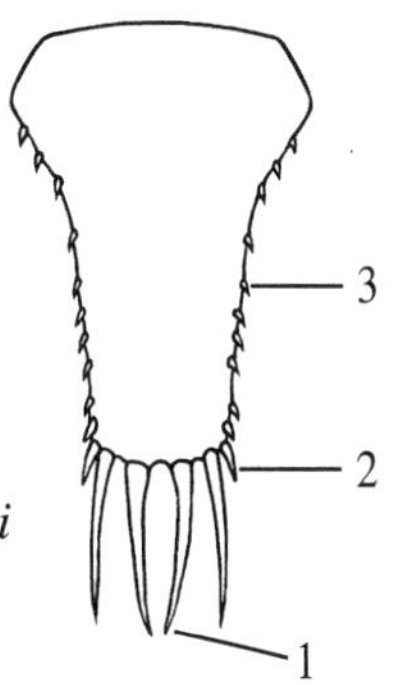

- telson:
  - terminal margin with 3 pairs of strong, widely spaced spines
  - central pair slightly longer than next outer pair (1)
  - corner pair short but longer than lateral margin spines (2)
  - 9-11 lateral spines on each side (3)
- very large second thoracic limb as in *A. bigelowi* and *A. alleni*

Americamysis bahia, formerly *Mysidopsis bahia*

Occurrence. *Americamysis bahia* occurs throughout the Gulf of Mexico and from eastern Rhode Island to Florida. It is found in open bay areas on sand and mud but is often associated with *Thalassia* seagrass areas in the Gulf. Salinity ranges of 2–54 are reported, with maximum abundances often near 30 psu. It appears to be restricted to shallow water, usually found in <2 m.

Biology and Ecology. Because of its wide occurrence, adaptability to laboratory conditions, and short generation time, *A. bahia* is widely used in toxicity bioassays. *Mysidopsis bahia* can reach maturity in 10–15 days, incubate and release broods after 6 or 7 days, and produce multiple broods during a lifetime under controlled laboratory conditions.

References. Gentile et al. 1982; Molenock 1969; Wortham-Neal and Price 2002.

Americamysis almyra, formerly *Mysidopsis almyra*

Occurrence. *Americamysis almyra* occurs from Maryland to Florida and throughout the Gulf, where it is found in seagrass beds and shallow marsh areas, mostly in depths <4 m. Although it is found from 0 to 32 psu, it is most commonly collected at <20 psu.

Biology and Ecology. Even in shallow water, *A. almyra* tends to remain just above the bottom during day and night. Analyses of gut contents of individuals in Florida revealed an omnivorous diet, including algae, copepods, and organic detritus. In Louisiana, it is an important food for anchovies, catfishes, seatrouts, Atlantic croaker, red drum, silver perch, and silversides.

References. Darnell 1961.

Mysidopsis furca

Occurrence. *Mysidopsis furca* occurs in the eastern and northern Gulf and, in the Atlantic, from North Carolina to southern Florida. This mysid occurs in high-salinity coastal areas and is not often reported from shallow waters along shorelines. *Mysidopsis mortenseni* occurs in shallow water off the west coast of Florida and a similar undescribed species, *Mysidopsis* sp. (*mortenseni* complex, Heard et al. 2006), occurs in shallow waters off the Carolinas.

Biology and Ecology. *Mysidopsis furca* may reproduce year-round in the southern end of its range. Unlike most other mysids, sexual dimorphism is seen in the shapes and spination of the telsons. Both *M. mortenseni* and the undescribed species have numerous short spines on the telson margin and do not display the sexual dimorphism characteristic of *M. furca*.

References. Bowman 1957.

Americamysis bahia ♀ to 8.0 mm, ♂ to 7.0 mm

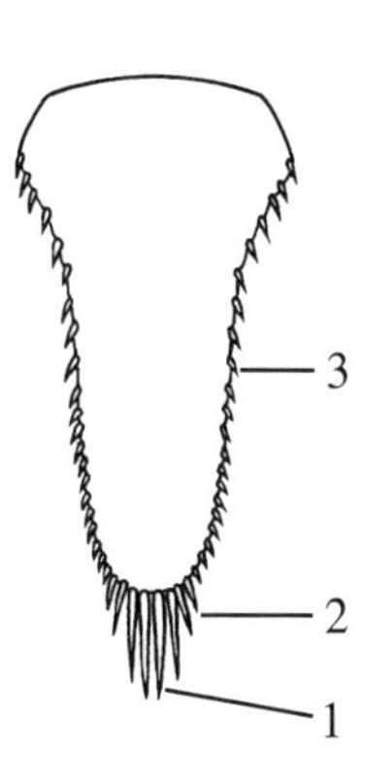

- telson:
  - terminal margin with 3 to 6 pairs of spines
  - central pair always longest (1)
  - lengths of outer pairs decrease abruptly (2)
  - 10-21 lateral spines on each side (3)
- inner uropod usually with 2 or 3 (sometimes 1 or 4) spines at base near statocyst (not shown)

Americamysis almyra ♀ to 10.0 mm, ♂ to 7.5 mm

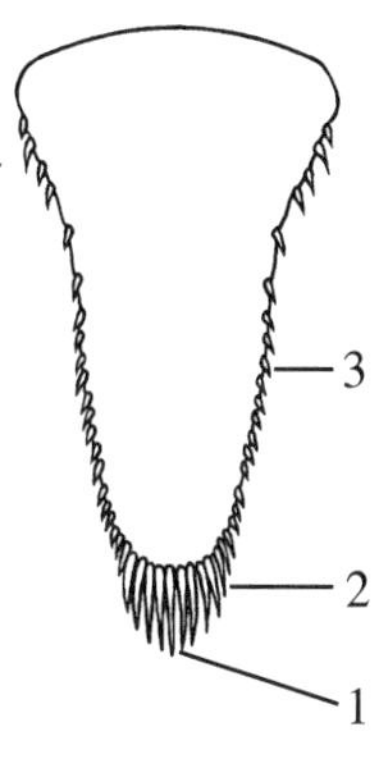

- telson:
  - terminal margin with 4 to 8 pairs of narrow spines
  - central pair longest (1)
  - lengths of other pairs decrease gradually (2)
  - 15-24 lateral spines of various sizes on each side (3)
- inner uropod always with 1 spine at base near statocyst (not shown)

Mysidopsis furca

- telson:
  - terminal margin with cleft and 2 pairs of spines
  - female with deep cleft; spines similar in length (1)
  - male with shallow cleft; central pair much longer (2)
- both sexes: inner uropod with spines along entire length (not shown)

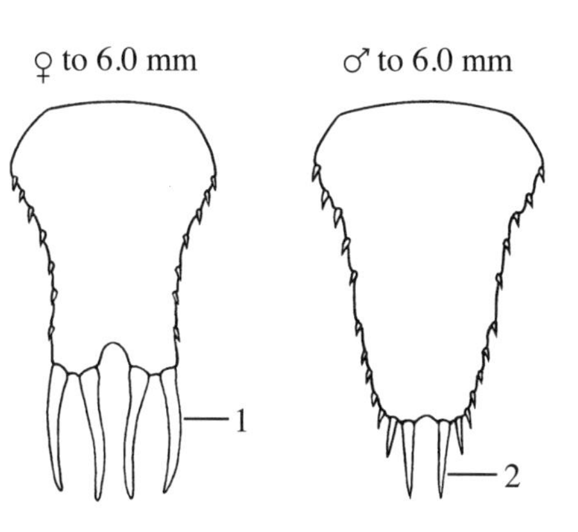

Metamysidopsis swifti

Occurrence. *Metamysidopsis swifti* is presently recognized as the only *Metamysidopsis* species occurring on the East Coast, where it is known from Delaware through Florida and throughout the Gulf. Previous reports of *M. munda* and *M. mexicana* refer to *M. swifti.* It occurs in high-salinity environments and is most often found in high-energy, sandy-bottom areas. High densities are often encountered in the surf zone of barrier beaches, especially at warm temperatures.

Biology and Ecology. Like many shallow water mysids, *M. swifti* aggregates to form schools and usually remains close to the bottom during the day. It is likely an important source of food for small fishes that inhabit the surf zone.

Chlamydopleon dissimile, formerly *Bowmaniella disimilis*

Occurrence. *Chlamydopleon dissimile* was previously known as *Bowmaniella dissimilis* and as *Coifmanniella dissimilis.* Records of both *Bowmaniella floridana* and *B. brasiliensis* are now attributable to *Ch. dissimile.* It is widely distributed from Delaware through Florida and throughout the Gulf. Although it has been collected at depths greater than 20 m, *Ch. dissimile* is found mostly in shallow waters and especially near beaches. *Chlamydopleon dissimile* often co-occurs with *Metamysidopsis swifti. Coifmanniella mexicana, Co. johnsoni,* and *Co. parageia*, which are similar to *Ch. dissimile*, are found along the Florida west coast.

Biology and Ecology. *Chlamydopleon dissimile* is a relatively large, strong swimmer that can quickly disappear into sandy bottoms and remain buried.

Reference. Wittmann 2009.

Promysis atlantica

Occurrence. *Promysis atlantica* occurs from North Carolina to Florida and throughout the northern Gulf. It is collected in open coastal waters and from estuarine areas >20 psu. Perhaps more so than many other coastal mysids, *P. atlantica* is found in the water column away from the bottom.

Biology and Ecology. This mysid occurs in large aggregations. In North Carolina, ovigerous female *Promysis* were collected during all but the winter months.

Metamysidopsis swifti

♀ to 6.0 mm, ♂ to 5.0 mm

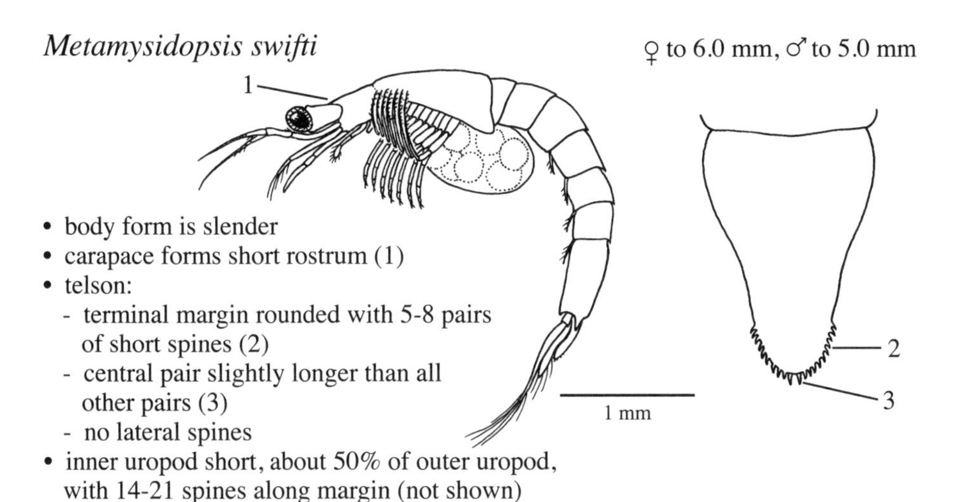

- body form is slender
- carapace forms short rostrum (1)
- telson:
  - terminal margin rounded with 5-8 pairs of short spines (2)
  - central pair slightly longer than all other pairs (3)
  - no lateral spines
- inner uropod short, about 50% of outer uropod, with 14-21 spines along margin (not shown)

Chlamydopleon dissimile

♀ to 12 mm, ♂ to 8 mm

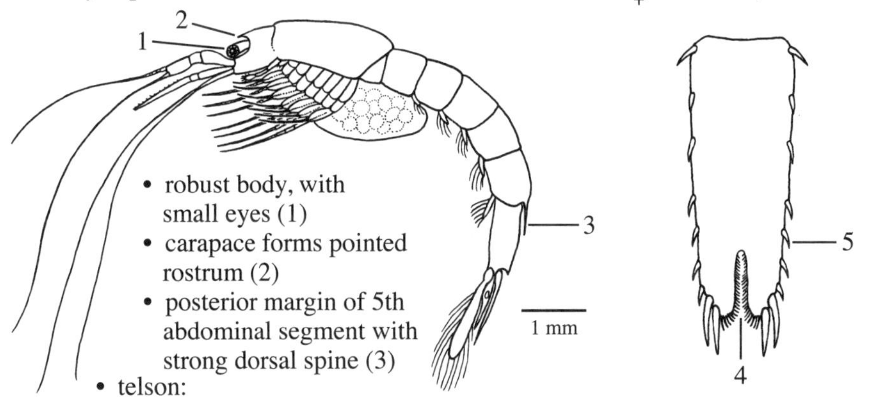

- robust body, with small eyes (1)
- carapace forms pointed rostrum (2)
- posterior margin of 5th abdominal segment with strong dorsal spine (3)
- telson:
  - terminal margin with deep, narrow cleft lined with very small spines (4)
  - 2 or more pairs of very strong spines flank cleft (5)

Promysis atlantica

♀ and ♂ to 8 mm

- body form narrow with very long eyestalks and blunt rostrum (not shown)
- telson:
  - terminal margin with deep cleft without internal spines (1)
  - 2 pairs of terminal spines slightly longer than lateral spines (2)
  - lateral spines along posterior half only (3)
- inner uropod inwardly curved with many thick spines (not shown)

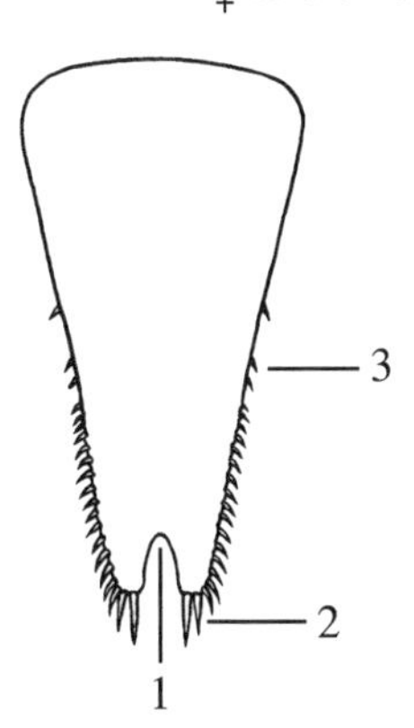

Heteromysis formosa

Occurrence. *Heteromysis formosa* occurs along the entire Atlantic Coast from Maine to Florida and has been reported from the Gulf but only in deeper coastal waters (40–200 m). In the Atlantic, it prefers high-salinity estuaries and the coastal ocean. Although it occurs mostly in channels and other deep areas of estuaries, it is sometimes found in shallow pools and margins of waterways. In the Gulf, at least five other species of *Heteromysis* are reported in the shallow waters of the southwestern Florida. *Heteromysis bredini* occurs in <5 m from Louisiana through Texas.

Biology and Ecology. *Heteromysis formosa* often appears in shallow-water collections near the bottom, usually in low densities. It has an affinity for dead, gaped bivalve shells, where small groups often swarm during the day. This mysid is more active after dark when it appears particularly vulnerable to prey by cusk eels, American eels, oyster toadfish, flounders, hakes, and searobins. Along the Atlantic Coast, *H. formosa* reproduces from spring through fall, and females carry from 7 to 32 young.

References. Allen 1982; Williams 1972.

Taphromysis bowmani and *T. louisianae*

Occurrence. *Taphromysis bowmani* occurs in the shallow ocean and estuaries from southern Florida to Texas at salinities ranging from 0 to 30. It sometimes occupies large intertidal pools on sandy flats. *Taphromysis louisianae* occurs from northern Florida through Texas and prefers low-salinity to freshwater areas of major rivers.

Biology and Ecology. *Taphromysis bowmani* reproduces through most of the year, with the larger adults occurring in the winter. Females typically carry less than 20 eggs or embryos.

References. Bacescu 1961; Banner 1953; Compton and Price 1979.

Brasilomysis castroi

Occurrence. *Brasilomysis castroi* occurs from South Carolina to Florida in the Atlantic and throughout the Gulf. It appears to prefer open estuarine sounds and bays to the coastal ocean. In tropical areas where it is more common, *B. castroi* is a nearshore mysid not usually seen in high-salinity (34–35) ocean waters.

Biology and Ecology. Little is known about the biology of *Brasilomysis*.

Heteromysis formosa

♀ and ♂ to 9.0 mm

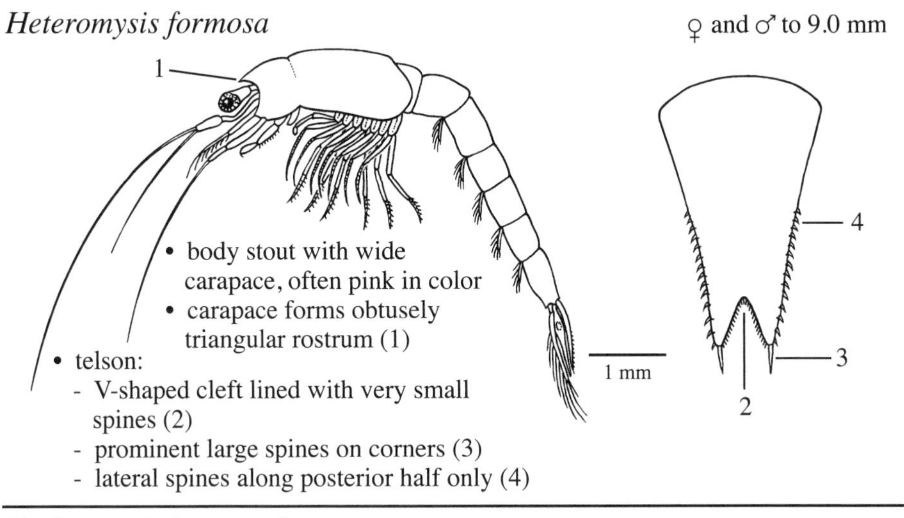

- body stout with wide carapace, often pink in color
- carapace forms obtusely triangular rostrum (1)
- telson:
  - V-shaped cleft lined with very small spines (2)
  - prominent large spines on corners (3)
  - lateral spines along posterior half only (4)

Taphromysis bowmani

♀ and ♂ to 9.0 mm

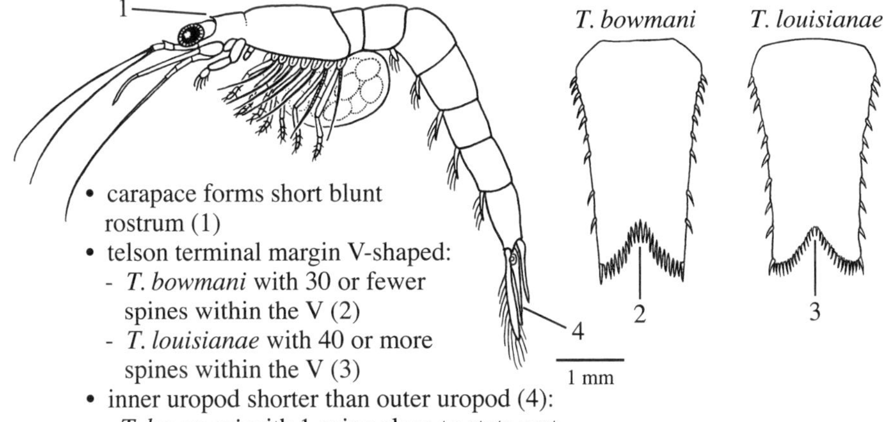

- carapace forms short blunt rostrum (1)
- telson terminal margin V-shaped:
  - *T. bowmani* with 30 or fewer spines within the V (2)
  - *T. louisianae* with 40 or more spines within the V (3)
- inner uropod shorter than outer uropod (4):
  - *T. bowmani* with 1 spine close to statocyst
  - *T. louisianae* with 1 spine along margin

Brasilomysis castroi

♀ and ♂ to 8.0 mm

- body slender with very long eyestalks and very long rostrum with rounded end (not shown)
- telson: terminal margin round with closely spaced flattened spines that extend about one half the length of the lateral margins (1)

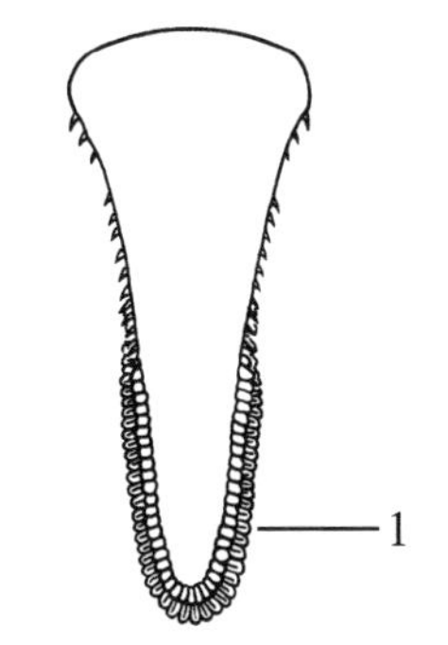

AMPHIPODS, ISOPODS, TANAIDACEANS, AND CUMACEANS

Amphipods, cumaceans, tanaidaceans, and isopods are primarily benthic or epibenthic, but many species make excursions into the water column, especially at night. Like the mysid crustaceans, amphipods, isopods, tanaidaceans, and cumaceans brood their young in a ventral brood pouch, or marsupium, made of thin overlapping plates. With the exception of some aberrant parasitic members of this group of crustaceans, there are no planktonic larval stages. Typically, young emerge from the brood pouch looking much like the adults. Basic differences in the morphology and biology of these peracarid orders are given next with additional details in the identification notes for individual taxa.

Amphipods are laterally compressed crustaceans without carapaces. Of the roughly 6,000 species, those occurring in plankton collections usually belong to one of three distinctive groups. As the name "amphipod" implies, they have several different kinds of appendages, including some with pincers. They typically lie on their sides in sample dishes. They are important sources of food for many coastal and estuarine fishes.

Gammaridean amphipods are among the most diverse and abundant benthic animals in virtually all brackish and marine habitats. Some are rapid swimmers, whereas others are closely associated with the bottom or live in tubes. Of the many gammarid species, a few make periodic nocturnal sojourns into the water column, and many others may be swept into the plankton by strong currents in shallow areas. Gammarideans are commonly reported in the plankton, but they are seldom identified to species in the literature.

Caprellid amphipods ("skeleton shrimps"), with their elongated, lightly armored bodies and simple appendages, hardly resemble the other amphipods. Caprellids typically cling tightly to and crawl slowly across substrates, but they may be swept into the plankton, especially in tidal creeks.

Hyperiid amphipods look like gammarideans but have huge compound eyes that cover much of the head. Unlike most other peracarids, hyperiids are entirely planktonic. All species are parasitic or commensal on gelatinous zooplankton, including jellyfishes, ctenophores, salps, and doliolids. Hyperiids are good swimmers and are caught swimming in the plankton as well as attached to jellyfish.

Isopods also lack a carapace and have a rigid, plated exoskeleton like amphipods. Unlike amphipods, they are often flattened dorsoventrally, and their seven pairs of legs are all similar. Most isopods are benthic and are not common in the plankton. However, some are good swimmers when they leave the bottom, primarily at night. In addition, the larval stages of parasitic isopods are occasionally caught in plankton samples.

Cumaceans are small (usually <5 mm) crustaceans with a large, "swollen" anterior carapace and a thin, forked "tail." Cumaceans are primarily benthic, but some, especially males, are accomplished swimmers and enter the water column at night, where pairing and mating occur.

Tanaidaceans are primarily benthic, but the nonfeeding males are capable of swimming and appear occasionally in plankton collections.

IDENTIFICATION HINTS

The morphological features noted above should distinguish the different basic groups. We have figured representative species commonly found in the plankton, but many more occur. Because all stages found in the plankton are either juveniles (immature) or adults, key characteristics of the species usually apply regardless of the size; however, many species exhibit sexual dimorphism. If you do not find a positive match among the species listed, consult one of the identification references in the reading list that follows.

USEFUL IDENTIFICATION REFERENCES

Barnard, J. L. 1969. The families and genera of the marine gammaridean Amphipoda. *US National Museum Bulletin* 271:1–535. (Ecological notes appear in the introductory chapter.)

Bousfield, E. L. 1973. *Shallow-Water Gammaridean Amphipoda of New England*. Comstock Publishing Associates, Cornell University Press, Ithaca, NY. 312 pp. (Also contains many of the most common southern and Gulf species.)

Bowman, T. E. 1973. Pelagic amphipods of the genus *Hyperia* and closely related genera (Hyperiidea: Hyperiidae). *Smithsonian Contributions to Zoology* 136:1–76.

Clark, S. T., Robertson, P. B. 1982. Shallow water marine isopods of Texas. *Contributions in Marine Science* 25:45–59.

Fox, R. S., Bynum, K H. 1975. The amphipod crustaceans of North Carolina estuarine waters. *Chesapeake Science* 16:223–237. (Planktonic occurrence is noted for many species.)

Heard, R. W. 1982. *Guide to Common Tidal Marsh Invertebrates of the Northeastern Gulf of Mexico*. Mississippi Alabama Sea Grant Consortium. MASGP-79-004. 82 pp.

Heard, R. W., Roccatagliata, T. D., Petrescu, I. 2007. *An Illustrated Guide to Cumacea (Crustacea: Malacostraca: Peracarida) from Florida Coastal and Shelf Waters to Depths of 100 m*. Florida Department of Environmental Protection, Bureau of Laboratories, Tallahassee. 175 pp. <www.dep.state.fl.us/labs/cgi-bin/sbio/keys.asp>. Accessed August 19, 2011.

Kensley, B., Schotte, M. 1989. *Guide to the Marine Isopod Crustaceans of the Caribbean*. Smithsonian Institution Press, Washington, DC. 308 pp.

LeCroy, Sara E. 2000. *An Illustrated Identification Guide to the Nearshore Marine and Estuarine Amphipoda of Florida*. Vol. 1. Florida Department of Environmental Protection, Bureau of Laboratories, Tallahassee. <www.dep.state.fl.us/labs/cgi-bin/sbio/keys.asp>. Accessed August 19, 2011.

LeCroy, Sara E. 2002–2007. *An Illustrated Identification Guide to the Nearshore Marine and Estua-*

rine Amphipoda of Florida. Vols. 2–4. Florida Department of Environmental Protection, Bureau of Laboratories, Tallahassee. <www.dep.state.fl.us/labs/cgi-bin/sbio/keys.asp>. Accessed August 19, 2011.

Menzies, R. J., Frankenberg, D. 1966. *Handbook on the Common Marine Isopod Crustacea of Georgia*. University of Georgia Press, Athens. 93 pp.

Richardson, H. 1905. *Monograph on the Isopods of North America*. Government Printing Office, Washington, DC. 727 pp. <www.archive.org/stream/monographsonisop00rich#page/iv/mode/2up>. Accessed August 19, 2011.

Schultz, G. A. 1969. *The Marine Isopod Crustacea*. William C. Brown, Dubuque, IA. 359 pp.

Steinberg, J. E., Dougherty, E. C. 1957. The skeleton shrimps (Crustacea: Caprellidae) of the Gulf of Mexico. *Tulane Studies in Zoology* 5:267–288.

Watling, L. 1979. *Marine Flora and Fauna of the Northeastern United States. Crustacea: Cumacea.* NOAA Technical Report NMFS Circular 423. 22 pp.

Zimmer, C. 1980. Cumaceans of the American Atlantic boreal coast region (Crustacea: Peracarida). *Smithsonian Contributions to Zoology,* No. 302. 29 pp.

SUGGESTED READINGS

Johnson, W. S., Allen, D. M., Ogburn, M. V., et al. 1990. Short-term predation responses of adult bay anchovies *Anchoa mitchilli* to estuarine zooplankton availability. *Marine Ecology Progress Series* 64:55–68. (Gammarideans and caprellids in the plankton are favored prey of anchovies.)

Johnson, W. S., Stephens, M., Watling, L. 2000. Reproduction and development of marine peracaridans. *Advances in Marine Biology* 39:107–260. (A review. Details on mating and the release of broods in the water column.)

Peebles, E. B. 2005. *An Analysis of Freshwater Inflow Effects on the Early Stages of Fish and Their Invertebrate Prey in the Alafia River Estuary*. Southwest Florida Water Management District, Brooksville. 147 pp.

<www.marine.usf.edu/documents/Peebles/AlafiaPlanktonReport_090705.pdf>. Accessed August 19, 2011.

Williams, A. B., Bynum, K. H. 1972. A ten-year study of the meroplankton in North Carolina estuaries: Amphipods. *Chesapeake Science* 13:175–192.

GAMMARIDEAN AMPHIPODS

These six representative gammarideans are among those most likely to appear in the plankton, but others are not infrequent. In a 10-year study of meroplankton in North Carolina estuaries, Williams and Bynum (1972) identified more than 50 species. Most *Gammarus* species and members of dozens of related genera share the same general look of the illustrated forms and require inspection of fine structures to identify correctly. Identification is complicated by differences between sexes and between mature and immature specimens. Under the microscope, they are usually seen lying on their sides, revealing a small head followed by six or more segments most of which have complex appendages. More so than for most other groups, very small juveniles (<1 mm) and large adults (>20 mm) that look very much alike can be found in the same collections.

ID hint: Most gammarideans are flattened laterally and lie on their sides.

Gammarus tigrinus

Gammarus tigrinus inhabits the upper reaches of estuaries from Canada south to Florida in the Atlantic and from southwestern Florida to Louisiana in the Gulf. While it is found in salinities as high as 25, it is more typical of lower salinities near the freshwater–saltwater interface, particularly at night. Unlike most members of *Gammarus*, *G. tigrinus* is a predator and may attack mysids and larval fishes. In turn, it is consumed by juvenile fishes.

References. Bailey et al. 2006; Stearns and Dardeau 1990.

Gammarus mucronatus

This brackish-water amphipod ranges from New England throughout the Gulf.
It is common in shallow water, in marsh pools, and in seagrass beds but is often taken in large numbers in plankton collections, especially at night.

References. Fredette and Diaz 1986; LaFrance and Ruber 1985.

Ameroculodes edwardsi and *Ameroculodes miltoni*

Ameroculodes edwardsi ranges from New England to at least the Mid-Atlantic. Earlier reports from more southerly waters were probably *A. miltoni*, which occurs from North Carolina southward along the Atlantic Coast and throughout the Gulf. Both species are euryhaline. *Monoculodes nyei* is often abundant over grassbeds and is widespread in moderate to high salinities. Both *Monoculodes* and *Ameroculodes* are good swimmers and are most common in the night plankton.

Reference. Foster and Heard 2002.

Gammarus tigrinus ♂ up to 14 mm

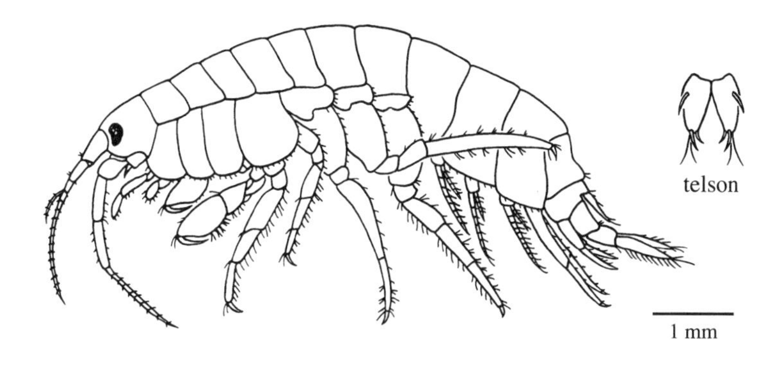

Gammarus mucronatus ♂ up to 13 mm

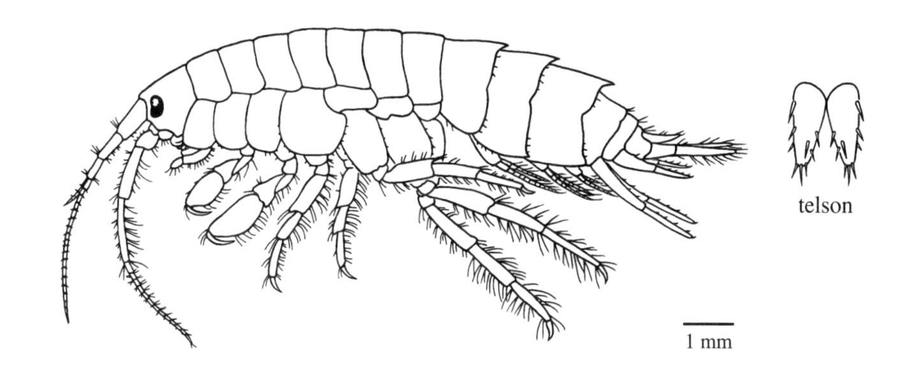

Ameroculodes edwardsi ♀ up to 9 mm

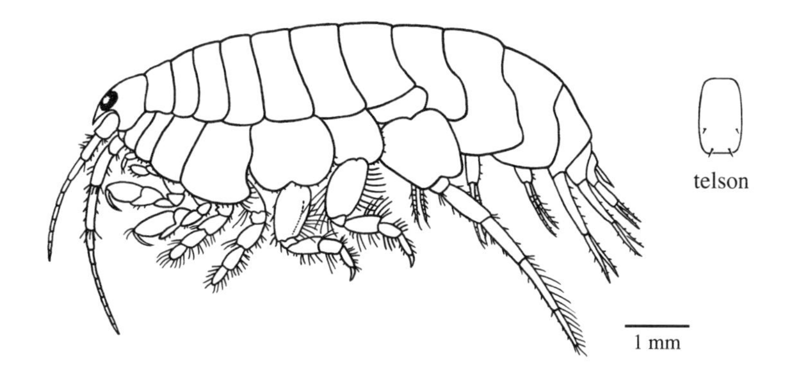

Calliopius laeviusculus

Calliopius laeviusculus is a northern, coastal species that ranges south to at least New Jersey. It is common near rocky coastlines, where it is often in the plankton, occasionally in high densities. These epibenthic-planktonic amphipods are omnivorous and can feed on phytoplankton, fish eggs, larval fish, and other zooplankton while in the plankton. Occasionally, they are common in the neuston.

References. Hudon 1983; Williams and Brown 1992.

Cerapus tubularis

This small benthic gammarid sometimes swims in the water column with large antennae extending from a tube that it constructs to cover most of its body. *Cerapus* is common in estuaries from Massachusetts south to at least South Carolina. It is also reported from the Gulf from Ten Thousand Islands, Florida, westward to eastern Louisiana. *Cerapus* occurs more regularly in the plankton than most gammarids and may occur in oligohaline areas in south Florida. Most tube-dwelling amphipods leave their tubes when they enter the water column, but *Cerapus* takes its tube along when it goes for a swim.

Apocorophium lacustre, formerly *Corophium lacustre*

Apocorophium lacustre is common in estuaries and bays from Canada to Indian River Lagoon, Florida, and westward in the Gulf from Apalachicola, Florida, to Louisiana or beyond. It may be in salinities as high as 25 but is most abundant in oligohaline waters. *Apocorophium louisianum* also occurs in the plankton and ranges from Indian River, Florida, to the Atlantic Coast, and throughout the Gulf in meso- and oligohaline areas. *Apocorophium* typically lives on the bottom in mucus-lined tubes, but they occasionally enter the plankton, where they are often caught near the surface, especially at night. Presumably, these nocturnal sojourns are associated with mating.

References. Stearns and Dardeau 1990.

Calliopius laeviusculus

♀ up to 14 mm

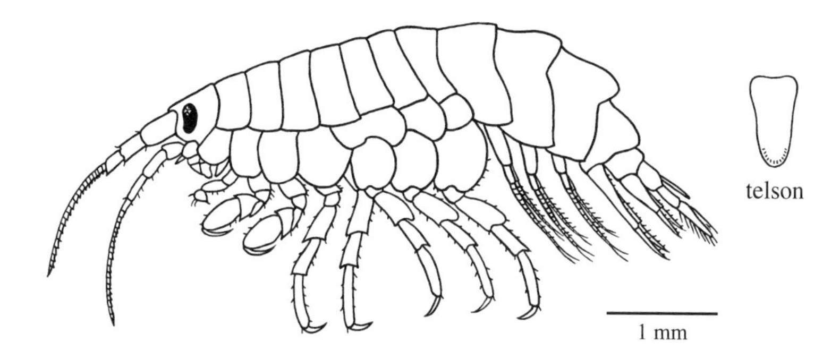

Cerapus sp.

♂ up to 5 mm

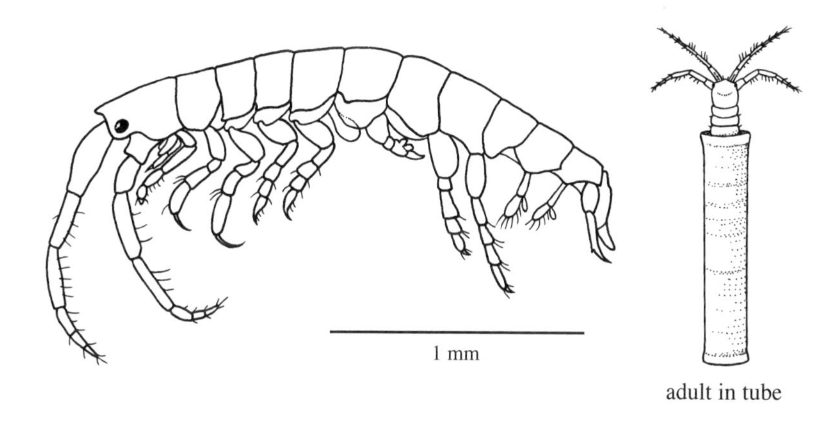

Apocorophium lacustre

♂ up to 4 mm

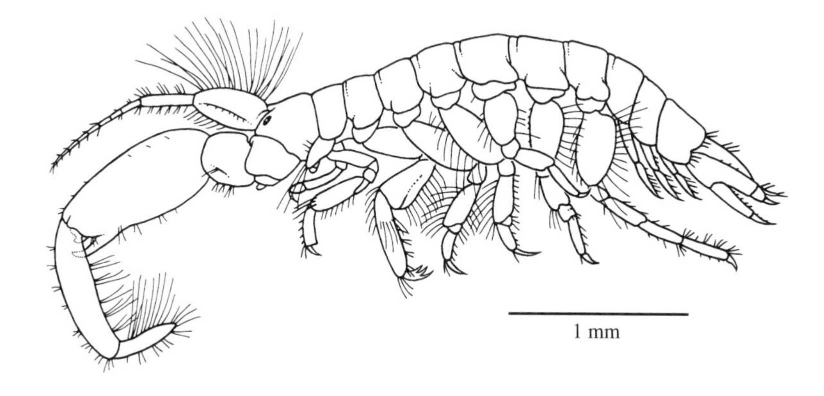

HYPERIID AMPHIPODS

Hyperia spp. and *Hyperoche* spp.

Hyperiid amphipods are cosmopolitan in the ocean but uncommon in less than full-strength salinity waters. The large-eyed, holoplanktonic *Hyperia* and *Hyperoche* are the hyperiids most likely to occur in nearshore samples. They may be free in the plankton or associated with gelatinous zooplankton, including jellyfishes, comb jellies, and salps. Females and their broods feed on their hosts. They probably do not inflict significant damage to the larger jellyfishes but frequently remove living tissue from salps and doliolids.

References. Cahoon et al. 1986; Dittrich 1987; Gasca et al. 2009; Harbison et al. 1977; Laval 1980.

CAPRELLID AMPHIPODS

Caprella spp.

Caprellids, with their elongate bodies, are unique among amphipods. Several similar-looking species of caprellids occur widely in coastal and estuarine areas from Massachusetts through the Gulf. These benthic amphipods are most common on hydroids, sea whips, and submerged aquatic vegetation. They typically fasten their rear appendages and wave their bodies to and fro with a slow and very characteristic motion. They are occasionally swept into the water column with algae and resuspended detritus.

TANAIDACEANS

Hargeria rapax

The euryhaline tanaidacean *Hargeria rapax* ranges from Canada to Texas (except possibly south Florida) in saltmarsh creeks, grassbeds and sandy bottoms in estuaries, and along the coast. It is especially abundant in the clear, spring-fed estuaries north of Tampa Bay. Typically a benthic tube dweller, *Hargeria* also occurs in the plankton, predominately at night. The pleopods of adult males, used in swimming, are especially well developed compared with pleopods in immature males and females. Presumably, enhanced swimming ability assists males in their nocturnal sojourns in search of females. *Hargeria* is largely protogynous, where most mature individuals begin as females and then switch to become males. Tanaidaceans can resemble isopods or elongate amphipods, but they can be distinguished from both of those groups because they lack a telson. Both males and females of many tanaidaceans can have unusually large chelae. Other genera of tanaidaceans may appear in the plankton.

References. Heard 1982; Modlin and Harris 1989.

Hyperia galba

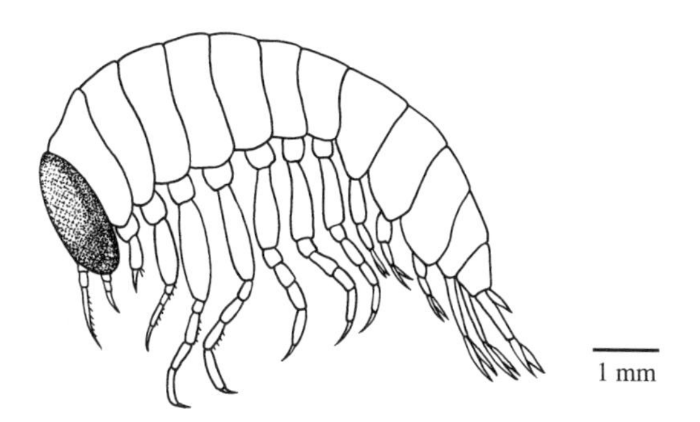

Caprella sp.

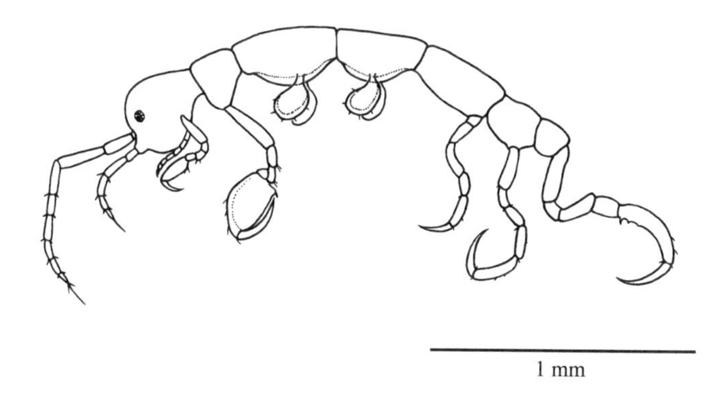

Hargeria rapax

♂ up to 7 mm

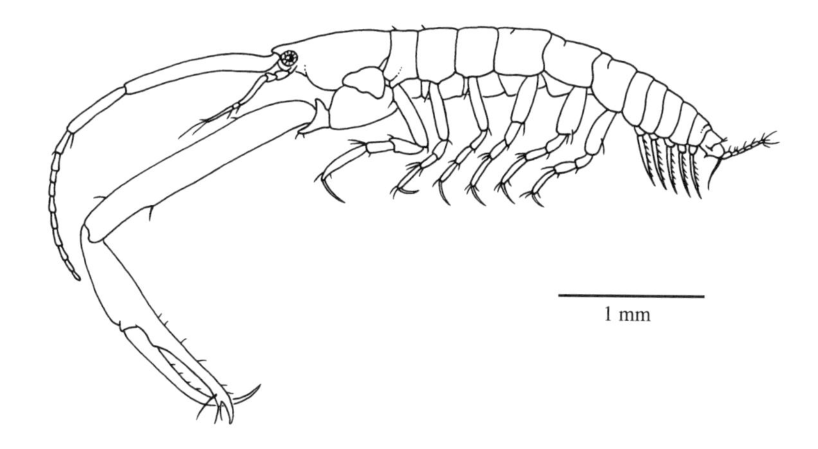

ISOPODS

ID hint: Isopods are flattened dorsoventrally and have seven thoracic segments behind the head, most of which have similar legs.

Idotea metallica and *I. balthica*

Idotea metallica is an active swimmer that inhabits coastal and estuarine waters from Massachusetts to Florida. Protruding lateral segments give *I. metallica* a characteristic saw-toothed appearance. *Idotea balthica* occurs from Massachusetts to Florida, often in estuaries. *Idotea balthica* has a smoother lateral profile, and its telson has a prominent point in the center rather than being square as in *I. metallica*. *Idotea* spp. are often found rafting on floating objects or floating patches of seaweed.

References. Gutow and Franke 2003; Gutow et al. 2006.

Eurydice littoralis

The very long antennae distinguish *Eurydice* from the next two similar species. Chromatophores usually give this isopod a dark appearance. *Eurydice* is particularly common in sandy areas from Georgia to Florida and on the Florida west coast. *Eurydice* is an active swimmer and may nibble on bathers in the surf. It also attaches to fishes as an ectoparasite.

Livoneca redmanii (gill louse), formerly *Lironeca ovalis*

Livoneca redmanii ranges from Massachusetts to Florida and in the Gulf west to Mississippi. It is common in both coastal and estuarine waters (>5 psu). The large eyes separate juvenile and male *Livoneca* from the two preceding species. The planktonic juvenile (drawn) reaches 3–7 mm and is light brown. *Livoneca* is a fish parasite found on a wide variety of fishes, including bluefish, striped bass, and menhaden. Newly released young swim actively in search of hosts and are not selective about species or where they attach. Newly attached *Livoneca* are easily dislodged and may reenter the plankton. Older isopods are typically found on the gills or branchial chamber. Like many parasitic isopods, *Livoneca* may undergo sex reversal from male to female. *Livoneca texana*, reported from the western Gulf, may be the same species. Newly released juveniles are identical to isopods previously assigned to the genus *Aegathoa*.

References. Marks et al. 1996; Sandifer and Kerby 1983.

Idotea metallica up to 38 mm

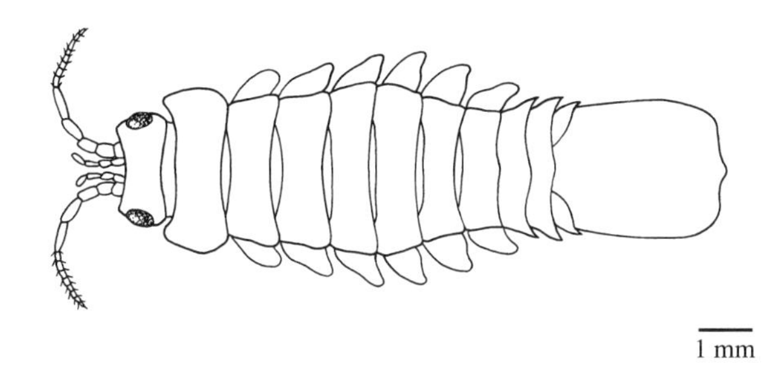

Eurydice sp. up to 6 mm

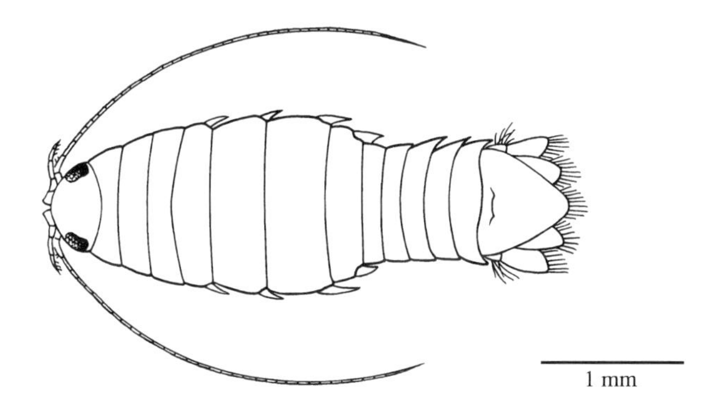

Livoneca redmanii up to 18 mm

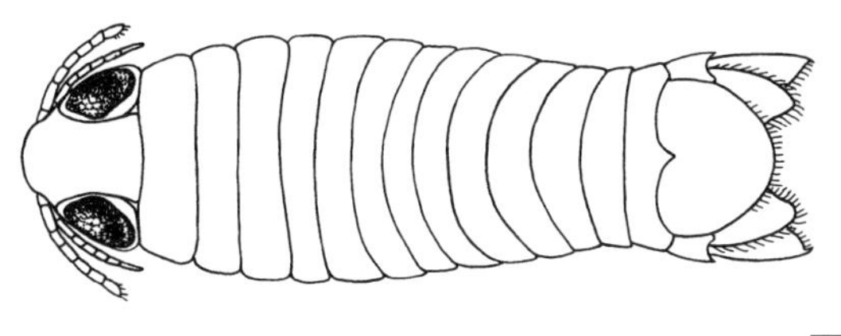

Politolana concharum, formerly *Cirolana concharum*

Politolana concharum ranges from Massachusetts to South Carolina, primarily along the coastline. These large isopods are good swimmers and may prey on other zooplankton, including copepods. They are also scavengers and will feed on fishes caught in gill nets.

Edotia triloba (=*E. montosa*)

Edotia triloba is very flat, has long legs, and lacks abdominal segments anterior to the telson. *Edotia* ranges from Massachusetts to Florida and in the eastern Gulf. Because the taxonomic designations of *E. triloba* and *E. montosa* have not been fully resolved, reports of both species on the Atlantic and in the Gulf remain unclear. *Edotia* is a weak swimmer that is common in the plankton from the swash zone along beaches to at least 10 km offshore, where it often swims just above the bottom. It also occurs in bays and creeks. In rivers on the Florida west coast, *E. triloba* typically occurs in distinct nocturnal swarms that extend along several kilometers of tidal-river reach. Other species of *Edotia* and *Chiridotea* are usually less common in the plankton.

Sphaeroma quadridentatum

This euryhaline isopod is common from Massachusetts to Florida and in the Gulf from Florida to Texas, occasionally in tidal freshwaters. *Sphaeroma terebrans* and *S. walkeri* also occur in the Gulf. Sphaeromatids as a group are generally most abundant in moderate to high salinities. *Sphaeroma* is usually associated with structures in shallow water and is distinguished by its ability to roll into a ball.

Politolana concharum up to 23 mm

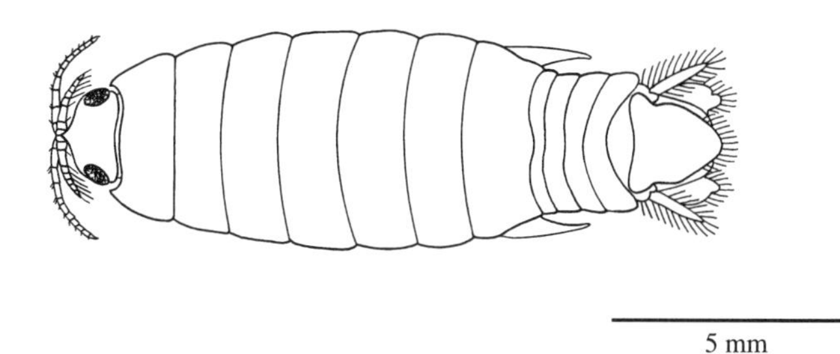

Edotia triloba up to 9 mm

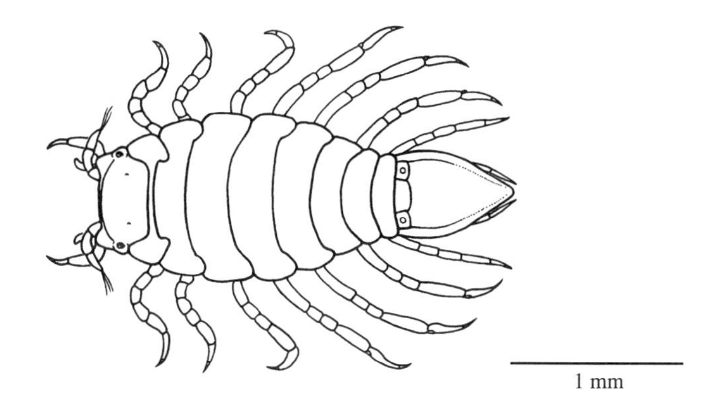

Sphaeroma quadridentatum up to 10 mm

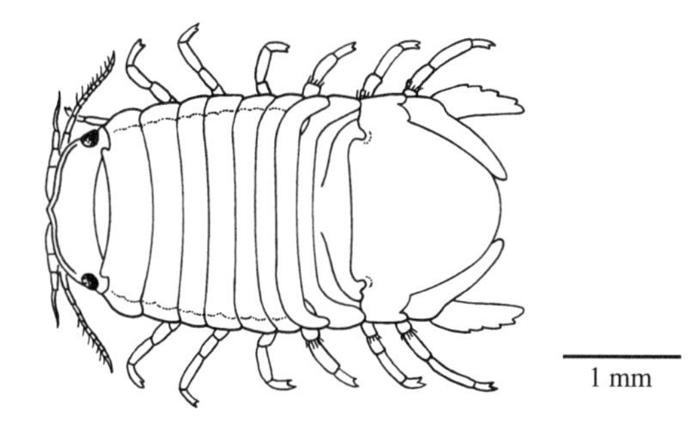

Uromunna reynoldsi

The brackish water isopod *Uromunna reynoldsi* occurs from North Carolina to Florida on the Atlantic Coast and from Florida to Texas in the Gulf. It is often overlooked because of its very small size. Unlike most isopods, males and females are morphologically distinct. It prefers shallow mesohaline and, especially, oligohaline bays, bayous, and marshes from 1 to 15 psu. It feeds on microalgae and detritus and is consumed by marsh killifishes. *Uromunna reynoldsi* is more common in the surface plankton at night, especially on flood tides.

References. Stearns and Dardeau 1990.

Cassidinidea ovalis

Cassidinidea ovalis occurs from New Jersey to north Texas in shallow brackish waters, usually from 1 to 20 psu. *Cassidinidea lunifrons*, which has been reported from the Atlantic and Gulf, is generally considered a synomyn of *C. ovalis*. *Cassidinidea* is a small, oval-shaped, benthic isopod that lives in intertidal marshes, and it is often colored to match its substrate. It has been reported as common in plankton collections in southwest Florida estuaries.

Cyathura polita

The euryhaline *Cyathura polita* is typically estuarine and is widely distributed from Massachusetts to Louisiana. It is especially common in lower salinities. This predominately benthic isopod is usually found buried in soft bottoms. Although adults are considered cryptic, juveniles occasionally swim up into the water column or attach to suspended pieces of detritus and algae. *Cyathura* often undergo a sex change with females transforming into males.

References. Burbanck 1962; Frankenberg and Burbanck 1963; Kruczynski and Subrahmanyam 1978.

Uromunna reynoldsi ♂ up to 1.6 mm

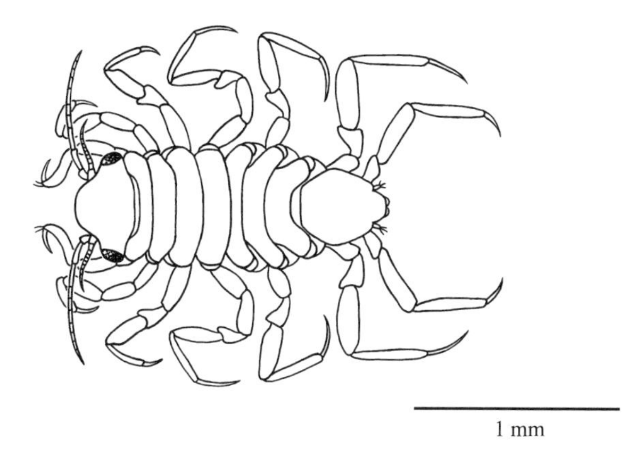

Cassidinidea ovalis up to 4 mm

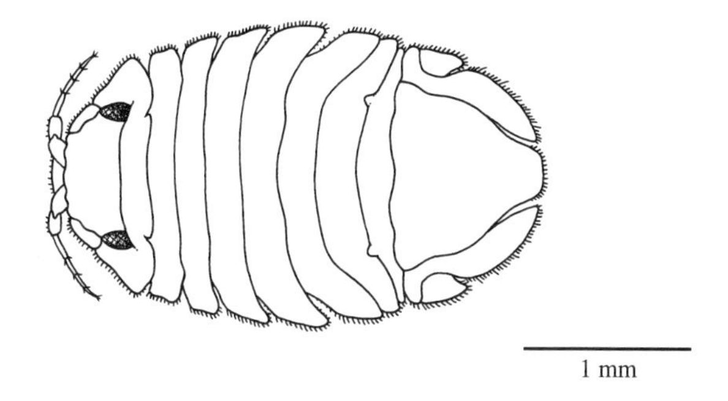

Cyathura polita up to 18 mm

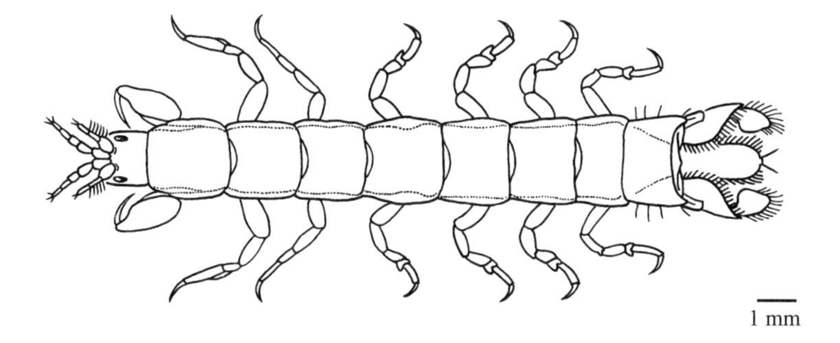

CUMACEANS

ID hint: The rigid carapace appears swollen compared with the slender posterior abdomen ("tail"). Use differences in the shape and the spination of the telson and uropods at the posterior tip of the abdomen (shown as inserts) to distinguish these six species. Male cumaceans differ from females of the same species. They are often larger, have longer antennae, and differ in other respects.

Diastylis polita and *D. sculpta*

The euhaline *Diastylis polita* occurs from Massachusetts to the Chesapeake Bay but is most common in New England. *Diastylis sculpta* occurs from Canada to Long Island, primarily over sandy bottoms. Most members of the genus *Diastylis* have numerous small spines on a pointed telson. Maximum size is about 14 mm. Several other species of *Diastylis* occur within our range, so consult the taxonomic literature listed just prior to this section for definitive species identifications. A European species (*D. rathkei*) is very mobile and considered an important component of the near-bottom zooplankton, especially at night.

References. Anger and Valentin 1976.

Oxyurostylis smithi

Oxyurostylis smithi occurs from Massachusetts south to Florida and throughout the Gulf from the coastal ocean to the mesohaline reaches of estuaries. This species is commonly collected in the water column. Arched dorsal carapace margins and prominent oblique ridges on the sides of the carapaces distinguish members of this genus; however, patterns vary considerably throughout the geographic range. This species has a lightly calcified integument, and its telson lacks an apical spine. Males grow to at least 9 mm. In Alabama, ovigerous *O. smithi* were collected from July through March. Females may brood 40 or more young at a time. Other species of *Oxyurostylis* are common but are seldom reported in the plankton.

Reference. Roccatagliata and Heard 1995.

Leucon americanus

Occurrence. *Leucon americanus* is reported from Massachusetts to northeast Florida and in the northeastern Gulf from Florida to Louisiana, generally in high salinities but occasionally from oligohaline areas. *Leucon* is frequently collected at the surface. In Mobile Bay, Alabama, it only occurred near the inlet, with ovigerous females present from September through March. This cumacean has a flat dorsal margin with weak serrations. It lacks an independent telson, and its inner uropod is shorter than the outer uropod. *Leucon* is uniformly brown and appears very slim when viewed from above or below. Most individuals are <6 mm.

References. Modlin and Dardeau 1987.

Diastylis sp. immature male

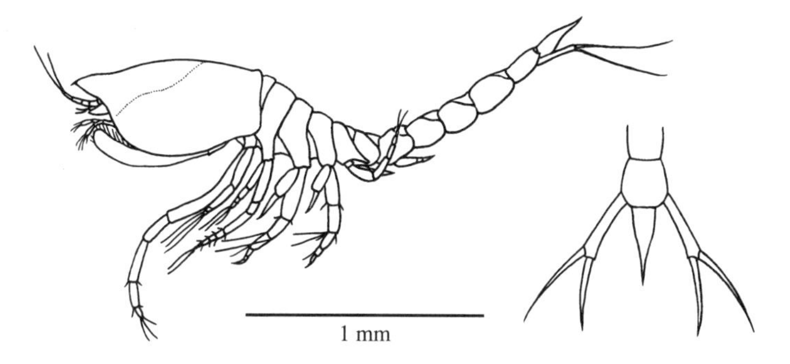

Oxyurostylis smithi male

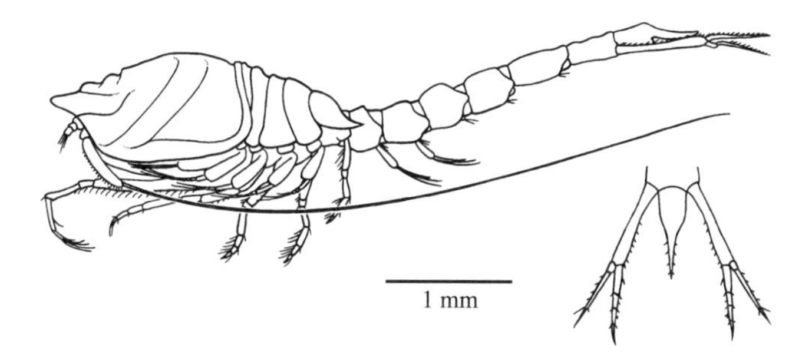

Leucon americanus immature female

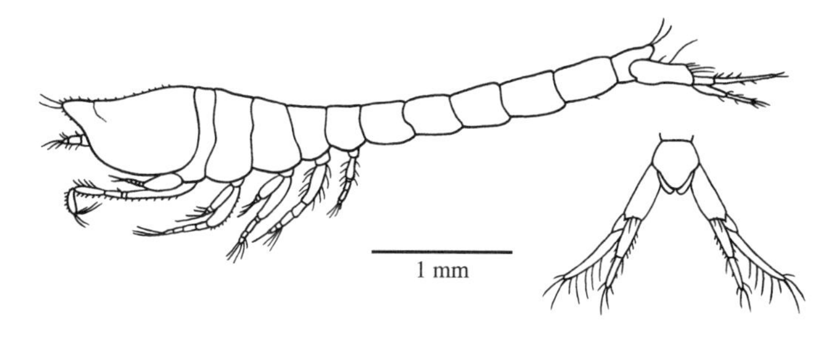

Cyclaspis varians

Occurrence. *Cyclaspis varians* ranges from New England south and through the Gulf of Mexico, primarily in shallow coastal waters or higher-salinity regions of estuaries. It is more common in the night plankton and can be the biomass-dominant animal in nighttime plankton samples in higher-salinity reaches of estuaries on the Florida west coast. Other species of *Cyclaspis* appear occasionally in the Gulf plankton.

References. Modlin and Dardeau 1987.

Mancocuma altera and *M. stellifera*

Mancocuma altera and *M. stellifera* occur along the eastern seaboard from Maine to the St. John's River, Florida. They are common along beaches just beyond the surf zone, especially in the night plankton. *Mancocuma altera* may be found in the mouths of sounds and estuaries in salinities between 15 and 37.

References. Gnewuch and Croker 1973.

Spilocuma watlingi and *S. salomani*

Spilocuma watlingi ranges from North Carolina through the Gulf while *S. salomani* seems restricted to the northern Gulf. In the Atlantic, both species are concentrated just seaward of open beaches. *Spilocuma salomani* prefers more exposed beaches with high wave energy. In the Gulf, *S. watlingi* appears to be confined to protected beaches inside of barrier islands. Like most cumaceans, they are more common in the night plankton.

References. Modlin 1992.

Cyclaspis varians

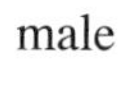
male

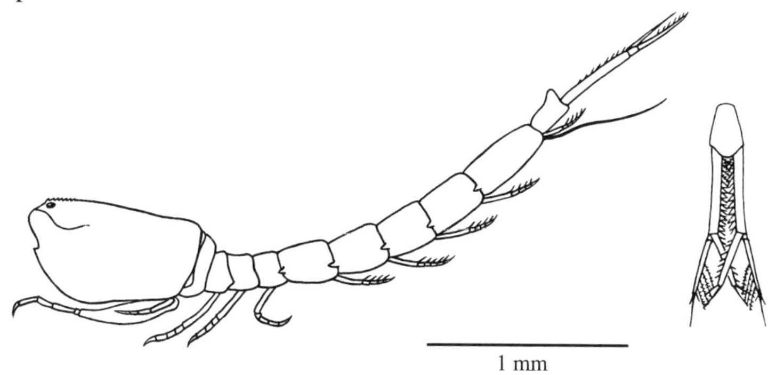

Manococuma altera

male

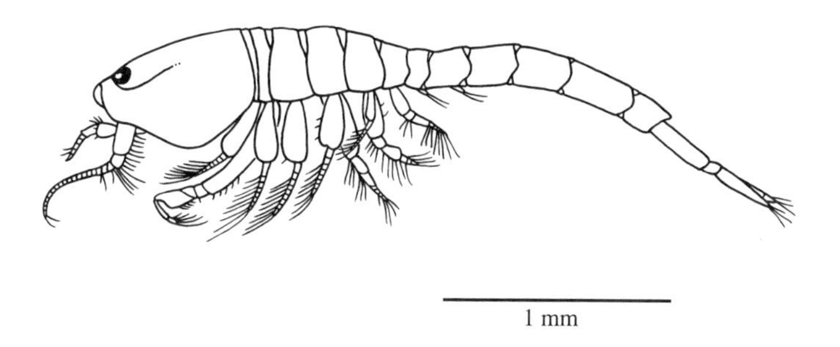

Spilocuma watlingi

male

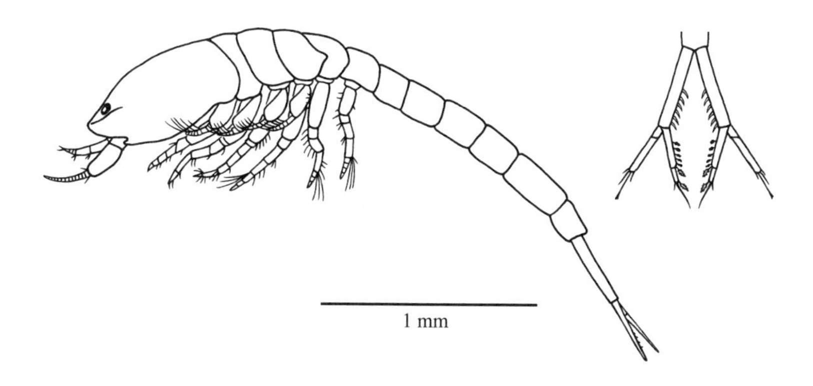

DECAPODS

Shrimps, Crabs, and Related Crustaceans

Although some decapods are planktonic for their entire lives, most spend their adulthood on the bottom. Their planktonic larvae bear little resemblance to the benthic adults, a source of confusion for many decades. Even now, the larval stages of many decapod species have not been described. This is not surprising because the only definitive way to match adults with larvae has required rearing of wild-caught larvae through all larval stages in the laboratory or hatching of eggs from known species of adults to determine their successive larval stages. Only of late has matching of gene sequences offered an alternative approach. Larval distribution and biology are well studied for commercially important species, but information remains scarce for many others.

Decapod life cycles typically involve mating followed by egg deposition and subsequent hatching of planktonic larval stages. Females use pheromones to attract males. Mating often occurs at the female's maturation molt when she is in the "soft shell" state. Once mated, most penaeid shrimps, including the familiar commercial shrimps, release their eggs directly into the water, and a planktonic developmental sequence begins with the nauplius and protozoea stages. In contrast, females of other decapods attach their developing eggs to the setae of their abdominal appendages and brood them until zoea are released into the plankton. Each molt results in an increase in size and in progressive development of appendages, with the more posterior appendages developing last (Fig. 23). As noted in Figure 2, life cycles often include long-distance transport and dramatic changes in habitat during the larval period. The notes on individual species illustrate the complexity of these life histories.

Although there are many variations in the sequences of developmental stages and in the names given to special stages in various decapod groups, the following stages are the most representative of decapods overall. Penaeid and sergestid shrimps have a free-living **nauplius** stage. This stage of development occurs within the egg in all other decapods. The only functional naupliar appendages, used for swimming, arise from segments (somites) of the head region. The nonfeeding decapod nauplius is usually found offshore and should not be confused with the nauplii of copepods or barnacles (see p. 52).

The most characteristic larval stage of decapods is the **zoea.** The now functional thoracic appendages are used in swimming and feeding. Each successive molt gives rise to a sequence of progressively larger larval stages designated zoea I, zoea II, and so on, as the larvae develop. The total number and longevity of zoeal stages for a species may differ,

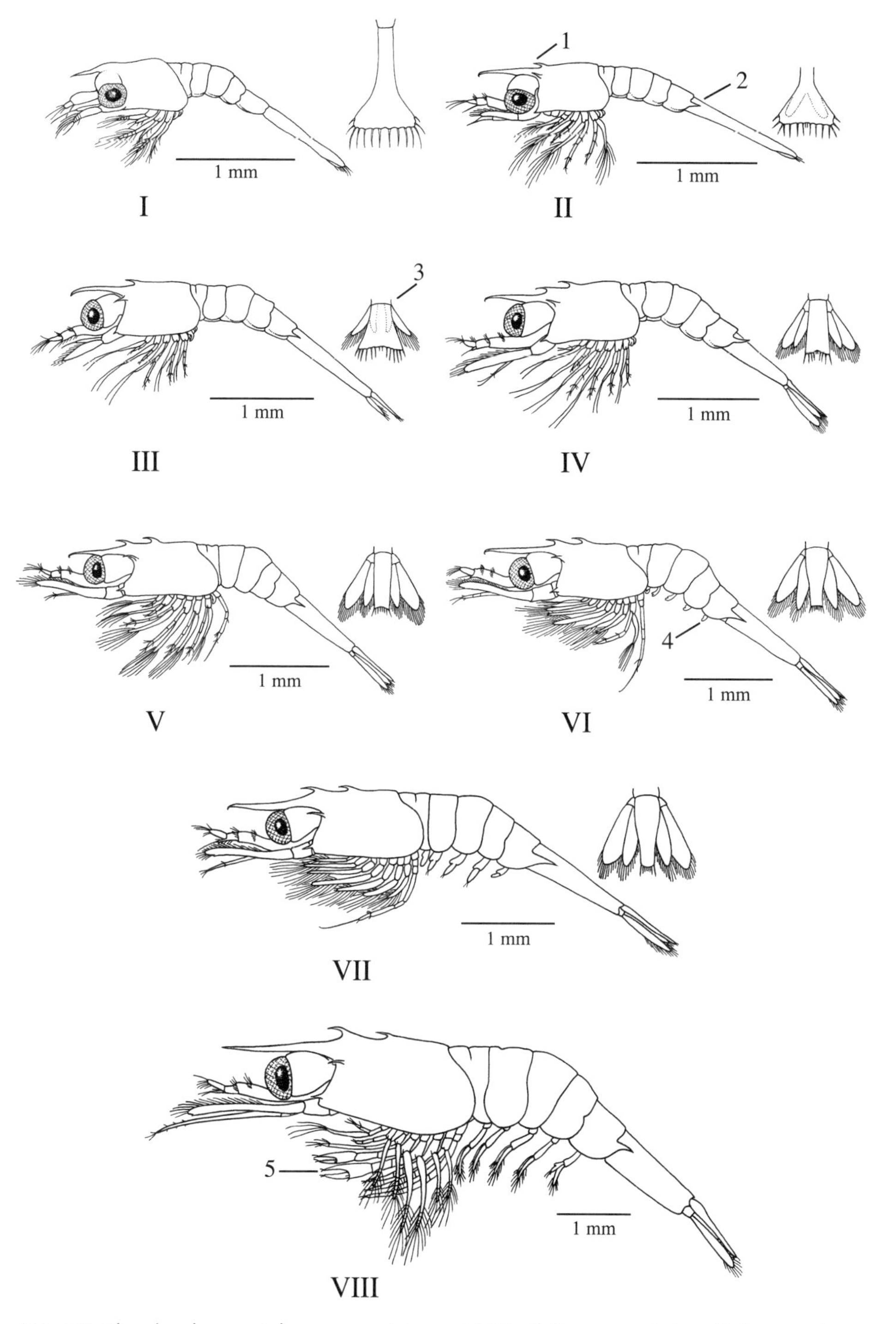

Fig. 23. The developmental sequence (stages I–VII) of the grass shrimp *Palaemonetes* sp. shows the incremental changes that occur in successive molts. In particular, note the addition of teeth on the rostrum (*1*), development of lateral spines on the fifth abdominal segment (*2*), the emergence of lateral tail fins (uropods) (*3*), the gradual development of swimmerets (pleopods) on the abdomen as the relative size of the abdomen increases (*4*), and the appearance of claws on some legs (*5*).

depending on environmental or laboratory culture conditions. The final larval stage is a transitional stage from planktonic to benthic lifestyles and usually bears a much closer resemblance to the adult than to earlier stages. This stage is technically the **decapodid** stage for all decapods, but it is sometimes called the postlarval stage. In contrast to the shrimps, most crabs show a dramatic metamorphosis from the zoeal stage to final larval stage as seen in Figure 24. Pincers, or claws, are used by the increasingly carnivorous decapodids to capture prey. In the Brachyura (true crabs), the last larval, or decapodid, stage is called a **megalopa** or megalops and that of the spiny and slipper lobsters is the **puerulus**. In this section, we use the term: (1) **nauplius** for all prezoeal stages of all decapods that produce them, (2) **zoea** for all but the final stage of all decapods, (3) **postlarva** for final stages of penaeid shrimps and the American lobster, (4) **megalopa** for the final stage of all crabs, and (5) **phyllosoma** for the prepuerulus stage of the spiny and slipper lobsters.

Many decapod larvae are accomplished swimmers. They respond to a variety of environmental stimuli and may undergo vertical migrations. Different groups and different stages often have characteristic swimming positions. For example, some snapping shrimps swim backward with the dorsal side down, while many crab zoeae swim leading with the dorsal spine as the larva faces down. In crabs, the thoracic appendages provide the locomotion for all stages, but abdominal flexions may be used when larvae are startled. In other decapods, abdominal appendages that appear in later larvae take over swimming duties.

Surprisingly few studies have examined prey capture or diets in field-caught larvae. Laboratory studies indicate that zoeal stages are generally omnivorous, wherein phytoplankton, detritus, and zooplankton are consumed. Although early penaeid larvae apparently catch small particles and microalgae in the 1–180 μm size range by suspension feeding, other zoeae and later larval stages grasp individual food particles with thoracic appendages. In some species, prey capture is effected as the abdomen flexes to pin the prey to the ventral side of the larvae near the feeding appendages. Items thus caught are examined and either consumed or released selectively. Decapod larvae become increasingly carnivorous at later stages of development and use their newly developed pincers to capture prey. These larvae are nonvisual feeders that attack virtually anything they contact. Diets include copepods, other decapod larvae, and fish eggs and larvae. Recent observations of feeding by blue crab megalopae when eating microzooplankton show the versatile use of both raptorial and suspension feeding.

Decapod larvae are at great risk from planktivorous fishes and exhibit a variety of antipredator defenses. Many larvae have long spines that may deter some predators. Others have special escape responses when attacked, and some remain close to the bottom during the day and rise at night to avoid visual plankton feeders. Still, many are consumed.

Shrimps and Prawns

Penaeid shrimps include virtually of all the commercial shrimps and prawns. (*Litopenaeus, Farfantepenaeus, Rimapenaeus*)

Sergestid shrimps are the only common holoplanktonic decapods in inshore waters. (*Acetes, Lucifer*)

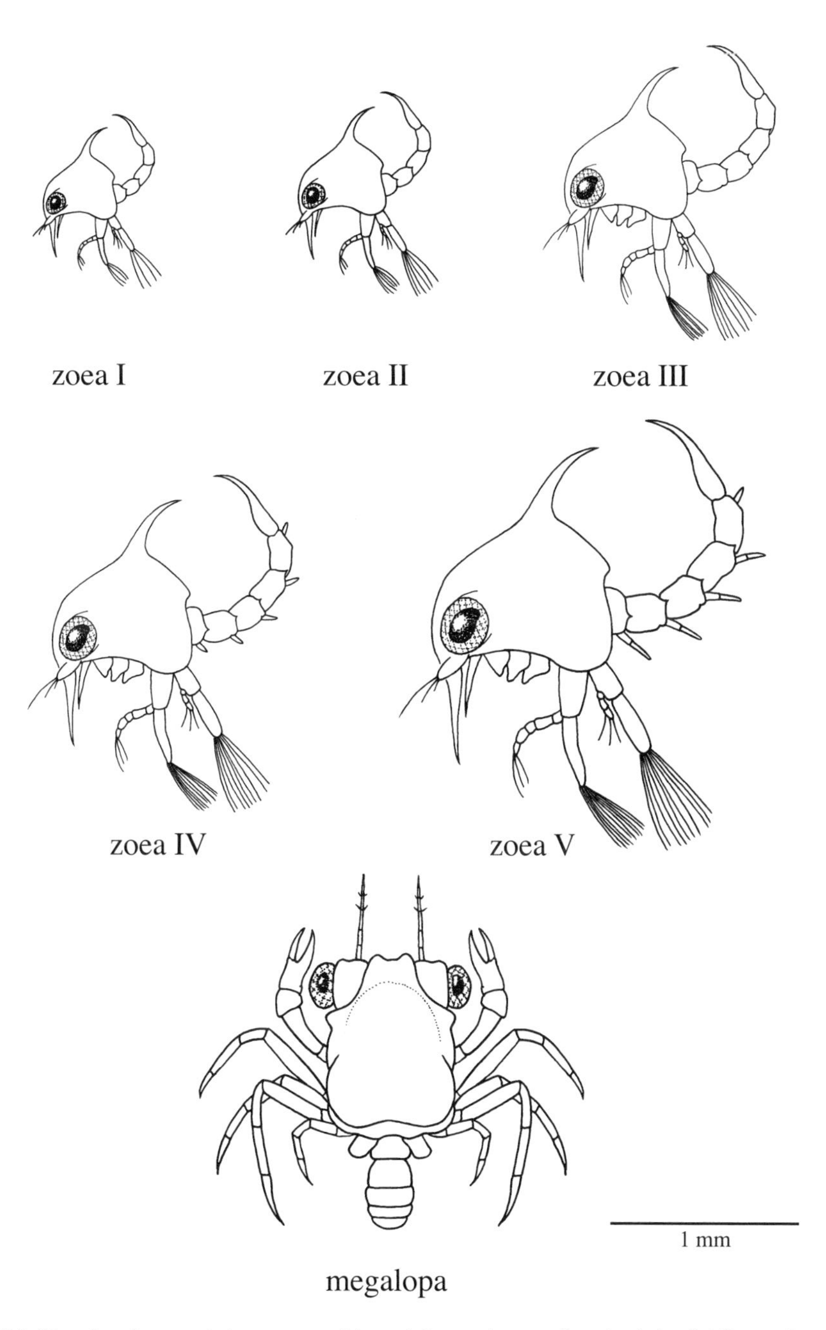

Fig. 24. The developmental sequence (stage I through megalopa) of the fiddler crab *Uca* sp., showing a gradual increase in size and slight increase in limb complexity over the five zoeal stages before the dramatic metamorphosis into the more crablike megalopa (postlarva).

Caridean shrimps comprise a diverse group of smaller shrimps including the snapping shrimps and the common sand and grass shrimps of coastal and estuarine waters. (*Palaemonetes*, *Crangon*, *Hippolyte*, *Alpheus*, *Lysmata*, *Thor*)

Mud and Ghost "Shrimps"

Upogebia, *Callichirus*, *Biffarius*, *Gilvossius*, and *Naushonia* are found in intertidal and shallow subtidal areas, where they make extensive subterranean burrows.

Lobsters

The **American lobster** is the cold-water lobster with large claws (*Homarus*).

Rock lobsters (spiny and slipper lobsters) are warm-water lobsters, lacking claws. (*Panulirus*, *Scyllarides*, *Scyllarus*)

Hermit Crabs, Porcelain Crabs, and Sand, or Mole, Crabs

These three rather different groups all have abdomens loosely joined to the thorax:

Hermit crabs use special hooks on their soft, fleshy abdomen to hold themselves securely inside their snail-shell homes. Usually, the megalopa stage locates and enters vacant snail shells. The final metamorphosis to the first crab stage may be delayed until a suitable shell is found. (*Clibanarius*, *Pagurus*)

Porcelain crabs are small, flattened, or subcylindrical intertidal or shallow water crabs found under rocks or in crevices. Some members have hard, white porcelain-like carapaces, hence the name, but most local species are drab. (*Euceramus*, *Petrolisthes*)

Sand, or **mole**, **"crabs"** are the burrowing crabs found in the surf zone of many beaches. (*Emerita*, *Albunea*, *Lepidopa*)

True Crabs (Brachyura)

This large group includes all of the common commercial crabs plus a variety of smaller ones. All have only a reduced abdomen that folds beneath the thorax. This diverse group includes spider crabs (*Libinia*), rock crabs (*Cancer*), blue crabs (*Callinectes*), mud and stone crabs (*Panopeus, Menippe*), fiddler crabs (*Uca*), and ghost crabs (*Ocypode*).

IDENTIFICATION HINTS

Is it a shrimp, a crab, a lobster, or something else? All decapod larvae are variations on the same basic body plan. Typically, a carapace covering the head and thorax, including the bases of their appendages, obscures segmentation. The posterior abdomen exhibits multiple segments (=somites), in which the last one supports a terminal telson (Fig. 25). Many decapod larvae have distinctive shapes that allow for quick identification to a general category. Some are shrimplike or crablike, whereas others are unique. If overall shape or the presence of certain unusual features, such as long spines, is not immediately apparent, then a more careful look at the following anatomical features illustrated in Figure 25 may be necessary.

If you encounter larvae that do not match those presented here, you may have

- An uncommon species that is not treated in the section. Dozens of additional species occur within our range. Those taxa judged most likely to be encountered are presented. Closely related species may be very similar or markedly different.
- A different developmental stage from a species that is treated here. With a few exceptions, the first stages of shrimps and crabs are presented. These are typically the stages collected nearshore. Look for and try to identify the smallest decapod larvae in your sample first. With a general understanding of how larvae change during development (see Figs. 23 and 24), you may be able to relate later stages to species identified in the guide.
- A regional or geographical variation of the larva illustrated here. Differences in size, spination, pigmentation, and other features have been observed between Atlantic and Gulf populations. Differences can occur within regions too.
- A shrimplike specimen that you are trying to identify could be a mysid shrimp rather than a decapod. See Figure 22 and the identification hints in each section for distinctions between these easily confused but very different groups of shrimps.

For many taxa, we provide notes on similar species and, sometimes, on differences between developmental stages to assist with identification. The technical references listed under the individual taxa and at the end of this section may help, but the larvae of many decapods are undescribed or unknown.

Sizes of decapod larvae are somewhat variable due to differences among populations, locations, and seasons; however, the size of the specimen of interest relative to that of other decapod larvae in the sample can be very useful in narrowing down your choices.

Color or specific pigmentation patterns may appear distinctive on fresh specimens, but color is not usually reliable. Exceptions where coloration is diagnostic are noted.

DIAGNOSTIC FEATURES OF THE HEAD, THORAX, AND CARAPACE

Carapace size should be compared to that of the abdomen. The extent to which the carapace is attached to the thoracic segments and its spination are features that should be inspected first. In some preserved specimens, the posterior portion of the carapace may be lifted away from the thorax, especially in *Pagurus* and *Upogebia.*

The **rostrum**, when present, extends from the front of the head. The length, thickness

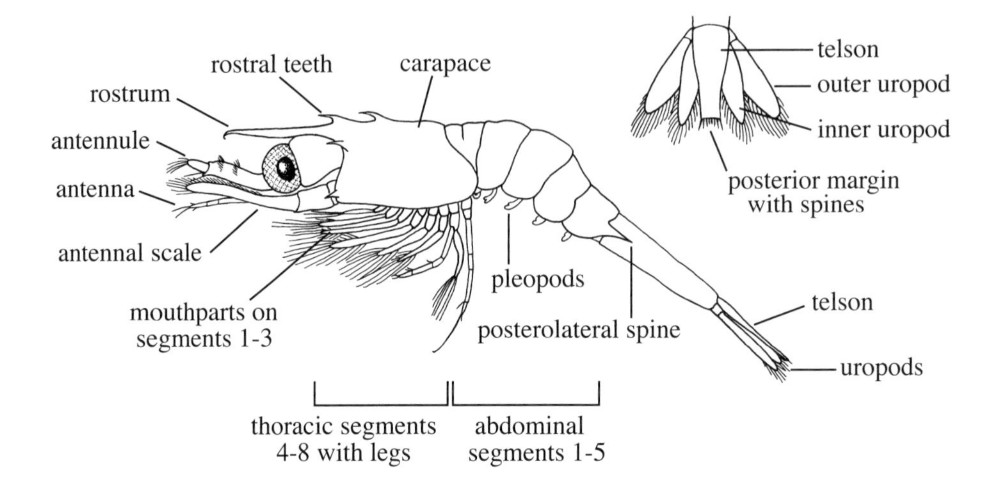

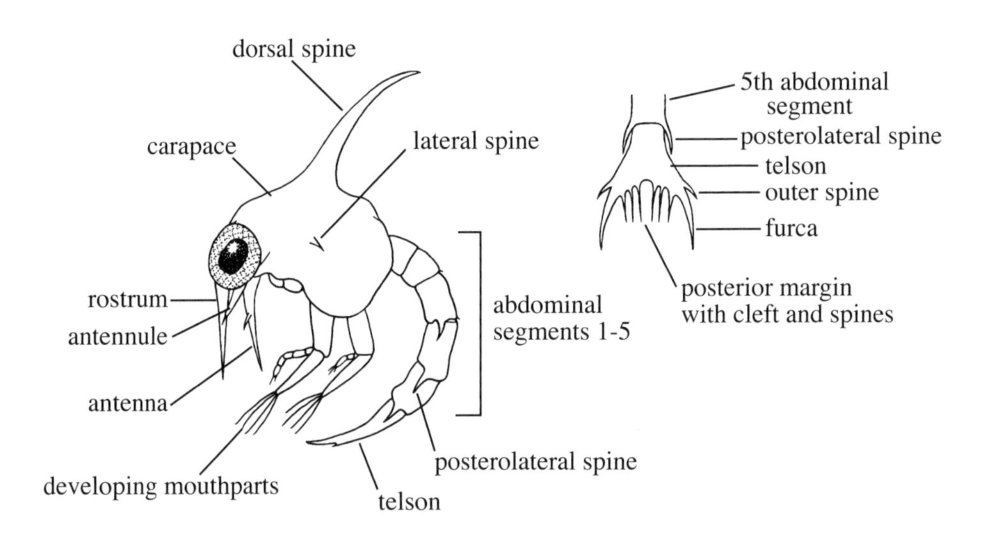

Fig. 25. Anatomical features useful for the identification of decapod larvae as shown in a caridean shrimp zoea (above) and a brachyuran crab zoea (below). Lateral views of both larval stages are supplemented with dorsal views of the telsons and associated structures.

of the base, curvature, and number of teeth may be important features. In some crab zoeae, the antennae appear similar to the rostrum in lateral figures, but note that the rostrum is an unsegmented extension of the carapace.

Antennae positions and lengths relative to the rostrum may be diagnostic.

Dorsal and/or **lateral spines** extending from the carapace are helpful in identifying crab larvae. Their presence or absence, shape, size, and position are often diagnostic.

Legs (off the thorax) are difficult features to use for identification because their numbers change from stage to stage and because they may be confused with developing mouthparts in some early stages. The presence or lack of **claws** on specific legs is especially useful in distinguishing groups of shrimplike larvae.

DIAGNOSTIC FEATURES OF THE ABDOMEN

In crab megalopae, the abdomen is small relative to the carapace and is usually tucked under it and thus not useful for identification. In contrast, the following features of the abdomen are extremely useful in identifying shrimp larvae and crab zoeae:

Dorsal or **posterolateral spines** are diagnostic according to their presence or absence on specific segments. When they are present, note their positions, sizes, and orientations.

Pleopods (abdominal appendages sometimes called swimmerets) appear on the underside of the segments, starting as buds in intermediate stages and becoming multisegmented in later zoeal and megalopal stages. The absence of abdominal appendages usually signals an early-stage larva in both shrimps and crabs.

Telson size, shape, and spination patterns are often species specific. We refer to the presence of large, pointed lateral telson extensions as **furca**. Note the shape of the inner (posterior) margin, the presence of a medial cleft on that margin, and the number and relative lengths of the spines or setae.[1] The shape and spination of the telson usually changes with each successive larval stage.

Uropods occur on both sides of the telson with which they comprise the tail fan of shrimps and lobsters. Each uropod has two branches so that it actually appears that there are two uropods on each side. Uropods usually appear late in the developmental series.

USEFUL IDENTIFICATION REFERENCES

Bullard, S. G. 2003. Larvae of anomuran and brachyuran crabs of North Carolina. *Crustaceana Monographs* 1:1–142.

Ditty, J. G. 2011. Young of *Litopenaeus setiferus*, *Farfantepenaeus aztecus* and *F. duorarum* (Decapoda: Penaeidae): A re-assessment of characters for species discrimination and their variability. *Journal of Crustacean Biology* 31:458–467.

---

1. We sometimes use the term *spine* (without a basal articulation) to refer to spinelike setae (with a basal articulation). This usage should not affect identification.

Kurata, H. 1970. *Studies on the Life Histories of Decapod Crustacea of Georgia.* Part 3, *Larvae of Decapod Crustacea of Georgia.* Final Report. University of Georgia Marine Institute, Sapelo Island. 274 pp. (Contains some rarer species not covered here.)

Maris, R. C. 1983. A key to the porcellanid crab zoeae (Crustacea: Decapoda: Anomura) of the north central Gulf of Mexico and a comparison of meristic characters of four species. *Gulf Research Reports* 7:237–246.

Martin, J. W. 1984. Notes and bibliography on the larvae of xanthid crabs, with a key to the known xanthid zoeas of the western Atlantic and the Gulf of Mexico. *Bulletin of Marine Science* 34:220–239. (Keys but no figures.)

Martin, J. W., Davis, G. E. 2001. An updated classification of the Recent Crustacea. Science series, *Natural History Museum of Los Angeles County* 39:1–124.

Pohle, G., Mantellato, F. L. M., Negreios-Fransozo, M. L., et al. 1999. Larval Decapoda (Brachyura). In: Boltovskoy, D., ed. *South Atlantic Zooplankton* (pp. 1281–1351). Vol. 2. Backhuys, Leiden, The Netherlands.

Sandifer, P. A. 1972b. Morphology and ecology of Chesapeake Bay decapod larvae. Ph.D. diss., University of Virginia, Charlottesville. 531 pp. (University Microfilms)

SUGGESTED READINGS

Anger, K. 2001. *Biology of Decapod Crustacean Larvae.* A. A. Balkema, Exton, PA. 419 pp.

Forward, R. B., Jr., Tankersley, R. A., Rittschof, D. 2001. Cues for metamorphosis of brachyuran crabs: An overview. *American Zoologist* 41:1108–1122.

Garrison, L. P. 1999. Vertical migration behavior and larval transport in brachyuran crabs. *Marine Ecology Progress Series* 176:103–113.

McConaugha, J. R. 1992. Decapod larvae—dispersal, mortality, and ecology: A working hypothesis. *American Zoologist* 32:512–523.

Morgan, S. G. 1987. Morphological and behavioral antipredatory adaptations of decapod zoeae. *Oecologia* 73:393–400.

Morgan, S. G., Christy, J. H. 1995. Adaptive significance of the timing of larval release by crabs. *American Naturalist* 145:457–479.

Queiroga, H., Blanton, J. O. 2004. Interactions between behavior and physical forcing in the control of horizontal transport of decapod larvae. *Advances in Marine Biology* 47:107–214.

Williamson, D. I. 1982. Larval morphology and diversity. In: Abele, L. G., ed. *The Biology of Crustacea: Embryology, Morphology, and Genetics.* Vol. 2. Academic Press, New York, 43–100.

Young, C. M., ed.; Sewell, M. A., Rice, M. E., assoc. eds. 2001. *Atlas of Marine Invertebrate Larvae.* Academic Press, New York. 656 pp.

PENAEID SHRIMPS

ID hint: All of the shrimps on this page were in the genus *Penaeus* until reclassification. Because the larvae of these three species are virtually indistinguishable, one example of each of three distinct larval stages is shown. Use the notes on seasonal occurrence, range, and habitat to arrive at a tentative species designation. See Ditty (2011) for assistance with identification to species.

Litopenaeus setiferus (white shrimp), formerly *Penaeus setiferus*

Occurrence. Adult white shrimp occur Massachusetts through Texas, but centers of abundance throughout South Carolina to northern Florida and in the central and southern Gulf.

Biology and Ecology. Spawning in white shrimp occurs from May through at least September in the Southeast region and begins as early as March in the Gulf. White shrimp spawn in the ocean but closer to shore than the other commercial shrimps, so the earlier larval stages may be in the nearshore plankton. The planktonic phase of larval development lasts about 10–12 days. Entrance of postlarvae (5–10 mm) into the bays and estuaries starts in June in the Carolinas and in April to May in Georgia and Florida. White shrimp postlarvae often migrate farther into low-salinity water than pink or brown shrimps, sometimes into <1 psu. Postlarvae prefer to settle in vegetated areas, either in submerged grass beds or in salt marshes.

References. Anderson et al. 1949; Baxter and Renfro 1967; Muncy 1984a, 1984c; Pattillo et al. 1997; Pérez-Farfante 1969; Williams 1984.

Farfantepenaeus duorarum (pink shrimp), formerly *Penaeus duorarum*

Occurrence. Adults occur from Virginia to Florida and throughout the Gulf, but they are only abundant along the Florida west coast. Relatively isolated concentrations occur near Beaufort, North Carolina (and the nearby sounds), and in hypersaline Texas lagoons.

Biology and Ecology. Most larval stages remain well offshore until the summer influx of postlarvae into inlets. Late larvae (3–7 mm) recruit to the Southeast estuaries from June to November, but abundances peak when temperatures exceed 24°C and salinities are above 12. Larvae preferentially settle in seagrasses. Development from nauplius to postlarva takes 15–20 days.

References. Bielsa et al. 1983; Dobkin 1961; Ewald 1965; Pattillo et al. 1997; Williams 1984.

Farfantepenaeus aztecus (brown shrimp), formerly *Penaeus aztecus*

Occurrence. Larvae are common in nearshore waters from Virginia to south Florida on the Atlantic Coast and the northern and western Gulf. Early larval stages are uncommon nearshore.

Biology and Ecology. Spawning occurs off the coast. In the Southeast, the large (8–12 mm) postlarvae recruit into bays and estuaries in spring well before white and pink shrimp postlarvae arrive. In the Gulf, maximum recruitment into coastal areas occurs in June and September. Brown shrimp postlarvae are less abundant in salinities <10.

References. Benfield and Downer 2001; Cook and Murphy 1971; Criales et al. 2007; Larson, et al. 1989; Lassuy 1983a; Pattillo et al. 1997; Rogers et al. 1993; Williams 1984.

penaeid shrimp nauplius

stage III (of V)

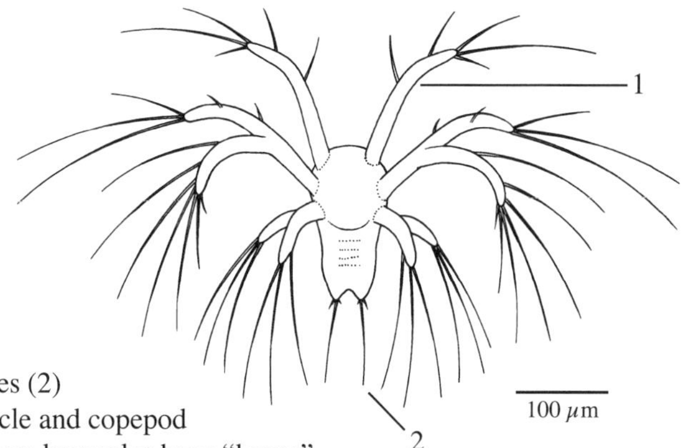

- 3 pairs of plumose appendages (1)
- prominent furcal spines (2)
- similar species: barnacle and copepod nauplii have shorter legs; barnacles have "horns"

---

penaeid shrimp zoea

stage III (of V)

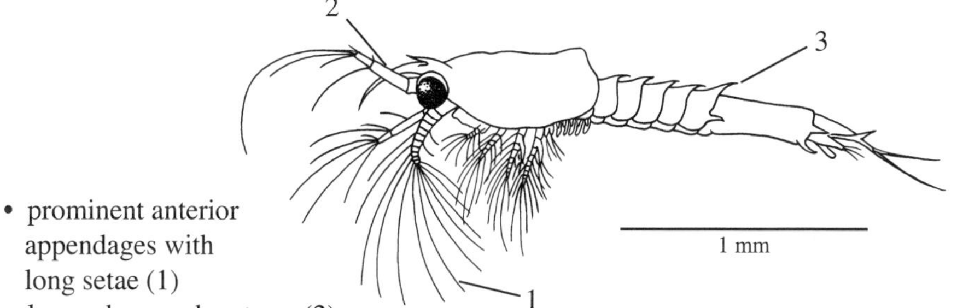

- prominent anterior appendages with long setae (1)
- large, decurved rostrum (2)
- dorsal processes on the abdominal segments (3)
- similar species: most other shrimp zoeae do not have such plumose antennae and mouthparts, a curved and toothed rostrum, and dorsal spines on abdominal segments

---

penaeid shrimp postlarva (*Litopenaeus setiferus*)

stage II (of II)

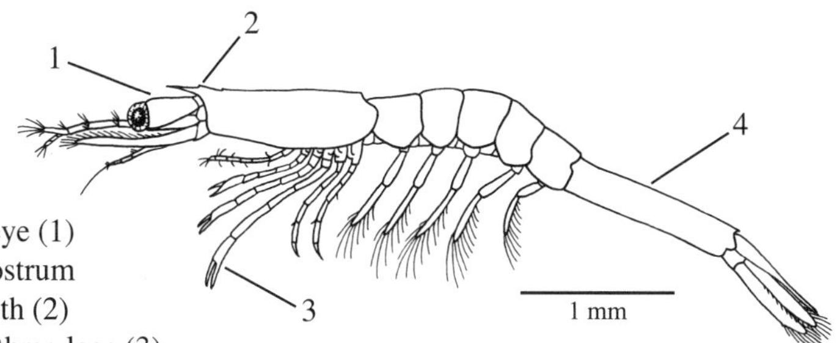

- long, stalked eye (1)
- straight thin rostrum with dorsal teeth (2)
- claws on first three legs (3)
- last abdominal segment very long (4)
- similar species: very difficult to distinguish the various penaeid shrimps without close inspection of spines (see Pearson 1939, Ringo and Zamora 1968)

Rimapenaeus constrictus (roughneck shrimp), formerly _Trachypenaeus constrictus_, and _R. similis_ (roughback shrimp)

Occurrence. Adults of both species occur from North Carolina to Texas with especially high larval abundances reported from Tampa Bay and from Mississippi Sound. Larvae are present from spring through fall.

Biology and Ecology. Spawning occurs offshore in spring and summer in the Carolinas but may be year-round in more southern areas. Peak recruitment of postlarvae in South Carolina occurs in midsummer, but they are found in high-salinity areas throughout the warm season. Larvae of these two species are virtually indistinguishable.

SERGESTID SHRIMPS

Acetes americanus carolinae

Occurrence. Adults occur from Virginia to Texas, although ocean currents may carry them to higher latitudes during summer and fall. They occur year-round in offshore waters but are especially abundant nearshore and in estuaries in warm seasons. Although *Acetes americanus carolinae* is typically coastal, it may be common around inlets and within estuaries with high salinities (>15). Densities are often greater in the lower portion of the water column.

Biology and Ecology. *Acetes* is a holoplanktonic shrimp. Adults spawn in the ocean, apparently during the warmest months. Unlike *Lucifer*, below, *Acetes* sheds its eggs. Larval stages generally resemble the penaeid shrimps but are smaller and less likely to be encountered in estuaries.

References. Oshiro and Omori 1996; Williams 1984.

Lucifer faxoni

Occurrence. Adults are found nearshore and in high-salinity inlets from Rhode Island to Texas. *Lucifer* occurs more commonly summer and fall in the northern part of its range and April to December in South Carolina. Occasionally, it occurs in salinities as low as 16, but it prefers salinities >30.

Biology and Ecology. While most penaeids shed their eggs directly into the water, *Lucifer* attaches its eggs to the third walking legs until nauplii hatch. The larval stages of this holoplanktonic species are typically farther offshore than adults but were collected in the lower Chesapeake Bay in the fall. Daily vertical migrations may be modified to effect shoreward migrations. In the laboratory, adults are predators, but nothing is known of their feeding habits in the field.

References. Brooks 1882; Harper 1968; Hashizume 1999; Woodmansee 1966.

Rimapenaeus constrictus

postlarva

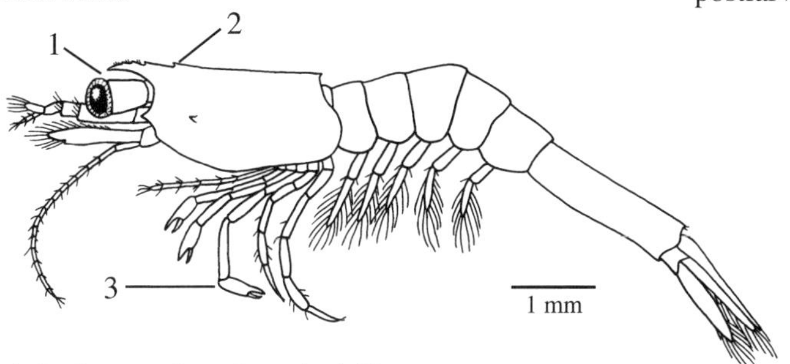

- stalked eye (1)
- rostrum stout, slightly decurved, and toothed (2)
- claws on front legs (3)
- similar species: *Rimapenaeus similis*; other penaeid postlarvae are not as robust and have longer abdominal segment 6

Acetes americanus carolinae

adult

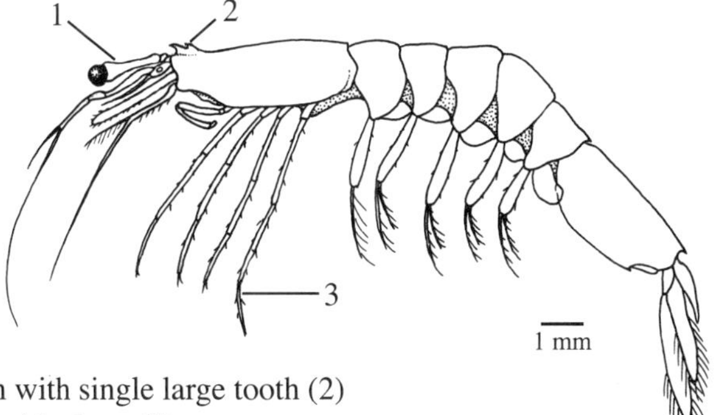

- very long stalked eye (1)
- very short rostrum with single large tooth (2)
- only 4 legs, none with claws (3)
- similar species: penaeid postlarvae with more slender abdominal segment 6, longer rostra, shorter eyestalks, and 5 legs, including some with claws

---

Lucifer faxoni

adult male

- carapace extremely long (1)
- body very thin and compressed laterally
- rostrum small and without teeth (2)
- spine on side of carapace near mouthparts (3)
- large ventral teeth on abdominal segment 6 in males; reduced in females (4)
- similar species: often confused with *Acetes*, which has deep carapace, toothed rostrum, and no spines on abdominal segment 6. Penaeid postlarvae have a long rostrum and deep carapace

CARIDEAN SHRIMPS

Palaemonetes spp. (grass shrimps)

Occurrence. *Palaemonetes pugio* and *P. intermedius* adults are among the most common and conspicuous shrimps in shallow coastal areas from Massachusetts to Texas. *Palaemonetes vulgaris* has a similar range but is not reported from the Florida Keys. Larvae of *Palaemonetes* species are widespread along the Atlantic and Gulf Coasts and in estuaries (2–35 psu), where they may be the dominant decapod shrimp larvae in summer in salinities >15.

Biology and Ecology. In larger Atlantic estuaries, all larval stages remain within the estuary. In some high-salinity marsh systems near the coast, early larvae are exported through inlets. They return as late-stage larvae as they recruit to the creeks and salt marshes. In tidal creeks, early stages are more abundant near the surface, especially after nocturnal high-tide hatches, whereas late stages are more common near the bottom. Variation in the numbers of larval stages and in their morphology makes identifying larvae to species difficult. Current work using DNA techniques may soon resolve taxonomic uncertainty within this genus. Refer to Figure 23 for the full larval sequence.

References. Anderson 1985; Broad 1957a; Pattillo et al. 1997; Sandifer 1973c; Williams 1984.

Macrobrachium spp. (river shrimps)

Occurrence. River shrimps are widely distributed in freshwater reaches of coastal rivers with access to brackish areas. The most abundant, *Macrobrachium ohione*, occurs from the James River, Virginia, southward along the Atlantic Coast and in the northern and western Gulf. *Macrobrachium olfersii* and *M. acanthurus* occur sporadically from North Carolina through the Gulf, whereas *M. carcinus* is found only in Gulf estuaries. Several other species of *Macrobrachium* have more limited distributions.

Biology and Ecology. River shrimps typically live in freshwater. In summer, gravid females usually migrate to the brackish waters to release their larvae since development seems most successful at salinities >15. Larvae swim most often on flooding tides, which prevents them from being swept out to sea and also assists with their migration back to fresh water as they grow. The only exception is *M. ohione*, which can complete its larval development in enclosed bodies of freshwater. In the laboratory, larvae thrive on brine shrimp larvae and are presumably largely carnivorous in the wild.

References. Bauer and Delahoussaye 2008; Dobkin 1971; Hughes and Richard 1973; Reimer et al. 1974; Rome et al. 2009; Truesdale and Mermilliod 1979.

Cuapetes americanus, formerly *Periclimenes* spp.

Occurrence. *Cuapetes americanus* is representative of a large group of related shrimps that is most diverse in south Florida, although a number of species occur from North Carolina south to Florida and through the Gulf, predominately in higher salinities. *Urocaris longicaudata* (formerly *Periclimenes longicaudatus*) is a year-round resident of estuaries in the Southeast to the Florida Keys, where it resides in high-salinity, subtidal bottoms. It also occurs in the Gulf.

Biology and Ecology. Adult *Cuapetes* are associated with coral reefs and with turtle grass beds. The planktonic stages have not been widely studied. Ovigerous females and larvae occur in summer.

Palaemonetes sp.

zoea stage I

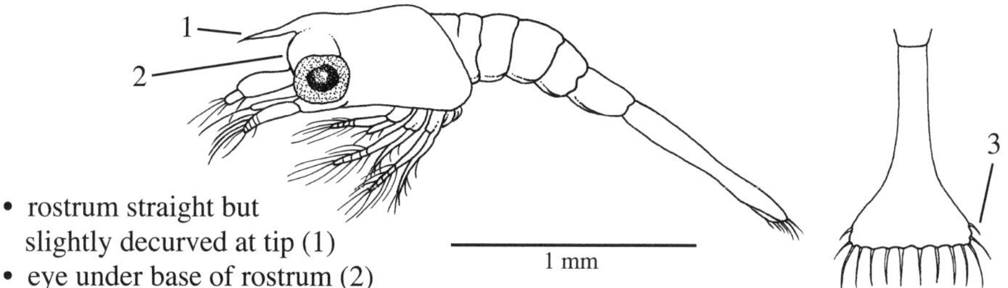

- rostrum straight but slightly decurved at tip (1)
- eye under base of rostrum (2)
- telson triangular with 10 similar setae along the straight posterior margin and two shorter spines at each corner (3)
- similar species: larvae of *Palaemonetes* species are very difficult to separate; relatively large overall size, long straight rostrum and lack of cleft on telson distinguishes stage I from most other shrimps
- see Figure 23 for entire larval sequence

Macrobrachium acanthurus

zoea stage I

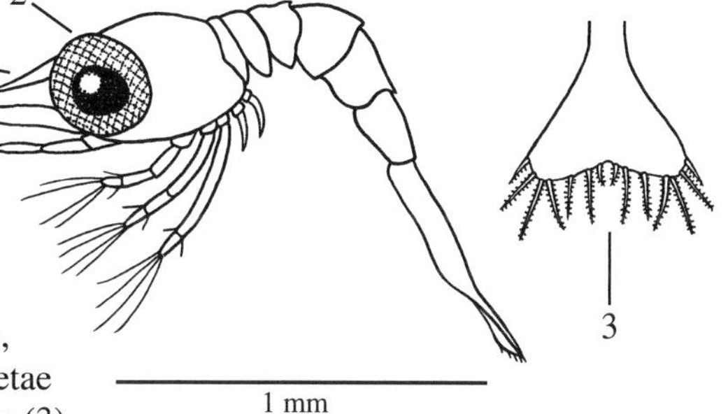

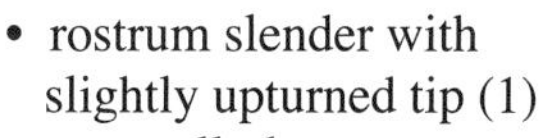

- rostrum slender with slightly upturned tip (1)
- unusually large eye obscures anterior detail (2)
- telson triangular with broad, slightly concave posterior margin, notched in the middle; plumose setae of various lengths and orientations (3)
- similar species: other *Macrobrachium* spp. are similar, but larvae have not been described for many species; *Palaemonetes*' telson has straight margin with uniform spines; telson spines on most other shrimps are not plumose

---

Cuapetes americanus

zoea stage I

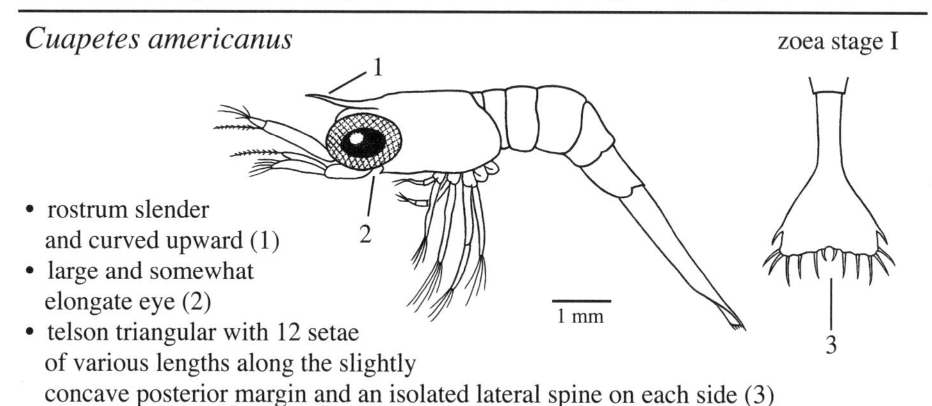

- rostrum slender and curved upward (1)
- large and somewhat elongate eye (2)
- telson triangular with 12 setae of various lengths along the slightly concave posterior margin and an isolated lateral spine on each side (3)
- similar species: *Palaemonetes* has straight rostrum and telson margin; other shrimps with slightly concave telson margins do not have distinctly isolated lateral spines

Hippolyte spp. (broken-back shrimps)

Occurrence. Of the several species of *Hippolyte* in the western Atlantic, *Hippolyte pleuracanthus* is most common from Massachusetts to south Florida. *Hippolyte obliquimanus* occurs from North Carolina to Florida. *Hippolyte zostericola* has the widest distribution, occurring from Massachusetts to Texas, with the possible exception of the Florida Keys.

Biology and Ecology. In North Carolina, ovigerous females of *H. pleuracanthus* occur from May to October with shorter spawning seasons to the north and longer ones to the south, where they may produce larvae all year long. *Hippolyte pleuracanthus* larvae are common in lower Chesapeake Bay from May to November at 14°–28°C and 15–31 psu. Early stages are more abundant in bottom than surface waters, and later stages occurred almost exclusively in bottom collections.

References. Shield 1978.

Alpheus spp. (snapping shrimps)

Occurrence. More than 20 species of *Alpheus* and other closely related shrimps occur in the Atlantic and Gulf. *Alpheus heterochaelis, A normanni,* and *A. packardii* are among the most common; they occur from Chesapeake Bay to Texas. Ovigerous females of both species occur throughout the year in all but the northern part of their ranges, where they occur from May to October. *Alpheus* spp. larvae occur from April through November in South Carolina estuaries.

Biology and Ecology. *Alpheus heterochaelis* has a variable number of zoeal stages, and larval development is designated as either extended or abbreviated (where some zoeal stages are lacking). The positively phototactic larvae swim backward (tail first) and ventral side up. Thoracic appendages provide the propulsion for normal swimming, but startled larvae respond with rapid backward movements produced by repeated abdominal flexions. In the lab, late larvae usually remain near the bottom. In both lower Chesapeake Bay and North Carolina estuaries, *Alpheus* larvae occur in salinities >20 in summer.

References. Knowlton 1973.

Crangon septemspinosa (sevenspine bay shrimp)

Occurrence. *Crangon* occurs along the Atlantic Coast from Canada to Florida but is most common from Virginia north. *Crangon* larvae may be the only decapods common in the winter plankton along the Mid-Atlantic Coast, where they occur from January to May. Farther north, larvae peak in spring and in summer. Larvae prefer salinities >20.

Biology and Ecology. Peak abundances sometimes exceed 700 m^{-3} in spring from the Mid-Atlantic north. Larvae have been collected every month in the lower Chesapeake Bay. The most common planktonic stage is the first zoea, often found near the surface. Adult *Crangon* play a major role in estuarine and coastal food webs, but the role of planktonic larvae is unknown.

References. Tesmer and Broad 1964; Wehrtmann 1991.

Hippolyte pleuracanthus

zoea stage I

- rostrum short with heavy base and no teeth (1)
- carapace anterior ventral margin serrated (2)
- prominent posterolateral spine on abdominal segment 5 (3)
- end of telson somewhat rectangular with long corner spines (4)
- widely reported to be richly colored (red)
- similar species: *Upogebia* has decurved rostrum and cleft telson

---

Alpheus heterochaelis

zoea stage II

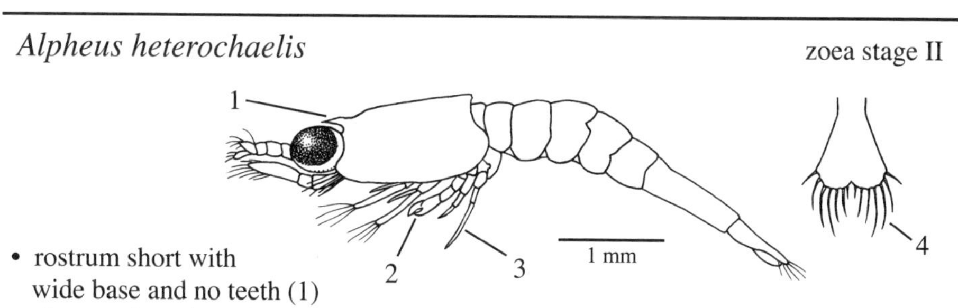

- rostrum short with wide base and no teeth (1)
- claw on front legs (2)
- branch on last leg unusually long (3)
- telson with 2 lobes that form medial cleft on posterior margin; longest setae in middle of lobes (4)
- similar species: the combination of short rostrum and claws at an early stage distinguishes *Alpheus* (and some closely related genera) from most other shrimps

---

Crangon septemspinosa

zoea stage I

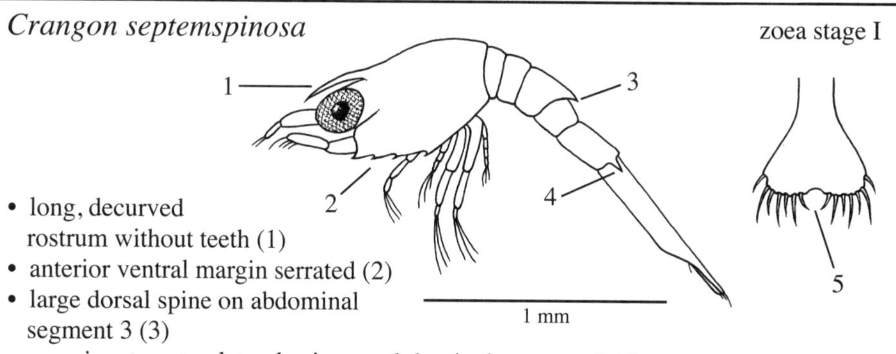

- long, decurved rostrum without teeth (1)
- anterior ventral margin serrated (2)
- large dorsal spine on abdominal segment 3 (3)
- prominent posterolateral spine on abdominal segment 5 (4)
- telson with short inwardly curved setae in center of posterior margin (5)
- similar species: ghost shrimps have a dorsal spine on the second abdominal segment; *Homarus* has spines on all segments

Lysmata wurdemanni (peppermint shrimp)

Occurrence. Adults occur from New Jersey to Texas along with several other less common species, especially around structures and rocky or shelly bottoms.

Biology and Ecology. Ovigerous females are most abundant during warm months. Reproduction may involve simultaneous hermaphroditism. Larvae are released at night and often swim just below the surface film. When disturbed, they dart away using rapid, repeated flexions of the abdomen.

References. Calado et al. 2004; Lin and Zhang 2001.

Thor spp.

Occurrence. Adults of *Thor floridanus*, *T. dobkini*, and *T. manningi* occur from North Carolina to Florida and along the Gulf Coast to Texas, especially in areas with extensive grass beds.

Biology and Ecology. Ovigerous females occur in North Carolina from May through July and increasingly later at more southern locations. Egg-bearing females occur year-round in some areas. Larvae have been raised from 11° to 31°C in the laboratory. Information on the distribution of larvae in the field is scarce.

References. Broad 1957b; Dobkin 1962, 1968; Sandifer 1972a.

HERMIT CRABS

Clibanarius vittatus (thinstripe hermit)

Occurrence. Adult *Clibanarius vittatus* occur from Virginia to Texas in shallow-water habitats. Larvae are common in the Gulf at least from Louisiana to Texas in warm seasons and probably occur throughout the adult range. Several other *Clibanarius* species are also common in the tropical part of this range.

Biology and Ecology. *Clibanarius* spawns in estuarine waters, but it may complete larval development offshore. The megalopae can delay final metamorphosis until a suitable snail shell is found. Total developmental time may be as long as three months.

References. Harms 1992; Harvey 1996; Truesdale and Andryszak 1983; Ziegler and Forward 2006.

Lysmata wurdemanni

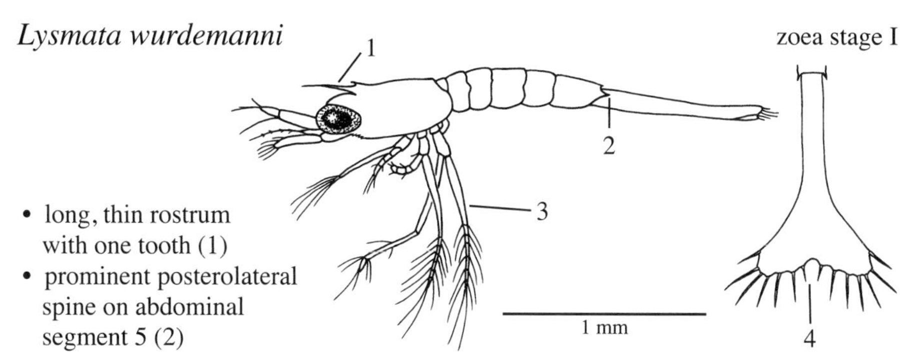

- long, thin rostrum with one tooth (1)
- prominent posterolateral spine on abdominal segment 5 (2)
- very long branches on last 2 legs (3)
- telson with very long neck; posterior margin with cleft and short straight setae in center (4)
- similar species: other *Lysmata* species; slender body and long hind legs are rather unique to this group

Thor sp.

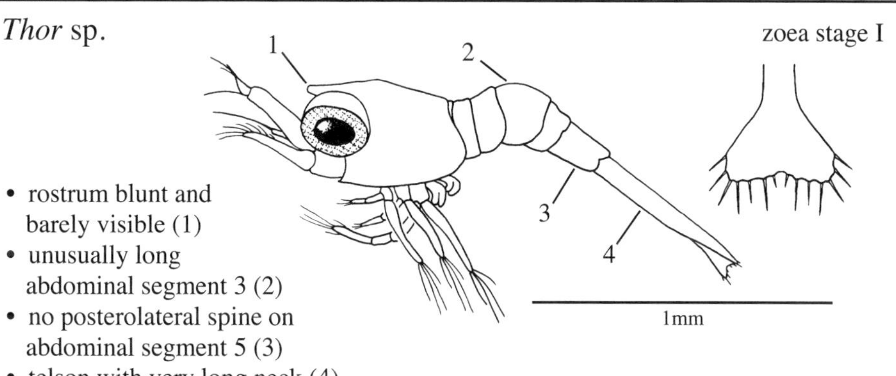

- rostrum blunt and barely visible (1)
- unusually long abdominal segment 3 (2)
- no posterolateral spine on abdominal segment 5 (3)
- telson with very long neck (4)
- similar species: other species of *Thor* and *Cuapetes* have a wide abdominal segment 3; other *Thor* species may have a rostrum

Clibanarius vittatus

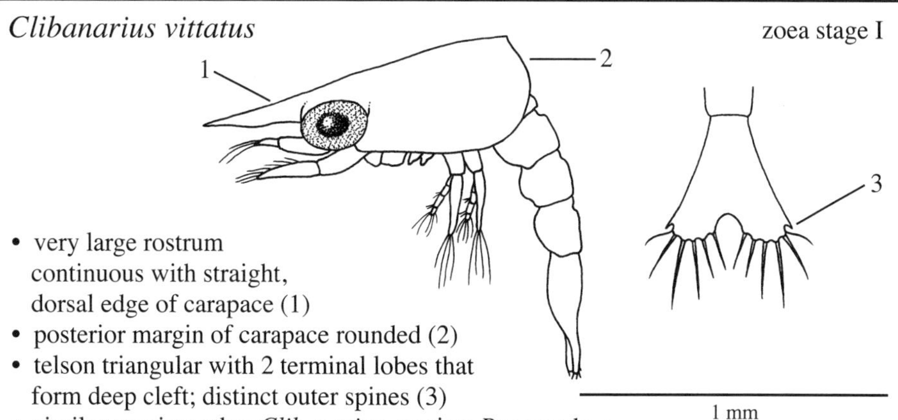

- very large rostrum continuous with straight, dorsal edge of carapace (1)
- posterior margin of carapace rounded (2)
- telson triangular with 2 terminal lobes that form deep cleft; distinct outer spines (3)
- similar species: other *Clibanarius* species; *Pagurus* has posterolateral spines on carapace and abdominal segments

Pagurus spp.

Occurrence. More than two dozen species of *Pagurus* and related genera occur within our range. Some have restricted distributions in cold or warm areas, but some are widely distributed. *Pagurus longicarpus* (bluespine hermit) is one of most abundant and widespread of the shallow water hermit crabs, occurring from Massachusetts to Texas. *Pagurus pollicaris* has a similar range.

Biology and Ecology. In lower Chesapeake Bay, larvae of *P. longicarpus* were found from May to November, with most collected at salinities >20 and near the bottom. The megalopa settles from the plankton and actively searches for vacant snail shells. If no suitable shell is found, metamorphosis may be delayed.

References. Harvey 1996; Nyblade 1970; Roberts 1970, 1971a, 1971b.

MUD SHRIMPS

Naushonia crangonoides (Naushon mud shrimp)

Occurrence: Although adults are uncommon compared with other ghost and mud shrimps in nearshore waters along our coasts, the larvae appear in summer plankton collections along the coasts and in higher-salinity (>25) reaches of estuaries from Massachusetts to the mouth of the Chesapeake Bay. Adults are reported as far south as Georgia.

Biology and Ecology. Because larvae have been raised in the laboratory on newly hatched *Artemia*, they are presumably carnivorous and may feed on small crustaceans in the wild.

Reference. Goy and Provenzano 1978.

Upogebia affinis (coastal mud shrimp)

Occurrence. Adults occur from Massachusetts to Texas. They are common in coastal lagoons and in estuaries in salinities as low as 15. In South Carolina estuaries, larvae are produced from April through November.

Biology and Ecology. Ovigerous females occur year-round in the southern waters. Larvae are common during the warmest months in the northern part of the range. In lower Chesapeake Bay, all stages have been collected, with most occurring near the bottom. Recent research on an African species of *Upogebia* found that zoea I is facultatively lecithotrophic (nourished by yolk alone). It can go without food until it molts but will feed if it has food, and this feeding accelerates development. Feeding and behavior in the field are unreported.

References. Ngoc-Ho 1981; Sandifer 1973a; Truesdale and Andryszak 1983.

Pagurus longicarpus

zoea stage I

- large rostrum uniformly deep along most of length; with distinct ventrally directed curve (1)
- carapace wide in dorsal view, with strong posterolateral spine at center of posterior margin (2)
- large spine on last abdominal segment (3)
- telson triangular with furcal spines and fairly straight posterior margin (4)
- similar species: many other *Pagurus* species; separated from *Clibanarius* species, which do not have spines on carapace or abdomen

Naushonia crangonoides

zoea stage I

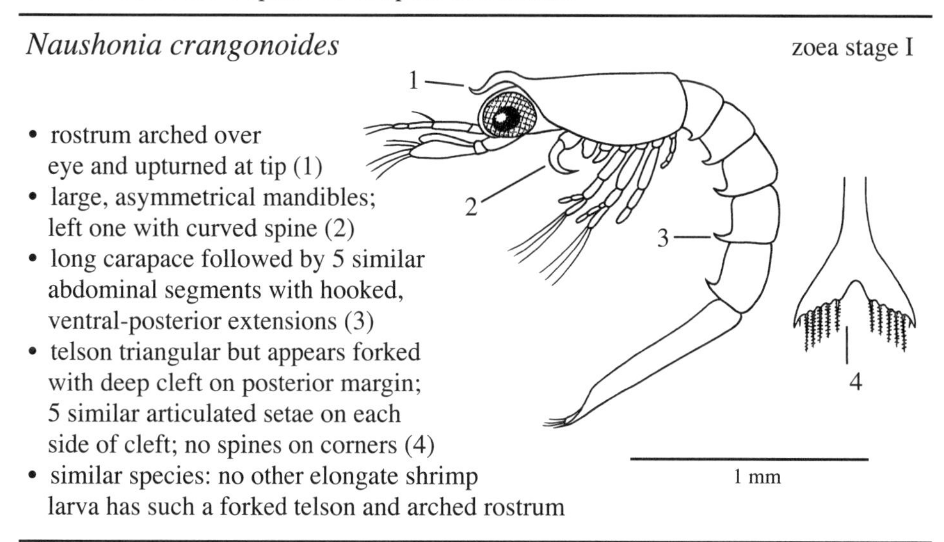

- rostrum arched over eye and upturned at tip (1)
- large, asymmetrical mandibles; left one with curved spine (2)
- long carapace followed by 5 similar abdominal segments with hooked, ventral-posterior extensions (3)
- telson triangular but appears forked with deep cleft on posterior margin; 5 similar articulated setae on each side of cleft; no spines on corners (4)
- similar species: no other elongate shrimp larva has such a forked telson and arched rostrum

Upogebia affinis

zoea stage I

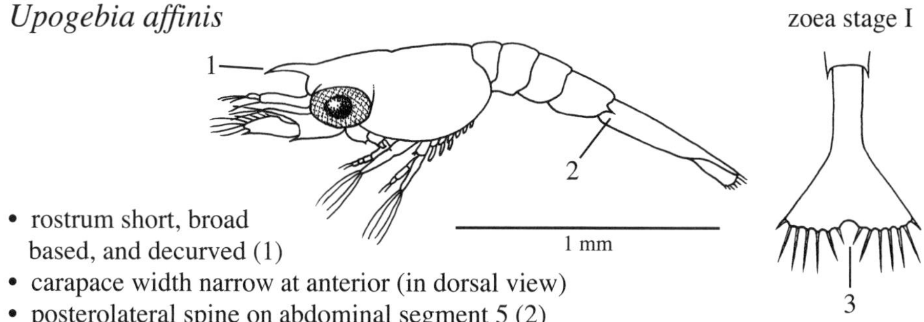

- rostrum short, broad based, and decurved (1)
- carapace width narrow at anterior (in dorsal view)
- posterolateral spine on abdominal segment 5 (2)
- telson with triangular end; posterior margin with cleft and short inwardly directed central setae (3)
- similar species: *Palaemonetes* is larger and stage I has no abdominal spines; *Hippolyte* has shorter rostrum, a serrated ventral margin, and heavy pigmentation

GHOST SHRIMPS

Biffarius biformis and other callianassids

Occurrence. Adults of multiple species of ghost shrimps occur in intertidal or shallow subtidal habitats, predominantly in high-salinity areas. The following species were, until recently, classified as different members of the genus *Callianassa. Gilvossius setimanus* occurs from Massachusetts to Florida and is probably the most common ghost shrimp on the Atlantic Coast. *Biffarius biformis* and *Callichirus major* occur from Massachusetts to Florida and into the Gulf. *Callichirus islagrande* inhabits the northern and western Gulf. In South Carolina, larvae belonging to this group of shrimps occur from April through October. The recently described larvae of the "lobster shrimp" *Axius serratus* resemble *Homarus* larvae and range from Canada south to Long Island (see Pohle et al. 2011).

Biology and Ecology. Larvae of most species are apparently more coastal than estuarine, preferring salinities >25 but occur rarely in 10. An exception is the strictly oligohaline ghost shrimp *Lepidophthalmus louisianensis*, which has a highly abbreviated larval life history. *Lepidophthalmus* larvae do not resemble those species formerly assigned to the genus *Callianassa* as figured to the right.

References. Nates et al. 1997; Ngoc-Ho 1981; Sandifer 1973b; Strasser and Felder 1999a, 1999b, 2000; Truesdale and Andryszak 1983.

LOBSTERS

Homarus americanus (American lobster)

Occurrence. Larvae occur in nearshore coastal waters from New England south to at least New Jersey, principally from May to July.

Biology and Ecology. Early stages are found near the surface until the postlarvae begin diving before settlement. A number of dives to explore the substrate may precede final settlement, with returns to the surface between dives. Larvae are voracious feeders that consume diatoms, copepods, other crustaceans, and fish eggs. Larval survival is best at salinities >20 and optimal near 30. The strong directional swimming of the postlarvae may assist in their recruitment to inshore waters.

References. Able and Cowen 2003; Cobb and Wahle 1994; Ennis 1995; Herrick 1978; Jackson and MacMillan 2000; Juinio and Cobb 1992; Katz et al. 1994; Pohle et al. 2011; Rooney and Cobb 1991.

Biffarius biformis

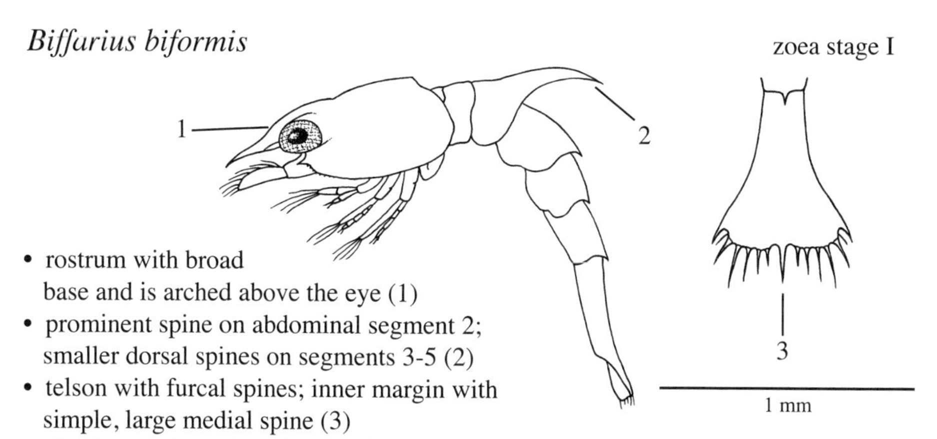

- rostrum with broad base and is arched above the eye (1)
- prominent spine on abdominal segment 2; smaller dorsal spines on segments 3-5 (2)
- telson with furcal spines; inner margin with simple, large medial spine (3)
- similar species: other ghost shrimps also have a dorsal spine on abdominal segment 2; *Crangon* has a dorsal spine on segment 3 and lacks a medial spine on the telson

Homarus americanus

zoea stage I

- rostrum heavy and fairly rectangular out to tip (1)
- robust claws on first legs (2)
- all abdominal segments with prominent dorsal and posterolateral spines (3)
- telson margin concave with large medial spine, multiple short setae, and strong furca (4)
- all legs with outer branches (5)
- similar species: large size and spination easily distinguish *Homarus* from other shrimp-like zoeae

postlarva

- resembles very small adult; wide carapace in dorsal view (1)
- large claws on first legs; small claws on second and third legs (2)
- well-developed pleopods and uropods (3)
- similar species: unlikely to be confused with late larval stages of most other shrimps or crabs; mud shrimps do not have claws on the third legs

Panulirus argus (Caribbean spiny lobster)

Occurrence. Adults range from North Carolina to Texas but are most common in southern Florida. Other species of *Panulirus* may occur as adults or larvae within our range. Larvae are more common in southern waters, but the Gulf Stream carries some as far north as New England.

Biology and Ecology. Adult females migrate into deeper water as the embryos attached to their abdomens develop. Embryos hatch directly into the phyllosoma stage, which may last 8–11 months or more as the larvae grow from about 2 mm to more than 30 mm. Later phyllosoma stages occur progressively farther from shore, predominately in surface waters. Metamorphosis produces the nonfeeding puerulus stage, a transitional stage between the planktonic phyllosoma and the benthic adults. The pueruli are active swimmers. As they move inshore, pueruli remain on the bottom during the day but are common in the surface plankton on the dark phases of the moon. Phyllosoma larvae of various species raised in captivity ate chaetognaths, larger crustaceans, ctenophores, hydromedusae, and even young fishes. Since larvae congregate in the top few centimeters of the water column, subsurface plankton nets sample them poorly.

References. Acosta et al. 1997; Acosta and Butler 1999; Calinski and Lyons 1983; Lewis 1951; Phillips and Sastry 1980; Yeung et al. 2001.

Scyllarides nodifer and *Scyllarus* spp. (slipper lobsters)

Occurrence. Adults of both genera occur from North Carolina to Texas.

Biology and Ecology. *Scyllarides nodifer* spawns in summer with a larval period of about 9–10 months. Larval development in *Scyllarus* is relatively short, about 4–6 weeks. Eggs develop directly into the phyllosoma larva. The transparent postlarval "nisto" stage becomes pigmented just before it molts into the first juvenile stage. The late nisto stage is predominately benthic rather than planktonic. Scyllarid phyllosoma larvae are sometimes associated with various jellyfishes. The exact nature of this association is unknown, but the young lobsters may hitchhike a ride as they are transported back toward shore with their jellyfish host.

References. Robertson 1968; Sandifer 1971.

Panulirus sp.

phyllosoma

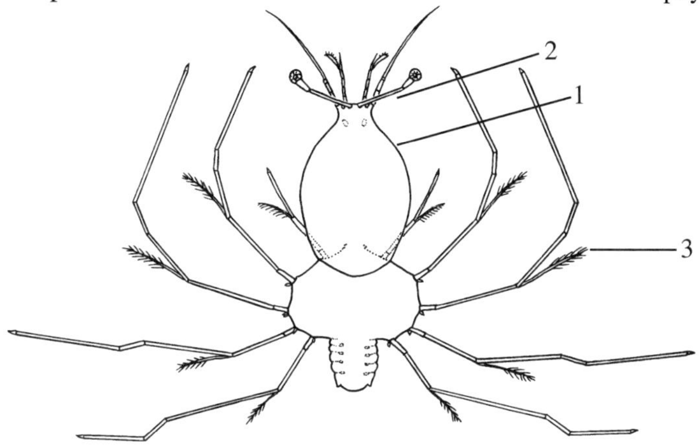

10 mm

- dorsoventrally flat and transparent
- head oval but elongate (1) and narrower than thorax
- long, stalked eyes not flanked by rostral horns (2)
- long legs with short plumose branches (3)
- similar species: other *Panulirus* and related lobsters; scyllarid lobster larvae with round head and rostral horns; very large and distinct from other decapod larvae

---

Scyllarides sp. or *Scyllarus* sp.

phyllosoma

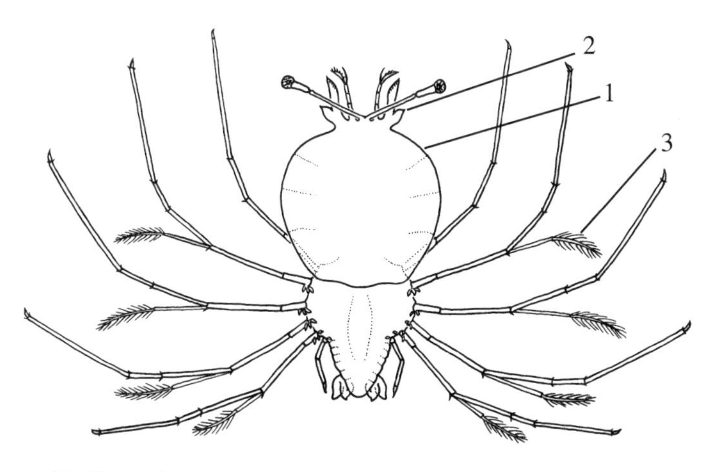

10 mm

- dorsoventrally flat and transparent
- head round (1) and wider than thorax
- long, stalked eyes flanked by rostral horns (2)
- long legs with short plumose branches (3)
- similar species: other species of both genera; *Panulirus* phyllosomes with more oval head and without rostral horns; very large and distinctive larvae

PORCELAIN CRABS

Petrolisthes spp. and related genera

Occurrence. *Petrolisthes* spp. zoeae and megalopae occur from North Carolina to Texas, and the zoeae may be especially abundant in summer in estuaries and lagoons from southwestern Florida to Texas. Members of other porcelain crab genera, including *Euceramus*, *Megalobrachium*, *Porcellana*, *Pachycheles*, and *Polyonyx,* produce similar distinctive zoeae treated in Maris (1983). All five of these genera have representatives on both the Atlantic and Gulf Coasts.

Biology and Ecology. Before 1994, *Petrolisthes armatus* was not known north of the Indian River Lagoon in Florida. This crab was first reported in South Carolina in 1995, at which time it was rare. It is now abundant, especially in areas with oyster reefs. Its range is probably limited by temperature because cold winters have resulted in almost total elimination, followed by reestablishment in the following years. Despite their spines, porcellanid zoeae are important food for many fishes. In turn, these large larvae are predators on larval fishes.

References. Gonor and Gonor 1973; Greenwood 1965; Hiller et al. 2006; Maris 1983; Roberts 1968; Sandifer 1972b; Tilburg et al. 2010.

MOLE CRABS AND SAND CRABS

Lepidopa websteri (mole crab)

Occurrence. Adults occur along beaches from Virginia to Florida and in the Gulf at least to Mississippi.

Biology and Ecology. Larval distribution is poorly known, but presumably they occur near the beaches where adults are found. Some were collected in salinities 22–35 at the mouth of the Delaware and Chesapeake Bays during summer.

References. Sandifer and Van Engle 1972.

Petrolisthes armatus

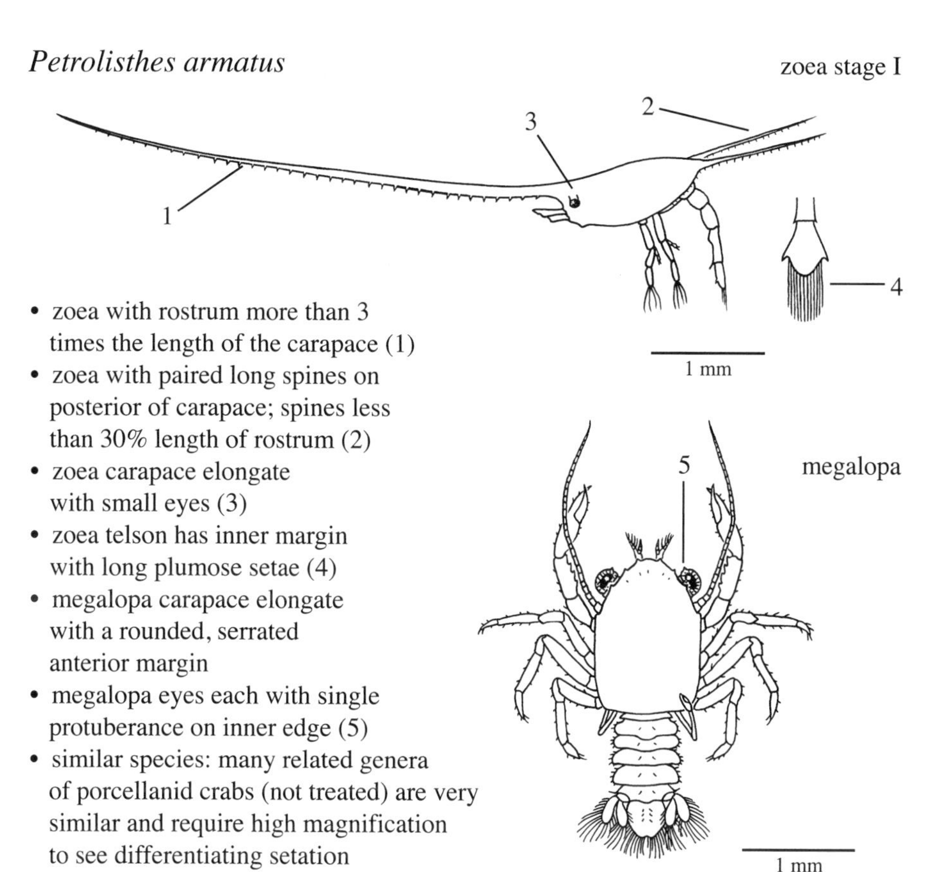

- zoea with rostrum more than 3 times the length of the carapace (1)
- zoea with paired long spines on posterior of carapace; spines less than 30% length of rostrum (2)
- zoea carapace elongate with small eyes (3)
- zoea telson has inner margin with long plumose setae (4)
- megalopa carapace elongate with a rounded, serrated anterior margin
- megalopa eyes each with single protuberance on inner edge (5)
- similar species: many related genera of porcellanid crabs (not treated) are very similar and require high magnification to see differentiating setation

---

Lepidopa websteri

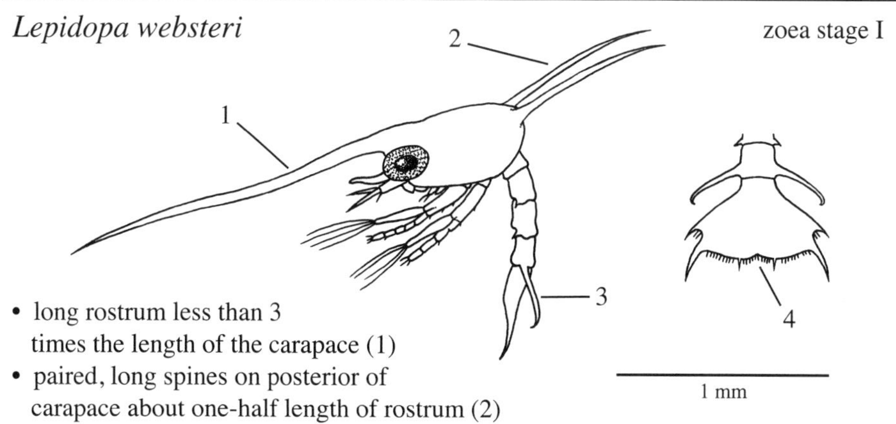

- long rostrum less than 3 times the length of the carapace (1)
- paired, long spines on posterior of carapace about one-half length of rostrum (2)
- long, dorsally directed spines on abdominal segment 5 (3)
- telson broad with strong furcal spines and fairly straight posterior margin with numerous small setae (4)
- similar species: porcellanid crab zoeae have a longer rostrum, narrower telson, and a short spine on abdominal segment 5

Emerita spp. (sand crabs)

Occurrence. The most common and widespread member of this genus, *Emerita talpoida* (Atlantic sand crab), occurs from Massachusetts to Texas, with *E. benedicti* reported from South Carolina to Texas. *Emerita portoricensis* occurs in southern and western Florida.

Biology and Ecology. Ovigerous female *E. talpoida* are seen from January to October in North Carolina and probably produce larvae all year in more southern waters. In lower Chesapeake Bay, some larvae were collected in open waters during summer and fall at salinities 28–32. Adults live in the surf on sandy beaches. Along the coast, larvae occur from the surf zone out at least several kilometers. Larvae may settle as stage V zoeae since megalopae are seldom collected in the plankton.

References. Amend and Shanks 1999; Harvey 1993; Johnson 1939; Rees 1959; Shield 1973.

Albunea spp. (mole crabs)

Occurrence. *Albunea gibbesii* (surf mole crab) adults occur from North Carolina to Texas and *A. catherinae* from Virginia to Texas. They are usually associated with sandy areas from the shore to shoals in the coastal ocean.

Biology and Ecology. Ovigerous females occur in North Carolina during summer. *Albunea* larvae are among the most abundant crab larvae in coastal Louisiana plankton in summer, especially in depths >21 m.

Emerita talpoida

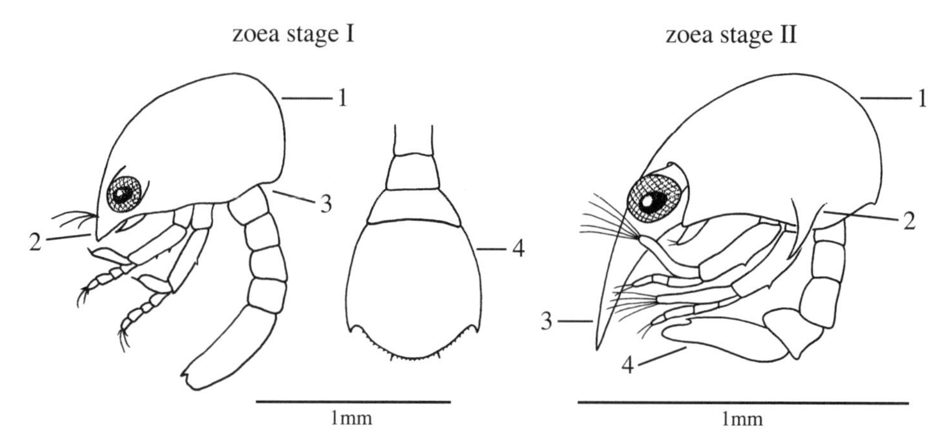

- rounded, helmet-shaped carapace without dorsal or lateral spines (1)
- anterior end of carapace pointed; not a true rostrum (2)
- abdomen typically held at a right angle to the thorax (3)
- telson broad; small furcal teeth and rounded posterior margin with very short spines (4)
- similar species: other *Emerita* spp. are similar and all later stages have large lateral spines; *Zaops* is one of the few other crab zoeae without a dorsal spine

- rounded, helmet-shaped carapace without dorsal spines (1)
- ventrally directed lateral spines (2)
- prominent, wide rostrum that elongates in the next 4 zoeal stages (3)
- rounded telson similar in all stages (similar to stage I) (4)
- similar species: *Albunea* is more elongate with a dorsal bulge and a wide telson

Albunea sp.

zoea stage I

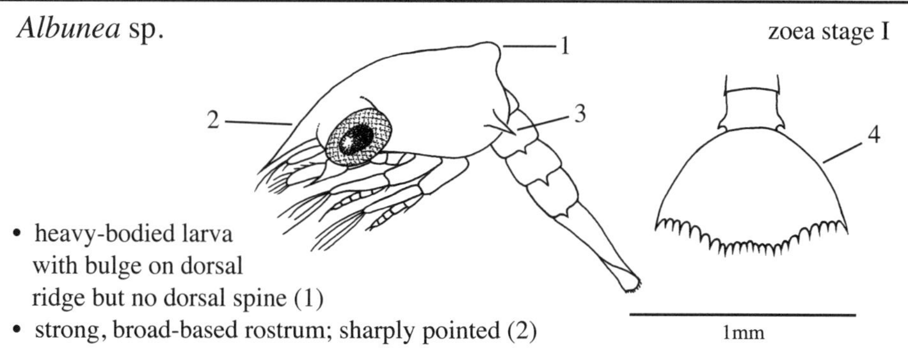

- heavy-bodied larva with bulge on dorsal ridge but no dorsal spine (1)
- strong, broad-based rostrum; sharply pointed (2)
- short posterolateral spine on carapace (3)
- round outer (lateral) margin forms semicircular telson; posterior margin with straight center section and concave outer sections; short spines throughout (4)
- similar species: both species of *Albunea* similar; *Emerita* and *Zaops* have less robust body forms, different telsons, and they lack lateral spines

PEA CRABS

Zaops ostreum (oyster pea crab), formerly *Pinnotheres ostreum*

Occurrence. *Zaops ostreum* occurs from Canada to south Florida in the Atlantic and along the southern Texas coast in the Gulf. Related pea crabs in the *genera Tumidotheres* (below) and *Pinnotheres* are found in the Gulf. Larvae are in the plankton in the warmer months. Another pea crab, *Dissodactylus mellitae*, ranges from North Carolina through the Gulf.

Biology and Ecology. Adults are commensal inside oysters and occasionally in the mussel *Geukensia demissa*. The megalops metamorphoses into an unusual planktonic first crab stage that has flattened third and fourth "walking legs" to assist swimming. The first crab stage is the infective stage. *Zaops* prefer to enter oyster spat (newly settled oysters) with larger oysters infected at lower rates. In South Carolina, planktonic first crabs occur from May through October. Planktivorous fishes rejected the planktonic fourth crab stage in the laboratory, suggesting that they are chemically protected. Zoeae are omnivorous. The similar *Gemmotheres chamae* adults have been found only in the corrugate jewelbox clam *Chama congregata* from North and South Carolina.

References: Christensen and McDermott 1958; Luckenbach and Orth 1990; Roberts 1975; Sandifer and Van Engle 1970; Sandoz and Hopkins 1947.

Tumidotheres maculatus (squatter pea crab), formerly *Pinnotheres maculatus*

Occurrence. *Tumidotheres* occurs from Massachusetts to Texas from the coastal ocean to salinities as low as 11.

Biology and Ecology. *Tumidotheres* is commensal inside mussels, scallops, and occasionally in tubes of larger polychaetes. Larvae can be raised on zooplankton.

References: Costlow and Bookout 1966c.

Pinnixa spp. (pea crabs)

Occurrence. Many commensal and free-living species of *Pinnixa* occur within our coverage area: *P. chaetopterana* (Massachusetts to Texas), *P. sayana* (Massachusetts to Texas), and *P. cylindrica* (Massachusetts to northwestern Florida). Zoeal release for most species on the Atlantic Coast occurs during the warmest months.

Biology and Ecology. These pea crabs are best known for their associations with worms, shrimps, and sea squirts, especially in estuaries. In lower Chesapeake Bay, zoeae of all three species mentioned occurred from June to September at salinities above 18 with zoeal densities much higher near the bottom.

References. Brookins and Epifanio 1985; McDermott 2005.

Zaops ostreum

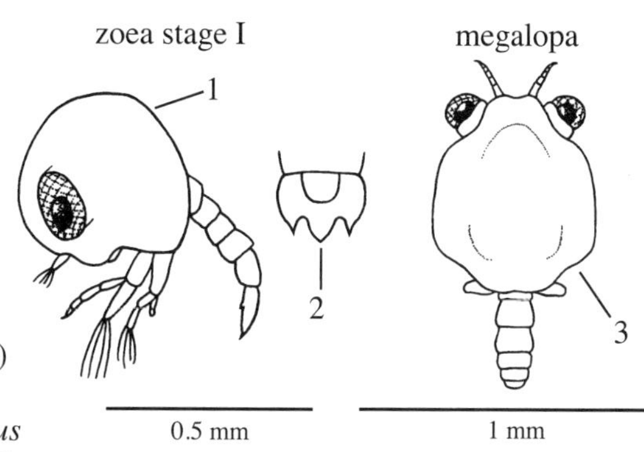

- zoea without rostrum or dorsal or lateral spines (1)
- zoea telson short with bifurcated posterior margin (2)
- megalopa carapace rectangular with rounded corners (3)
- megalopa rostrum pointed ventrally (not visible from above)
- similar species: *Gemmotheres chamae* is similar but *T. maculatus* has large spines; *Emerita* also lacks a dorsal spine but has a pointed anterior end and broadly rounded telson

---

Tumidotheres maculatus

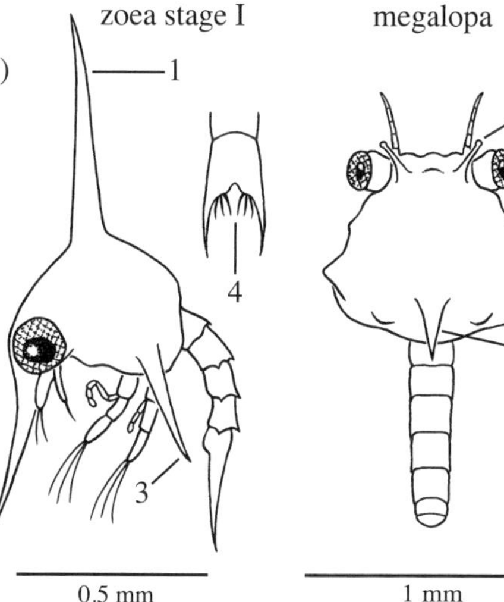

- zoea with long, straight dorsal spine (1)
- zoea with straight or slightly curved rostrum, shorter than dorsal spine (2)
- zoea with lateral spines, long and ventrally curved, shorter than dorsal spine (3)
- zoea with rectangular telson with long furca, concave posterior margin with notch in middle, and 6 similar inner setae (4)
- megalopa carapace wide and has large spine near posterior margin (5)
- megalopa with narrow orbital spines (6)
- similar species: xanthids and other zoeae with long dorsal spines have short lateral spines; *Pinnixa* spp. have very different telsons

---

Pinnixa sp.

zoea stage I

- rostrum long with broad base (1)
- dorsal spine straight and about equal to rostrum in length (2)
- lateral spines pointed ventrally and almost opposite direction of the dorsal spine (3)
- distinctive wing-like projections from abdominal segment 5 overlaps telson (4)
- similar species: other *Pinnixa* species share most features but telsons are slightly different; *Zaops* and *Gemmotheres* have similar telsons but no dorsal spines

1 mm

SPIDER CRABS

Libinia spp. (spider crabs)

Occurrence. Although the adults of many genera and species of spider crabs occur along the Atlantic and Gulf Coasts, *Libinia dubia* and *Libinia emarginata* are among the most frequently encountered spider crabs within our range. Both occur from Massachusetts to Texas and are found on all kinds of bottoms from the shoreline to the coastal ocean. Larvae occur offshore, but scattered reports from Rhode Island to Louisiana indicate that they also occur sporadically in nearshore coastal waters in summer and fall.

Biology and Ecology. Ovigerous females occur during summer and sometimes in spring and fall. In lower Chesapeake Bay, zoeal stages occurred from June to October at 16–32 psu. The juvenile spider crabs attached to the large jellyfish *Stomolophus meleagris*, *Rhopilema verrilli*, and *Aurelia aurita* probably first attach during the crab's planktonic larval stage.

References. Johns and Lang 1977; Sandifer and Van Engle 1971; Shanks 1998.

CANCRID CRABS

Cancer borealis (Jonah crab) and *C. irroratus* (Atlantic rock crab)

Occurrence. Both species occur from Massachusetts to Florida but are far more abundant in coastal ocean and estuarine waters north of Virginia.

Biology and Ecology. Ovigerous females occur all year in northern areas, and larvae are typically released offshore. In the lower Chesapeake Bay, zoeae were collected from May to October, with highest densities occurring in the spring and fall when temperatures were below 18°C. Studies on Pacific species of *Cancer* indicate the importance of zooplankton as essential food for the first zoeal stage but also suggest that the ability to feed on heterotrophic protozoans may be important when preferred prey are absent.

References. Clancy and Cobb 1997; Sastry 1977a, 1977b; Sulkin et al. 1998.

Carcinus maenas (green shore crab)

Occurrence. Adults occur from Massachusetts to Virginia over a wide range of salinities.

Biology and Ecology. This European native has been on the U.S. coastlines since at least the early 1800s. Ovigerous females occur in winter and spring. Other than the collection of larvae from Delaware Bay from June to August, little is reported on the occurrence of larvae in the plankton. Vertical migrations may be linked to endogenous tidal rhythms and selective tidal transport.

References. Anger et al. 1998; Factor and Dexter 1993; Williams 1968; Zeng and Naylor 1996a, 1996b; Zeng et al. 1997.

Libinia dubia

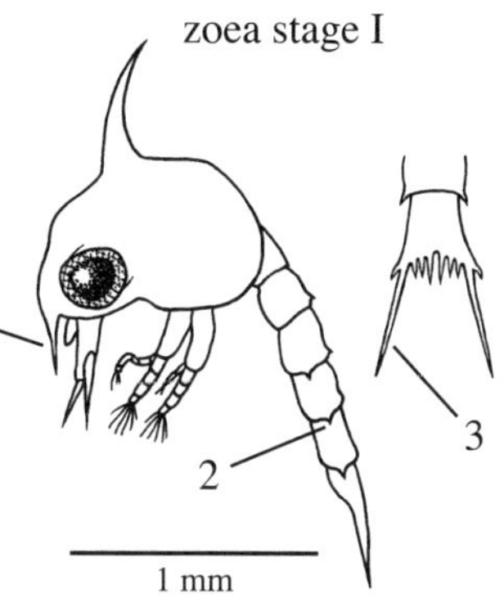

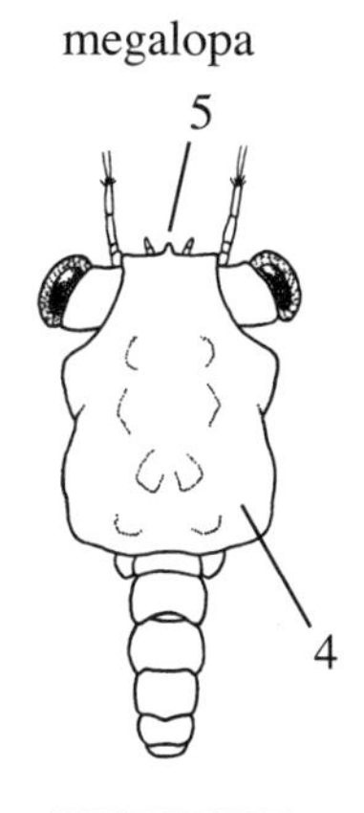

- zoea with short rostrum, medium dorsal and no lateral spines (1)
- zoea with short posterolateral spines on abdominal segments 3-5 (2)
- zoea telson with very long furcal spines, short corner spines, and straight posterior margin with deep notch and 6 stout setae (3)
- megalopa with very irregular surface on carapace (4)
- megalopa with very small rostrum (5)
- similar species: zoeae of all spider crabs are very similar; telson shape separates them from other large crab zoeae

---

Cancer irroratus

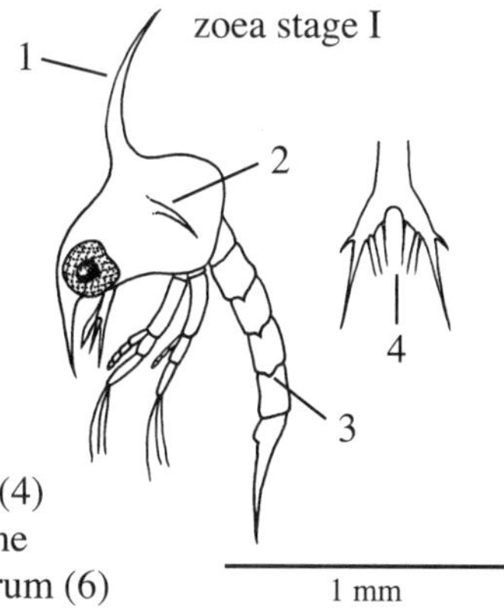

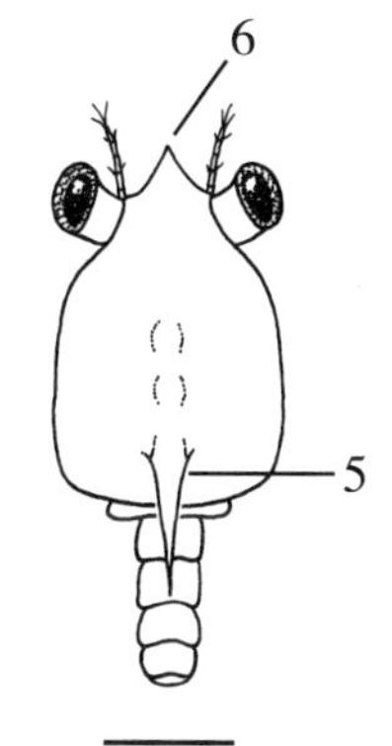

- zoea with very long and strongly curved dorsal spine (1)
- zoea with lateral spine with origin under base of dorsal spine (2)
- zoea with posterolateral spines on all abdominal segments (3)
- zoea telson with strong furca; posterior margin with deep cleft (4)
- megalopa with strong dorsal spine on carapace (5) and pointed rostrum (6)
- similar species: other species of *Cancer*; some xanthid zoeae (especially *Panopeus*) but they have a much longer rostrum

---

Carcinus maenas

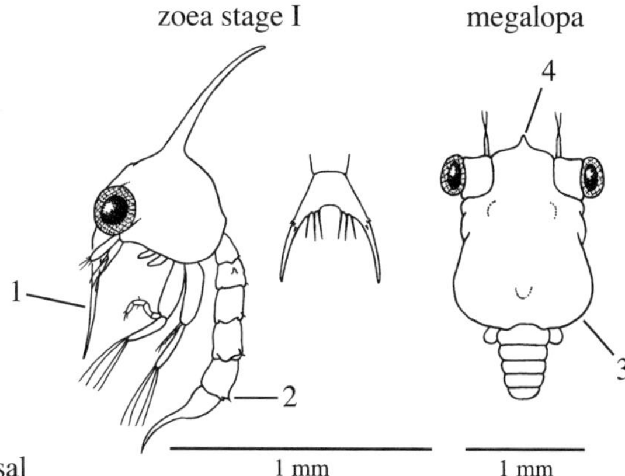

- zoea with tip of long rostrum curved slightly anteriorly (1)
- zoea without lateral spine
- zoea with small dorsal but no posterolateral spine on abdominal segment 5 (2)
- megalopa with elongate rectangular carapace (3)
- megalopa with short pointed rostrum (4)
- similar species: xanthid and most other zoeae with long dorsal spines have lateral spines

XANTHID CRABS (MUD CRABS)

Panopeus spp.

Occurrence. Adults of *Panopeus herbstii* range from Massachusetts southeastern Florida, where they and are abundant in the mid- to high-salinity reaches of estuaries. *Panopeus simpsoni* (sometimes included with *P. herbstii*) occurs throughout the Gulf. *Panopeus occidentalis* and *P. obesus* occur from North Carolina south and in the Gulf. Other species may occur on both coasts of Florida and in the Gulf.

Biology and Ecology. Ovigerous females occur from spring to fall along the Atlantic Coast and during most of the year farther south. Zoeae occurred in the lower Chesapeake Bay throughout the warm season at salinities from 2 to 34, although they were most abundant at 15–25 psu.

References. Andrews et al. 2001; Costlow and Bookhout 1961b; Costlow et al. 1962; Dittel et al. 1996; Harvey and Epifanio 1997; Rodriguez and Epifanio 2000.

Dyspanopeus spp.

Occurrence. *Dyspanopeus sayi* occurs along the entire Atlantic Coast in the mid- to high-salinity areas of estuaries. *Dyspanopeus texanus* may be the dominant species from the west coast of Florida to Texas. The taxonomy of the group is not clear; other species or subspecies may occur in the Southeast and Gulf. Zoeae can be abundant for short periods in summer and fall. The similar *Neopanope packardii* occurs in the Gulf.

Biology and Ecology. Ovigerous females of *D. sayi* occur in the Carolinas from April to October. In lower Chesapeake Bay, zoeae were collected at salinities from 11 to 32 and were evenly distributed throughout the water column. Large concentrations of zoeae are sometimes associated with tidal fronts.

References. DeVries et al. 1991; Epifanio 1987; Forward 1989; Van Montfrans et al. 1990.

Hexapanopeus angustifrons

Occurrence. Adults occur from Massachusetts to Texas in estuaries and along the coast. Other species and subspecies may occur in the Southeast and Gulf.

Biology and Ecology. Ovigerous females occur during the warmer months on the Atlantic Coast, and spawning occurs during most of the year in southern waters. Larval development takes about two to three weeks. *Hexapanopeus angustifrons* zoeae were very abundant in the lower Chesapeake Bay from June to October at salinities from 14 to 32. All larval stages were present; late stages were more abundant in bottom waters.

References. Costlow and Bookhout 1966b; Steppe and Epifanio 2006.

Panopeus herbstii

zoea stage I megalopa

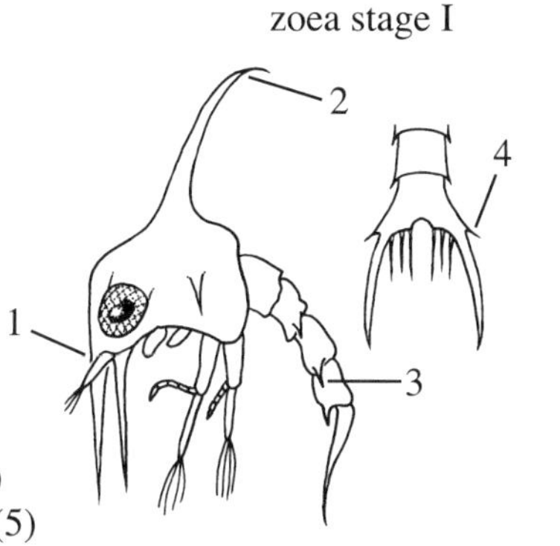

- zoea with long rostrum and straight forehead (1)
- zoea with dorsal spine tip curved; about same length as rostrum (2)
- zoea with spines on last 3 abdominal segments (3)
- zoea telson with long furca and prominent outer spines (4)
- megalopa is broadly rounded (5)
- lateral horns flank blunt rostrum (6)
- similar species: except for *Eurypanopeus*, other xanthid zoeae have straight dorsal spines

Dyspanopeus texanus

zoea stage I megalopa

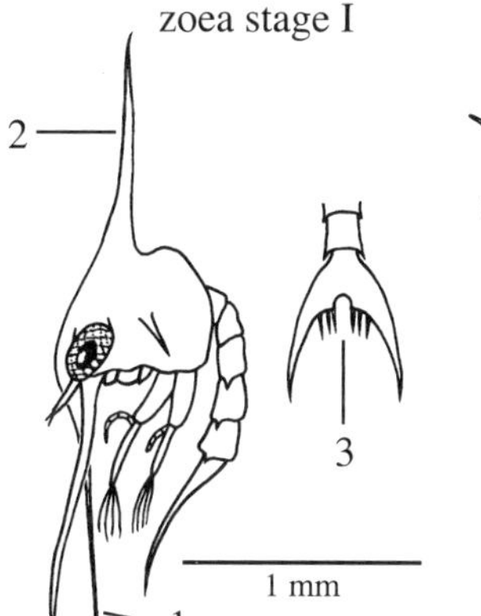

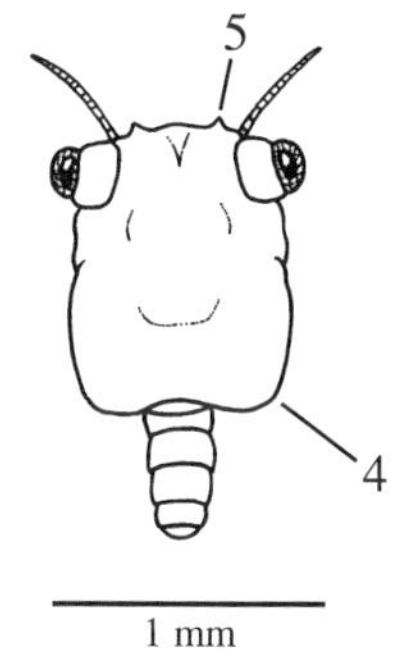

- zoea with very long rostrum directed slightly under the body (1)
- zoea with long and very straight dorsal spine (2)
- zoea telson with wide, strongly curved furca and deep cleft (3)
- megalopa rectangular but elongate (4)
- megalopa with very small rostral horns (5)
- similar species: other *Dyspanopeus* and xanthid species; *Eurypanopeus* has a curved dorsal spine tip and all abdominal segments have spines and knobs

Hexapanopeus angustifrons

zoea stage I megalopa

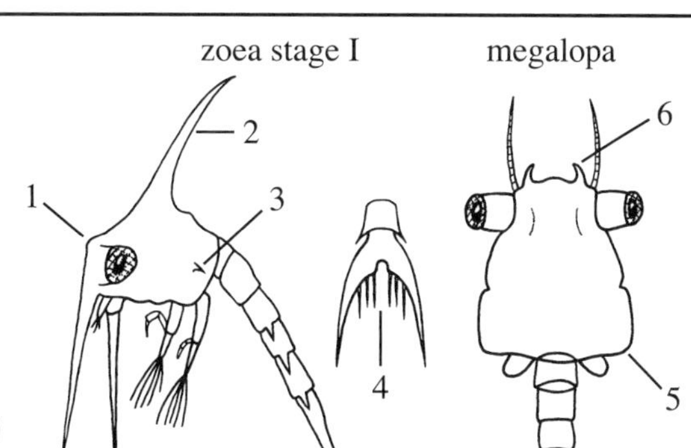

- zoea with prominent forehead at base of rostrum (1)
- zoea with slightly curved dorsal spine (2)
- zoea with short posteriorly directed lateral spine (3)
- zoea telson with wide furca and deep cleft (4)
- megalopa widest at posterior (5)
- megalopa with prominent rostral horns (6)
- similar species: other *Hexapanopeus* and xanthid zoeae; *Panopeus* and *Eurypanopeus* have more curved dorsal spine tips and the lateral spines are in different positions

Rhithropanopeus harrisii (estuarine mud crab)

Occurrence. Adults occur from Massachusetts to Texas from almost fresh to marine waters, especially within estuaries.

Biology and Ecology. Females spawn from spring to fall in the middle Atlantic, and the species probably spawns through most of the year in southern waters. All larval stages are retained in estuaries, particularly in low salinities (0–10; less common to 20). In the lower Chesapeake Bay, all zoeal stages were more abundant in the lower portion of the water column, which may minimize the potential for export from the low-salinity estuarine areas preferred for settlement.

References. Chen et al. 1997; Costlow et al. 1966; Fitzgerald et al. 1998; Forward 1989, 2009; Forward and Bourla 2008.

Eurypanopeus depressus (flatback mud crab)

Occurrence. *Eurypanopeus* is one of the most common crabs on both Atlantic and Gulf Coasts and is especially abundant in estuaries, sometimes in low salinities. Other species of *Eurypanopeus* occur in the Gulf.

Biology and Ecology. Ovigerous females occur from spring to fall in the Carolinas. Zoeae were collected from May to October in North Carolina sounds at salinities as low as 12. Larvae occurred from April to October in northeastern Florida.

References. Costlow and Bookhout 1961a; Levine and Sulkin 1984; Sulkin et al. 1983.

Menippe mercenaria (Florida stone crab) and *M. adina* (Gulf stone crab)

Occurrence. *Menippe mercenaria* occurs from North Carolina to Louisiana and *M. adina* from throughout the Gulf in estuaries and along the coast. Because these two species hybridize where they overlap, the ranges of each on the Florida west coast are uncertain.

Biology and Ecology. During all but the coldest months, females produce up to six successive egg masses of up to 1 million eggs each, without intervening molts. *Menippe adina* larvae appear to tolerate salinities of 10–15. In the Gulf, all larval stages occur in nearshore coastal and estuarine (15–30 psu) areas. Both species support important commercial trap fisheries. The taxonomy of the genus has been revised leaving some confusion about old reports.

References. Brown et al. 1992; Guillory et al. 1995; Krimsky et al. 2009; Martin et al. 1988; Ong and Costlow 1970; Porter 1960; Stuck and Perry 1992.

Rhithropanopeus harrisii

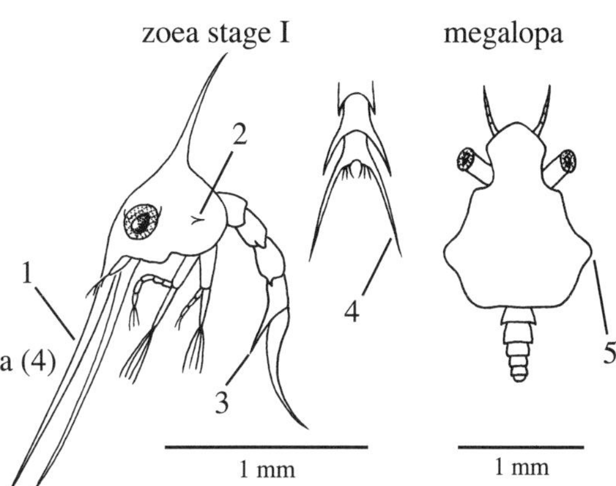

- zoea with very long rostrum somewhat upturned (1)
- zoea with very short lateral spine directed posteriorly (2)
- abdominal segment 5 with a prominent ventrally directed spine (3)
- zoea telson with long, narrow furca (4)
- megalopa carapace with projected anterior margin and widest in center (5)
- similar species: other xanthid zoeae have longer rostrums but do not have a long spine on the last abdominal segment

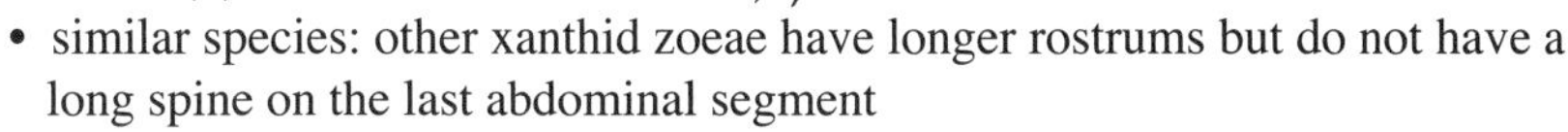

Eurypanopeus depressus

zoea stage I megalopa

- zoea with long rostrum pointed in almost opposite direction of dorsal spine (1)
- zoea with hooked tip on dorsal spine; about equal to rostrum in length (2)
- zoea with posterolateral spines and/or knobs on posterior abdominal segments (3)
- megalopa carapace square with blunt rostrum (4)
- megalopa carapace with many blunt spikes on carapace (not visible in illustration)
- similar species: other *Eurypanopeus* species; other xanthid zoeae are without spines or knobs on abdominal segment 2

1 mm 1 mm

Menippe mercenaria

zoea stage I megalopa

- zoea with heavy dorsal spine equal to rostrum in length (1)
- zoea with large lateral spine directed ventrally (2)
- zoea with long posterolateral spines on abdominal segments 3-5 (3)
- zoea telson with wide furca and shallow cleft (4)
- megalopa square with irregular lateral margins and notched anterior margin (5)
- similar species: other *Menippe* species; other xanthid zoeae are without stout dorsal or multiple long abdominal spines

1 mm 1 mm

SWIMMING (PORTUNID) CRABS

Callinectes spp. (blue crabs)

Occurrence. *Callinectes sapidus* occurs from Massachusetts to Texas and is the only *Callinectes* species common in the nearshore plankton north of Cape Hatteras. The first zoeal stage and the megalopa are the most common stages in nearshore and estuarine areas (>20 psu), principally in warmer months. Several other species of *Callinectes*, including *C. ornatus* and *C. similis*, occur in the Southeast and Gulf regions. Both *C. sapidus* and *C. similis* are abundant in the northeastern Gulf, with *C. similis* larvae more common in higher salinities, especially from Louisiana west, during winter and spring. Other swimming crabs, especially *Portunus* spp., occur within our region of coverage.

Biology and Ecology. Adult *C. sapidus* spawn near mouths of estuaries in spring to summer, and early zoeae are carried out to coastal areas, where larval development through the seven zoeal stages continues during summer and fall. Transport of larvae out of estuaries and their subsequent recruitment back to estuaries and inshore waters involves vertical migration and use of both wind-driven and tidal currents (see Fig. 2). Megalopae are abundant in Atlantic coastal waters and reenter estuaries borne on surface currents. Once they enter estuarine waters, megalopae swim only at night and are especially abundant on rising tides. Megalopae settle preferentially in vegetated areas. On the Gulf Coast, zoeal stages I and II and megalopae are common inshore with *C. sapidus* most abundant spring to fall and with *C. similis* in winter to early spring. Entrance into Gulf passes seems most pronounced on nocturnal flood tides. In northern areas, megalopae are eaten by the sevenspine bay shrimp *Crangon septemspinosa*. Larvae of swimming crabs are difficult to distinguish based on morphological features. Genetic markers are being developed to identify larvae of several co-occurring species.

References. Blackmon and Eggleston 2001; Costlow and Bookhout 1959; Epifanio 1995, 2007; Etherington and Eggleston 2003; Forward et al. 2001, 2003; Johnson and Perry 1999; Lukenbach and Orth 1992; McConnaugha 2002; Mense and Wenner 1989; Ogburn et al. 2007; Perry et al. 1995, 2003; Perry and McIlwain 1986.

Ovalipes spp. (lady crabs)

Occurrence. *Ovalipes ocellatus* is probably the only *Ovalipes* species common north of Virginia. *Ovalipes stephensoni* becomes increasingly dominant from North Carolina to Biscayne Bay, Florida. *Ovalipes floridanus* occurs in the Gulf from southwest Florida to Texas.

Biology and Ecology. Reports of ovigerous females vary throughout the range, but larvae were collected in lower Chesapeake Bay and Delaware Bay from June to October at salinities from 25 to 32. Because these crabs are most abundant from the surf and coastal ocean, larvae might not be expected to be common in estuaries.

References. Costlow and Bookhout 1966a.

Callinectes sapidus

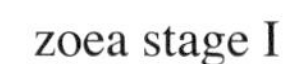

megalopa

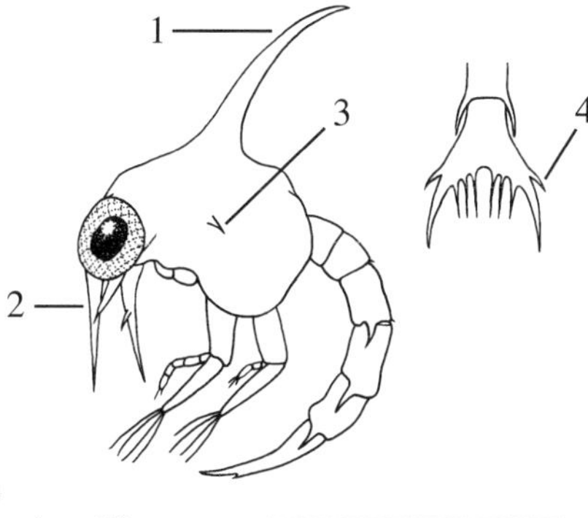

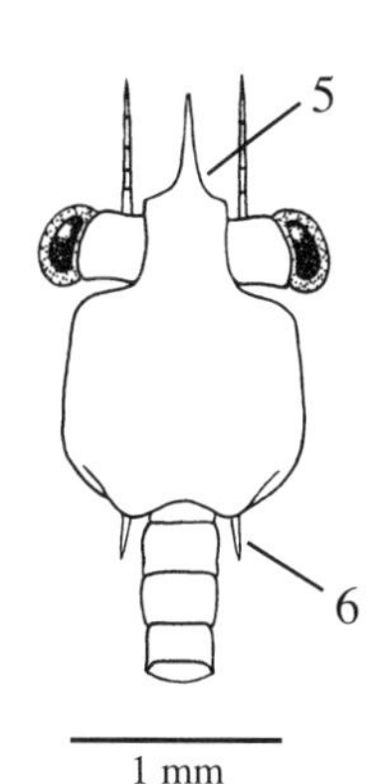

- zoea very small with strong curved dorsal spine (1)
- zoea with rostrum shorter than dorsal spine (2)
- zoea with very short lateral spine located in center of carapace (3)
- zoea telson with slightly curved furca and strong outer spines (4)
- megalopa with strong rostrum on anteriorly projected margin (5)
- megalopa with small spines projecting from posterior margin of carapace (6)
- similar species: difficult to separate zoeae and megalopae from other *Callinectes* and portunid crab species

---

Ovalipes ocellatus

zoea stage I

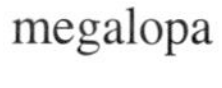

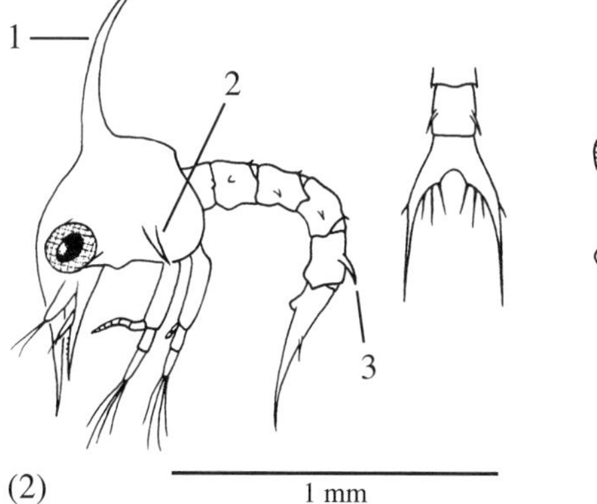

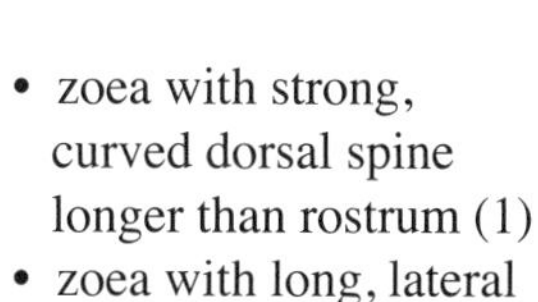

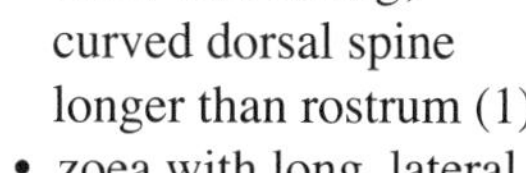

- zoea with strong, curved dorsal spine longer than rostrum (1)
- zoea with long, lateral spine directed ventrally (2)
- zoea with large, posteriorly directed, hooked blunt spines on abdominal segment 5 (3)
- megalopa somewhat square with centrally located projections on sides of carapace (4)
- megalopa with sharply pointed rostrum (5)
- similar species: other *Ovalipes* species; other portunid crab zoeae are without long, hooked knobs on abdominal segment 5

SHORE AND MARSH (GRAPSOID) CRABS

Pachygrapsus transversus (mottled shore crab)

Occurrence. *Pachygrapsus transversus* ranges from North Carolina south and throughout the Gulf.

Biology and Ecology. Larvae are released at high tide.

References. Flores et al. 2007; Moreira et al. 2007; Morgan and Christy 1996.

Hemigrapsus sanguineus (Asian shore crab)

Occurrence. This introduced species was first reported from the intertidal of Cape May, New Jersey, in 1988. It has since spread north to Maine and south to North Carolina.

Biology and Ecology. Reproduction appears to occur during summer. The early larvae are tolerant of salinities down to 15, but the later stages require high salinities consistent with the adult habitat along coastal shorelines. Exudates from the adult and juvenile crabs promote settlement and the final metamorphosis of the planktonic megalopae to the first crab stage. The rapid spread of this crab, presumably by larval dispersal, from Cape May northward against the net southerly flow of coastal water raises interesting questions.

References. Epifanio et al. 1998; Hwang et al. 1993; Kopin et al. 2001; Steinberg et al. 2007.

Armases cinereum and *Sesarma reticulatum* (marsh crabs)

Occurrence. *Sesarma reticulatum* occurs from Massachusetts to Texas and *Armases cinereum* from Maryland to Texas, but neither species occurs on the southern tip of the Florida peninsula. Adults are typically found on shorelines or on salt marshes. Additional species occur in the Gulf.

Biology and Ecology. Ovigerous females of both species occur from April to October in the Carolinas and over longer periods farther south. Zoeal stages have been collected from 12 to 35 psu in estuaries and in the coastal ocean. Because the genus name change from *Sesarma* to *Armases cinereum* is recent, most literature references are to *Sesarma*.

References. Costlow and Bookhout 1960, 1962; Hovel and Morgan 1997; Staton and Sulkin 1991; Zimmerman and Felder 1991.

Pachygrapsus transversus

- zoea with symmetrically shaped and straight dorsal spine (1)
- zoea with broad-based rostrum about as long as dorsal spine (2)
- zoea with rectangular telson with short furca, concave margin, and 6 plumose setae (3)
- megalopa carapace pear shaped with straight anterior margin slightly notched in middle (4)
- similar species: other first stage zoeae without lateral spines lack a stout rostrum and straight dorsal spine; later stages have lateral spines

zoea stage I

megalopa

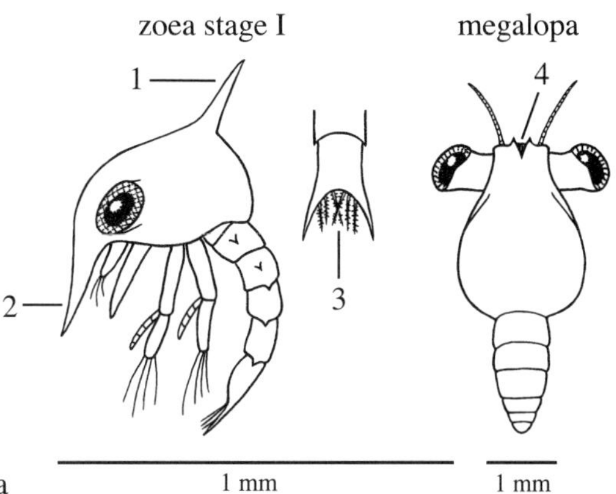

Hemigrapsus sanguineus

- zoea with broad-based, curved or straight dorsal spine; longer than rostrum (1)
- zoea with ventrally directed lateral spine (2)
- zoea with wide telson; long, straight, or curved furca; and a posterior margin with very shallow cleft (3)
- megalopa with rectangular but elongate carapace; cleft on anterior margin (4)
- similar species: *Armases*, *Sesarma*, and *Uca* are without lateral spines

zoea stage I

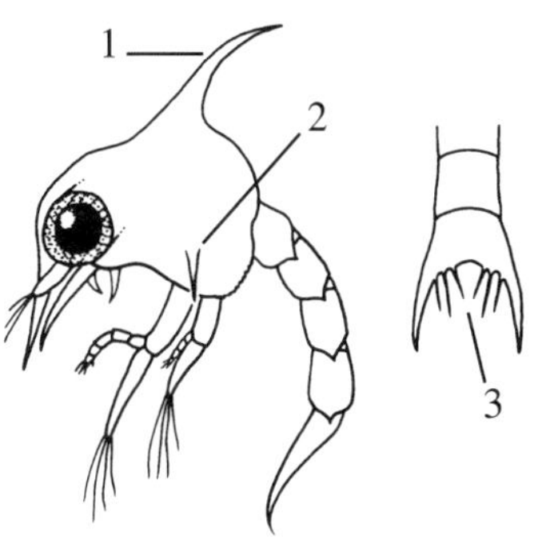

megalopa

Sesarma reticulatum

- zoea with rostrum longer than dorsal spine; directed under carapace (1)
- zoea with no lateral spine
- zoea telson with evenly curved lateral margins; posterior margin with roughly parallel setae of equal lengths (2)
- megalopa carapace narrow between eyes; with prominent rostrum (3)
- similar species: other *Armases* and *Sesarma* species; *Uca* zoeae are similar but smaller

zoea stage I

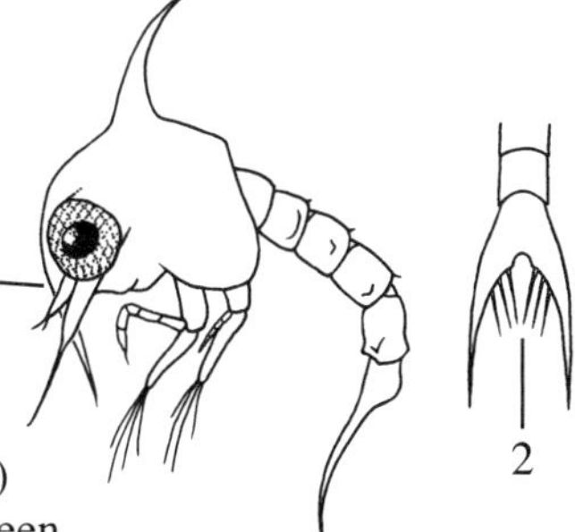

megalopa

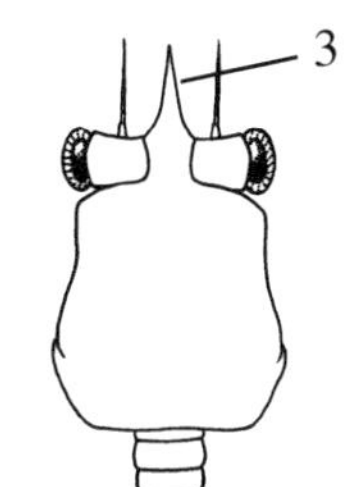

FIDDLER AND GHOST CRABS

Uca spp. (fiddler crabs)

Occurrence. Many species of *Uca* co-occur in marshes and mangroves along the Atlantic and Gulf Coasts in salinities as low as 5. High densities of zoeae are present in many tidal systems throughout the warmest months. At least 16 *Uca* spp. are known from the Atlantic and Gulf Coasts. The following list identifies distributions of some of the most commonly encountered species:

U. pugnax (Atlantic marsh fiddler) Massachusetts to northeastern Florida
U. pugilator (Atlantic sand fiddler) Massachusetts west Florida
U. minax (redjointed fiddler) Massachusetts to Florida and Gulf of Mexico (low salinity)
U. panacea (gulf sand fiddler) northern Gulf of Mexico
U. speciosa (longfinger fiddler) Gulf from Florida Keys to Mexico
U. spinicarpa (spined fiddler) Pensacola, Florida, to western Gulf
U. longisignalis (gulf marsh fiddler) northwestern Florida to Texas

Biology and Ecology. Zoeal releases are synchronized with the highest tides in the lunar cycle (spring rather than neap), and they occur almost exclusively at night. Zoeae of many species are known from a wide range of salinities. In larger estuaries, the entire larval development may occur within the estuary. In coastal systems, larvae enter the ocean and return as megalopae. In South Carolina marsh creeks, megalopae have pronounced tidal and diel rhythms, with peak abundance on nocturnal flooding tides on the highest spring tides of the month. Presumably, this facilitates recruitment back into the estuary and onto the marshes for settlement. Since megalopae are favored food for zooplanktivorous fish, they may reduce their exposure by favoring the bottom during daylight. The timing and locations of megalopa settlement may be guided by olfactory cues from adult fiddler crabs. Larvae of the various species are almost indistinguishable based on morphology; however, genetic markers may eventually allow us to establish species-specific larval distributions.

References. Christy 1982; Forward and Rittschof 1994; Grimes et al. 1989; Herrnkind 1968; Hovel and Morgan 1997; Little and Epifanio 1991; O'Connor and Van 2006; Petrone et al. 2005; Shanks 1998; Tankersley et al. 1995.

Ocypode quadrata (Atlantic ghost crab)

Occurrence. Adults occur from Rhode Island to Texas. These semiterrestrial crabs are restricted to sandy beaches throughout their range.

Biology and Ecology. Spawning occurs from spring to fall in the Carolinas, but ovigerous females occur almost all year in the Gulf. *Ocypode* larvae are surprisingly rare in plankton collections despite the abundance of adults. Perhaps the larvae develop in the surf zone, which is seldom sampled in our region. A few larvae were collected from the mouth of Chesapeake Bay.

References. Diaz and Costlow 1972; McDermott 2009.

Uca sp.

zoea stage I — megalopa

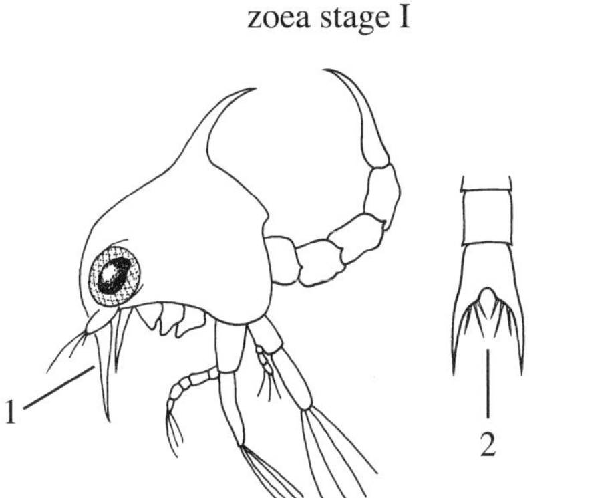

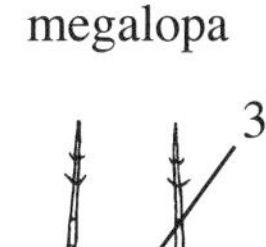

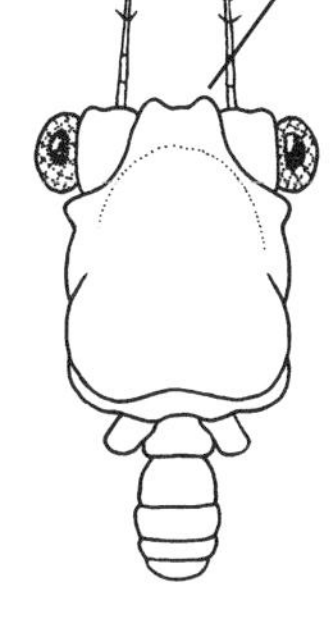

0.5 mm — 1 mm

- zoea with rostrum slightly longer than dorsal spine (1)
- zoea without lateral spine
- zoea telson narrow with irregularly curved lateral margins; posterior margin with setae of irregular lengths and orientations (2)
- megalopa carapace rectangular with anterior margin composed of 2 blunt knobs (3)
- similar species: other *Uca* species; *Armases* zoeae are similar but larger
- See Figure 24 for entire larval sequence

Ocypode quadrata

zoea stage I — megalopa

1 mm — 1 mm

- zoea carapace large and heavy (1)
- zoea with thick rostrum and dorsal spine of equal lengths (2)
- zoea with distinct protuberance on central frontal region (not visible in illustration) (3)
- zoea with long, lateral knobs on abdominal segments 4 and 5 (4)
- megalopa carapace oval with straight anterior margin (5)
- similar species: other *Ocypode* species; distinct robustness and prominent abdominal knobs distinguish *Ocypode* from other crab zoeae

STOMATOPODS

Mantis Shrimps

ID hint: The relatively large stomatopod larvae typically have oversized and spiny carapaces, conspicuous claws, and are unlike those produced by any other crustacean group.

Squilla empusa and other mantis shrimps

Occurrence. *Squilla empusa* is the most common stomatopod in nearshore waters, both along the coast and in higher-salinity reaches of estuaries from Maine to Texas. Other stomatopods, including *Neogonodactylus* spp., are widespread from North Carolina south to Texas. *Cloridopsis dubia* occurs in Southeast estuaries. Larvae are most abundant in warmer months.

Biology and Ecology. The larvae are formidable predators and use their claws to strike their victims with amazing speed. Larval development in *Squilla* includes nine stages and a postlarva, all relatively similar. Because of their size and mobility, larvae are best collected using large nets with at least 500 μm mesh. Stomatopod larvae are little studied compared with the adults, and relatively few have even been described. As a result, we lack definitive information about the abundances and distributions of the individual species and know virtually nothing of their biology in the field. In laboratory cultures, *S. empusa* can be raised on brine shrimp: early stages on nauplii and later stages on adults. Although the mantis shrimps share many features with the decapods, they constitute a separate crustacean group.

References. Diaz 1998; Morgan 1980; Morgan and Provenzano 1979; Pyne 1972.

Mantis shrimp antizoea larvae

Occurrence: Some mantis shrimps (Superfamily Lysiosquilloidea) have a distinctive first larval stage, the **antizoae**, which is not seen in *Squilla* or other mantis shrimps. These larvae appear in nearshore waters along the Mid-Atlantic Coast but probably range much more widely in both Atlantic and Gulf waters. The most common nearshore adults in this group, *Lysiosquilla scabricauda* and *Coronis scolopendra*, are reported from the Carolinas southward in the Atlantic and in the Gulf. *Bigelowina biminiensis* occurs in Gulf nearshore waters and is reported from North Carolina.

Biology and Ecology. Few antizoea from the Atlantic and Gulf have been described or studied. The zoeal stages following the antizoea resemble those of *Squilla*.

References. Rodrigues and Manning 1992; Young 2001.

Squilla empusa

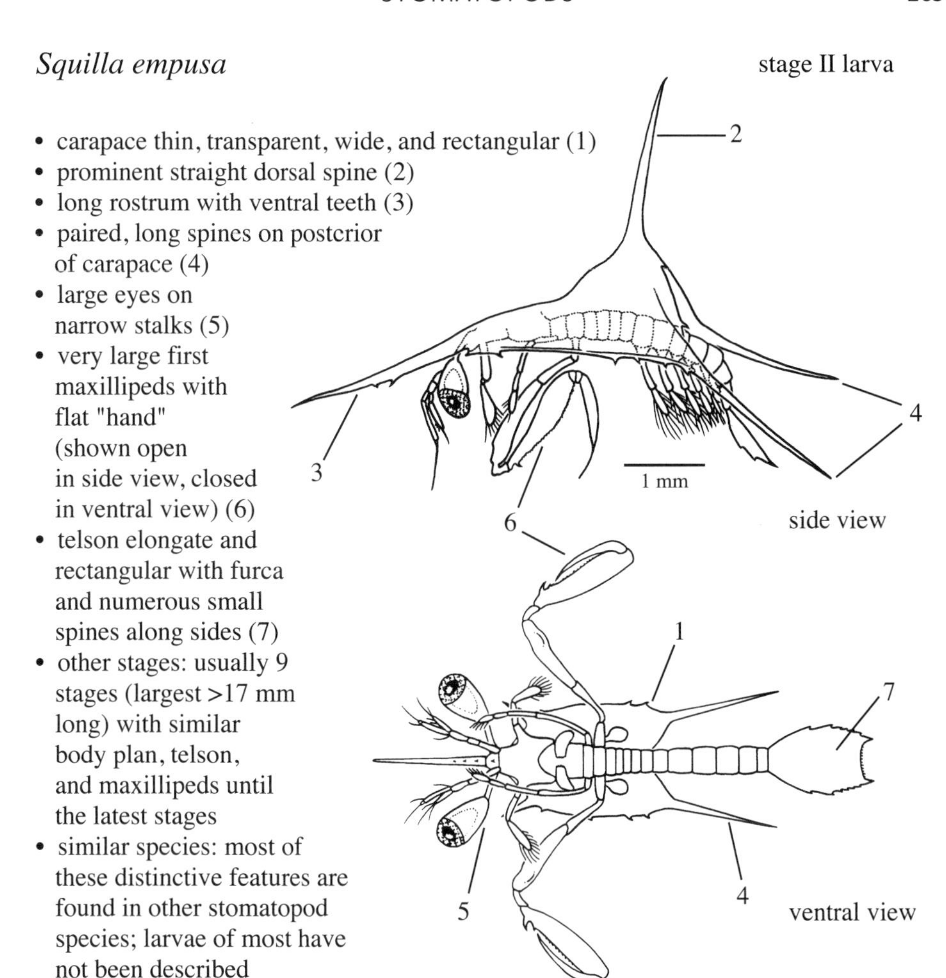

- carapace thin, transparent, wide, and rectangular (1)
- prominent straight dorsal spine (2)
- long rostrum with ventral teeth (3)
- paired, long spines on posterior of carapace (4)
- large eyes on narrow stalks (5)
- very large first maxillipeds with flat "hand" (shown open in side view, closed in ventral view) (6)
- telson elongate and rectangular with furca and numerous small spines along sides (7)
- other stages: usually 9 stages (largest >17 mm long) with similar body plan, telson, and maxillipeds until the latest stages
- similar species: most of these distinctive features are found in other stomatopod species; larvae of most have not been described

Coronis sp.

antizoea

- carapace longer than broad with dorsal spine slightly curved down (1)
- eyes not stalked (2)
- posteriolateral spines straight but slightly ventrally directed (3)
- rostrum smooth and ventrally directed (4)
- 5 pairs of biramous thoracic appendages and no abdominal appendages (5)
- similar species: other stomatopod larvae have long, spined carapaces that are not fused to most thoracic segments, but only antizoeae are lacking abdominal appendages and claws

1mm

SEA SPIDERS, MITES, AND INSECTS

This section contains two groups of primarily benthic arthropods that are occasionally swept off the bottom and then caught in plankton samples plus insects often incidentally collected at the water's surface. These animals are particularly common in saltmarsh creeks and other shallow inshore habitats.

SEA SPIDERS (PYCNOGONIDS)

Sea spiders occur in high-salinity areas throughout our range, especially in saltmarsh creeks or on hard substrates such as docks, jetties, or oyster reefs. These bizarre-looking, eight-legged creatures are usually associated with hydroids or benthic fouling communities. Although they are not true spiders, pycnogonids resemble spiders sufficiently to make them recognizable.

MARINE MITES (FAMILY HYDRACHNIDAE)

The mites resemble diminutive (<2 mm) seagoing ticks. Unlike most terrestrial mites, the most common marine mites are predators on minute organisms rather than parasites.

INCIDENTAL INSECTS

Most insects in plankton collections are terrestrial insects that happened into the net before sampling. However, flying insects sometimes land on the surface film and are caught as nets are deployed or retrieved. Among the more common insects are various dipterans and their larvae. We show an adult ceratopogonid, one of the more common biting "no-see-ems," or gnats, found on salt marshes. Midge (chironomid) larvae are swept into the plankton on occasion.

pycnogonid
or sea spider

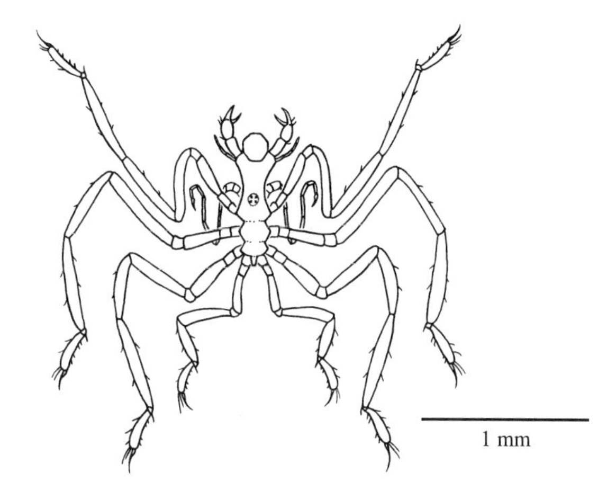

marine mite

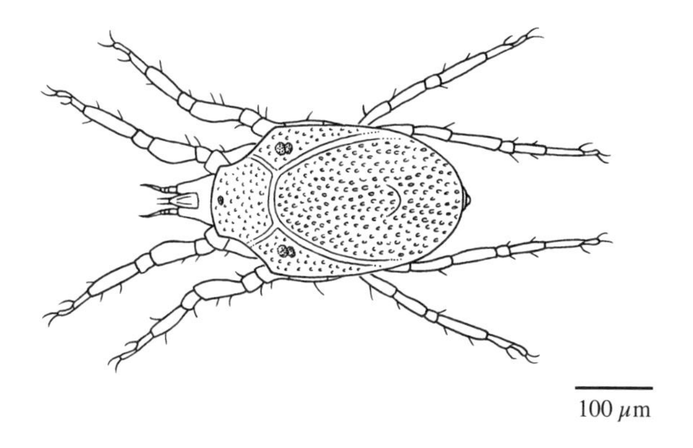

winged insect

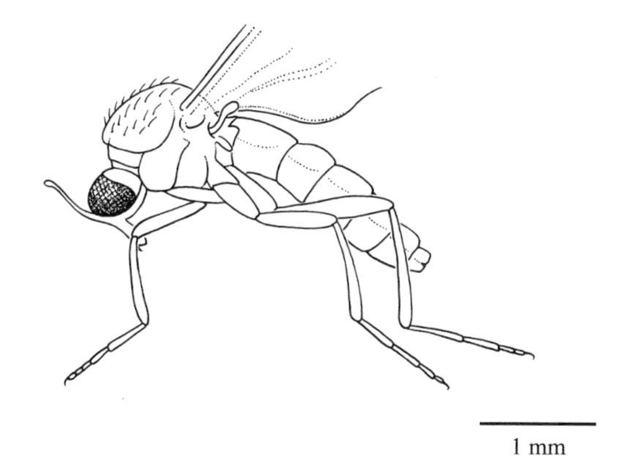

SUGGESTED READINGS

Biology and Identification References

Arnaud, F., Bamber, R. N. 1987. The biology of Pycnogonida. *Advances in Marine Biology* 24:1–96.

Child, C. A. 1992. Shallow-water Pycnogonida of the Gulf of Mexico. *Memoirs of the Hourglass Cruises* 9(1):1–86.

McCafferty, W. P. 1998. *Aquatic Entomology: The Fisherman's and Ecologist's Illustrated Guide to Insects and Their Relatives*. Jones & Bartlett, Boston. 448 pp. (Information on identification and the biology of marine mites and midge larvae.)

McCloskey, L. R. 1973. *Marine Flora and Fauna of the Northeastern United States: Pycnogonida.* NOAA Technical Report NMFS Circular 386. 12 pp.

Newell, I. M. 1947. A systematic and ecological study of the Halicaridae of eastern North America. *Bulletin of the Bingham Oceanographic Collection* 10:1–266.

Weiss, H. M. 1995. *Marine Animals of Southern New England and New York.* State Geological and Natural History Survey of Connecticut, Department of Environmental Protection, Hartford. 344 pp. (Contains an illustrated guide to identification of marsh and beach insects and to sea spiders, with some notes on ecology and habitat.)

ANNELIDS

Segmented Worms and Nematodes

POLYCHAETES

The most common and familiar annelids (segmented worms) in marine waters are the polychaete worms. Several species of polychaetes are holoplanktonic, spending their entire life cycles in the plankton, whereas many of the more than 9,000 benthic species currently described have planktonic larvae. Segmentation in this group is usually obvious with paired fleshy appendages called **parapodia** located on each body segment. The parapodia are commonly tipped with fine bristles called **chaetae** (formerly called setae).

The larvae of benthic polychaetes are among the most morphologically variable of any invertebrate group. Each of the roughly 80 polychaete families has a different larval appearance, bearing only a slight resemblance to corresponding benthic adults. Many benthic polychaetes increase fertilization success through synchronous spawning, resulting in episodic pulses of larvae in zooplankton samples. The earliest larval stage is the ciliated trochophore. The early trochophore larva most often illustrated in zoology texts is representative of the Families Serpulidae and Phyllodocidae. These polychaete trochophores are similar to the mollusc trochophore (see Fig. 26). Segmentation begins in the late trochophore as chaetae develop and cilia disappear. Later, larvae use their parapodia for locomotion as the ciliary bands disappear and segments are added sequentially to the posterior of the now clearly wormlike stages. Some polychaetes have nonfeeding larvae, which live entirely on stored yolk, whereas others have **planktotrophic** larvae that feed on phytoplankton and other small particles. *Streblospio benedicti* is unusual because individual females can brood their young, or they can release either feeding or nonfeeding larvae. Larval feeding type in this species is genetically (not environmentally) determined. Late, presettlement polychaete larvae often test several different locations before settling to benthic environments.

Three families of benthic polychaetes (Nereidae, Syllidae, and Eunicidae) enter the plankton for mating, which they accomplish during massive aggregations at the surface. In some nereids (e.g., *Alitta*, formerly *Nereis* or *Neanthes*), the reproductive planktonic adults (often called **epitokes**[1]) are specially modified and have larger eyes, reduced head appendages, and parapodia altered for swimming. Mating aggregations of some species are synchronized so that the time of swarming is highly predictable within days or hours of the event. A variety of environmental cues are involved, including phases of the moon or tide, water temperature, and photoperiod. Epitokes may be numerous in plankton tows made during the spawning period, especially at night or early morning. Some syllid polychaetes (e.g., *Myrianida*) have complex life cycles that result in planktonic reproductive stages called stolons.

Holoplanktonic polychaetes have special adaptations for living in the water column, including large and complex eyes, long sensory antennae, and flattened and, often, transparent bodies. These polychaetes prefer the clear waters of the open sea. The most common nearshore representatives are usually encountered in ephemeral parcels of offshore water. These pelagic polychaetes are predators.

OLIGOCHAETES (EARTHWORMS, ETC.)

The larger and more familiar oligochaetes are predominately terrestrial or freshwater, but many less conspicuous oligochaetes are abundant in estuarine or marine sediments. They are often indicators of polluted areas. These small (<1 cm), segmented benthic worms are occasionally resuspended in the water column and collected in plankton nets. Oligochaetes lack the parapodia of polychaetes.

HIRUDINEA (LEECHES)

Leeches are annelids with a sucker at each end. Although leeches are segmented worms, the pseudosegmentation visible on the outside of the body does not reflect true segmentation found internally. Most leeches are good swimmers and are frequently caught in the plankton. Many marine species have associations with other animals, including fishes (especially in boreal and Arctic waters), sea turtles, and a variety of crustaceans, such as blue crabs. There is no free-swimming larval stage. Leeches deposit eggs in cocoons that may be attached to the host or to the substrate. Newly hatched leeches swim immediately and readily attach to their preferred host on contact. Because leeches may leave a host if disturbed, individuals of all sizes may be in the plankton.

USEFUL IDENTIFICATION REFERENCES

Appy, R. G., Dadswell, M. J. 1981. Marine and estuarine piscicolid leeches (Hirudinea) of the Bay of Fundy and adjacent waters with a key to species. *Canadian Journal of Zoology* 59:183–192.

Bhaud, M., Cazaux, C. 1987. Description and identification of polychaete larvae: Their implications in current biological problems. *Oceanis* 13:595–753. (A guide to European polychaetes; should be useful to the family level for local species.)

Cook, D. G., Brinkhurst, R. O. 1973. *Marine Flora and Fauna of the Northeastern United States. Annelida: Oligochaeta.* NOAA Technical Report NMFS Circular 374. 23 pp.

---

1. Some authorities reserve the term **epitoke** for the posterior of the reproductive worm in contrast to the anterior part of the worm, the atoke.

Lacalli, T. C. 1980. A guide to the marine flora and fauna of the Bay of Fundy: Polychaete larvae from Passamaquoddy Bay. *Canadian Technical Report, Fisheries and Aquatic Sciences* 940:1–27. (This guide to Canadian Atlantic fauna is the closest available to our area.)

Larink, O., Westheide, W. 2011. *Coastal Plankton: Photo Guide for European Seas*. 2nd ed. Pfeil Verlag, Munich, Germany. 191 pp. (The polychaete section of this European guide is useful to identify families and genera of western Atlantic polychaetes.)

Plate, S., Husemann, E. 1994. Identification guide to the planktonic polychaete larvae around the island of Helgoland (German Bight). *Helgoländer Meeresuntersuchengen* 48:1–58. (Many of the families, genera, and species also occur on this side of the Atlantic.)

Sawyer, R. T., Lawler, A. R., Overstreet, R. M. 1975. Marine leeches of the eastern United States and the Gulf of Mexico with a key to the species. *Journal of Natural History* 9:633–667. (Contains behavioral and ecological notes.)

Shanks, A. L., ed. 2001. *An Identification Guide to the Larval Marine Invertebrates of the Pacific Northwest*. Oregon State University Press, Corvallis. 314 pp. (Useful to the family level for Atlantic and Gulf fauna.)

SUGGESTED READINGS

Planktonic Annelids and Polychaete Larvae

Giangrande, A. 1997. Polychaete reproductive patterns, life cycles, and life histories: An overview. *Oceanography and Marine Biology: An Annual Review* 35:323–386.

Levin, L. A., Caswell, H., DePatra, K. D., et al. 1987. Demographic consequences of larval development mode: Planktotrophy vs. lecithotrophy in *Streblospio benedicti*. *Ecology* 68:1877–1886.

Sawyer, R. T., Lawler, A. R., Overstreet, R. M. 1975. Marine leeches of the eastern United States and the Gulf of Mexico with a key to the species. *Journal of Natural History* 9:633–667.

Young, C. M., ed.; Rice, M. E., Sewell, M., assoc. eds. 2001. *Atlas of Marine Invertebrate Larvae*. Academic Press, New York. 656 pp. (See chapter on polychaetes by Pernet et al.)

POLYCHAETE LARVAE

ID hint: Polychaetes are distinguished by bundles of chaetae coming from the body segments or parapodia and by the characteristic peristaltic movement of annelids. Later-stage larvae usually have definitive heads with eyes, palps, or other structures. The stages shown are snapshots from a continuous process, with changes occurring over hours. The number, size, and shape of segments continue to change as the larva develops.

Late Trochophore Larva

The trochophore larva is recognized by its ciliary bands. The example shown is typical, but there is considerable variability in appearance among different polychaete families. The ciliary bands gradually disappear. Late trochophores show the beginnings of segmentation. Some trochophores, like this one, have a strong resemblance to mollusc trochophores while others are quite different.

Early Segmented Larva

The presence of segmentation with early parapodia indicates a transitional phase following the late trochophore stage. The ciliary bands characteristic of the trochophore usually disappear as the body becomes segmented, but these features co-occur in some larvae. This intermediate stage is sometimes called a metatrochophore.

Presettlement Larva

As larval development continues, many additional segments appear and the parapodia usually become more prominent and complex. At later stages, the larvae usually have distinct heads with eyes, palps, or tentacles characteristic of adults. Again, late larval stages of the many species of polychaetes co-occurring in shallow waters vary greatly in size, complexity, and behavior. Presettlement larvae are usually motile enough to be able to sample and finally select suitable benthic environments for settlement and development into juvenile forms.

References. Blake 1969; Dales 1950; Dean 1965; Levin and Creed 1986; Levin et al. 1987.

late trochophore larva

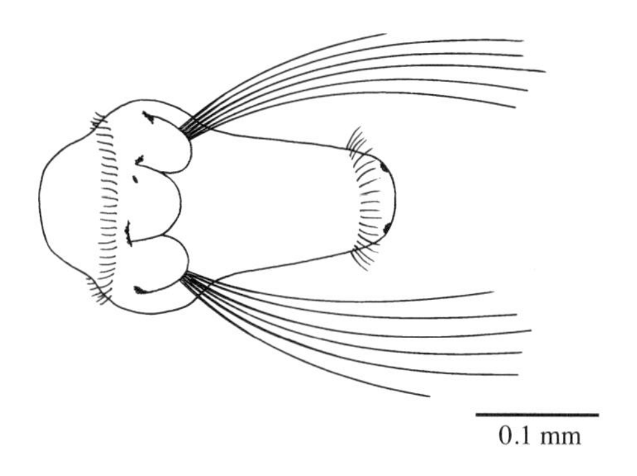

early segmented larva

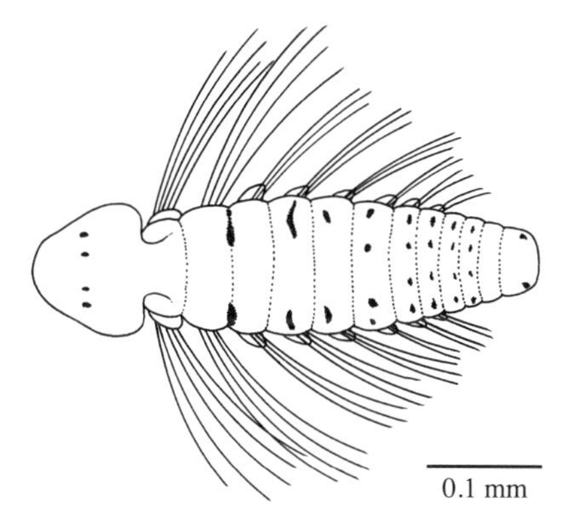

presettlement larva

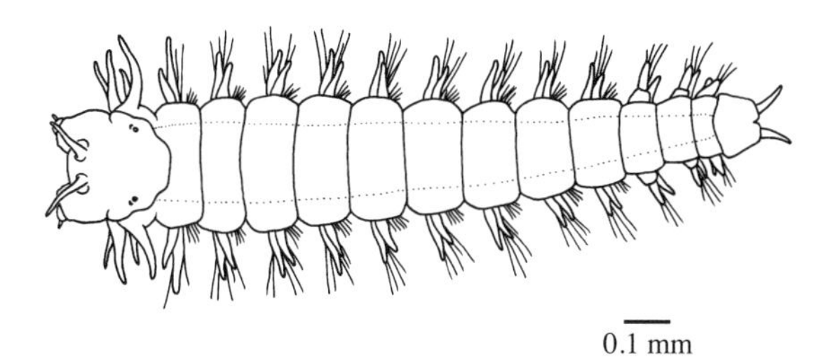

POLYCHAETES: REPRODUCTIVE ADULTS

Alitta spp. (clamworms), formerly *Nereis (Neanthes)* spp.

Occurrence. The clamworm *Alitta virens* (formerly *Nereis virens*) is common in northern, high-salinity coastal systems. Another clamworm, *A. succinea* (formerly *N. succinea*), is abundant across a wide range of salinities in Atlantic and Gulf Coast marshes and estuaries.

Biology and Ecology. Nocturnal or dawn-breeding aggregations of the specially modified reproductive worms of *A. succinea* take place periodically at the surface of both coastal and estuarine waters throughout our region. At breeding time, the benthic worms transform into a planktonic reproductive form often called an epitoke in which the posterior region is filled with eggs or sperm. Once at the surface, these fast-swimming worms with enlarged parapodia release their gametes. The resulting larvae may be locally abundant in the plankton. The precise timing of the aggregations is in response to lunar cycles. In *A. virens* only, the males become planktonic and mate with females on the bottom. Both *A. succinea* and *A. virens* were recently in the genera *Nereis* and *Neanthes*. Much of the literature on these species can be found under those genus names.

References. Bass and Brafield 1972; Chatelain et al. 2008; Dales 1950; Ram et al. 2008; Wilson and Ruff 1988.

Myrianida spp., formerly *Autolytus* spp.

Occurrence. Many species of syllid polychaetes (Family Syllidae) occur in shallow water systems along the Atlantic and Gulf Coasts. Reproductive individuals are commonly encountered in plankton collections in the warmer months, especially in high-salinity regions near the coast. Different species of *Myrianida* seem to be the most common syllids in nearshore plankton collections.

Biology and Ecology. Juveniles living on the bottom become male or female "stolons" through an extraordinary form of asexual budding. Stolons are active swimmers and leave the bottom to enter the water column where reproduction occurs. Mature planktonic female stolons carry their eggs in a ventral pouch, which bursts to release free-swimming young.

References. Franke 1999; Gidholm 1965; Schiedges 1979a, 1979b.

Alitta succinea epitoke

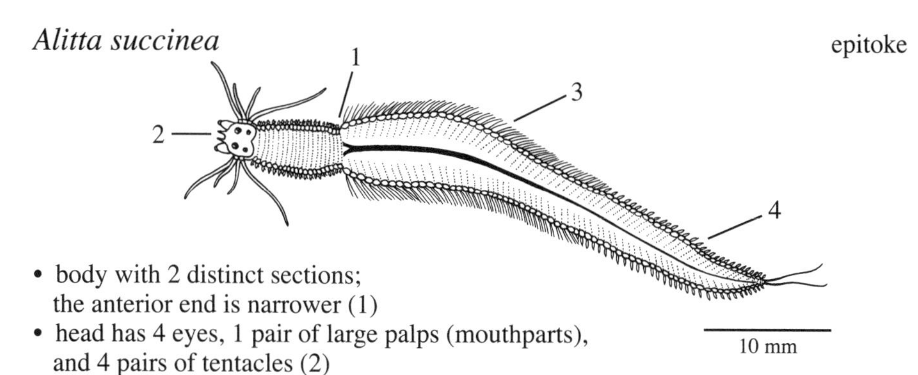

- body with 2 distinct sections;
  the anterior end is narrower (1)
- head has 4 eyes, 1 pair of large palps (mouthparts), and 4 pairs of tentacles (2)
- posterior section is tapered with large chaetae on the front half (3) and short chaetae along the tapered end (4)
- similar species: other species of clamworms and other types of polychaetes produce similar epitokes, and many are difficult to distinguish from one another

---

Myrianida sp. stolon

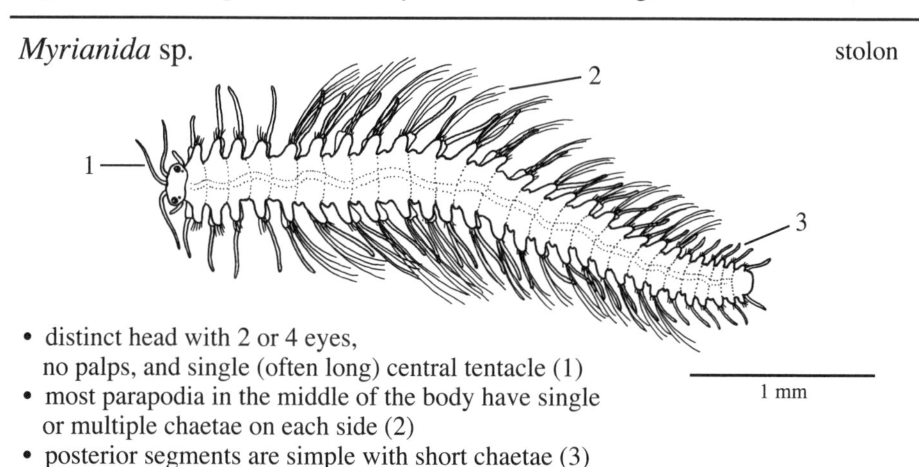

- distinct head with 2 or 4 eyes,
  no palps, and single (often long) central tentacle (1)
- most parapodia in the middle of the body have single or multiple chaetae on each side (2)
- posterior segments are simple with short chaetae (3)

Myrianida sp. stolon with eggs

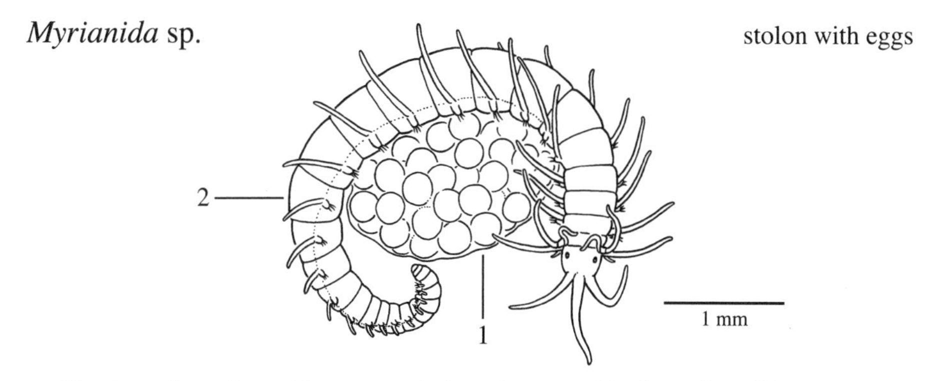

- side view of a stolon with eggs carried on the ventral body surface (1)
- body is usually seen curled around egg mass (2)
- similar species: many species of syllids produce free-swimming stolons, and they are difficult to identify to the species level

Polygordius spp. *larvae*

Occurrence. *Polygordius jouinae* ranges from Massachusetts south to at least Chesapeake Bay in coastal bays and harbors and along the coast out to the continental shelf. Other species may occur farther south.

Biology and Ecology. Adults live in coarse, sandy sediments and release their gametes into the water column where fertilization occurs. This genus exhibits two types of planktonic larval forms. The exolarva of *P. jouinae* begins without segments as an early trochophore, but in later stages, the worm trunk is gradually elongated by addition of new segments to the posterior end. *Polygordius* spp. are unusual because both larvae and adults lack parapodia and chaetae typical of polychaetes. *Polygordius* larvae have been raised successfully on diatoms. Multiple larval stages and young juveniles of *P. jouinae* are most commonly found in the plankton from the end of May to September.

References. Cowles 1903; Ramey 2008.

HOLOPLANKTONIC POLYCHAETES

Tomopteris spp.

Occurrence. *Tomopteris helgolandica* is a cold-water species found from Maine to the Chesapeake Bay, especially in colder months. *Tomopteris septentrionalis* ranges from Maine to South Carolina. Although it is most often found in clear ocean water, *Tomopteris* may occur in open coastal bays, in salinities as low as 24. *Tomopteris* is frequently caught but never abundant.

Biology and Ecology. These truly pelagic polychaetes are often transparent, with broad, flattened bodies. *Tomopteris* is a fast-swimming predator that uses its eversible proboscis to capture fish larvae, chaetognaths, larvaceans, and siphonophores.

References. Jiménez-Cueto and Suárez-Morales 1999; Rakusa-Suszczewski 1968.

Polygordius jouinae exolarva

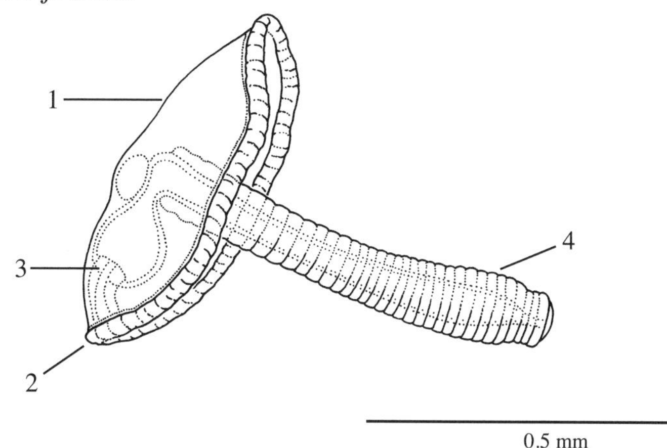

- umbrella-shaped form with stalk becoming longer as the larva develops (1)
- reddish-brown margin around the translucent cap (2)
- cap not attached to stalk; organs visible under cap (3)
- distinct segmentation on trunk (4)
- similar species: distinctive larval form among temperate coastal polychaetes

---

Tomopteris helgolandica adult

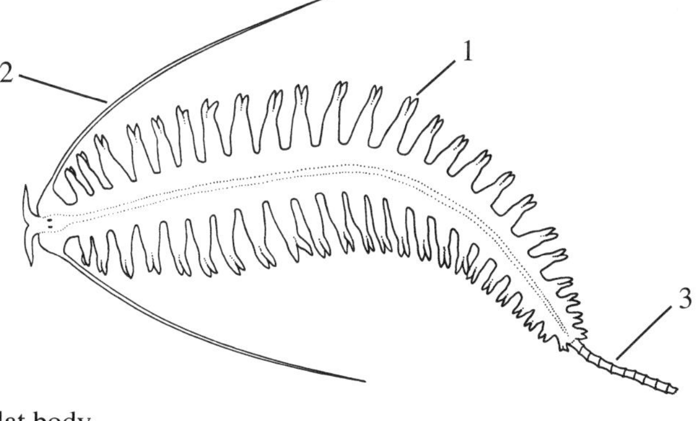

- wide and flat body with broad parapodia (1)
- one pair of very long tentacles extend from the head (2)
- posterior segments are simple, and they are often lost in collections (3)
- body transparent, although prey are often visible in the axial gut

Leeches

ID hint: Leeches appear segmented and have a sucker at either end. They move by extension of the body and alternating attachment of the anterior and posterior suckers.

Occurrence. *Cystobranchus vividus*, formerly *Calliobdella vivida*, occurs from Massachusetts to the Gulf Coast, primarily in the coldest months and especially in salinities of 5–20. *Mysidobdella borealis* is most common in colder months from New England to southern New Jersey. *Myzobdella lugubris* occurs from Massachusetts to the Gulf in brackish and freshwater. Many other species of small leeches are known from the Atlantic and Gulf Coasts.

Biology and Ecology. A variety of free-living and parasitic leeches may be in the plankton, but they are seldom numerous. Those listed next are strong swimmers and are often taken in plankton nets, especially during the coldest months. *Cystobranchus* occurs on a variety of fish hosts but is especially common in or on the mouths of menhaden. It leaves the host to deposit its eggs on oyster beds, especially near the green seaweed *Ulva*. *Cystobranchus* has chromatophores and can change color rapidly. *Mysidobdella borealis* occurs in the plankton, especially at night and may be collected near the bottom with epibenthic sleds. It is common in areas with an abundance of the mysids *Neomysis americana* and *Mysis stenolepis*, the only known hosts for this ectoparasitic leech. *Myzobdella lugubris* leaves its fish hosts to deposit its egg cocoons on crustaceans, including blue crabs and grass shrimps. Fish hosts include mullets, killifishes, and catfishes.

References. Burreson and Allen 1978; Burreson and Zwerner 1982; Daniels and Sawyer 1975; Sawyer and Hammond 1973; Sawyer et al. 1975.

Oligochaetes

Marine oligochaetes are small, cylindrical segmented worms; however, segmentation may not be obvious. Parapodia are absent. Careful examination at high magnification reveals minute chaetae imbedded in the body wall. Peristaltic movement and a tendency to become twisted distinguish them from the similar roundworms.

Biology and Ecology. Oligochaetes are widespread, often abundant, free-burrowing benthic worms. They are commonly caught in shallow nearshore and estuarine waters, where the currents sweep them from their benthic habitats. Marine oligochaetes do not produce larval stages, and juveniles resemble adults.

Reference. Attrill et al. 2009.

Nematodes

Although nematodes are not annelids, they look much like oligochaetes and some polychaetes. However, they are not segmented and move with a characteristic lashing motion rather than with peristalsis. Their cuticles have no chaetae. Most are less than 3 mm and are long and thin. The posterior end has a firm, sharp point. Unlike oligochaetes and polychaetes, nematodes are often colorless. Identification to species is not practical.

Biology and Ecology. Although best known as internal parasites, most roundworms are free living. These minute and abundant benthic worms are frequently swept into the plankton, especially in shallow areas with high turbulence, including saltmarsh creeks. Only adults are likely to be found in plankton collections.

Cystobranchus vividus

adult

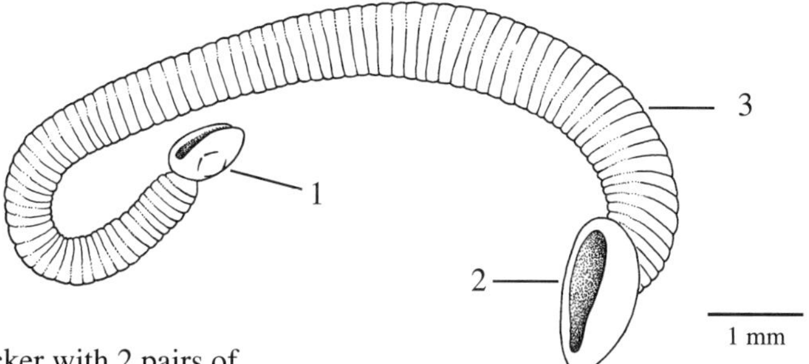

- small, oral sucker with 2 pairs of small, slit-shaped eyes on dorsal surface (1)
- large, caudal sucker is joined to the last body segment by a narrow neck (2)
- segmentation is less conspicuous in living specimens (3)

oligochaete

adult

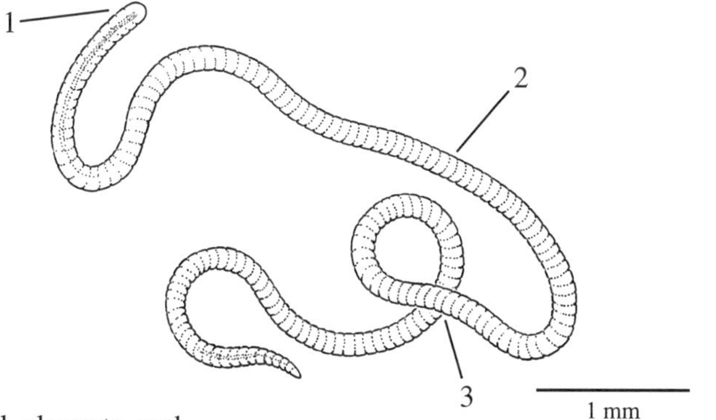

- body cylindrical, elongate, and pigmented with head end less pointed than posterior (1)
- symmetrical segments usually visible; very fine chaetae are not usually seen (2)
- body can extend and twist in life; usually contorted in preserved specimens (3)

---

nematode

adult

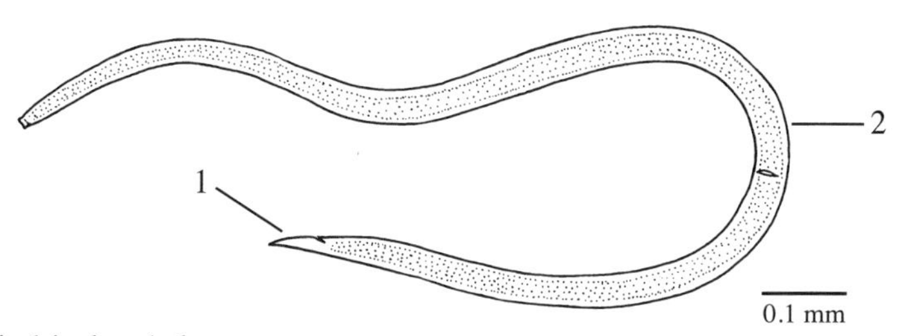

- body cylindrical and elongate, usually with firm, sharply pointed posterior end (1)
- smooth cuticle and no segmentation (2)
- limited flexibility possible but no twisting or complex convolutions possible
- body often colorless or transparent

MOLLUSCS

Gastropods, Bivalves, and Cephalopods

Many of the familiar coastal and estuarine gastropods (snails) and bivalves (clams, oysters, and scallops) have planktonic larvae that are often abundant in nearshore plankton samples. Less familiar are the holoplanktonic gastropods seen nearshore when pulses of offshore ocean water come ashore. Although most cephalopods (squid) are fast enough to escape plankton nets, juvenile squid appear sporadically in estuarine and nearshore plankton collections.

MOLLUSCAN LARVAE

Trochophore and **veliger** larvae are typical of bivalves and snails and common in the plankton (Fig. 26). Although the molluscan and polychaete trochophores are quite similar, recent evidence suggests that the similarity may be more due to coincidence than common ancestry. Trochophore larvae spin as the band of cilia on the anterior end propels them through water. The trochophore usually transforms into the veliger stage in less than 24 hours. The veliger is a uniquely molluscan larval type, characterized by two (sometimes four) ciliated lobes (the velum) that provide locomotion. When feeding, the cilia and mucus of the velum catch small particles. Veligers soon develop eyes and statocysts, both useful in vertical positioning. At the appearance of the foot, the larvae (now called **pediveligers**) often alternate between swimming and crawling as they investigate the substrate before settlement and metamorphosis. Comb jellies, jellyfishes, and many larval fishes eat veligers.

Gastropods (snails) show a variety of reproductive modes. In some snails, including whelks (*Busycon*), mating and internal fertilization lead to production of elaborate egg cases from which tiny snails emerge, thus eliminating planktonic larval stages. In other snails, the trochophore stage takes place within egg cases or gelatinous egg masses, and larvae first appear in the plankton as veligers. For example, veliger larvae of moon snails (*Polynices*) emerge from the sandy, collar-shaped masses often seen on beaches. Gastropod larvae are episodic and may be among the most abundant zooplankton in some nearshore samples.

Bivalves typically shed their gametes into the water where fertilization and complete larval development ensue. The majority of trochophores in the plankton are those of bi-

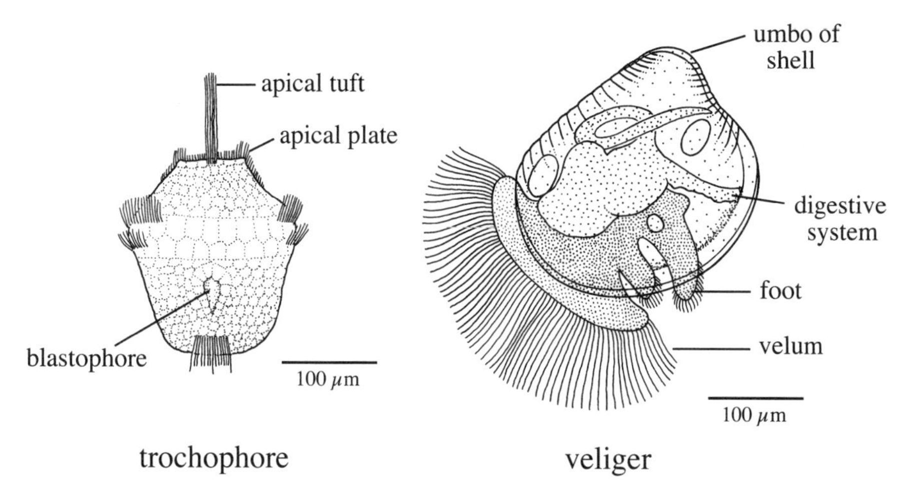

Fig. 26. Typical molluscan larval stages. A ciliated molluscan trochophore larva and a bivalve veliger.

valve molluscs. In less than a day, the trochophores transform into veligers with a shell and velum. The first shelled larva is called the straight hinge larva. As the larva grows into the "umbo stage," the umbo gradually obscures the hinge. The first larval foods are small pico- and nanoplankton (0.5–3 μm) and even bacteria. As the larvae grow, they may shift to slightly larger food (3–10 μm). The final veliger stage (the pediveliger) is often epibenthic, alternating between swimming and searching for a substrate. In some species, habitat selection involves a complex set of behaviors involving both physical and chemical testing. The newly developed foot is used in locomotion as the larva crawls on the bottom. If a suitable substrate is not found, metamorphosis may be delayed as searching continues. After metamorphosis, many bivalves reenter the plankton during a secondary planktonic dispersal phase before final settlement. Bivalve larvae are common in the nearshore plankton, especially in the spring and early summer.

HOLOPLANKTONIC SNAILS: PTEROPODS (SEA BUTTERFLIES)

The pteropod gastropods, sometimes called sea butterflies, include both shelled (Thecosomata) and shell-less (Gymnosomata) orders of holoplanktonic snails. The traditional term pteropod remains a convenient informal grouping for Thecosomata and Gymnosomata. Adaptations to planktonic life include loss or reduction of the shell to retard sinking and the conversion of a portion of the foot and body wall into lateral "wings" for swimming. Pteropods literally flap themselves through the water. The shelled thecosomes use a large external mucous net to trap phytoplankton and protozooplankton, including tintinnids and forams. The shell-less gymnosomes are specialized predators, often feeding on their thecosome relatives. Fishes, baleen whales, and seabirds also eat thecosomes. Sea butterflies are widespread in the open oceans but are rare nearshore, except when brought shoreward in oceanic water masses.

CEPHALOPODS (SQUID AND OCTOPUS)

The first posthatching stage of cephalopods is called the **paralarva**, which is often found in near-surface waters during the day. Occasionally, newly hatched individuals of commercial squid (*Illex* or *Doryteuthis*) are captured within a few kilometers of shore. The only squid found regularly in nearshore areas is the brief squid (*Lolliguncula brevis*), common along the Southeast and the Gulf Coasts and estuaries. Although the brief squid is an agile swimmer, juveniles (<15 mm) are vulnerable to capture by larger plankton nets. The entirely transparent octopus paralarvae are rarely collected in nearshore waters.

IDENTIFICATION HINTS

Most of the molluscan larvae encountered will be veligers, and the shape of the developing shell is the key to identification. The coiled shell and ciliated velum of the gastropods readily distinguish them from other plankton. Some of the shelled sea butterflies also have coiled shells, but they swim with muscular flaps rather than cilia. These distinctions are more easily made in living specimens. Identification of snail and bivalve larvae to genus or to species is generally only practical for late larvae where the shell resembles that of the adult. Earlier stages of many local gastropods are not yet described.

Bivalve molluscs can be confused with several crustacean groups (cladocerans, ostracods, and barnacle cyprid larvae) that also have bivalved shells. These crustaceans have thinner, more flexible shells, although ostracod shells may be brittle like those of developing bivalves. All of the crustaceans have jointed appendages that may protrude from their shells, whereas only soft tissue extends from bivalve veligers. Look for sculpturing of the shell to distinguish ostracods from bivalve larvae.

Because both size and morphology change rapidly during larval development, identification of most gastropod and bivalve larvae to species is difficult, requiring practice, patience, and a compound microscope. Dozens of other species occur in our range, and the larval stages of many are either undescribed or difficult to distinguish. We present only one stage of some familiar species of bivalves.

USEFUL IDENTIFICATION REFERENCES

Larval Molluscs

Chanley, P., Andrews, J. D. 1971. Aids for identification of bivalve larvae of Virginia. *Malacologia* 11:45–119.

Garland, E. D., Zimmer, C. A. 2002. Techniques for the identification of bivalve larvae. *Marine Ecology Progress Series* 225:299–310.

Lebour, M. V. 1937. The eggs and larvae of the British prosobranchs with special references to those living in the plankton. *Journal of the Marine Biological Association of the UK* 22:105–166.

Scheltema, R. S. 1962. Pelagic larvae of New England intertidal gastropods. I. *Nassarius obsoletus* Say and *Nassarius vibex* Say. *Transactions of the American Microscopy Society* 81:1–11.

Scheltema, R. S., Scheltema, A. H. 1965. Pelagic larvae of New England intertidal gastropods. III. *Nassarius trivittatus. Internationale Revue für Gesamten Hydrobiologie* 25:321–329.

Thiriot-Quievreux, C. 1980. Identification of some planktonic prosobranch larvae present off of Beaufort, North Carolina. *Veliger* 23:1–10.

Thiriot-Quievreux, C., Scheltema, R. C. 1982. Pelagic larvae of New England intertidal gastropods. V. *Bittium alternatum, Triphora nigrocincta, Cerithiopsis emersoni, Lunatia heros* and *Crepidula plana. Malacologia* 23:37–46.

Holoplanktonic Gastropods and Cephalopods

Castellanos, I., Suárez -Morales, E. 2001. Heteropod molluscs (Carinariidae and Pterotracheidae) of the Gulf of Mexico and the western Caribbean Sea. *Anales del Instituto de Biologia Universidad National Autonoma de Mexico Serie Zoologia 2001* 72:221–232.

McConnathy, D. A., Hanlon, R. T., Hixon, R. F. 1980. Chromatophore arrangements of hatchling loliginid squids (Cephalopoda, Myopsida). *Malacologia* 19:279–288.

Noji, T. T., et al. 1997. Notes on the identification and speciation of Heteropoda (Gastropoda). *Zoologische Mededelingen, Leiden* 47:545–560.

Sweeney, M. J., Roper, C. F. E., Mangold, K. M., et al. 1992. *"Larval" and Juvenile Cephalopods: A Manual for Their Identification.* Smithsonian Contributions to Zoology 513. Smithsonian Institution Press, Washington, DC. 282 pp.

Tesch, J. J. 1948. The Gymnosomata. I. *Dana Reports*, No. 30, 1–45.

Tesch, J. J. 1950. The Gymnosomata II. *Dana Reports*, No. 36, 1–55.

van der Spoel, S. 1976. *Pseudothecosomata, Gymnosomata and Heteropoda (Gastropoda).* Bohn, Scheltema & Holkema, Utrecht, The Netherlands. 484 pp.

van der Spoel, S., Newman, L., Estep, K. W. 1997. *Pelagic Molluscs of the World.* ETI Expert Centre for Taxonomic Identification. CD-ROM. UNESCO Publishing, Amsterdam.

Vecchione, M., Roper, C. F. E., Swenney, M. J. 1989. *Marine Flora and Fauna of the Eastern United States—Mollusca: Cephalopoda.* NOAA Technical Report NMFS 73. 23 pp.

SUGGESTED READINGS

Baker, P., Mann, R. 1997. The postlarval phase of bivalve mollusks: A review of functional ecology and new records of postlarval drifting of Chesapeake Bay bivalves. *Bulletin of Marine Science* 61:409–430.

Carriker, M. R. 1990. Functional significance of the pediveliger in bivalve development. In: Morton, B., ed. *The Bivalvia: Proceedings of a Memorial Symposium in Honour of Sir Charles Maurice Yonge, Edinburgh, 1986.* Hong Kong University Press, Hong Kong, 267–282.

Huxham, M., Richards, M. 2003. Can postlarval bivalves select sediment type during settlement? A field test with *Macoma balthica* (L.) and *Cerastoderma edule* (L.). *Journal of Experiment Biology and Ecology* 288:279–293.

Lalli, C. M., Gilmer, R. W. 1989. *Pelagic Snails: The Biology of Holoplanktonic Gastropod Mollusks.* Stanford University Press, Palo Alto, CA. 259 pp.

Vecchione, M., Roper, C. F. E., Sweeney, M. J., et al. 2001. *Distribution, Relative Abundance and Developmental Morphology of Paralarval Cephalopods in the Western North Atlantic Ocean.* NOAA Technical Report 152. US Department of Commerce, Seattle, WA. 54 pp.

Villanueva, R., Norman, M. D. 2008. Biology of the planktonic stages of benthic octopuses. *Oceanography and Marine Biology: An Annual Review* 46:105–202.

Webber, H. H. 1977. Gastropoda: Prosobranchia. In: Giese, A. C., Pearse, J. S., eds. *Reproduction in Marine Invertebrates*. Vol. 4, *Molluscs: Gastropods and Cephalopods*. Academic Press, New York, 1–97.

Young, R. E., Harman, R. F. 1988. "Larva," "paralarva," and "subadult" in cephalopod terminology. *Malacologia* 29:201–207.

BIVALVE VELIGER LARVAE (SELECTED REPRESENTATIVES)

Crassostrea virginica (eastern oyster)

Occurrence. Eastern oyster larvae are most common during summer in estuaries or embayments from Cape Cod to Texas in salinities from 7 to 35.

Biology and Ecology. Trochophore larvae appear 6–9 hours after fertilization and veligers after 24–48 hours. Larvae remain in the plankton two to three weeks. *Crassostrea* veligers feed optimally on 3–5 μm phytoplankton but also eat bacteria and small protozooplankton. Pediveligers use the foot to crawl on the substrate before settling. Both chemical and physical factors influence final settlement choice. Newly settled oysters are called spat.

References. Baldwin and Newell 1995; Chanley and Andrews 1971; Kennedy 1996; Newell and Langdon 1996; Tamburri et al. 1992, 1996; Zimmer-Faust et al. 1996.

Mytilus edulis (blue mussel) and others

Occurrence. Blue mussels occur from Massachusetts to South Carolina, where they are predominately coastal and particularly abundant in areas with rocky coastlines. Larvae prefer high salinities, usually 15–30. Ribbed mussels (*Geukensia demissa*) live on salt marshes throughout the region and hooked mussels (*Ischadium recurvum*) occur from Virginia through the Gulf. The scorched mussel (*Brachidontes exustus*) is common on intertidal rocks or oyster shells from New Jersey to Florida and through the Gulf.

Biology and Ecology. The planktonic trochophores last about 24 hours and do not feed. Veliger larvae feed primarily on phytoplankton <10 μm during the 15–35 day larval period. Younger larvae tend to aggregate near the surface. A hard surface with filamentous algae or hydroids seems preferred for initial settlement. The larvae metamorphose into a plantigrade form and attach to the substrate using sticky filaments (byssus threads). The plantigrades later free themselves and reenter the plankton, using long byssal threads as underwater sails (byssal drifting). Plantigrades settle preferentially around adult mussels and then reattach by means of byssus threads.

References. Bayne 1976; de Schweinitz and Lutz 1976; Newell 1989; Wang and Widdows 1991.

Mercenaria spp. (hard clams)

Occurrence. Of the many clams in our area, the hard clam (also known as quahog) is among the most widespread and conspicuous. *Mercenaria mercenaria* is abundant from Massachusetts to South Carolina and less common farther south to Florida. *Mercenaria campechiensis* occurs from Cape May, New Jersey, south and is the primary species found in the Gulf. Hard clam veligers are abundant in the summer plankton, especially in coastal embayments. Larvae occur in estuaries in salinities as low as 5 but are more common above 20 psu.

Biology and Ecology. Larvae feed on small (<10 μm) phytoplankton during the nearly 30-day planktonic phase. Pediveligers descend to the bottom, where they alternate between swimming and crawling as they investigate the substrate before settling. Juveniles may again become suspended in the plankton before final settlement, although they can no longer swim.

References. Hard clams: Carriker 2001; Eversole 1987; Gallager 1988; Stanley and DeWitt 1983. Other clams and scallops: Abraham and Dillon 1986; Fay et al. 1983c; Garrison and Morgan 1999; LaSalle and de la Cruz 1985; Sastry 1965.

Crassostrea virginica veliger

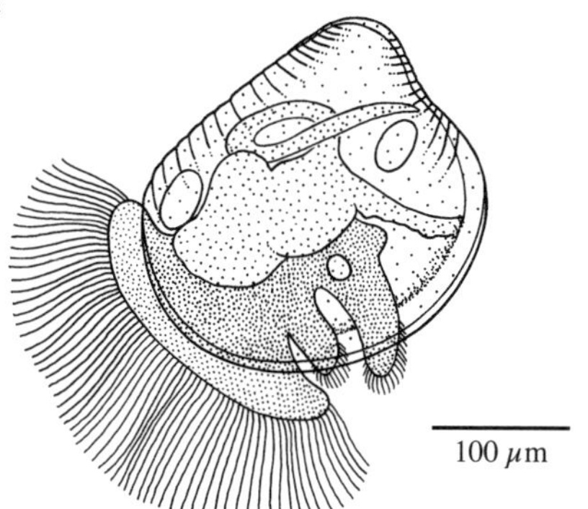

Mytilus edulis veliger

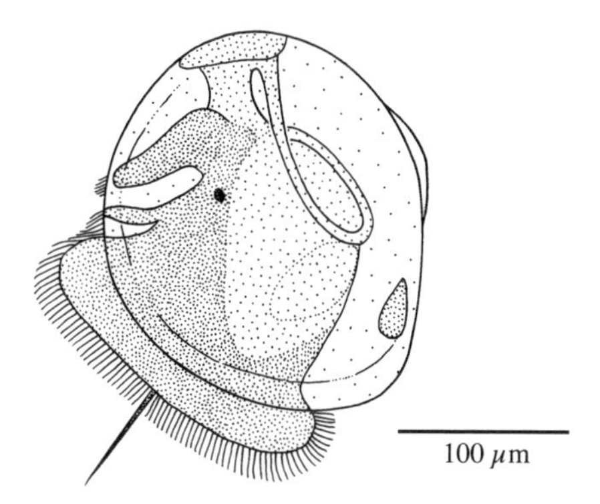

Mercenaria mercenaria veliger

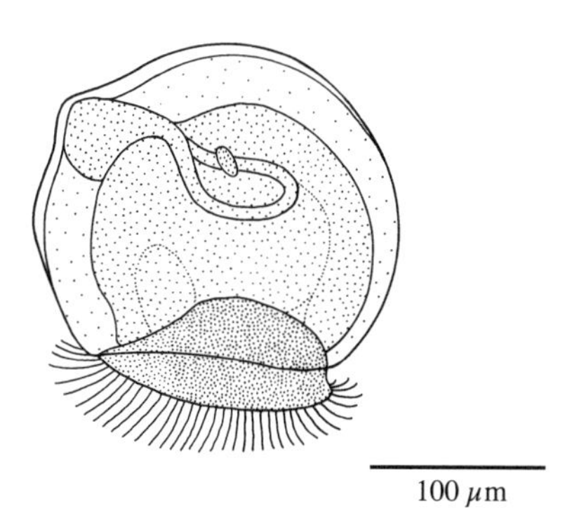

GASTROPOD LARVAE (SNAILS)

ID hint: Gastropod larvae, with their coiled shells, resemble tiny snails. Sometimes larvae can be identified by matching their shells to local adults.

Occurrence. On occasion, gastropod larvae are among the most abundant nearshore zooplankton, but the occurrence of any given species is likely to be local and brief.

Biology and Ecology. Studies on gastropod larval ecology are sparse compared with those on bivalves.

References. Hadfield et al. 2000; Russell-Hunter et al.1972; Scheltema 1961; Webber 1977.

HOLOPLANKTONIC GASTROPODS: SEA BUTTERFLIES

Limacina spp.

Occurrence. *Limacina retroversa* is widely distributed in cold boreal and north temperate waters as far south as Virginia in winter and spring. *Limacina trochiformis* occurs in Gulf Stream water and in the Gulf. Both are found nearshore in occasional pulses of clear, oceanic waters.

Biology and Ecology. Minute phytoplankton are caught with a large mucous web. Cilia transport the entire web and its food to the mouth for ingestion, and a new web is set out. *Limacina* is near the surface at night and migrates to deeper water during the day. When attacked by the predatory pteropod *Clione* in the laboratory, *Limacina* displayed several escape responses, all to little avail. *Limacina* is a protandric hermaphrodite. Swimming veligers emerge from floating egg capsules. Young of a common fish known as spot consume *Limacina trochiformis* in the northern Gulf.

References. Gilmer and Harbison 1986; Lalli and Gilmer 1989; Noij et al. 1997.

Clione limacina

Occurrence. Limited information suggests that this open-water gastropod comes close to shore only sporadically. *Clione* may be more abundant near the surface at night.

Biology and Ecology. *Clione* preys almost exclusively on *Limacina* spp. Prehensile tentacles catch the prey, which are then extracted from their shells with chitinous hooks and swallowed whole. *Clione* is hermaphroditic and deposits its eggs in gelatinous masses. The veliger larvae develop a shell, which is retained for only a few days. During early development, veligers feed on phytoplankton. Soon after shell loss, *Clione* veligers begin feeding on *Limacina* veligers.

References. Gilmer and Harbison 1986; Hermans and Satterlie 1992; Lalli 1970; Lalli and Gilmer 1989.

gastropod larva

veliger

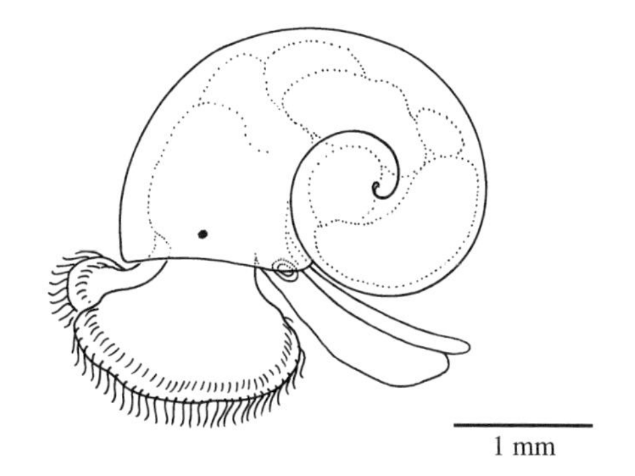

Limacina retroversa

adult

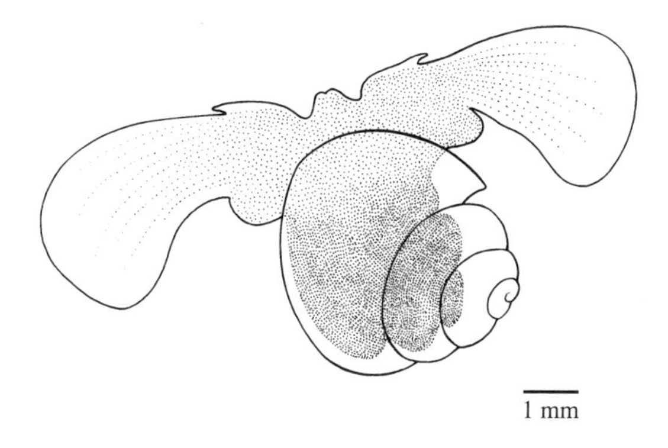

Clione limacina

adult

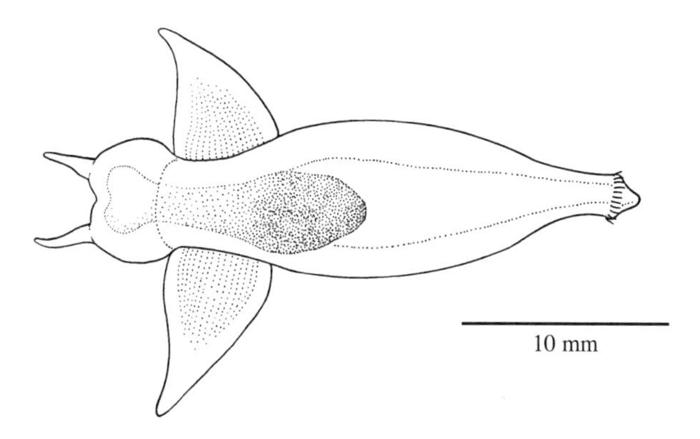

CEPHALOPODA

Lolliguncula brevis (brief squid)

Occurrence. Diminutive brief squids (also called the bay squid) occur on the Atlantic Coast from New Jersey south to Florida, but they are most abundant from Virginia south and throughout the Gulf in warm seasons. *Lolliguncula* is abundant both along the coast and in high-salinity reaches of estuaries (usually >18 psu), especially in deeper channels.

Biology and Ecology. The brief squid is the only squid common in brackish water in our region. Larvae hatch from benthic egg cases at 2–4 mm. Adults are good swimmers and usually avoid small plankton nets, but juveniles less than 15 mm are more vulnerable. This squid has a rich behavioral repertoire associated with rapid changes in color patterns produced by its chromatophores. Both its behavior and chromatophore physiology are currently under active investigation. The paralarvae of brief squid are more common in high salinities and near the bottom. Juvenile squids feed in the plankton where copepods often fall victim. Fins of fully developed brief squids are rounded and extend over less than half of the mantle's length. Very small squids are difficult to separate into species.

References. Bartol et al. 2001, 2002; Laughlin and Livingston 1982; Pattillo et al. 1997; Vecchione 1982, 1991a, 1991b; Vecchione et al. 2001.

Doryteuthis pealeii, formerly *Loligo pealeii* (longfin inshore squid)

Occurrence. Occasionally, early juvenile longfin squid are caught in nearshore plankton collections. These squid are common in the ocean from Massachusetts to North Carolina and less frequent south to Florida and in the Gulf. Unlike the brief squid, they are rare in brackish waters. Shortfin squid (*Illex illecebrosus*) paralarvae are associated with the Gulf Stream–slope water frontal zone from Canada to Florida and may be most common nearshore in the Atlantic off south Florida, where the Gulf Stream passes closest to shore.

Biology and Ecology. Newly hatched young, or paralarvae, begin hunting right away. In the laboratory, *Doryteuthis opalescens* waits with open arms as copepod prey approach and then jet swiftly forward as the arms snap shut. Identification of very young squid is difficult. *Doryteuthis pealeii* is the primary commercial squid south of Cape Cod. Note: Fins in hatchlings are not well developed and may not show the triangular shape or relative length characteristic of adults.

References. Chen et al. 1996; Dawe and Beck 1985; Pattillo et al. 1997; Vecchione 1981; Vecchione et al. 2001.

Lolliguncula brevis (lateral view)

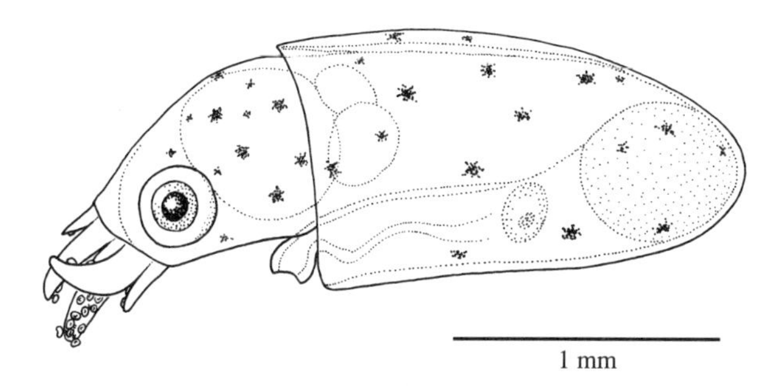

Doryteuthis pealeii (ventral view)

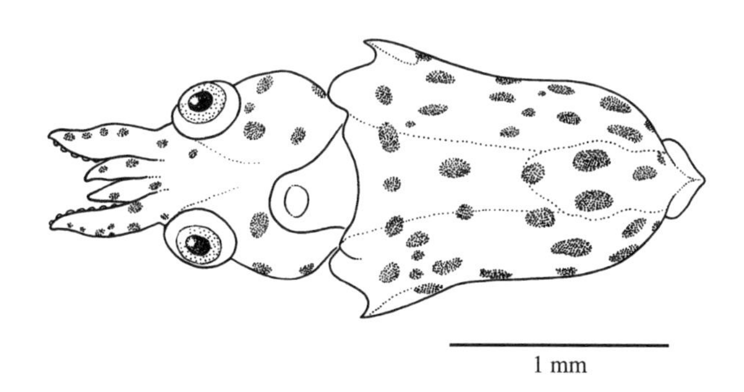

CHAETOGNATHS

Arrow Worms

The Chaetognatha constitutes a distinctive phylum of about 100 marine species ubiquitous in the world's oceans. These flattened, transparent-to-opaque creatures are easily distinguished from all other taxa. They may reach 1–2 cm, but most inshore arrow worms are smaller. Arrow worms are common along the coast and in the mouths of major estuaries during warm seasons. Most arrow worm species are so sensitive to minor changes in salinity that oceanographers can identify water masses as oceanic or coastal in origin by noting which specific arrow worms are present. Although their relationship to other invertebrate phyla is uncertain, new information suggests that the closest arrow worm relatives are roundworms (Phylum Nematoda).

Arrow worms are hermaphroditic, with the male gonads developing first. Although self-fertilization is possible in the lab, cross-fertilization may be more common in nature. Fertilized eggs are often encased in a gelatinous matrix that may float at the surface or be attached to submerged vegetation. The eggs develop directly (without metamorphosis) into juvenile arrow worms. Several generations a year may occur.

Most of the arrow worms likely to be encountered in nearshore waters were all conveniently in the genus *Sagitta* until Tokioka (1965) revised the phylum and divided *Sagitta* into nine genera. More recently, workers in the Gulf of Mexico have expanded on Tokioka's revisions, resulting in name changes for most of the local "*Sagitta*" species. Some, but not all, workers have accepted the newer classification. Both old and new names are listed in the identification section.

Arrow worms often remain quiescent while waiting to ambush prey. When passive, they sink for a period and then swim upward in rapid short bursts as undulating waves pass from head to tail. Stiffness in arrow worms is caused by pressure in their internal "hydroskeleton." They swim by alternating contractions of dorsal and then ventral longitudinal muscles. Many species undertake diel vertical migrations.

Arrow worms are nonvisual predators. Arrays of sensory cilia detect hydrodynamic vibrations of prey, triggering sudden attack when victims are within range. A flick of the tail and flex of the body propel the arrow worm forward as it unsheathes its chitinous grasping hooks (or spines) around the mouth to capture the prey, primarily copepods. A venom (tetrodotoxin) associated with the smaller teeth subdues the prey, which are then swallowed whole. Newly hatched arrow worms feed in the same fashion as the adults but on smaller prey, especially copepod nauplii. Other common foods include tintinnids, barnacle larvae, and, on occasion, fish larvae and other arrow worms. When they are abundant in coastal waters, arrow worms can have a significant effect on copepod populations. Predaceous copepods, such as *Candacia* and *Oncaea*, and larger decapod larvae, fishes, squid, ctenophores, and jellyfishes eat arrow worms.

IDENTIFICATION HINTS

Arrow worms and some elongate larval fishes appear similar on first inspection, especially if the fish's eyes have been lost (not uncommon). Arrow worms are distinguished from fishes by their anterior hooks and the presence of identical lateral fins on both sides (Fig. 27). Because the teeth and spines are often enclosed within the hood, they are often not visible through a dissecting microscope. Instead, use the following features to identify arrow worms to species.

The location of the **ovaries** relative to the fins is usually an excellent characteristic in mature specimens. Variations in the lengths of the strings of eggs are common among individuals. Inspect multiple specimens before making a firm identification.

The position of the **ventral ganglion** is often diagnostic. It is most easily seen from the side view, where it appears as a swelling on the ventral surface.

Note the presence or absence of a **fork at the anterior end of the intestine**. This feature may not be visible in opaque specimens.

The presence or absence of a **collarette**, a thin membrane on either side of the anterior trunk. Note its length and width. Collarettes are often rubbed off during collection, but remnants may still be found. Presence of a collarette is thus a more reliable characteristic than its absence. You may need to alter lighting to see the collarettes.

Whether the **rays on the lateral fins** are complete (the rays extend from the outside of the fin to the body) as opposed to being restricted to the outer fin margin (incomplete) may be diagnostic.

Sometimes the **shape of the tail section**, possession of seminal vesicles, or presence of **hairs along the trunk** can be useful identification features.

If these features do not suffice, then count teeth and spines under a compound microscope using Michel (1984) as a guide.

USEFUL IDENTIFICATION AND CLASSIFICATION REFERENCES

Bieri, R. 1991a. Six new genera in the chaetognath family Sagittidae. *Gulf Research Reports* 8(3):221–225.

Bieri, R. 1991b. Systematics of the Chaetognatha. In: Bone, Q., Kapp, A. C., Pierrot-Bults, A. C., eds. *The Biology of the Chaetognatha*. Oxford University Press, Oxford, 122–136. (Reviews the history of the taxonomic flux within *Sagitta*.)

Casanova, J.-P. 1999. Chaetognatha. In: Boltovskoy, D., ed. *South Atlantic Zooplankton*. Vol. 2. Backhuys, Leiden, The Netherlands, 1353–1374.

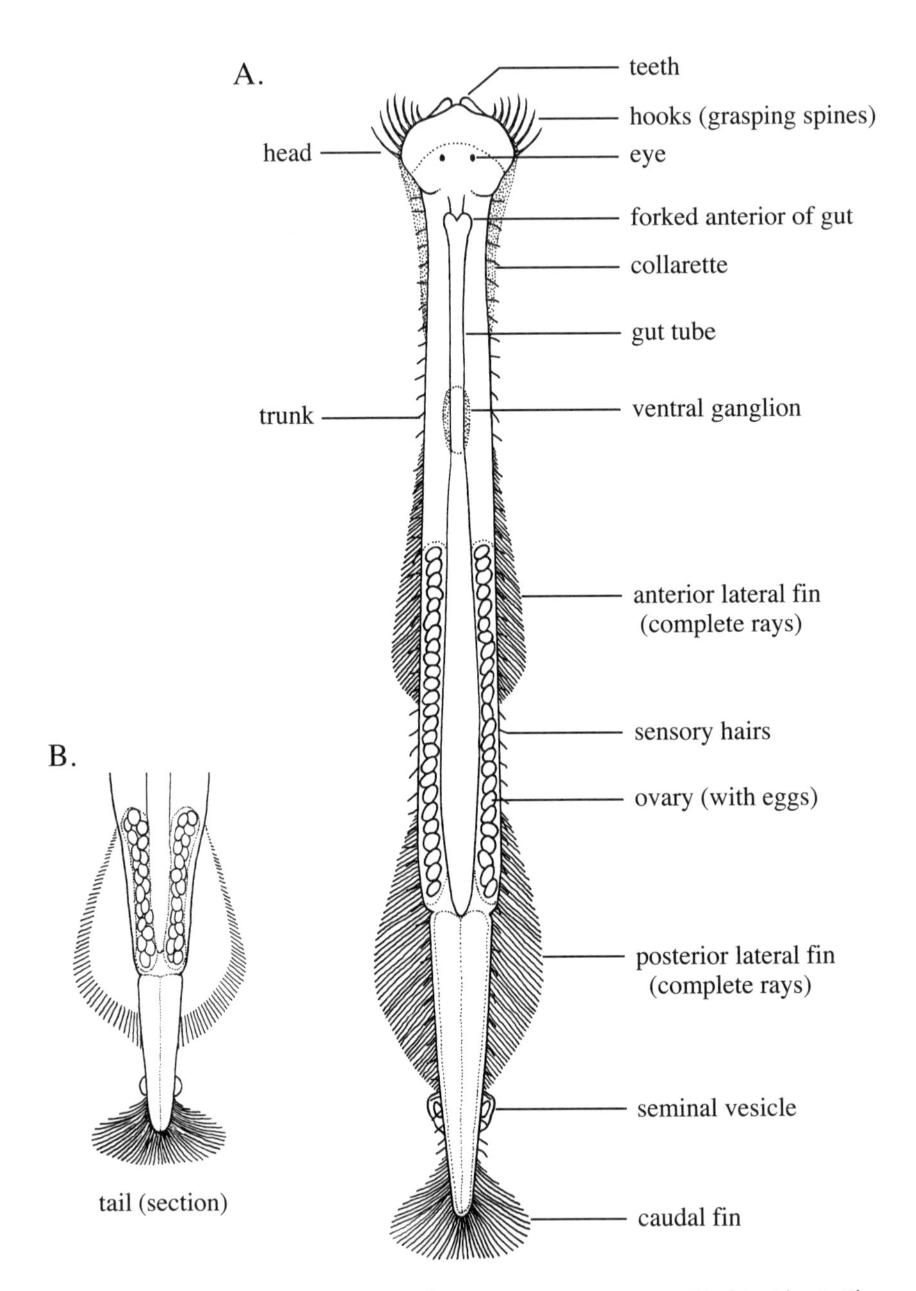

Fig. 27. *A*, The basic anatomical features of an arrow worm, *Ferosagitta hispida*. *B*, The posterior of *Flaccisagitta enflata*, showing fins with incomplete rays.

Grant, G. C. 1963. Investigations of the inner continental shelf waters off lower Chesapeake Bay. Pt. 4, Descriptions of the Chaetognatha and a key to their identification. *Chesapeake Science* 4:107–119.

McLelland, J. A. 1989. An illustrated key to the Chaetognatha of the northern Gulf of Mexico with notes on their distribution. *Gulf Research Reports* 8:145–172.

Michel, H. B. 1984. *Chaetognaths of the Caribbean Sea and Adjacent Areas*. NOAA Technical Report NMFS 15. US Department of Commerce, Washington, DC. 33 pp. (Identification details for all of the Gulf and southeast Atlantic species are here.)

Southeastern Regional Taxonomic Center (SERTC). 2004. Chaetognaths of the South Atlantic Bight and the Northern Gulf of Mexico. *SERTC Taxonomic Information and Educational Resources*. <www.dnr.sc.gov/marine/sertc/Chaetognath%20key/Chaetognath%20key.htm>. Accessed August 20, 2011.

SUGGESTED READINGS

Bieri, R., Thuesen, E. V. 1990. The strange worm *Bathybelos*. *American Scientist* 78:542–549. (Begins with an introduction to chaetognath biology.)

Bone, Q., Kapp, H., Pierrot-Bults, A. C., eds. 1991. *The Biology of Chaetognaths*. Oxford Scientific Publications, Oxford. 173 pp. (Includes chapters on feeding, locomotion, and tetrodotoxin venom. This source summarizes and cites several earlier reviews.)

Coston-Clements, L., Waggett, R. J., Tester, P. A. 2009. Chaetognaths of the United States South Atlantic Bight: Distribution, abundance, and potential interactions with newly spawned larval fish. *Journal of Experimental Marine Biology and Ecology* 373:111–123.

Grant, G. C. 1991. Chaetognatha from the central and southern Middle Atlantic Bight—species composition, temperature-salinity relationships, and interspecific associations. *Fishery Bulletin* 89:33–40.

Thuesen, E. V., Kogure, K., Hashimoto, K., et al. 1988. Poison arrow worms: A tetrodotoxin in the marine phylum Chaetognatha. *Journal of Experimental Marine Biology and Ecology* 116:249–256.

ID hint: In the three species covered on this page, the ovaries in mature specimens are confined to the posterior of the trunk and do not reach the anterior lateral fins.

Parasagitta elegans, formerly *Sagitta elegans*

Occurrence. *Parasagitta elegans* is one of the most common nearshore arrow worms from Cape Hatteras north to Massachusetts. This northern species is more common (and larger) in the southern parts of its range in winter. It occurs in the mouths of the Delaware and Chesapeake Bays, chiefly in salinities >20. Small specimens reach salinities as low as 11.

Biology and Ecology. In deeper waters, *P. elegans* is usually near the surface but has been reported as abundant just off the bottom (epibenthic zone) in shallow water.

References. Pearre 1973; Saito and Kiørboe 2001; Sullivan 1980; Tonnesson and Tiselius 2005.

Parasagitta tenuis, formerly *Sagitta tenuis,* and *Parasagitta friderici*

Occurrence. *Parasagitta tenuis* is common from Delaware Bay south through the Gulf, where it is considered an indicator of lower-salinity water along the coast. It is tolerant of slightly brackish waters of coastal bays, sounds, and inlets. It is often the most common nearshore arrow worm within its range. In the Gulf, *P. friderici* is often confused with and may even be a form of *P. tenuis.*

Biology and Ecology. The diet is more than 90% copepods in the field. Confusion over the status of *P. friderici* from 1951 to the present has plagued anyone trying to use the literature.

Reference. Canino and Grant 1985; Coston-Clements et al. 2009.

Flaccisagitta enflata, formerly *Sagitta enflata*

Occurrence. This is one of the most widespread and common chaetognaths in both the Atlantic and Gulf. It ranges north at least to Massachusetts. *Flaccisagitta enflata* has been collected nearshore and in higher-salinity areas of North Carolina and Mississippi sounds and inlets.

Biology and Ecology. Several authors have suggested a possible competition between *F. enflata* and *P. tenuis* with very little overlap between them. *Flaccisagitta enflata*'s diet includes other arrow worms.

Reference. Bushing and Feigenbaum 1984.

Parasagitta elegans

reaches 20 mm

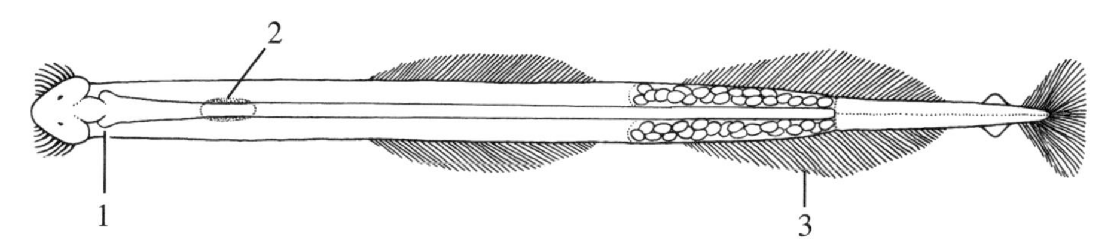

- moderately wide body without collarette
- anterior of gut forked (1)
- ventral ganglion slightly closer to head than to first pair of fins (2)
- lateral fins fully rayed (3)

Parasagitta tenuis

reaches 8 mm

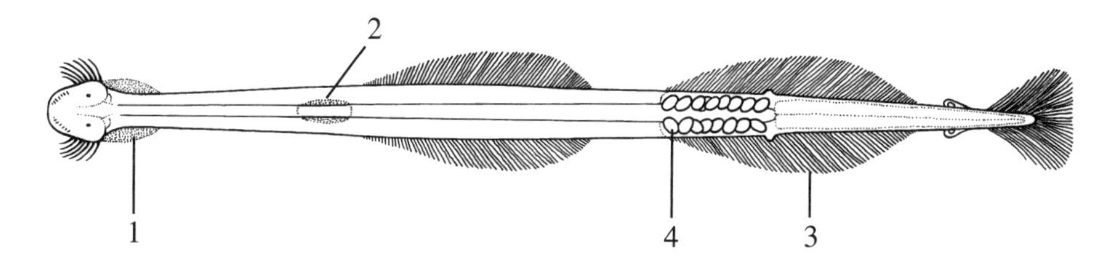

- narrow body with short collarette (1)
- ventral ganglion near origin of first pair of lateral fins (2)
- anterior of gut not forked
- lateral fins fully rayed (3)
- individual ova large (4)

Flaccisagitta enflata

reaches 20 mm

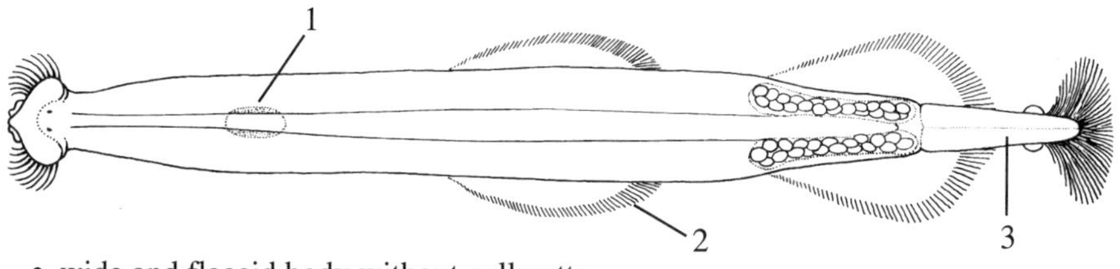

- wide and flaccid body without collarette
- ventral ganglion about halfway between head and first pair of lateral fins (1)
- anterior of gut not forked
- lateral fins not fully rayed (2)
- tail section short (3)

ID hint: In the three species covered on this page, the ovaries in mature specimens extend anteriorly at least to the base of the anterior-lateral fins.

Ferosagitta hispida, formerly *Sagitta hispida*

Occurrence. *Ferosagitta hispida* occurs in Atlantic waters from Chesapeake Bay south and in the Gulf, typically within a few kilometers of shore. It is common in sounds and estuaries of North Carolina and the Florida west coast. *Ferosagitta hispida* is common in shallow water, especially in grass beds.

Biology and Ecology. This important coastal predator feeds heavily on copepods. It spends much time near the bottom and sometimes attaches its eggs to the substrate. *Ferosagitta hispida* often shows pronounced vertical migrations into the water column at night, apparently mediated by light levels. *Ferosagitta hispida* is seldom taken in the company of *Sagitta helenae* or *Flaccisagitta enflata*, which are typically further from shore.

References. Alvarez-Cadena 1993; Baier and Purcell 1997; Coston-Clements et al. 2009; Fulton 1984; Sweatt and Forward 1985.

Sagitta helenae

Occurrence. *Sagitta helenae* occurs in Atlantic and Gulf subtropical waters, ranging north to Cape Hatteras and occasionally to Delaware. Like *Flaccisagitta enflata*, and unlike the other species noted previously, *S. helenae* is typically associated with higher-salinity continental shelf waters, several kilometers or more offshore.

Sagitta bipunctata

Occurrence. *Sagitta bipunctata* is another widespread temperate and tropical species more typical of oceanic than inshore waters.

Biology and Ecology. Its primary food is copepods. *Sagitta bipunctata* also feeds on other arrow worms, as does *Flaccisagitta enflata*.

Ferosagitta hispida reaches 15 mm

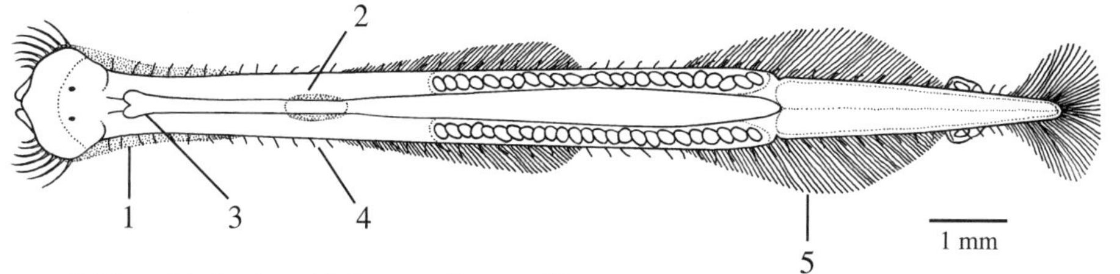

- fairly wide body with long collarette (1)
- ventral ganglion slightly anterior to origin of first pair of lateral fins (2)
- anterior of gut forked (3)
- body sparsely covered with short, fine hairs (4)
- lateral fins fully rayed (5)

Sagitta helenae reaches 15 mm

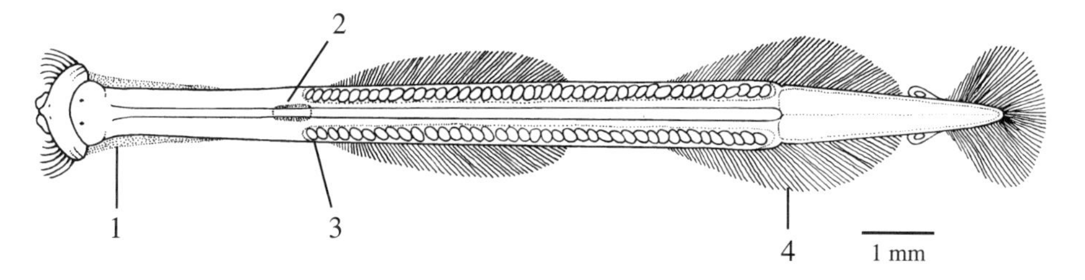

- moderately wide body with medium collarette (1)
- ventral ganglion slightly anterior to origin of first pair of lateral fins (2)
- anterior of gut not forked
- ovaries extend to near posterior end of ventral ganglion (3)
- lateral fins fully rayed (4)

Sagitta bipunctata reaches 19 mm

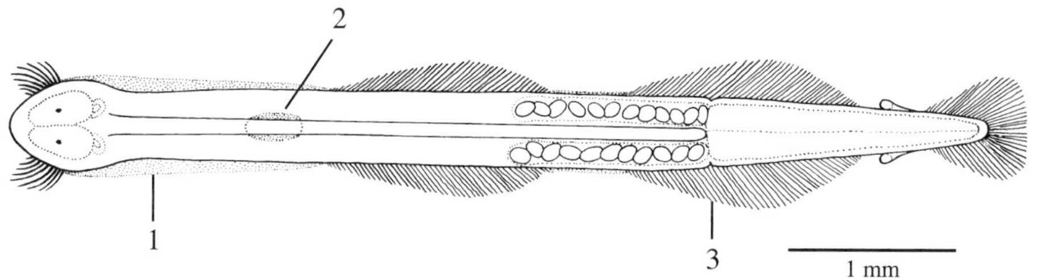

- wide body with very long collarette (1)
- ventral ganglion anterior to origin of first pair of lateral fins (2)
- anterior of gut not forked
- lateral fins fully rayed (3)

ECHINODERM LARVAE

Starfishes and Sea Urchins

The Phylum Echinodermata contains starfishes, brittle stars, sea urchins, and several additional classes. Most of the familiar coastal and estuarine echinoderms have planktonic larvae that are occasionally common in nearshore plankton but rare in brackish waters. Fertilization and larval development occur in the water column. Each class of echinoderms has one or more characteristic ciliated larval forms. Some echinoderms produce nonfeeding larvae that subsist on food reserves in the egg. More typically, the larvae feed on phytoplankton. Cilia provide currents for both locomotion and feeding. Many larvae are undescribed, and information on echinoderm larval occurrence and ecology along our coasts is spotty at best. Adult echinoderms often spawn synchronously, releasing clouds of gametes. Echinoderm larvae may be locally abundant for brief periods after spawning events. Identification to species is generally not feasible because larvae of many species have not been described. No systematic surveys of echinoderm larval distribution are currently available.

PLUTEUS LARVA OF BRITTLE STARS, SEA URCHINS, AND SAND DOLLARS

The rather similar echinopluteus larva of sea urchins and sand dollars and the ophiopluteus larva of brittle stars typically have eight arms supported by rigid, brittle skeletal rods. A dramatic metamorphosis follows shortly after they settle.

BIPINNARIA LARVA OF SEA STARS

The bipinnaria are variable in morphology; they change during development, and there may be noticeable differences between species. The bipinnaria is perhaps the most common of several different larval stages in the typical sea star developmental sequence. The similar brachiolaria larva is a later stage in this larval continuum rather than a separate larval type. The brachiolaria uses its slightly longer larval arms to test the substrate before settling. At metamorphosis, the larval arms are resorbed, and the adult arms form independently.

SUGGESTED READINGS

Emlet, R. B. 1994. Body form and patterns of ciliation in nonfeeding larvae of echinoderms: Functional solutions to swimming in the plankton. *American Zoologist* 34:570–585.

Emlet, R. B., McEdward, L. R., Strathmann, R. R. 1987. Echinoderm larval ecology viewed from the egg. In: Jangoux, M., Lawrence, J. M., eds. *Echinoderm Studies*. Vol. 2. A. A. Balkema Press, Rotterdam, 55–136.

Hart, M. W. 1991. Particle captures and the methods of suspension feeding by echinoderm larvae. *Biological Bulletin* 180:12–27.

Strathmann, R. R. 1975. Larval feeding in echinoderms. *American Zoologist* 15:717–730.

Strathmann, R. R. 1978. Larval settlement in echinoderms. In: Chia, F. S., Rice, M. E., eds. *Settlement and Metamorphosis of Marine Invertebrates*. Elsevier, New York, 235–246.

pluteus

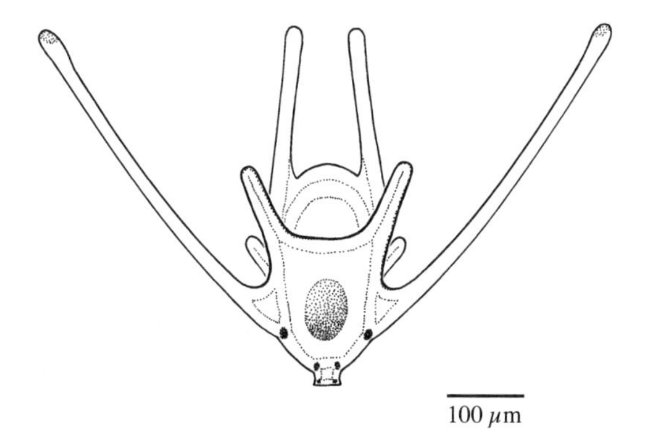

bipinnaria

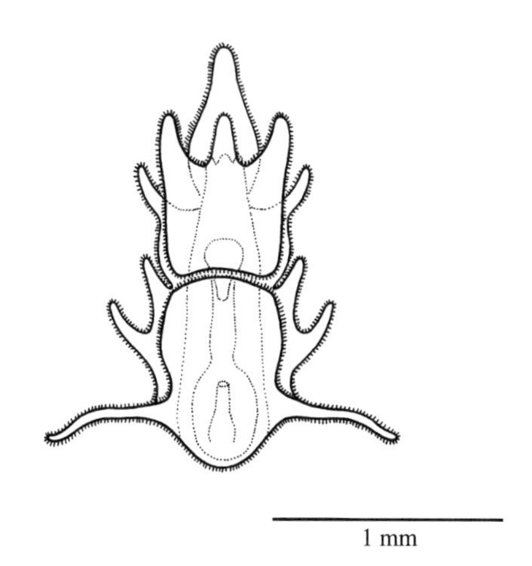

LESS COMMON CILIATED INVERTEBRATE LARVAE

Many benthic invertebrate groups produce ciliated larvae that are minor constituents of the plankton. Although uncommon, they are interesting creatures and worthy of observation when alive. Larvae of Phoronida, Brachiopoda, Echiura, Entoprocta, and Sipunculida are rare in the nearshore plankton of our geographic range and are not treated here. Identification beyond the phylum level is rarely feasible for the groups covered below.

Pilidium larva of ribbon worms (nemertea)

Ribbon worms (nemerteans) produce two types of ciliated planktonic larvae: an elongated wormlike larva and the helmet-shaped pilidium larva. Both larvae feed on small phytoplankton. Ribbon worm larvae are widespread but generally rare.

Reference. Riser 1974.

Müller's larva of flatworms (platyhelminthes)

The most common larva of the nonparasitic flatworms is the somewhat variable Müller's larva. Few, if any, feed during their several-day to several-week planktonic stage. Müller's larvae are widespread but uncommon. The specialized planktonic stages of parasitic flukes (trematodes) and tapeworms (cestodes) are seldom recognized in plankton samples.

References. Henley 1974; Ruppert 1978.

Tornaria larva of acorn worms (hemichordata)

With an apical ciliary tuft and a band of cilia encircling the body, the tornaria resembles a mollusc or polychaete trochophore. The posterior position of this band (opposite end from the apical tuft) is diagnostic. Known as the telotroch, this band provides propulsion while the more anterior looping band of cilia is used for feeding. Tornaria development has been intensively studied to provide insights into chordate origins.

Reference. Strathmann and Bonar 1976.

Cyphonautes larva of ectoprocts (bryozoa)

Cyphonautes are generally uncommon but not rare. The flattened and distinctive shape of the paired shells separates the cyphonautes from other ciliated forms in the plankton. Cyphonautes larvae are good swimmers, using a corona of cilia to provide currents for propulsion and feeding. Reported foods include phytoplankton and tintinnids. Cyphonautes may be in the plankton for a month or more before settlement. Some ectoprocts (bryozoans) release nonfeeding pseudocyphonautes or coronate larvae that are only in the plankton a few hours.

References. Burgess et al. 2009; Maki et al. 1989; McEdward and Strathmann 1987; Ryland 1976; Strathmann and McEdward 1986; Wendt 1998.

SUGGESTED READING

Also see invertebrate zoology texts for general information.

Christy, J. H., Stancyk, S. E. 1982. Timing of larval production and flux of invertebrate larvae in a well-mixed estuary. In: Kennedy, V. S., ed. *Estuarine Comparisons*. Academic Press, New York, 489–503.

Young, C. M., ed.; Rice, M.E., Sewell, M., assoc. eds. 2001. *Atlas of Marine Invertebrate Larvae*. Academic Press, New York. 656 pp.

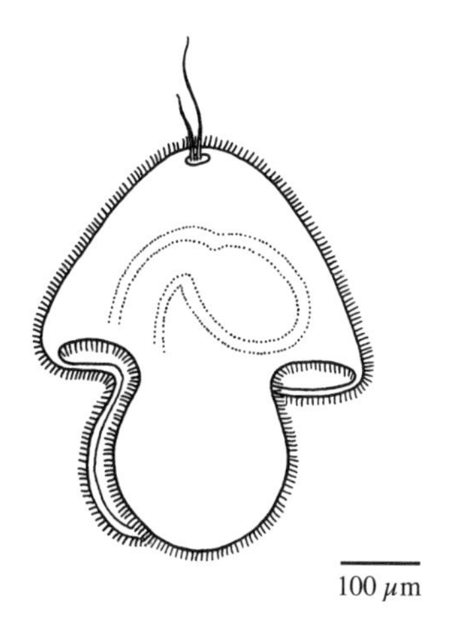

pilidium

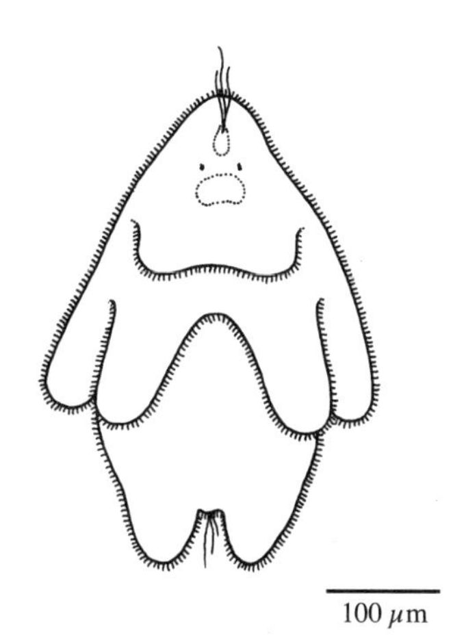

Müller's larva

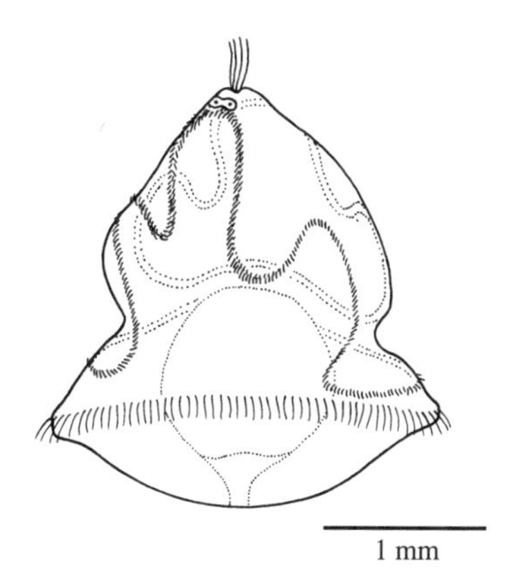

tornaria

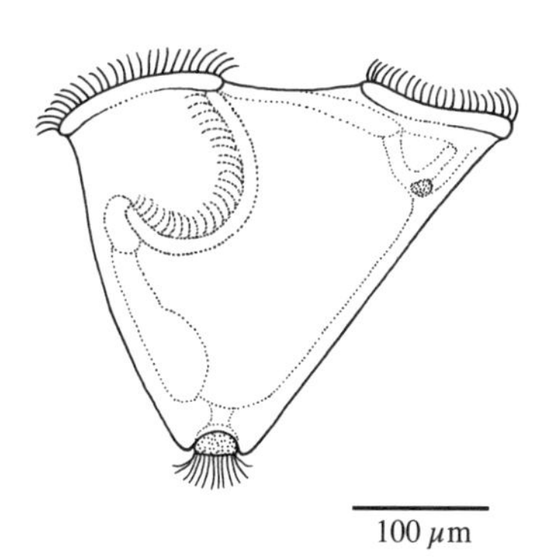

cyphonautes

LOWER CHORDATES

Larvaceans, Sea Squirts, Salps, Doliolids, and Lancelets

Phylum Urochordata contains the benthic sea squirts (or ascidians), most of which have planktonic "tadpole" larvae. The phylum also contains some holoplanktonic groups (larvaceans, salps, and doliolids), all highly modified for planktonic life and bearing little resemblance to the more familiar sea squirts. The planktonic urochordates are filter feeders that pump water through a mucous sieve to catch small, suspended particles. They can form an important link in the food web by transferring energy contained in bacteria and nanoplankton to higher trophic levels.

LARVACEANS (APPENDICULARIA)

Larvaceans are one of the two types of holoplanktonic urochordates. Their name is derived from their resemblance to the tadpole larvae of sea squirts. These small (1–5 mm) urochordates are common in coastal waters where they are often abundant. There is no definitive head. The body consists of an amorphous-looking trunk and a characteristic, curved tail. The most remarkable feature of larvaceans is their external gelatinous "house" containing a series of filters (Fig. 28). The empty mucous houses contribute to a particular form of detritus known as **marine snow**, an important part of coastal and offshore food webs. Planktivorous fishes (including many fish larvae), arrow worms, and jellyfishes prey on larvaceans.

SALPS AND DOLIOLIDS (THALIACEA)

Salps and doliolids represent the other group of holoplanktonic urochordates. They are hollow, barrel-shaped, gelatinous animals common in the open sea. Both salps and doliolids are nonselective feeders on particles ranging in size from bacteria to diatoms. The filter is composed of two mucous nets secreted by the endostyle, one fine and one of coarser mesh. Water enters the oral opening or siphon, passes through the mucous filters, and then exits the atrial opening at the opposite end of the barrel. The fine net with its trapped food particles is ingested, and another is secreted. Although somewhat larger particles are ingested, particles 1–4 μm are removed with maximum efficiency. Most other suspension feeders cannot remove such small cells, so salps and doliolids, along with larvaceans, constitute a

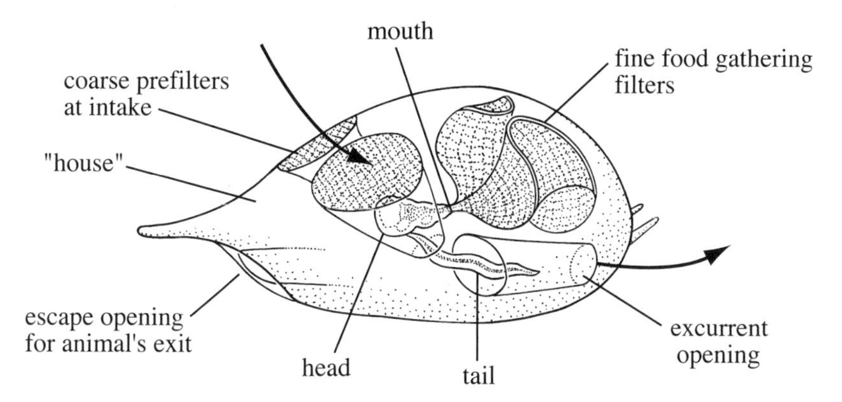

Fig. 28. Larvaceans, such as *Oikopleura*, secrete complex gelatinous houses. The tail propels the "house" through the water and simultaneously creates a water current through the house. As water flows through the fine filter, small particles become trapped in mucus and are eventually ingested. The houses are disposable and are abandoned when the filters become clogged or if the larvacean is threatened.

key link between nanoplankton and larger consumers. In salps, peristaltic contractions of the muscular bands that encircle their bodies (Fig. 29) produce a water current through the barrel and its filter and provide propulsion. In contrast, doliolids use cilia to provide both the feeding current and limited locomotion.

Both salps and doliolids have complex life histories, with alternation of sexual and asexual generations. Asexual budding results in rapid colony formation. In addition, salps and doliolids have some of the fastest generation times recorded for metazoans, resulting in massive "swarms" that can extend for hundreds of kilometers when food is plentiful. An individual salp or "zooid" initiates asexual reproduction by producing a stolon (Fig. 29) that gives rise to a chain or colony of many individuals. Members of the colony reproduce sexually with internal fertilization that leads to development inside the parental individual until young zooids are released through the atrial opening. This entire cycle may take only 50 hours. The doliolid life cycle is even more complex, with considerable specialization in the "zooids" not seen in salps. Doliolids produce a short-lived tadpole larva.

ASCIDIAN TADPOLE LARVAE (UROCHORDATA)

Several features of the swimming tadpole larva suggest an evolutionary link between sea squirts and the vertebrates. The nonfeeding larvae seem specifically adapted for dispersal; statocysts and ocelli are common. Initial photopositive behavior followed by photonegativity before settlement occurs in the lab. Chemical cues facilitate settlement on or near adults of the same species. At settlement, adhesive papillae cement the tadpole in place. A complex metamorphosis into the sessile adult form follows promptly. Most solitary sea squirts shed their gametes, leading to external fertilization and a planktonic phase of sev-

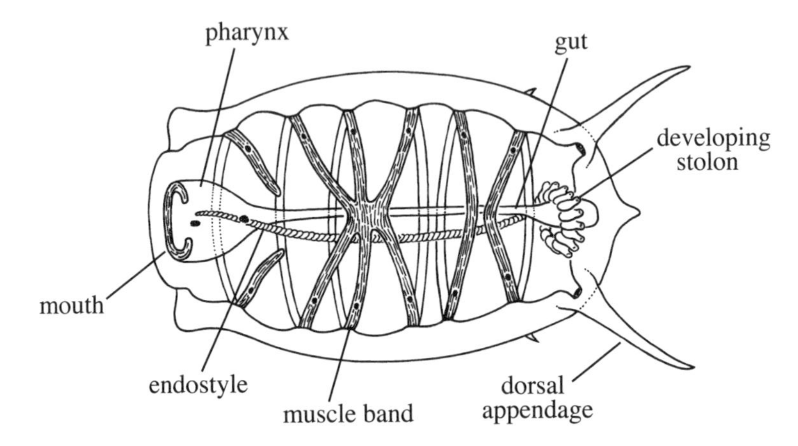

Fig. 29. Salps, such as *Thalia*, feed by filtering particles from water taken into the mouth. Water exits through the atrial opening at the other end of the barrel-shaped body. The mucous filters, located at the anterior end of the gut, are produced by the endostyle. Muscle bands enable locomotion.

eral weeks. In contrast, most colonial sea squirts brood their young, and the larval phase lasts only a day or two. Two common genera, *Molgula* and *Styela,* release nonswimming larvae that settle after a brief dispersal phase.

Ascidian tadpole larvae can occur anywhere along the coast, especially where there are hard substrates for the adults. Samples taken just down current from marinas or floating docks often yield tadpole larvae. Most ascidians require higher salinities, but *Molgula* is tolerant of more brackish areas and is abundant in estuaries, especially on oyster reefs. Our treatment of different species of sea squirt larvae is restricted to just two among the dozens that occur within our range.

LANCELETS (CEPHALOCHORDATA OR "AMPHIOXUS")

Adult lancelets live in sediments, but the larvae, which resemble the adults, are planktonic. When feeding, branchial cilia bring small particles (primarily phytoplankton) into the atrial chamber, where they become trapped in mucus and transported to the stomach. *Branchiostoma*, the only genus of these small, fishlike chordates found within our range, was formerly classified as *Amphioxus* and is featured in discussions of chordate evolution.

USEFUL IDENTIFICATION REFERENCES

Boltovskoy, D., ed. 1999. *South Atlantic Zooplankton.* Vols. 1, 2. Backhuys, Leiden, The Netherlands. (See chapters on Appendicularia by Esnal and on salps and doliolids by Esnal and Daponte.)

Bullard, S. G., Whitlatch, R. 2004. *A Guide to the Larval and Juvenile Stages of Common Long Island Sound Ascidians and Bryozoans.* Connecticut Sea Grant, Groton, CT. 39 pp.

Fenaux, R. 1998. The classification of Appendicularia. In: Bone, Q., ed. *The Biology of Pelagic Tunicates*. Oxford University Press, Oxford, UK, 295–306.

Fraser, J. H. 1982. *British Pelagic Tunicates*. Cambridge University Press, Cambridge, UK. 57 pp.

Godeaux, J. 1998. The relationships and systematics of the Thaliacea, with keys for identification. In: Bone, Q., ed. *The Biology of Pelagic Tunicates*. Oxford University Press, Oxford, UK, 274–294.

SUGGESTED READINGS

Alldredge, A. L., Madin L. P. 1982. Pelagic tunicates: Unique herbivores in the marine plankton. *BioScience* 32:655–663. (Reviews life cycles and ecology as well as feeding.)

Boltovskoy, D., ed. 1999. *South Atlantic Zooplankton*. Vols. 1, 2. Backhuys, Leiden, The Netherlands. (See chapters on Appendicularia by Esnal and on salps and doliolids by Esnal and Daponte.)

Bone, Q., ed. 1998. *The Biology of Pelagic Tunicates*. Oxford University Press, New York. 340 pp. (A comprehensive treatment.)

Stokes, M. D., Holland, N. D. 1998. The lancelet. *American Scientist* 86:552–560.

Svane, I., Young, C. M. 1989. The behavior and ecology of ascidian larvae. *Oceanography and Marine Biology: An Annual Review* 27:45–90.

Vargas, C. A., Madin, L. P. 2004. Zooplankton feeding ecology: Clearance and ingestion rates of salps *Thalia democratica*, *Cyclosalpa affinis* and *Salpa cylindrica* on naturally occurring particles in the Mid-Atlantic Bight. *Journal of Plankton Research* 26:827–833.

UROCHORDATES: LARVACEANS (APPENDICULARIANS)

ID hint: In larvaceans, the tail is attached to the trunk at a 90° angle.

Oikopleura spp.

Occurrence. *Oikopleura dioica* is abundant from Massachusetts to Texas in warm seasons. *Oikopleura vanhoeffeni* is a more northern species. The similar but smaller *Appendicularia sicula* occurs along the Atlantic and Gulf Coasts.

Biology and Ecology. *Oikopleura* feed extensively on nanoplankton of 1–2 μm. A high concentration of larger particles can clog their filters causing them to produce new houses and filters (Fig. 28). Generation times may be as short as one day. Predaceous copepods eat *Oikopleura* eggs and larvae.

References. Acuña et al. 1996; Bedo et al. 1993; Costello and Stancyk 1983; Deibel and Lee 1992; Galt and Fenaux 1990; Heron 1972; López-Urrutia et al. 2004; Troedsson et al. 2009; Vargas and Madin 2004.

Fritillaria borealis

Occurrence. *Fritillaria borealis* ranges widely in ocean waters from the poles to the tropics. It is common over the continental shelf and occasionally in nearshore waters.

Biology and Ecology. *Fritillaria* can retain particles from 0.2 to 30 μm. When the coarse filter becomes clogged, the water flow can be reversed to clear it.

References. Bone et al. 1979; Flood 2003; Purcell et al. 2005.

SALPS AND DOLIOLIDS (THALACEANS)

ID hint: Circular muscle bands distinguish salps and doliolids from other gelatinous zooplankton. Doliolids are thin and delicate as opposed to the thicker and more rigid salps. Both groups undergo morphological changes during their life cycles.

Thalia democratica

Occurrence. *Thalia* occurs in warm ocean waters occasionally appear inshore.

Biology and Ecology. *Thalia* feed on bacteria and small phytoplankton (mostly 1–4 μm). Under favorable conditions, they reproduce rapidly to form swarms of colonies at the surface. Hyperiid amphipods attack them and hollow out the barrels. Seabirds, sea turtles, fishes, heteropod molluscs, ctenophores, and jellyfishes all eat salps.

References. Deibel and Paffenhöfer 2009; Heron 1972; Kremer and Madin 1992; Madin 1974, 1990; Madin and Harbison 1977.

Salpa fusiformis

Occurrence. *Salpa fusiformis* is typically oceanic with periodic incursions into nearshore waters, often in large swarms.

Biology and Ecology. *Salpa fusiformis* often prefers deeper waters during the day. Its biology is similar to *Thalia*, occurring in both solitary and aggregate forms.

Oikopleura sp. adult (without house)

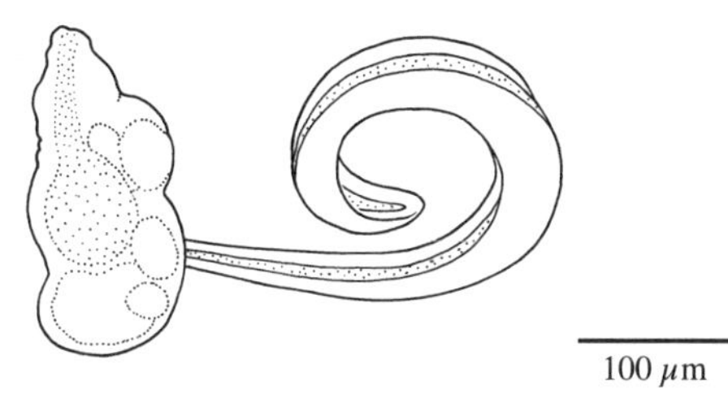

Fritillaria borealis adult (without house)

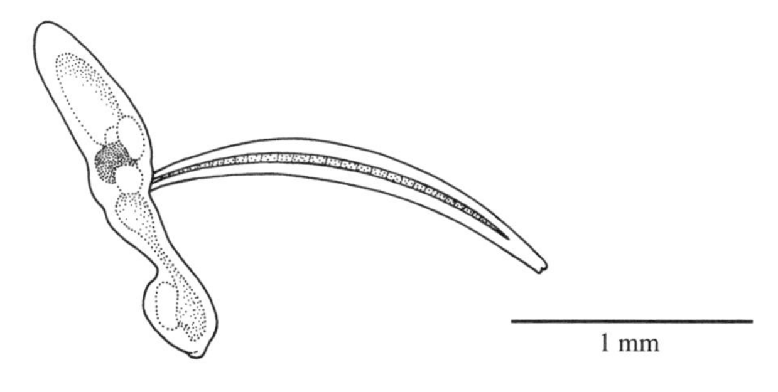

Thalia democratica solitary form

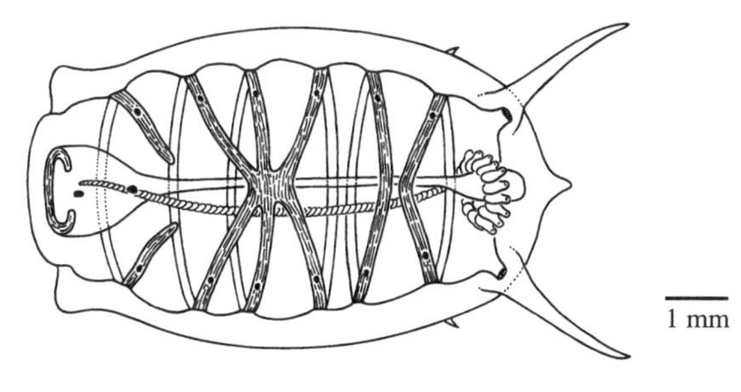

Salpa fusiformis

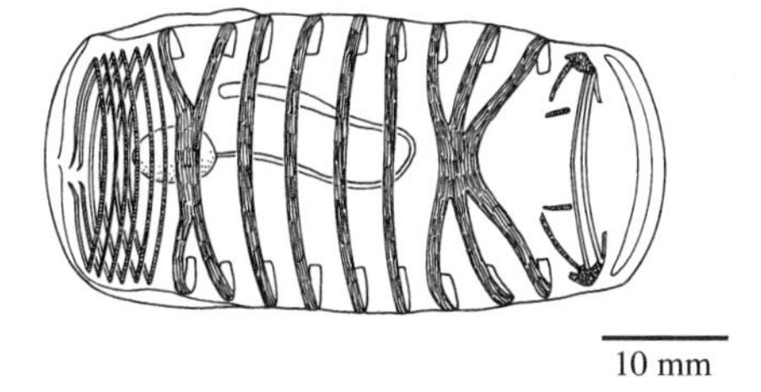

solitary form

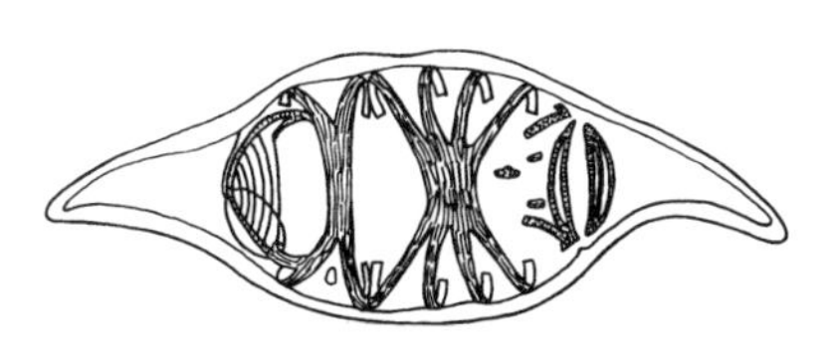

aggregate form

ASCIDIAN TADPOLE LARVAE (ASCIDIACEANS)

ID hint: The tail extends straight behind the posterior of the trunk, along the same axis.

Molgula spp. (sea grapes)

Occurrence. *Molgula* is the only ascidian common in brackish waters. *Molgula manhattensis* is common from Massachusetts southward in the Atlantic and in the northeastern Gulf. *Molgula citrina* is a northern species occurring south to the Mid-Atlantic. *Molgula occidentalis* occurs from North Carolina southward in the Atlantic and on the west coast of Florida.

Biology and Ecology. Larvae are in the plankton for only minutes to a few hours and rarely settle more than a few meters from where they were released. The tadpoles lack definitive attachment organs, but the general body surface seems to be adhesive.

References. Durante 1991; Osman and Whitlatch 1995; Vázquez and Young 2000.

Botryllus schlosseri (star ascidian)

Occurrence. *Botryllus schlosseri* is common from Canada south to at least the Chesapeake Bay and less commonly south to the Gulf.

Biology and Ecology. Many larvae settle within minutes, although some may postpone settlement for up to a few days. Most travel only a few meters before settling.

References. Caicci et al. 2010; Graham and Sebens 1996; Grosberg 1987.

LANCELETS (CEPHALOCHORDATES)

ID hint: Adult and larval lancelets lack the eyes, jaws, and opercula of fish larvae. Their tail lacks the characteristic shape seen in arrow worms (Chaetognatha). *Branchiostoma* reaches 3–5 mm and is pale pink.

Branchiostoma spp.

Occurrence. *Branchiostoma* spp. occur from Chesapeake Bay south through the Gulf and are particularly abundant in the plankton of Tampa Bay. Multiple species occur, but their precise distributions have not been determined. *Branchiostoma lanceolatum* (=*caribaeum*) occurs from the Chesapeake Bay to Texas. It is common in shallow, sandy areas, especially near barrier islands. *Branchiostoma floridae* occurs in the Gulf, while *B. virginiae* occurs from Virginia south at least to Florida. *Branchiostoma longirostrum* may be the common species off Louisiana.

Biology and Ecology. Newly hatched young use epidermal cilia for swimming or hovering. At 60–72 hours, a transition to swimming using muscular undulations takes place. Recent studies show that both juvenile and adult lancelets may spend more time in the night plankton than previously suspected.

References. Boschung and Shaw 1988; Riisgård and Svane 1999; Stokes 1996, 1997; Stokes and Holland 1995a, 1995b; Webb 1969.

Molgula manhattensis

ascidian (tadpole) larva

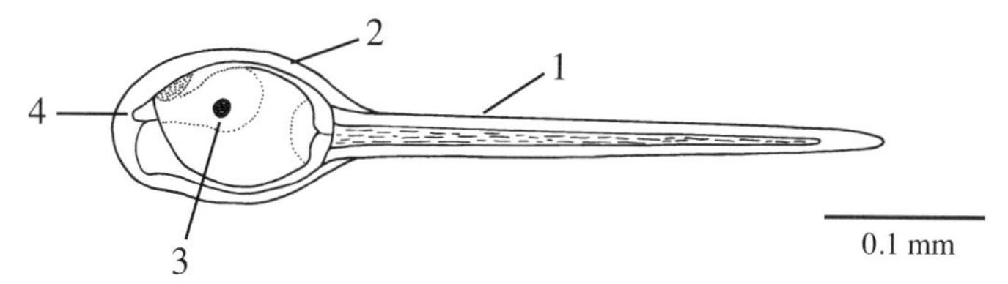

- tadpole-like form with slightly oblong body and narrow tail (1)
- well-developed capsule encloses entire body, which is usually pigmented (2)
- one ocellus (eyespot) is present near center (3)
- papillae do not extend outside of capsule (4)
- similar species: other ascidian larvae have a similar body shape, but most are larger and have either a less conspicuous ocellus or two of them

---

Botryllus schlosseri

ascidian (tadpole) larva

0.5 mm

- tadpole-like form with oval to oblong body and long tail (1)
- well-developed adhesive glands that project in front of the body (2)
- anterior portion of body transparent; posterior portion pigmented (3)
- tail is opaque or white (4)
- similar species: other ascidians with similar shape, but among mid-sized larvae it is most like *Ciona*, which has a narrower body

---

Branchiostoma floridae

lancelet or amphioxus larva

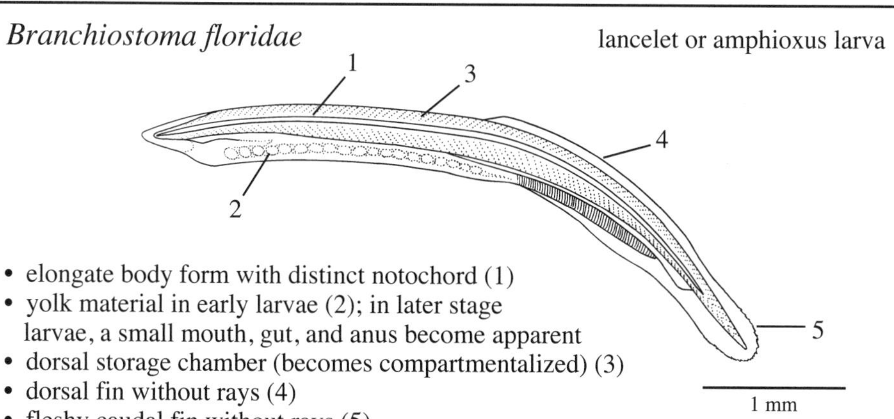

- elongate body form with distinct notochord (1)
- yolk material in early larvae (2); in later stage larvae, a small mouth, gut, and anus become apparent
- dorsal storage chamber (becomes compartmentalized) (3)
- dorsal fin without rays (4)
- fleshy caudal fin without rays (5)
- similar species: other cephalochordates similar; adults have conspicuous external gills; larval fishes have eyes, pectoral and anal fins, and pigment spots; chaetognaths are bilaterally symmetrical

FISH LARVAE

Ichthyoplankton, fish eggs and larvae, are common in estuaries and nearshore areas. Protected coastal regions and especially estuarine areas are critical spawning and/or nursery areas for most of the common commercial, sport, and forage fishes along our coasts. The larval stage is one of the most crucial and is, perhaps, the least understood of the developmental stages in the life history of most fishes.

SPAWNING PATTERNS

Most local bony fishes (teleosts) release eggs into the water where fertilization and subsequent development occur. Pipefishes, seahorses, and sea catfishes are notable exceptions since the males brood the young until their release as juveniles. Although the fish species that occur within the limits of our coverage area exhibit many reproductive strategies, most generally fall into one of the following general spawning patterns:

1. Adults spawn at sea, and early larvae are transported inshore by behaviorally mediated responses to currents as they grow (e.g., spot, left-eyed flounders, menhaden, pinfish).
2. Adults spawn close to the mouths of estuaries or in bays, and the larvae enter and grow in estuarine waters (e.g., red drum, weakfish).
3. Anadromous marine species migrate from the sea to spawn in fresh or almost freshwater areas of estuaries. The larvae move slowly into brackish waters as they grow (e.g., herrings, striped bass).
4. Catadromous freshwater fishes migrate as adults to the sea to spawn. Within our coverage area, the American eel is the singular representative of this pattern.
5. Resident estuarine fishes, including gobies and blennies, spawn in the middle reaches of estuaries. In larger estuaries, some residents undergo spawning migrations within the estuary itself (e.g., white perch to freshwater and hogchoker to more saline areas).
6. Resident species spawn across a wide range of salinities and conditions and complete larval development locally (e.g., mummichog, silversides).

Spawning of most species is seasonal and is strongly influenced by changes in water temperature. The result is a temporal succession of larval fish abundance in each location with a predictable assemblage of species in each season. Of course, the timing and duration of spawning and larval recruitment varies between the northern and southern parts of many species' ranges, but general patterns of succession and species associations are recognized in most of the Atlantic and Gulf of Mexico. In winter to early spring, larvae of offshore spawners arrive in nearshore estuarine waters; this is known as larval ingress or settlement. Menhaden, spot, croaker, pinfish, and flounders (summer, southern, and gulf) exemplify this group. In summer, shallow coastal spawners, especially anchovies, gobies, and drums, produce larvae that dominate most estuarine ichthyoplankton collections. Each location along the coast has its own unique temporal sequence due to broad-scale latitudinal patterns of species occurrence, current conditions, local bathymetry (e.g., deep channels, shallow sounds, and broad intertidal flats), and habitat distribution (e.g., presence of marshes, oyster reefs, and sea grasses). In estuaries, the oligohaline interface between fresh- and saltwater is populated by anadromous fishes and by some larvae that migrate upstream from more saline spawning areas (e.g., gobies, hogchoker, and some drums). Anchovy, goby, and silverside larvae usually dominate middle reaches. The mouths of estuaries, coastal lagoons, and surf areas have their own characteristic and usually more diverse seasonal assemblages. The one constant feature of larval abundances in warm, temperate estuaries throughout the region is that in summer bay anchovy and goby larvae usually outnumber all other species combined in all but the freshest areas.

Early larval stages of fishes are usually dispersed from spawning areas to other locations during developmental periods that often last from weeks to months. Because most larvae are relatively feeble swimmers, their redistributions are primarily determined by prevailing currents. As larvae grow and acquire more capacity to move relative to currents, they can facilitate their transport to more favorable locations. Behaviorally determined movements, including endogenous rhythms and responses to changing environmental conditions, have been noted for many larval fishes. Vertical movements can facilitate both transport and retention. The difficulties of studying microscopic larvae and physical characteristics of the water column at such a fine scale impede our understanding of the mechanisms of larval ingress and settlement. For instance, the relative roles of passive and active transport of larval fishes across the ocean shelf are largely unknown. Some of the general hydrological factors involved in the distribution, ingress, and settlement of larvae are covered in the introductory section, and movements for individual fish taxa are discussed in the text entries for identification.

A few species attach sticky eggs to objects on the bottom, but most simply release their eggs into the water column; some are buoyant, and some sink to the bottom (demersal eggs). Fish eggs are usually spherical, from 0.5 to 2 mm. Their transparency distinguishes them from the typically opaque eggs of invertebrates. Recently spawned eggs have oil droplets and yolk that support early development. Development of an embryo is usually rapid, and often the outline of the young larva is seen in the egg within hours of spawning. At hatching, the now-free larva still has a yolk sac attached to support early development (Fig. 30).

The change from the larval planktonic phase to the juvenile and adult lifestyle often involves "remodeling" of the basic larval body plan, resulting in major changes in morphology and functional capabilities. The newly hatched yolk-sac larvae grow and develop using nutrients provided by the yolk. As the larvae grow, they add functional mouths,

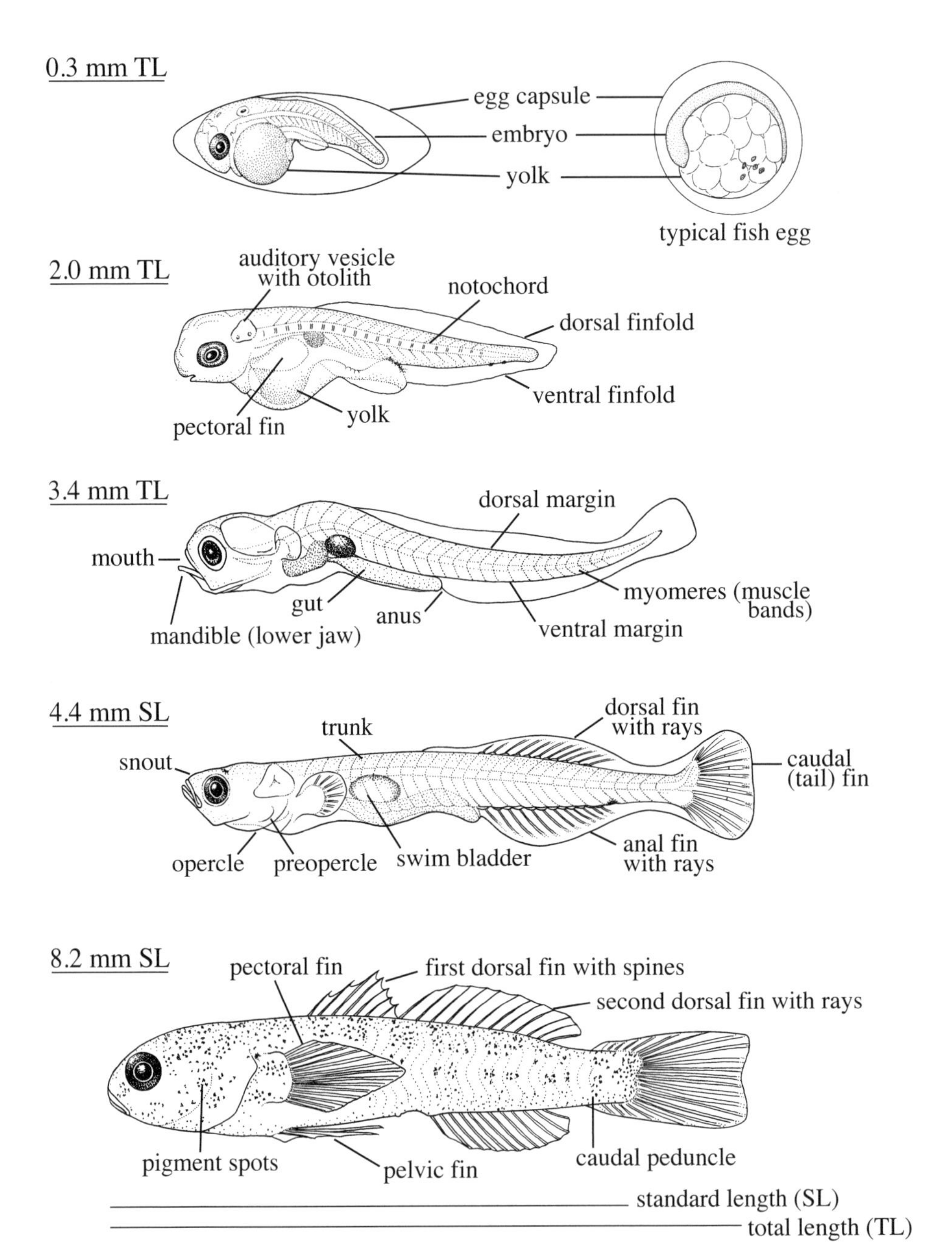

Fig. 30. Anatomy of larval fishes and eggs according to features seen in developmental stages of the naked goby *Gobiosoma bosc*. An elongate 0.3 mm TL (total length) goby egg is compared with a more typical, nearly spherical bony fish egg. At 2.0 mm TL, the mouth is not developed and the larva still relies on yolk for energy; this stage is referred to as a *yolk-sac larva*. At about 3.4 mm TL, the mouth and gut become functional, the body and fin-folds elongate, and a tail begins to develop. At 4.4 mm SL (standard length), the developing trunk muscu-

musculature, fin rays, scales, and sensory capabilities. Newly hatched larvae are feeble swimmers, but swimming capabilities improve rapidly as larvae grow. Some larvae show schooling behavior at a relatively early age. As swimming speed and endurance increase, some larvae show distinctive patterns of vertical migration that may be associated with predator avoidance, feeding migrations, or selective tidal transport. Early larvae are usually transparent with pigmentation increasing during development.

When the yolk sac is resorbed, the larvae enter the critical first feeding stage where they must find and capture prey in the plankton or perish. First foods are usually quite small: rotifers, ciliates, and copepod nauplii. Vision develops early, and most larval fishes are visual feeders. Prey are perceived at close range, usually within one fish body length. Success rate in capturing prey improves as both vision and coordination develop. Prey detection and selection seem to be based on prey size more than on any other factor, but motion and color may come into play. Feeding stops in complete darkness but may continue in moonlight. Larvae can see and feed using ultraviolet light, a capability lost in many adults. When prey are sighted, menhaden, herring, and anchovy larvae assume a quiescent S-shaped posture with the tail bent to one side. They then use undulating motions of the fins to maintain this position as they approach their target. More vigorous species, such as mackerel, approach first and then cock the tail (C-start position). When a targeted prey is within range, a flick of the tail shoots the larva forward as it strikes with mouth open. As larvae grow and become faster and more coordinated, the poised striking from a C- or S-start diminishes, and feeding motions become integrated into normal swimming behavior. Different feeding strategies emerge with some fish cruising continuously while feeding, whereas others rely on a pause-travel motion. Larvae typically switch to progressively larger foods, especially larger copepod larvae and adult copepods. Successful feeding and rapid growth greatly enhance survival, so the density, size, and type of prey available during the larval phase may be crucial. Entire year-classes of fishes can fail if the timing of the availability of required foods and the occurrence of larvae do not coincide; starvation can occur in a matter of hours.

Fish eggs and larvae often encounter risks from different suites of predators as they grow and change habitats. Planktonic predators include chaetognaths, decapod crustacean larvae, predatory copepods, gelatinous predators (e.g., hydromedusae, scyphomedusae, siphonophores, and ctenophores), and zooplanktivorous fishes (including juveniles of the same species). Should eggs or larvae settle near the bottom, various benthic predators await. Eggs and small larvae with limited escape abilities may go unnoticed by visual feeders because of their small size and transparency. Remaining motionless also reduces

lature and the fins and tail (now with rays) enable more mobility. At 8.2 mm SL, adult-like features, including skin with scales, pigment patterns, and a full complement of fin spines and rays, are present. The lines below the 8.2 mm larva show that SL is measured from the snout to the base of the central caudal (tail) rays and TL is measured from the snout to the posterior end of the caudal fin. Because the tails of small fishes tend to erode in collections, length measurements of fishes starting with larval stages with caudal rays are usually based on SL.

hydrodynamic signals that could attract attention. Escape rate improves steadily as larvae grow. This is partially due to an increased ability to detect predators using visual and hydrodynamic cues as well as to stronger and better coordinated evasive responses. Larger larvae may also gain a "size refuge" from some of the smaller predators.

IDENTIFICATION HINTS

Many hundreds of species of fishes occur in shallow coastal systems of the Atlantic and Gulf Coasts. Our coverage of larval fishes is limited to those species or groups of species that are most likely to be collected in the shallow ocean, in coastal embayments, and at various locations along estuarine salinity gradients. The morphological features of larvae identified in Figure 30 represent a minimum set of key characteristics useful for distinguishing taxa. Larval fish biologists use many more features and terms to help with this often challenging task. Other sources listed in the references at the end of this section should be consulted for additional technical terms. The following features are particularly useful for larval fish identification:

Size. There are many ways to represent the lengths of larval fishes. In this guide, we provide total lengths (TL) for stages that have not yet developed caudal (tail) rays as shown in Figure 30. The fleshy ends of larval fish bodies tend to erode during capture and handling. Because length is often critical for identification, larval fish biologists use standard length (SL) as a more stable measure for specimens that have developing tail fins. SL is the length from the tip of the snout to the end of the body at the base of the central caudal rays. We provide a TL or SL for each illustrated fish. We also provide size ranges from the total length of the smallest hatchlings to the standard length of the most advanced larvae. If your specimen is within the stated range and generally resembles the illustration, look more closely for specific identifying features. If the specimen has similar characteristics but is considerably smaller or larger, you probably need to look for other alternatives.

Shape. When viewing a larva from the side, the degree to which the body tapers from the head to tail and the relative width of the caudal peduncle (see Fig. 30) are good features for differentiating some taxa. The differences between especially thin (laterally compressed), broad (round in cross-section), and flat (dorsally ventrally depressed) larvae are useful for identification. The relative size of the head and shape of the anterior margin can also be useful.

Fins. The location of the origins of developing dorsal and anal fins relative to one another can be very useful. The position of the anus relative to the dorsal or anal fin is an important characteristic in differentiating some otherwise similar-looking larvae. Numbers of fin rays and spines can be important features for identifying all but the youngest larvae. We refer readers to references listed here for information about stage and species-specific counts of fin rays during development.

Mouth. The size of the mouth relative to the head or body, its angle relative to the axis of the body, and its location (e.g., at or below the snout) are sometimes diagnostic.

Especially important is the position of the corner (angle) of the jaws relative to the eye. It may be anywhere from totally anterior to below and behind the eye. Experts often use the numbers, shapes, and sizes of inconspicuous teeth for definitive identifications, but we only refer to conspicuous teeth.

Pigmentation. In some instances, the presence or absence of a pigment spot (or dash, band, or cluster) in a specific location can provide certain identification to the species level. More often pigment patterns overlap, and the pattern observed must be considered along with other features. Variations over the developmental sequence and among individuals of the same size complicate interpretation. Some pigment may not be visible under certain types of lighting or with different preservation techniques.

Other features that are useful for some of the taxa treated here include size, shape and placement of the pectoral fins, number of muscle bands (myomeres), spines on the head or gill covers (opercles), length, shape and extent of coiling of the gut, and presence or absence of a swim bladder. Some of these are apparent in the illustrations provided. Check other sources for myomere and fin ray counts and other features. Illustrated stages are snapshots along a continuum, and major changes in features can occur before the larva grows another millimeter. Because the rate of development is so rapid in larval fishes, the illustrated stage may only exist for a matter of hours. Many of the resources in the "Useful Identification References" illustrate a more complete sequence of larval development.

USEFUL IDENTIFICATION REFERENCES

Of the many general guides to larval fish identification, the ones listed below are the most recent, most complete, and contain the most information on distribution and biology. Additional identification materials exist as in-house reports and keys at local universities, marine laboratories, state agencies, and environmental consulting firms. Also note the multivolume series *Development of Fishes of the Mid-Atlantic Bight: An Atlas of Egg, Larval, and Juvenile Stages* available from U.S. Fish and Wildlife Service FWS/OBS-789/12.455, Government Printing Office, Washington, DC.

Able, K. W., Fahay, M. P. 1998. *The First Year in the Life of Estuarine Fishes in the Middle Atlantic Bight.* Rutgers University Press, New Brunswick, NJ. 342 pp. (Contains detailed notes on ecology and distribution of larvae, but the focus is more on juvenile than on larval fishes.)

Ditty, J. G., Shaw, R. F. 1994. *Preliminary Guide to the Identification of the Early Life History Stages of Sciaenid Fishes from the Western Central Atlantic.* NOAA Technical Memorandum NMFS-SEFSC-349. 118 pp.

Ditty, J. G., Shaw, R. F., Fuiman, L. A. 2005. Larval development of five species of blenny (Teleostei: Blenniidae) from the western central North Atlantic, with a synopsis of blennioid family characters. *Journal of Fish Biology* 66:1261–1284.

Fahay, M. P. 2007. *Early Stages of Fishes in the Western North Atlantic Ocean (Davis Strait, Southern Greenland and Flemish Cap to Cape Hatteras).* Vol. 1, *Acipenseriformes through Syngnathiformes* (pp. 1–931). Vol. 2, *Scorpaeniformes through Tetraodontiformes* (pp. 932–1696). Northwestern Atlantic Fisheries Organization, Dartmouth, Nova Scotia. Available online: <www.nafo.int/publications/fahay/pdfs.html>. Accessed March 5, 2012.

Richards, W. J., ed. 2005. *Early Life History of the Fishes of the Western Central North Atlantic.* CRC Press, Boca Raton, FL. 2672 pp.

Wang, J. C. S., Kernehan, R. J. 1979. *Fishes of the Delaware Estuaries: A Guide to the Early Life*

Histories. Ecological Analysts, Towson, MD. 410 pp. (A pictorial guide featuring illustrations of many larval stages. Out of print, but photocopies can be purchased from EA Engineering, Science, and Technology, Inc., 15 Loveton Circle, Sparks, MD 21152.)

SUGGESTED READINGS

Able, K. W., Fahay, M. P. 2010. *Ecology of Estuarine Fishes: Temperate Waters of the Western North Atlantic*. Johns Hopkins University Press, Baltimore. 566 pp. (Contains coverage of the larval ecology of each or the 87 species covered.)

Able, K. W., Fahay, M. P., Witting, D. W., et al. 2006. Fish settlement in the ocean vs. estuary: Comparison of pelagic larval and settled juvenile composition and abundance from southern New Jersey, U.S.A. *Estuarine, Coastal and Shelf Science* 66:280–290.

Bailey, K. M., Houde, E. D. 1989. Predation on eggs and larvae of marine fishes and the recruitment problem. *Advances in Marine Biology* 25:1–83.

Bailey, K. M., Nakata, A., Van der Veer, H. W. 2005. The planktonic stages of flatfishes: Physical and biological interactions in transport processes. In: Gibson, R. N., ed. *Flatfishes: Biology and Exploitation*. Blackwell, Oxford, UK, 94–119.

Blaxter, J. H. S. 1988. Pattern and variety in development. In: Hoar, W. S., Randall, D. J., eds. *Fish Physiology*. Vol. 11, *The Physiology of Developing Fish*. Academic Press, San Diego, CA, 1–58.

Boehlert, G. W., Mundy, B. C. 1988. Roles of behavioral and physical factors in larval and juvenile fish recruitment to estuarine nursery areas. *American Fisheries Society Symposium* 3:51–67.

Collette, B. B., Klein-MacPhee, G., eds. 2002. *Bigelow and Schroeder's "Fishes of the Gulf of Maine."* 3rd ed. Smithsonian Institution Press, Washington, DC. 748 pp. (This updated classic reference also includes many southern fishes, with summaries of spawning patterns and larval biology.)

Fahay, M. P. 2007. *Early Stages of Fishes in the Western Atlantic Ocean (Davis Strait, Southern Greenland and Flemish Cap to Cape Hatteras)*. Vols. 1 and 2. Northwestern Atlantic Fisheries Organization, Dartmouth, Nova Scotia. (Contains an introduction to larval fish biology.)

Fuiman, L. A., Magurran, A. E. 1994. Development of predator defenses in fishes. *Reviews in Fish Biology and Fisheries* 4:145–183.

Fuiman, L. A., Werner, R. G., eds. 2002. *Fishery Science: The Unique Contributions of Early Life Stages*. Blackwell Science, Oxford, UK. 352 pp.

Hare, J. A., Thorrold, S., Walsh, H., et al. 2005. Biophysical mechanisms of larval fish ingress into Chesapeake Bay. *Marine Ecology Progress Series* 303:295–310.

Hernandez, F. J., Jr., Powers, S. P., Graham, W. M. 2010. Detailed examination of ichthyoplankton seasonality from a high resolution time series in the northern Gulf of Mexico during 2004–2006. *Transactions of the American Fisheries Society* 139:1511–1525.

Houde, E. D., Schekter, R. C. 1980. Feeding by marine fish larvae: Developmental and functional responses. *Environmental Biology of Fishes* 5:315–334.

Jakobsen, T., Fogarty, M. J., Megrey, B. A., eds. 2009. *Fish Reproductive Biology: Implications for Assessment and Management*. Wiley-Blackwell, Oxford, UK. 429 pp. (Chapters on recruitment are especially noteworthy.)

Pattillo, M. E., Czapla, T. E., Nelson, D. M., et al. 1997. *Distribution and Abundance of Fishes and Invertebrates in Gulf of Mexico Estuaries*. Vol. 2, *Species Life History Summaries*. ELMR Report No. 11. NOAA/NOS Strategic Environmental Assessments Division, Silver Spring, MD. 377 pp. (A synopsis of the biology and distribution of most of the larval fishes in the Gulf.)

Webb, J. F. 1999. Larvae in fish development and evolution. In: Hall, B. K., Wake, M. H., eds. *The Origin and Evolution of Larval Forms*. Academic Press, New York, 109–157. (An overview of adaptations of larval fishes, including sensory adaptations.)

Anguilla rostrata (American eel)

Occurrence. Leptocephalus larvae of the American eel are oceanic and generally do not occur close to shore. Glass eels and elvers occur in Atlantic and Gulf estuaries in winter and spring but are more common north of North Carolina.

Biology and Ecology. The American eel has a catadromous life cycle. Adults live in freshwater or brackish water then migrate to the Sargasso Sea to spawn. The flattened leptocephalus larvae remain in the plankton for 6–18 months and grow to 40–70 mm during the long migration back to the coasts. As they near the coasts, the eels metamorphose into round, unpigmented "glass eels" (50–65 mm) and then into pigmented elver-stage eels (>65 mm); both stages resemble adults. Glass eels arrive later and at progressively older ages from North Carolina north. Glass eels and elvers are collected in the water column within sounds and tidal creeks. Little is known regarding feeding or means of navigation of eels during their planktonic stages.

References. Facey and Van Den Avyle 1987; McCleave and Kleckner 1982; Pfeiler 1986; Powles and Warlen 2002; Sorensen 1986; Sullivan et al. 2009.

Myrophis punctatus (speckled worm eel)

Occurrence. Speckled worm eel larvae are caught from New Jersey to Texas usually 2 to 4 months after leaving oceanic spawning grounds. Late-stage leptocephali and shrinking metamorphic stages are especially abundant in the winter ichthyoplankton in estuaries in the Carolinas but can occur in spring and summer farther north. In Alabama coastal waters, they are present from November to March. Larvae occasionally occur in mid to low salinities. Other similar eel larvae may occur in inshore samples in southern areas, including whip or sooty eels (*Bascanichthys* spp.) and the shrimp eel (*Ophichthus gomesii*). The similar-looking leptocephalus larva of the conger eel (*Conger oceanicus*) occurs in shallow-water collections from the Atlantic and northern Gulf Coasts.

Biology and Ecology. The larval ecology of this very abundant yet cryptic fish is not well known, but spawning takes place in the deep ocean. The resulting leptocephali eat larger protozooplankton, including ciliates.

References. Able et al. 2011; Fahay and Oberchain 1978; Govoni 2010; Leiby 1979.

Syngnathus spp. (pipefishes) and *Hippocampus* spp. (seahorses)

Occurrence. Adults of these closely related fishes are most common in estuaries and bays from Massachusetts to the Gulf, especially near seagrass beds or other structure. The northern pipefish (*Syngnathus fuscus*) occurs from Massachusetts to Florida over a wide salinity range. The dusky pipefish (*S. floridae*) occurs in the Southeast as well as the Gulf. The chain pipefish (*S. louisianae*) and the Gulf pipefish (*S. scovelli*) are common throughout the Gulf. The lined seahorse (*Hippocampus erectus*) occurs from Massachusetts through the Gulf. Several additional pipefishes and seahorses range along the southeastern United States and in the Gulf, including *Hippocampus reidi* and *H. zosterae*.

Biology and Ecology. Male pipefishes and seahorses brood eggs in a ventral marsupium. Juveniles are about 8–12 mm long when they are released into the plankton during the warm months. Newly freed young cluster near the surface, often associated with floating seaweed or seagrasses.

References. Campbell and Able 1998; Van Wassenbergh et al. 2009.

Anguilla rostrata American eel elver 81 mm (> 65 mm) SL

- long, cylindrical body, small head (1)
- dorsal fin origin anterior to anal fin; both continuous with caudal fin (2)
- round pectoral and no pelvic fin (3)
- elver stage (> 65 mm) with dark brown pigmentation (4)
- intermediate, cylindrical glass eel stage (48 - 65 mm) is transparent
- earlier leptocephalus stage (58 - 80 mm SL) is laterally flat, transparent, with chevron-shaped myomeres (not shown)
- similar species: leptocephalli in estuaries are more likely *Myrophis* and *Conger*; glass eels and elvers are more difficult to differentiate among these species and others in high salinity, but in low salinity, *Anguilla* is more likely; *Anguilla* elver has a projecting lower jaw

---

Myrophis punctatus speckled worm eel leptocephalus 51 mm (< 84 mm) SL

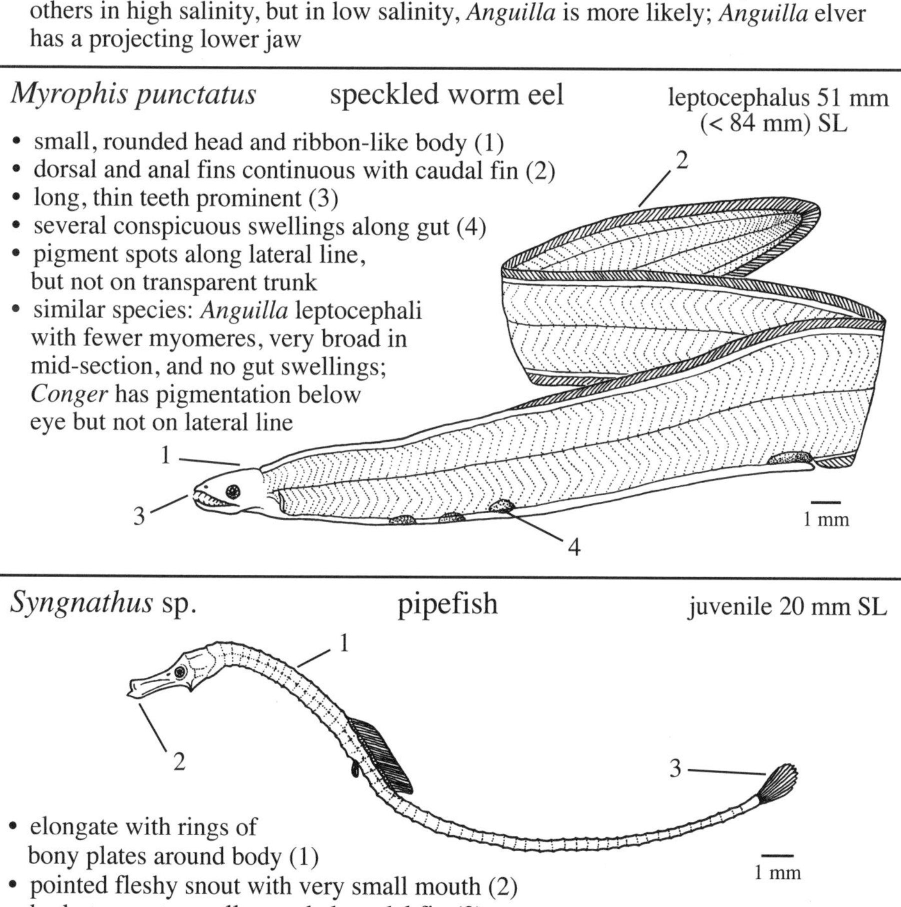

- small, rounded head and ribbon-like body (1)
- dorsal and anal fins continuous with caudal fin (2)
- long, thin teeth prominent (3)
- several conspicuous swellings along gut (4)
- pigment spots along lateral line, but not on transparent trunk
- similar species: *Anguilla* leptocephali with fewer myomeres, very broad in mid-section, and no gut swellings; *Conger* has pigmentation below eye but not on lateral line

---

Syngnathus sp. pipefish juvenile 20 mm SL

- elongate with rings of bony plates around body (1)
- pointed fleshy snout with very small mouth (2)
- body tapers to small, rounded caudal fin (3)
- increasingly pigmented after release from male pouch at 8 - 12 mm
- similar species: many *Syngnathus* species are difficult to separate as small juveniles; *Hippocampus* spp. (sea horses) have distinct crested heads and curled tails

Alosa spp. (river herrings and shads)

Occurrence. River herrings and shads spawn in freshwater streams along the Atlantic Coast and in some Gulf estuaries. The blueback herring (*Alosa aestivalis*), American shad (*A. sapidissima*), and hickory shad (*A. mediocris*) occur from Massachusetts to northern Florida. The alewife (*A. pseudoharengus*) occurs from Massachusetts to South Carolina. These herrings and shads spawn in winter or early spring. Early larvae occur in freshwater and oligohaline areas, especially near the surface. Older herring larvae gradually move into more saline areas. The positions of various sizes, ages, and species vary along the salinity gradient.

Biology and Ecology. All shads and herrings in the genus *Alosa* are anadromous. Adults remain at sea until they are mature and return for a spring spawning run to their natal rivers. Spawning migrations are temperature dependent and occur later in the northern part of the range. Early larvae feed on zooplankton, especially tintinnids and larval copepods.

References. Chambers et al. 1976; O'Connell and Angermeier 1997.

Brevoortia tyrannus (Atlantic menhaden) and other *Brevoortia* spp.

Occurrence. Atlantic menhaden larvae occur in coastal and estuarine areas from Massachusetts to Florida. Maximum abundance is in winter or spring, but smaller numbers appear in the summer and fall as larvae can originate from both local and remote spawning areas. The Gulf menhaden (*Brevoortia patronus*) occurs throughout the Gulf, and the yellowfin menhaden (*B. smithi*) is in the eastern Gulf; both species occur but are uncommon in the Southeast. The fine-scale menhaden (*B. gunteri*) occurs from Louisiana to Texas.

Biology and Ecology. Menhaden spawn over the continental shelf, but postlarvae move into estuaries where they prefer low salinities. Most spawning occurs in winter, but some spawning occurs in summer and fall closer inshore, resulting in multiple cohorts of larvae in some areas. Early larvae feed on small zooplankton, including bivalve larvae and the dinoflagellate *Prorocentrum*. Later, larvae eat larval and adult copepods before becoming suspension feeders on phytoplankton as juveniles and adults. The larvae and juveniles spend the summer in low-salinity and brackish nursery areas before migrating out to sea in the fall. Larvae are often near the surface during the day.

References. Fitzhugh et al. 1997; Forward et al. 1996; Govoni and Stoecker 1984; Hettler 1984; June and Carlson 1971; Kjelson et al. 1975; Lassuy 1983b; Lozano 2011; Powell 1993; Rogers and Van Den Avyle 1983; Warlen 1988; Warlen et al. 2002.

Clupea harengus (Atlantic herring)

Occurrence. The Atlantic herring is a northern ocean species, with larvae occurring nearshore and in high-salinity estuaries from Delaware north. Young larvae occur in Mid-Atlantic estuaries from November to August, with highest densities usually occurring from February to May. Larvae often occur near the surface.

Biology and Ecology. Unlike other herrings, *Clupea harengus* spawns in North Atlantic shelf or coastal waters rather than in estuaries. In late summer or fall, females deposit ribbons of adhesive eggs that attach to hard-bottom substrates. Larvae feed first on copepod and bivalve larvae and then switch to adult copepods and larger prey as they grow. The planktonic larval stage lasts three months or more.

References. Blaxter and Fuiman 1990; Checkley 1982; Collette and Klein-MacPhee 2002; Fuiman and Gamble 1988; Kelly and Moring 1986; MacKenzie and Kiørboe 1995.

Alosa aestivalis blueback herring 13.4 mm (3 - 16 mm) SL

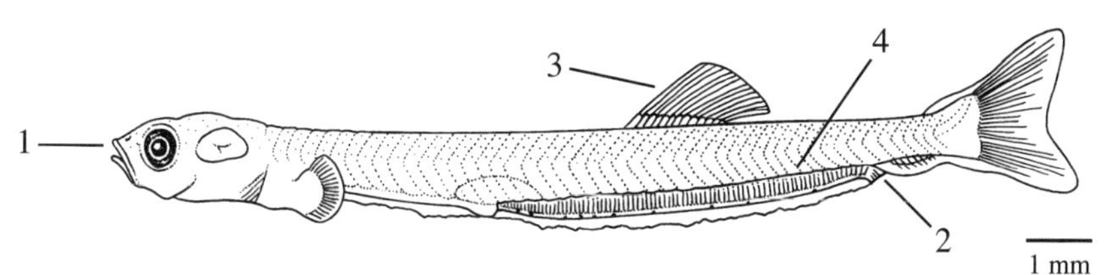

- very elongate body with pointed snout (1)
- muscle-band striations along posterior half of gut; anus close to caudal fin (2)
- short dorsal fin is well forward of the anal fin (3)
- light, scattered pigmentation especially on the gut and ventral body margin anterior to the anal fin (4)
- similar species: other *Alosa* spp. and *Dorosoma* spp. in freshwater very similar; *A. sapidissima* is larger at the same stage; *Brevoortia*'s anus is at posterior end of dorsal fin; *Anchoa*'s anus is below the dorsal fin

---

Brevoortia tyrannus Atlantic menhaden 17.9 mm (2.5 - 28 mm) SL

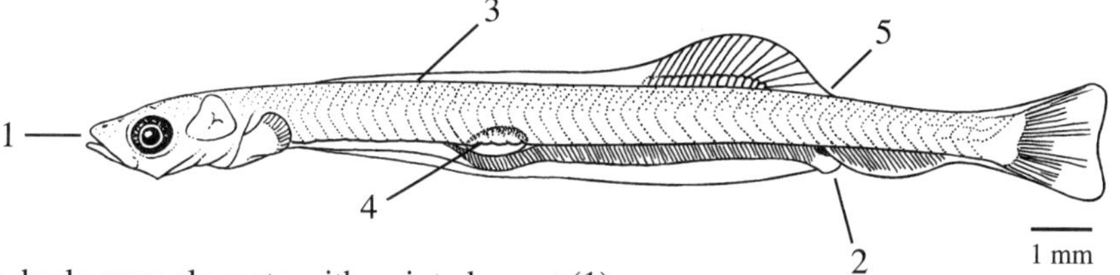

- body very elongate with pointed snout (1)
- muscle-band striations along most of gut; anus at posterior end of dorsal fin (2)
- moderate dorsal pigmentation and spots along gut (3)
- swim bladder evident by 11 mm (4)
- origin of anal fin is below dorsal fin (5)
- similar species: other *Brevoortia* spp. are very similar; *Alosa* spp. have anus closer to the caudal than dorsal fin; *Clupea*'s anus is midway to the caudal; *Anchoa*'s anus and anal fins are below the dorsal fin

---

Clupea harengus Atlantic herring 18.6 mm (4 - 42 mm) SL

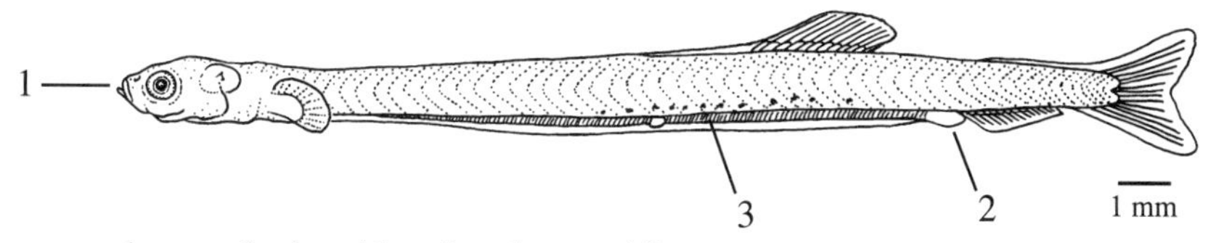

- very elongate body with pointed snout (1)
- gut narrow with muscle-band striations along most of length; anus about midway between dorsal and caudal fins (2)
- dark pigment spots above ventral margin (3)
- swim bladder not apparent until > 25 mm
- similar species: other elongate herring larvae are similar; *Brevoortia*'s anus is below the posterior end of the dorsal fin; *Anchoa*'s anus is below the dorsal fin; *Ammodytes* has dorsal and anal fins across from one another

Anchoa mitchilli (bay anchovy) and other *Anchoa* spp.

Occurrence. *Anchoa mitchilli* larvae often dominate the summer ichthyoplankton over the entire area of coverage and may exceed 90% of all fish eggs and larvae collected during warm seasons from nearshore coastal areas to the upper estuaries. They are common in summer in the Atlantic and much of the year in the Gulf. Striped anchovy (*A. hepsetus)* larvae also occur in estuaries but are usually more common in the coastal and deep ocean than the bay anchovy. Striped anchovy larvae are especially abundant in the frontal zone at the edge of the Mississippi River plume. Larvae of the dusky anchovy (*A. lyolepis*) and other coastal ocean anchovy species may co-occur in the Gulf and Southeast. Larvae of the silver anchovy (*Engraulis eurystole*), which is primarily a continental shelf species, sometimes occur in estuaries from North Carolina north and in the coastal ocean areas of Southeast and the northern Gulf of Mexico.

Biology and Ecology. Along the Southeast Coast, *Anchoa mitchilli* spawns from April through November in estuaries and the shallow ocean, but spawning starts later and is abbreviated from Virginia north. In the Gulf of Mexico, *A. mitchilli* can reproduce during all but the coldest periods. Bay anchovies spawn daily and mostly during the evening. Eggs hatch in about 24 hours at summer temperatures. In estuaries in South Carolina, larvae were collected at all stages of the tide, time of day, and at the surface and bottom. In the Suwannee River in Florida, bay anchovies preferentially spawned at the edge of the freshwater plume at the fresh/salt interface, where their favored prey, the copepod *Acartia tonsa* are concentrated. Eggs and larvae are vulnerable to predation by medusae and ctenophores; however, research in Chesapeake Bay suggests that *A. mitchilli* spawning is most intense in areas of high zooplankton and low ctenophore abundance.

References. Castro and Cowen 1991; Chesney 2008; Dorsey et al. 1996; Houde and Schekter 1980; Johnson et al. 1990; MacGregor and Houde 1996; Morton 1989; North and Houde 2004; Peebles 2002; Peebles and Tolley 1988; Purcell et al. 1994; Rilling and Houde 1999; Robinette 1983; Schultz et al. 2000.

Anchoa mitchilli bay anchovy 2 - 12 mm SL

3.0 mm

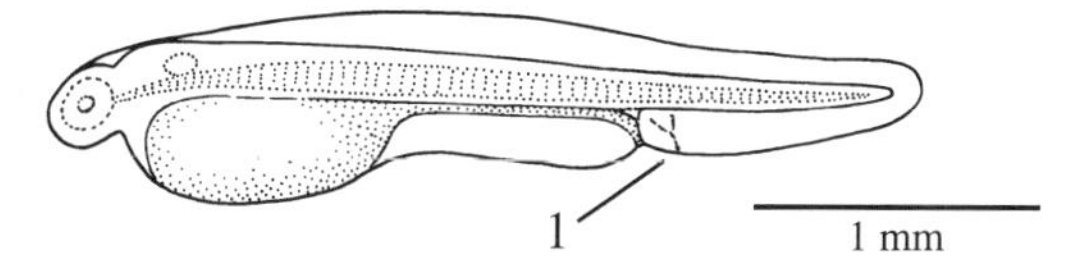

- yolk-sac larva (2.0 - 4.0 mm) is tadpole-like with unpigmented eyes and Y shaped set of pigment spots posterior to anus (1)

4.8 mm

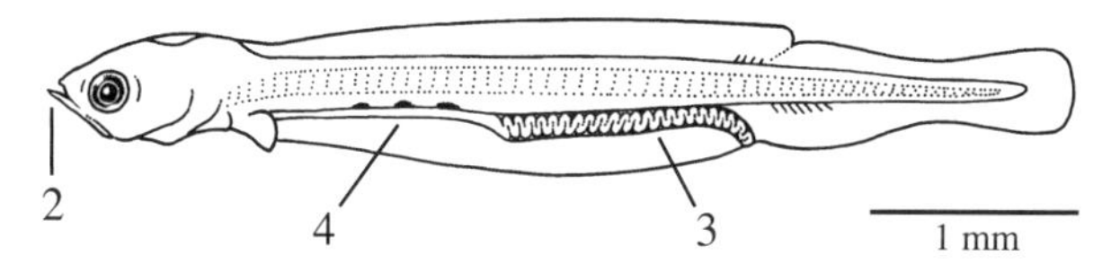

- starting at about 4.2 mm, elongate body with pointed snout and large mouth (2)
- gut with muscle striations in posterior half (3)
- pigment dashes along dorsal edge of anterior half of gut cavity (4)

8.2 mm

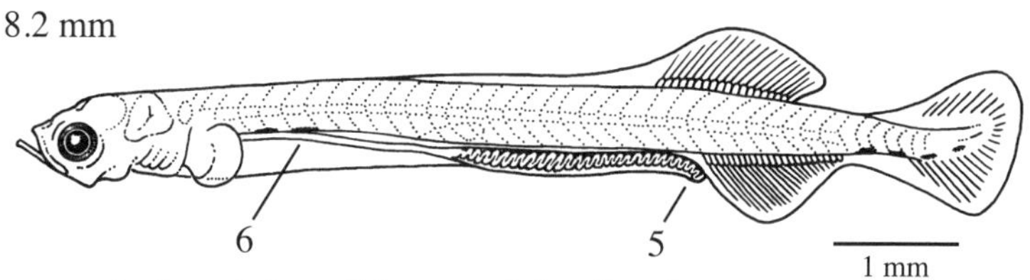

- starting at about 5.0 mm, origin of anal fin is under center of dorsal fin (5)
- pigment dashes along dorsal edge of gut cavity persist thoughout larval development (6)

10.9 mm

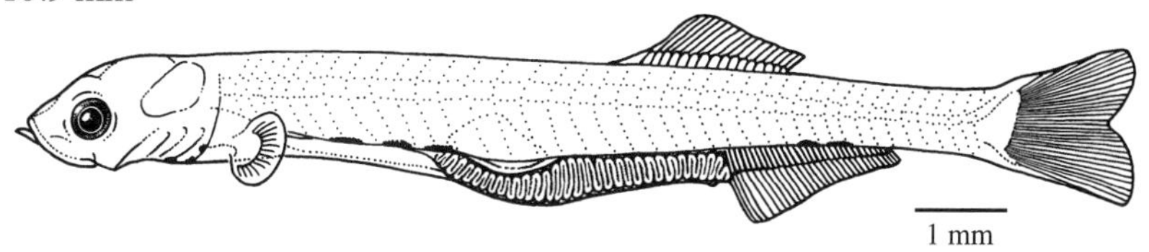

- similar species: in herrings, the anus is at or beyond the posterior end of the dorsal fin; *A. hepsetus* has an anal fin origin posterior to the midpoint of the dorsal fin; *Engraulis* spp. have an anal fin origin at the posterior end of the dorsal fin

Menidia spp. and *Membras martinica* (silversides)

Occurrence. Atlantic silverside (*Menidia menidia*) larvae occur from Canada to northern Florida in summer. In estuaries, they sometimes occur in 1–14 psu but are more prevalent in higher salinities along the coastline. Inland silverside (*M. beryllina*) larvae occur in warm months from Maine to Texas. They are among the most abundant fish larvae in parts of the northern Gulf, often found in low to mid-salinities. The tidewater silverside (*M. peninsulae*) is common in Gulf estuaries and bays, especially around the Florida peninsula. Larvae are associated with higher salinities and water temperatures of 12°–30°C. Rough silverside (*Membras martinica*) larvae occur from the coastal ocean to low salinity reaches of estuaries in summer and fall from New York through the Gulf.

Biology and Ecology. *Menidia menidia* spawns during daytime high tides in the upper intertidal zone of marshes and along shorelines from March to August, depending on latitude. Eggs of all of the silverside species are attached to the bottom (including vegetation) and hatch during nocturnal high tides. Silverside larvae are often caught near the surface.

References. *Menidia menidia:* Fay et al. 1983b; Lindsay et al. 1978; Middaugh 1981; *M. beryllina*: Gleason and Bengtson 1996.

Ammodytes spp. (sand lances)

Occurrence. Sand lance larvae occur from Massachusetts to Virginia and may dominate the nearshore ichthyoplankton from Maine to northern New Jersey from November to May. In the north, early larvae are euryhaline and may occur in low salinities.

Biology and Ecology. Sand lances spawn in winter and early spring from the inner continental shelf to high-salinity inlets. During their two to three month larval stage, sand lances feed first on phytoplankton and copepod nauplii and later switch to adult copepods. In the laboratory, individuals begin schooling at 25–30 mm and burying at 35–40 mm. Two other *Ammodytes* species have been distinguished, and the taxonomy of the group is unresolved.

References. Monteleone and Peterson 1986; Monteleone et al. 1987; Norcross et al. 1961; Smigielski et al. 1984.

Elops saurus (ladyfish)

Occurrence. Ladyfish larvae occur from the Chesapeake Bay south and throughout the Gulf. Larvae appear in the coastal waters of Georgia and the northern Gulf from March to October (April–May peak). Late-stage larvae were caught in low salinities during late spring in South Carolina estuaries.

Biology and Ecology. Ladyfish spawn offshore and larvae disperse to inshore waters, typically arriving at the coast as leptocephali. This is followed by a period of size decrease as the larvae metamorphose to acquire juvenile characteristics before larval growth resumes. Metamorphic larvae continue to migrate into estuaries and late-stage larvae are most common in lower salinities. Leptocephali have fanglike teeth, but details of their feeding in the plankton are lacking.

References. Gehringer 1959; Govoni and Merriner 1978; McBride et al. 2001.

Menidia menidia Atlantic silverside 9.8 mm (3.8 - 13.7 mm) SL

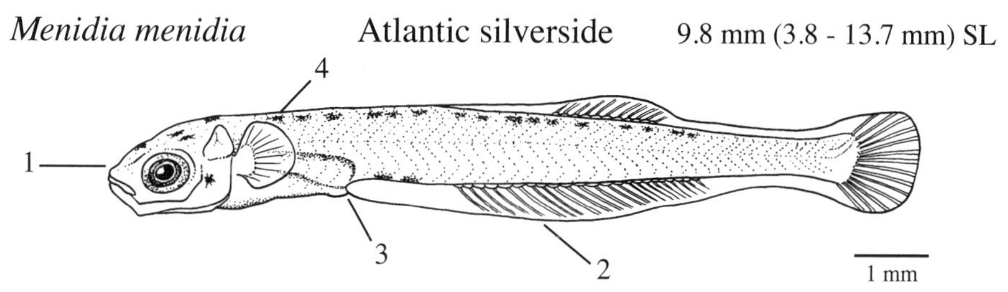

- elongate body with relatively blunt snout (1)
- anal fin begins anterior to dorsal fin origin; anal fin is longer (2)
- short gut and anus are far anterior to dorsal fin origin (3)
- one row of long pigment spots on each side of the dorsal edge (viewed from above) and large spots on head (4)
- similar species: *M. beryllina* does not have dorsal pigment spots, and the spots on the head are small; *Membras martinica* has a single row of spots along the dorsal edge; other elongate larvae have an anus that is more posterior

---

Ammodytes americanus sand lance 12.6 mm (3 - 14 mm) SL

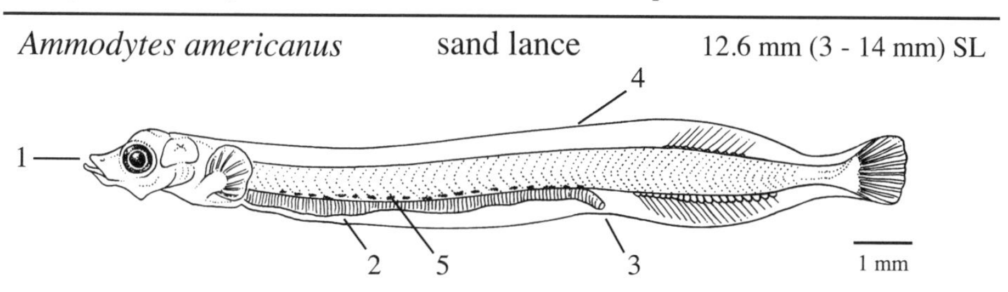

- elongate with duck-like snout; lower jaw is longer than upper jaw (1)
- long gut with internal folds (2)
- anus well anterior of the dorsal fin and exits to side rather than at ventral margin (3)
- dorsal and anal finfolds are nearly opposite (4)
- pigment spots along ventral margin above gut (5)
- similar species: herrings and anchovies have an anus posterior to or just below the dorsal fin

---

Elops saurus ladyfish 26.0 mm (5 - 34 mm) SL

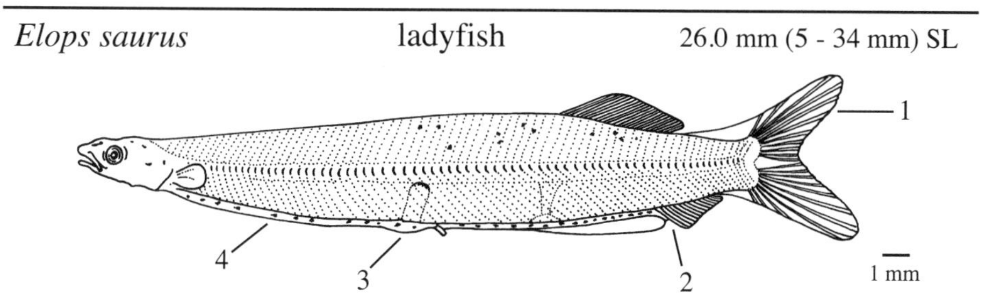

- leptocephalus larva with small head, elongate and tall body, and forked tail (1)
- anal fin origin under posterior end of dorsal fin (2)
- elongate, vertical swim bladder nearly perpendicular to ventral edge at mid-body (3)
- row of almost evenly spaced pigment spots along gut (4)
- similar species: eel leptocephali do not have forked tails; tarpon (*Megalops atlantica*) has dorsal and anal fins that are almost vertically aligned; bonefishes (*Albula* spp.) have an anal fin origin far posterior to the end of the dorsal fin

Gobiosoma bosc (naked goby) and other gobies

Occurrence. Naked gobies (*Gobiosoma bosc*; formerly, *G. bosci*) inhabit estuaries and protected waters from Connecticut through the Gulf, especially around oyster shell habitats. They spawn from May to September in temperate areas and all year in the Gulf. Larvae are abundant in 1–12 psu in north and Mid-Atlantic estuaries. In the Southeast and Gulf, they occur in more saline or even hypersaline waters, where they may be the most abundant fish larvae with densities sometimes exceeding 200 m^{-3}. The seaboard goby (*G. ginsburgi*) co-occurs in high-salinity areas from Massachusetts to Georgia. Larvae of other species of *Gobiosoma*, *Microgobius*, *Gobionellus*, *Ctenogobius*, *Bathygobius*, and related genera also occur in shallow-water collections in the Gulf or the Southeast.

Biology and Ecology. Gobies attach egg clusters to bivalve shells in nests made and defended by males. Eggs hatch after four to five days at summer temperatures, and the larvae often move to lower salinities. Larvae spawned in lower salinities may hover in shoals close to the reef substrate during the day and rise to the surface at night. Larvae are in the plankton for two to three weeks and feed on zooplankton, including oyster larvae, which are generally present at the same time. Goby larvae are susceptible to predation by jellyfishes.

References. *Gobiosoma bosc:* Breitburg 1990, 1991; Breitburg et al. 1994; Harding 1999; Hendon et al. 2000; Holt and Strawn 1983; Shenker et al. 1983; *Gobiosoma ginsburgi:* Dahlberg and Conyers 1973; Duval and Able 1998; Harding and Mann 2000; Hendon et al. 2000; *Gobiosoma robustum:* Pattillo et al. 1997; *Gobionellus* and *Ctenogobius*: Wyanski and Targett 2000.

Gobiosoma bosc naked goby 2.0 - 8.5 mm SL

2.0 mm

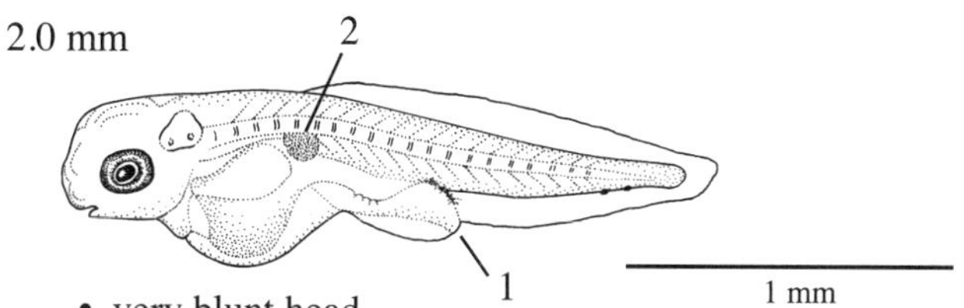

- very blunt head
- yolk-sac/gut is wide and convoluted with anus near midpoint of body (1)
- darkly pigmented swim bladder (2)

3.4 mm

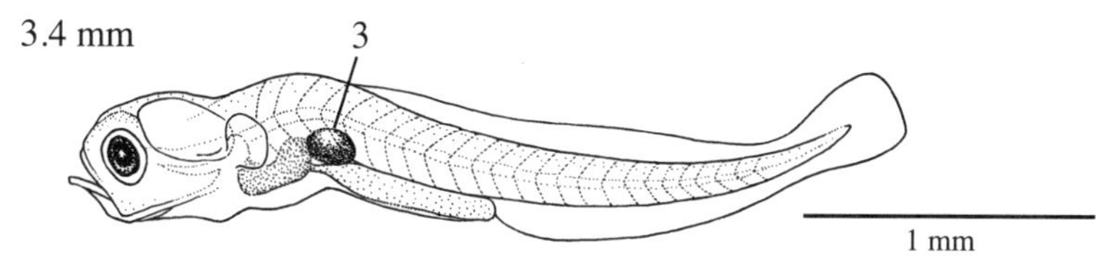

- starting at about 2.5 mm, body is elongate and slender
- oval and darkly pigmented swim bladder is very conspicuous throughout development (3)

4.4 mm

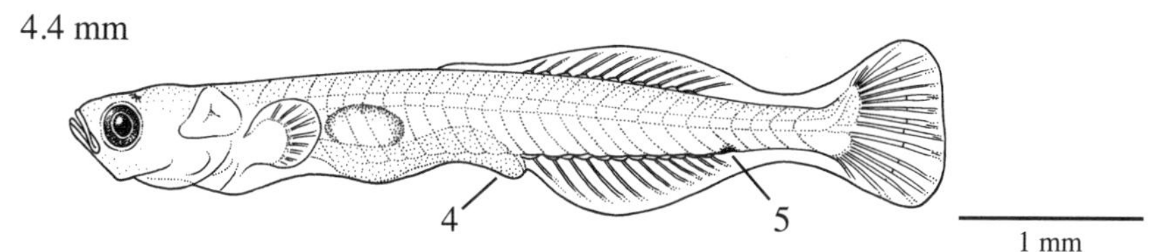

- gut remains wide, fairly straight, uncurled, and extended to midbody throughout development (4)
- few pigment spots along midventral region from anus to caudal fin; one spot usually larger than rest (5)

8.2 mm

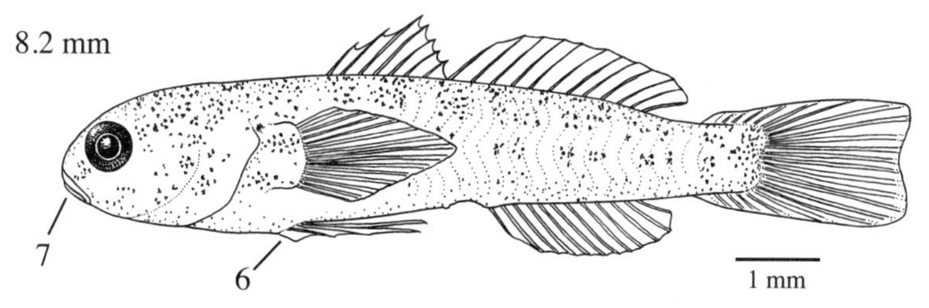

- starting at about 6.0 mm, pelvic fins emerge (6)
- distinctively oblique (almost vertical) mouth becomes smaller (7)
- similar species: many other goby species are similar; among elongate larvae, gobies typically have the most prominent swim bladders and oblique mouths; *G. ginsburgi* is very similar but has midventral dashes rather than spots; *Microgobius* spp. are more slender and heavily pigmented

Hypsoblennius hentz (feather blenny)

Occurrence. Feather blennies occupy estuarine and coastal areas from Massachusetts to northern Florida in the Atlantic and from western Florida through Texas in the Gulf. Larvae are most common in estuaries in brackish to high-salinity waters but have also been collected in shallow-ocean areas. Other closely related species, including the freckled blenny (*Hypsoblennius ionthas*), crested blenny (*Hypleurochilus geminatus*), and striped blenny (*Chasmodes bosquianus*), occur along most of the feather blenny's range, and larvae of these and other blenny species can co-occur.

Biology and Ecology. Spawning occurs in warmer months: typically from May through August in the northern end of its range, April through October in the Southeast, and in all but the coldest periods in the Gulf. Females deposit eggs in empty bivalve shells around oyster reefs or rocky bottoms. Males guard nests containing as many as 3,700 eggs. The larvae usually settle to the bottom at a length of about 15 mm.

References. Hildebrand and Cable 1938; Olney and Boehlert 1988.

Fundulus spp. (killifishes)

Occurrence. The mummichog (*Fundulus heteroclitus*) is euryhaline and ranges from Canada to Florida. Larvae tend to remain in flooded portions of the marsh surface and are incidental catches in the plankton of saltmarsh creeks and occasionally in open waters of estuaries. The similar gulf killifish (*Fundulus grandis*) is common throughout the Gulf, where larvae prefer 5–18 psu. The striped killifish (*F. majalis*) ranges from New England through the Gulf, with larvae especially common in marsh creeks in salinities >20. At least seven other *Fundulus* spp. occur in various regions of the Atlantic and Gulf Coasts, and all *Fundulus* species spawn in warmer months.

Biology and Ecology. Spawning usually shows lunar periodicity in *F. heteroclitus*, with adults laying eggs during high spring tides, although this periodicity may be reduced in northern populations. Both the mummichog and gulf killifish attach eggs to the substrate or to the bases of plants in the intertidal zone of salt marshes. Striped killifishes lay adhesive eggs over shallow sandy bottoms but typically not in marshes.

References. Able and Fahay 1998; Lang et al. 2011; Lopez et al. 2011; Pattillo et al. 1997; Petersen et al. 2010.

Morone saxatilis (striped bass)

Occurrence. Striped bass spawn in tidal freshwater reaches of Atlantic and Gulf Coast estuaries associated with large coastal rivers when temperatures rise above 12°C in spring (February in Florida; June in New England). The Chesapeake Bay and the Hudson River are primary spawning areas. Larvae tend to stay in freshwater or oligohaline water. Within each estuary, larvae often occur for a brief period. Juveniles gradually move into higher salinities (5–15).

Biology and Ecology. Unlike the eggs of the white perch, striped bass eggs are semibuoyant. Transport of newly hatched larvae to the "estuarine turbidity maximum" zone appears critical to survival. Larvae (5 mm) begin feeding at six to eight days after hatching. First foods are rotifers, water fleas (*Bosmina*), and copepod larvae, especially *Eurytemora affinis*. Larvae >14 mm eat adult copepods (*Eurytemora*) and water fleas.

References. Dunning et al. 2009; Fay et al. 1983a; Grant and Olney 1991; Limburg et al. 1997, 1999; McGovern and Olney 1996; Monteleone and Houde 1992; North and Houde 2001, 2006; North et al. 2005; Smith and Kernehan 1981.

Hypsoblennius hentz feather blenny 4.7 mm (2.4 - 10.5 mm) SL

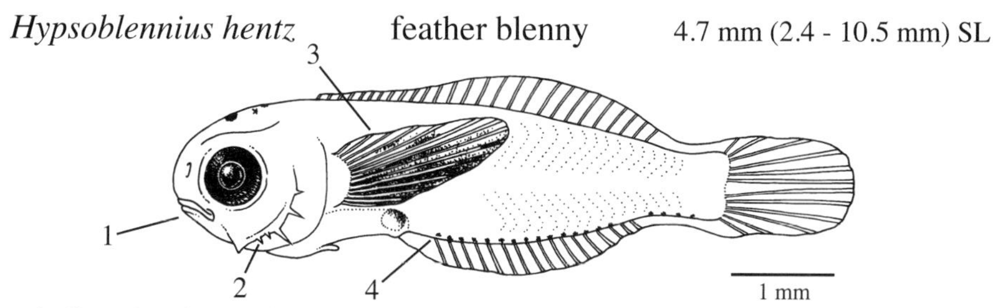

- bulbous head and robust trunk
- blunt snout, subterminal mouth, and very large eyes (1)
- multiple spines on operculum (2)
- large pectoral fin with heavy pigmentation (3)
- row of pigment spots along base of anal fin (4)
- similar species: many other blenny species are similar; drum species do not have large, pigmented pectoral fins and rounded foreheads

---

Fundulus heteroclitus mummichog 8.5 mm (5 - 11 mm) SL

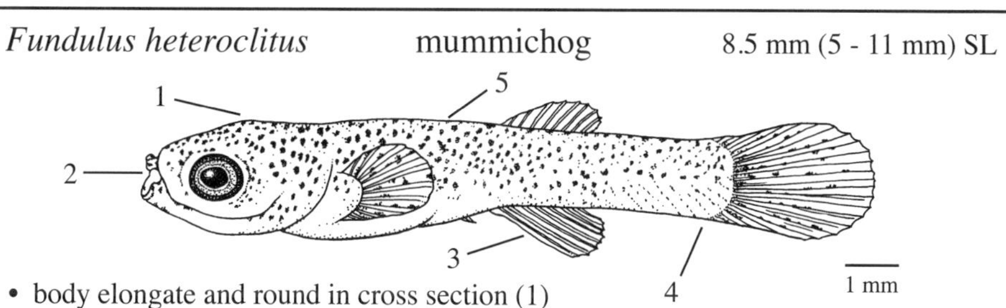

- body elongate and round in cross section (1)
- small mouth with short jaws (2)
- short dorsal fin has origin slightly posterior to origin of anal fin (3)
- broad caudal peduncle and long rounded caudal fin (4)
- dense pigmentation scattered, but sometimes short vertical bars are evident (5)
- similar species: many other *Fundulus* spp. are very similar and difficult to distinguish from one another; *F. majalis* is more elongate and has a pointed snout, less pigmentation, and bars that extend the full depth of the body; *F. diaphanus* has a dorsal fin origin anterior to the origin of the anal fin

---

Morone saxatilis striped bass 7.4 mm (3.5 - 13 mm) SL

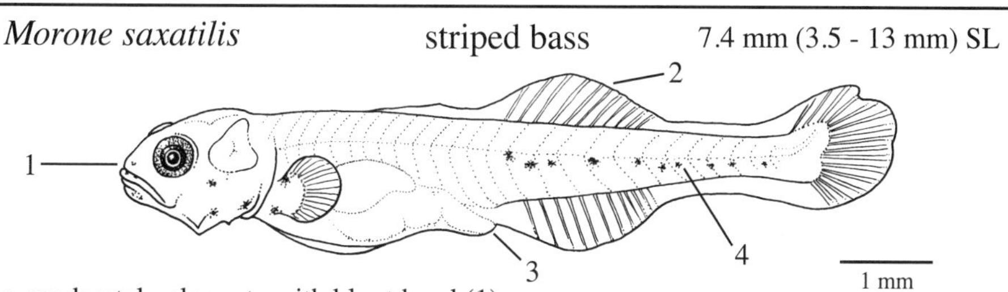

- moderately elongate with blunt head (1)
- dorsal and anal fins similar and opposite one another (2)
- anus is below the anterior origin of the dorsal fin (3)
- line of unevenly spaced pigment spots just below the lateral line (4)
- teeth present at early stages
- similar species: *M. americana* is much more developed at same length, has evenly spaced spots, and no teeth

Morone americana (white perch)

Occurrence. Within the coverage area, white perch range from Massachusetts to South Carolina but are most abundant in estuaries from the Chesapeake Bay north. Eggs and early larvae are primarily in upper reaches of estuaries and are often concentrated at the estuarine turbidity maximum. Larvae spawned in the spring move down the salinity gradient with juveniles in brackish areas (<10 psu).

Biology and Ecology. White perch undergo a semianadromous spawning migration from lower reaches of the larger estuaries to spawn upstream in <4 psu. Eggs adhere to substrates and usually hatch after two to six days. As is the case for the striped bass (*Morone saxatilis*), the larvae are closely associated with the estuarine turbidly maximum, where their primary prey, larval and adult copepods (*Eurytemora affinis*), are also concentrated.

References. Limburg et al. 1997, 1999; Mansueti 1964; North and Houde 2001, 2006; Setzler-Hamilton et al. 1982; Shoji et al. 2005; Smith and Kernehan 1981; Stanley and Danie 1983.

Morone americana white perch 3.0 - 9.5 mm SL

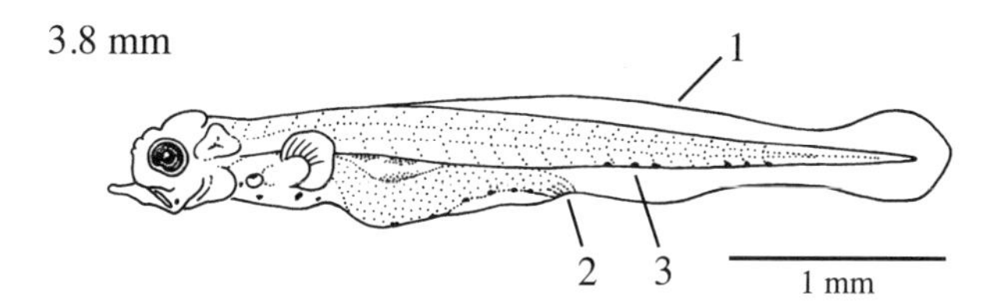

- elongate body of almost uniform height (1)
- yolk-sac/gut is wide, convoluted, but not coiled, and ends midbody (2)
- midventral row of small, evenly spaced pigment spots from the anus to the caudal fin (3)

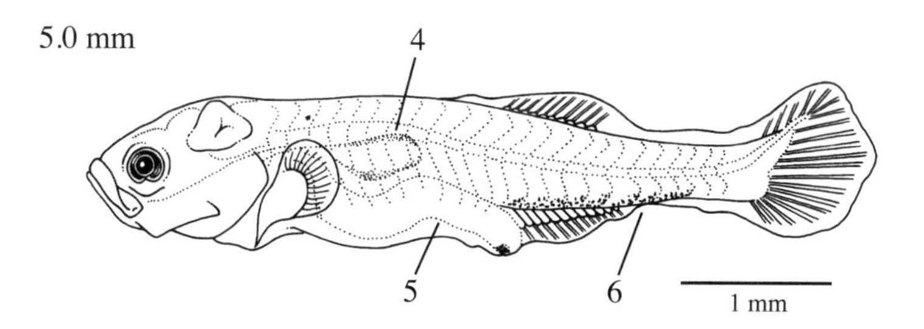

- starting at about 4.0 mm, swim bladder is larger and darkly pigmented (4)
- gut remains wide and uncoiled (5)
- small pigment spots become numerous on ventral edge over and posterior to the anal fin (6)

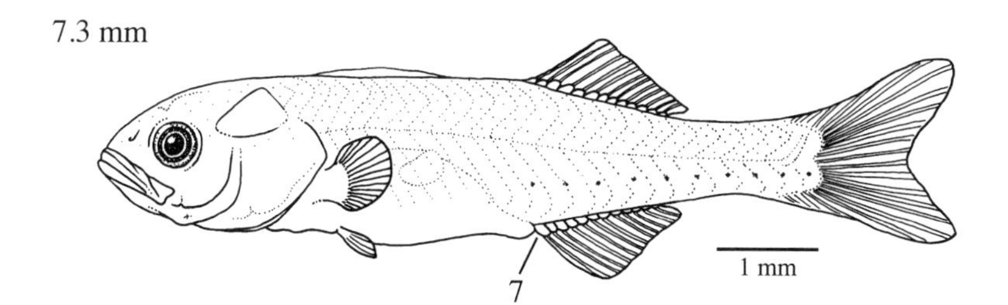

- starting at about 4.5 mm, origin of the anal fin is almost directly below origin of the dorsal fin (7)
- similar species: herrings and *Anchoa mitchilli* have an anus located posterior to the midpoint of the body; *Gobiosoma* spp. lack midventral pigment spots; *M. saxatilis* is less developed at the same length, more slender, lacks pigment on the swim bladder, and has large, unevenly spaced midventral spots

DRUMS (Family Sciaenidae: The Next Eight Species)

ID hint: Drum larvae have relatively large heads, large oblique mouths, truncate bodies, and, often, species-specific pigmentation patterns. Drums have long, continuous dorsal fins, usually with at least twice as many rays as the anal fin.

Leiostomus xanthurus (spot)

Occurrence. The spot is perhaps the most abundant demersal estuarine fish from Virginia to Florida and through the Gulf. Although larvae have been reported in Rhode Island waters and New Jersey, they are much more abundant from Virginia south. Larvae occur in Mid-Atlantic and Southeast estuaries from October to June with peaks in winter. Ingress to northern estuaries comes increasingly later in spring. In Alabama coastal waters, larvae occur from August to May with a peak in November. In South Carolina estuaries, arriving spot are usually at least 10 mm SL. Larger larvae are most common near the bottom from December through March. In estuaries with salinity gradients, late larval stages enter moderate salinity areas but seldom penetrate far up estuaries before becoming juveniles.

Biology and Ecology. Atlantic coast spot spawn primarily in midcontinental shelf areas south of Chesapeake Bay in late fall and winter. Larval spend about two months at sea before moving inshore and entering estuaries and inlets as late-stage larvae. Within each estuary, multiple cohorts of larvae can be identified over the three to five month period of ingress. Copepods make up the bulk of the larval diet.

References. Flores-Coto and Warlen 1993; Govoni 1993; Govoni and Chester 1990; Govoni et al. 1983; Lyczkowski-Shultz et al. 1990; Phillips et al. 1989.

Micropogonias undulatus (Atlantic croaker)

Occurrence. Larvae occur in the Atlantic from New York to central Florida. In South Carolina estuaries, larvae from 5 to 11 mm SL are most common from October to December but can occur until April. Atlantic croaker occurs throughout the Gulf, and, in Alabama coastal waters, larvae appear from August to April with a peak in October–November. Atlantic croaker larvae occur up-estuary to oligohaline tidal creeks. Atlantic croaker and spot often codominate larval fish collections in fall and winter, but the Atlantic croaker occurs earlier in the fall on both coasts.

Biology and Ecology. On the Atlantic Coast, spawning occurs over large areas of the continental shelf south of Delaware Bay. In the northern Gulf, croakers spawn close to inlets from August to April. Spawning in the western Gulf appears to be completed before January. Once in bays or estuaries, larvae tend to remain near the bottom in deep channels. Larvae feed selectively on small zooplankton: bivalves, copepod larvae, and small cyclopoid copepods before switching to microbenthic invertebrate prey as juveniles.

References. Govoni et al. 1983, 1986; Lassuy 1983c; Norcross 1991; Poling and Fuiman 1997; Schaffler et al. 2009; Sogard et al. 1987.

Leiostomus xanthurus spot 12.3 mm (2 - 16 mm) SL

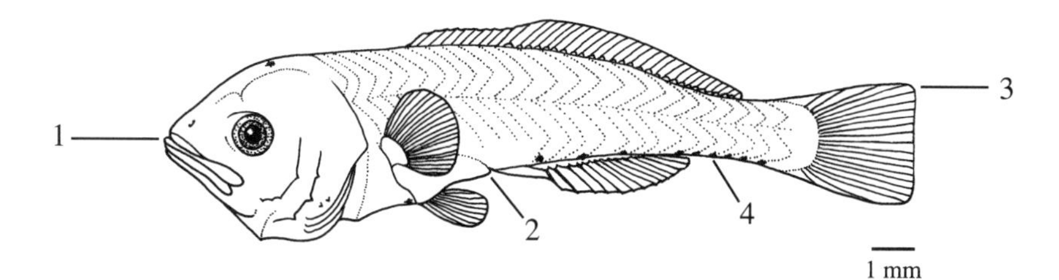

- short, somewhat pointed head with large eyes (1)
- anus near midpoint of body (2)
- caudal fin is rectangular becoming square (never pointed) (3)
- multiple pigment spots at base of anal fin and on ventral surface of caudal peduncle (4)
- similar species: other drums have similar fin and anus positions, opercular spines, and pigment above the lateral line; *Micropogonias* is most similar but has darker ventral pigmentation, fewer anal rays (8 or 9), and long, central and ventral caudal fin rays, giving the tail a low, pointed rather than a square shape

---

Micropogonias undulatus Atlantic croaker 8.1 mm (3 - 14 mm) SL

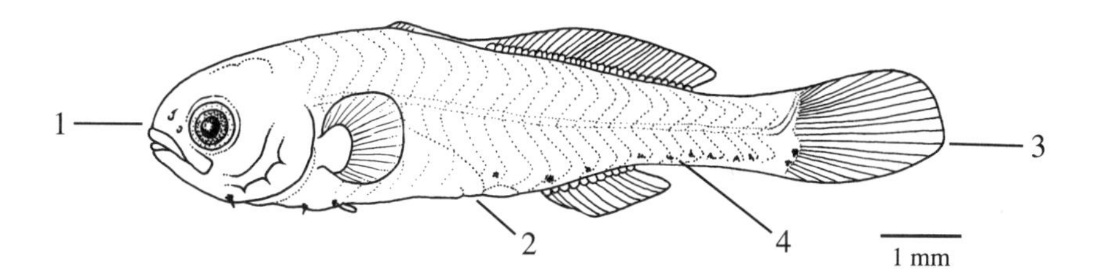

- short, somewhat rounded head with medium eyes (1)
- anus near midpoint of body (2)
- caudal fin pointed with ventral rays longest (3)
- pigment spots at base of anal fin and on ventral surface of caudal peduncle (4)
- no pigment along dorsal half of body
- no pigment on front of the gut when operculum lifted
- similar species: other drums have some pigment spots in the mid or dorsal regions of the trunk; *Leiostomus* has more anal fin rays (12 or 13) and caudal fin rays that are mostly of similar length

Bairdiella chrysoura (silver perch)

Occurrence. Silver perch are among the most abundant fish larvae in high-salinity areas in the Gulf. Larvae also occur regularly in spring and summer along the Atlantic Coast north to the Chesapeake Bay and sporadically north to New Jersey. In South Carolina estuaries, larvae from 3 to 6 mm are collected from May to August. Larvae may occur in waters from 1 to 37 psu although most occur at salinities >10.

Biology and Ecology. In both the Atlantic and Gulf, larvae are produced in the deepest parts of estuaries and close to inlets; they are rarely encountered very far from estuaries. Larvae are especially common in seagrass beds and saltmarsh creeks. Silver perch is a euryhaline species, with adults in spawning condition found from 14 to 26 psu in Gulf estuaries. Peak spawning along most of the Atlantic Coast is from May through August. In the Gulf, silver perch spawn from March to October with peak spawning occurring April to June. Unlike most drums, silver perch are capable of spawning daily. In Mississippi, early stage silver perch larvae feed on copepods, but mysids become increasingly important in the diets of larvae >10 mm SL.

References. Grammer et al. 2009; Ocaña-Luna and Sánchez-Ramírez 1998; Rooker et al. 1998; Waggy et al. 2007.

Bairdiella chrysoura silver perch 1.8 - 14 mm SL

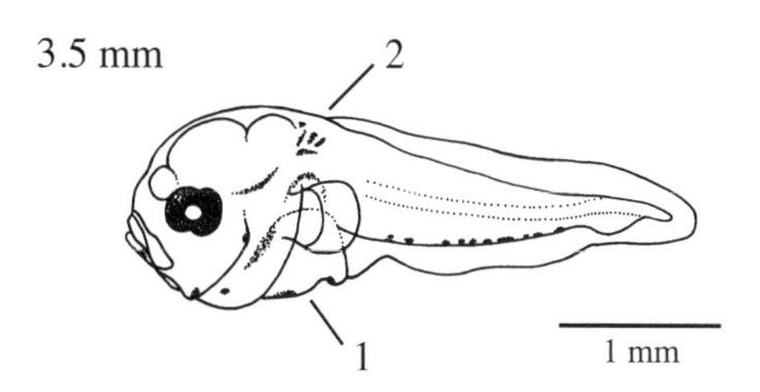

- very large head and short gut cavity (1)
- pigment spots posterior to the head (2)

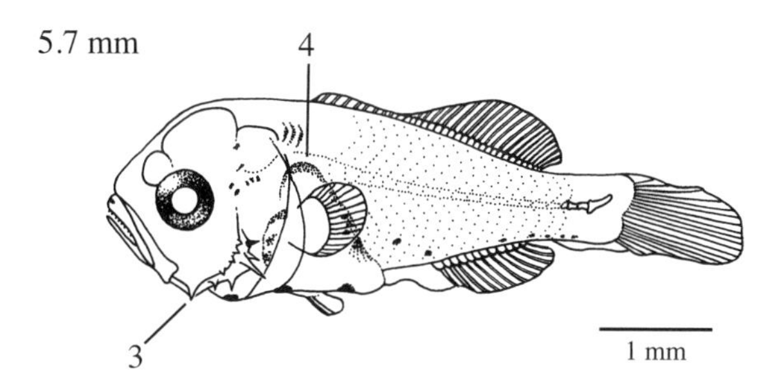

- starting at about 4.0 mm, strong preopercular spines occur (3)
- irregular band of pigment posterior to the head extends acrosss the gut, and dorsal edge of swim bladder is heavily pigmented throughout larval development (4)

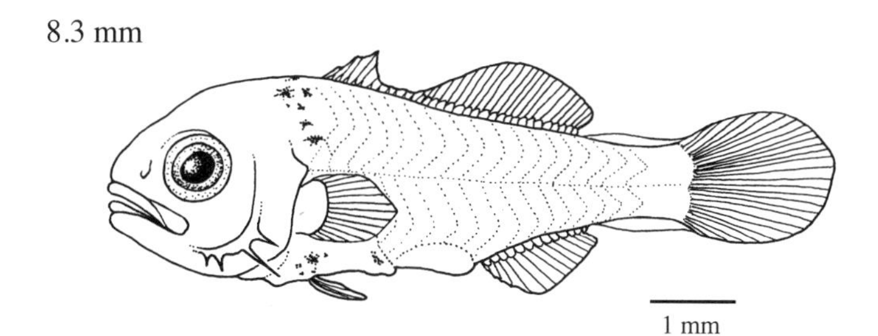

- large head and strong preopercular spines persist
- similar species: *Menticirrhus* spp. and *Pogonias chromis* have widespread and heavier pigmentation; *Cynoscion regalis* has multiple vertical bands; *Sciaenops ocellatus* and *Cynoscion nebulosus* are more elongate and have heavy pigment along the lateral line

Sciaenops ocellatus (red drum)

Occurrence. Red drum larvae occur in estuaries and enclosed bays and sounds from Chesapeake Bay south, but maximum abundances are recorded in the Gulf and the Southeast in fall, typically in mid to high salinities. Yolk-sac larvae occur near the bottom, and feeding stages occur near the surface.

Biology and Ecology. In both the Southeast and Gulf, red drum form spawning aggregations with individuals producing multiple batches of eggs, mostly between August and November. Spawning occurs close to barrier islands, in open sounds, or in the mouths of estuaries. The eggs and early larvae develop in shallow protected areas, especially seagrass beds. Late-stage larvae move into lower-salinity areas. Calanoid copepods form the primary diet of larval red drum, supplemented by bivalve, gastropod, and barnacle larvae. Larvae >10 mm feed extensively on larger crustaceans, especially mysids.

References. Duffy et al. 1997; Faria et al. 2009; Fuiman and Cowan 2003; Fuiman et al. 2006; Holt and Holt 2000; Pattillo et al. 1997; Pérez-Domínguez et al. 2006; Regan 1985.

Cynoscion nebulosus (spotted seatrout)

Occurrence. Spotted seatrout larvae appear in the plankton from New Jersey south and throughout the Gulf but are most common from Chesapeake Bay south. The larvae are euryhaline, often occurring in low salinities. They are usually found near the bottom and particularly near vegetation. In South Carolina, larvae occur from May to September, and in the northern Gulf from February through October; peaks in both areas are usually between June and August. Other seatrouts that produce similar larvae during the summer include the silver seatrout (*Cynoscion nothus*) predominately in the Gulf and the sand seatrout (*C. arenarius*) exclusively in the Gulf.

Biology and Ecology. Spotted seatrout spawn in high- to medium-salinity coastal areas, in or near bays, lower estuaries, and lagoons, especially in deep, fast-moving water at dusk. Females can release successive batches of eggs, several days apart. Larvae then disperse over wide areas and into low salinities. Calanoid copepods and bivalve larvae are the primary prey of larval seatrout.

References. Banks et al. 1991; Brown-Peterson et al. 2002; Holt and Holt 2000; Lassuy 1983d; McMichael and Peters 1989; Pattillo et al. 1997; Peebles and Tolley 1988; Pryor and Epifanio 1993.

Cynoscion regalis (weakfish)

Occurrence. Weakfish larvae occur from Rhode Island to Georgia. They can be abundant in the nearshore and surf zone areas of New Jersey, especially from June through August. In South Carolina estuaries, larvae occur from April to September.

Biology and Ecology. Weakfish spawn in late spring and summer, usually in or near large bays or estuaries at salinities >25. Adults return to natal estuaries and multiple spawning peaks can occur as aggregations spawn at different locations and times outside and within an estuary. In Delaware Bay, a major spawning area, the larvae move into upper estuarine areas, probably by using subsurface currents to transport them up-estuary. Feeding is presumably similar to that of spotted seatrout.

References. Able et al. 2006; Brown-Peterson et al. 2002; Connaughton and Epifanio 1993; Connaughton et al. 1994; Goshorn and Epifanio 1991; Mercer 1989; Pryor and Epifanio 1993.

Sciaenops ocellatus red drum 6.6 mm (2 - 10 mm) SL

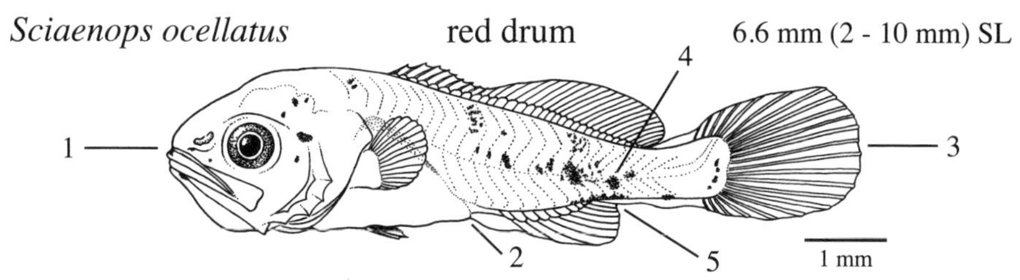

- head is large relative to elongate trunk (1)
- anus near midpoint of body (2)
- mid-caudal fin rays about same length; fin with evenly rounded shape (3)
- pigment spots (often 3) along dorsal margin and more numerous along lateral line and ventral margin (4)
- in later stages, prominent spot near termination of anal fin base (5)
- similar species: other drums similar; *Pogonias* does not have pigment spot under origin of soft dorsal fin; *Cynoscion* spp. have increasingly longer central caudal fin rays as they develop and pigment on roof of mouth

---

Cynoscion nebulosus spotted seatrout 8.9 mm (2 - 11 mm) SL

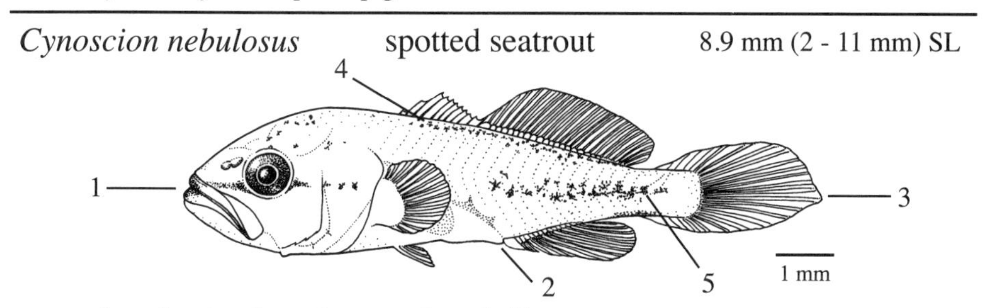

- very large head and evenly tapered trunk (1)
- anus posterior to midline of body but close to anal fin (2)
- mid-caudal fin rays long, giving tail a pointed appearance (3)
- long line of pigment spots along dorsal margin (4)
- broken line of pigment extends through eye and along the lateral line (5)
- similar species: other drums similar; many drum lack long central caudal fin rays; other *Cynoscion* spp. are not as elongate or as heavily pigmented along the midline; *C. regalis* has irregular vertical bands of pigment

---

Cynoscion regalis weakfish 8.7 mm (2 - 10 mm) SL

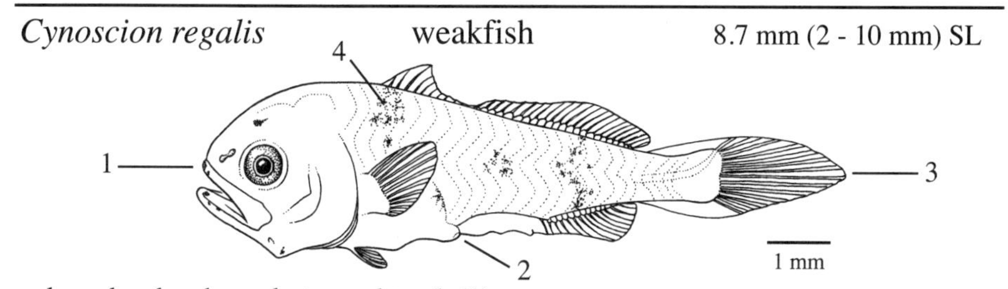

- large head and evenly tapered trunk (1)
- anus anterior to middle of body and well anterior to the anal fin origin (2)
- mid-caudal fin rays long, giving the tail a pointed appearance (3)
- clusters of pigment spots form bands across the trunk (4)
- similar species: other drums and especially other *Cynoscion* spp. similar, but *C. nebulosus* has much heavier pigmentation along mid-trunk; *Bairdiella* has a vertical band of pigment, but the anus is near the origin of the anal fin

Pogonias cromis (black drum)

Occurrence. Black drum larvae are most common from Delaware to Texas. Larvae are often transported from high-salinity spawning areas in bays and sounds to the upper regions of estuaries in 0–6 psu. In New Jersey and South Carolina, most larvae occur from May to July. In the Gulf, abundance peaks from February to May. Thus, throughout its range, the black drum's larvae are typically found earlier in the year than other co-occurring species in the drum family.

Biology and Ecology. Calanoid copepods form the primary diet of larval black drum, supplemented by bivalve, gastropod, and barnacle larvae.

References. Cowan et al. 1992; Sutter et al. 1986; Thomas and Smith 1973.

Menticirrhus saxatilis (northern kingfish) and other *Menticirrhus* spp.

Occurrence. Northern kingfish occur along the Atlantic from Massachusetts south and throughout the Gulf with larvae prevalent along beaches and inlets. Larvae appear in summer and fall. They occur in estuaries, generally at salinities above 10. Adult southern kingfish (*Menticirrhus americanus*) and Gulf kingfish (*M. littoralis*) are the most common species in the Southeast and Gulf and appear less frequently along the Atlantic Coast to Virginia so their larvae would be expected in this range. In South Carolina estuaries, *Menticirrhus* spp. larvae occur from May to October and in every month except December in Alabama coastal waters.

Biology and Ecology. All kingfishes reproduce in shallow water, including the surf zone throughout their ranges. Northern kingfish spawn from May to September north of Chesapeake Bay. In the Southeast and Gulf, all species spawn over extended periods but peak during the spring and summer.

Reference. Harding and Chittenden 1987; Miller et al. 2002.

Lophius americanus (goosefish)

Occurrence. Goosefish larvae occur in spring from Massachusetts to North Carolina. They are among the more common fish larvae taken off New Jersey beaches from May to July.

Biology and Ecology. The goosefish, sold in markets as monkfish, spawn in coastal waters and produce pinkish mucous sheets up to 12 m long, containing a single layer of floating eggs. Initially, larvae float at the surface through the yolk-sac stage and then remain in the plankton two to three months before settling to the bottom when they're ~40–50 mm long. Absence of reports of larvae in nearshore plankton may reflect the paucity of collections near sandy beaches rather than a lack of larvae.

Reference. Collette and Klein-MacPhee 2002.

Pogonias cromis black drum 6.6 mm (3 - 8.5 mm) SL

- large head and large, rounded pectoral fin (1)
- anus near midpoint of body close to origin of the anal fin (2)
- caudal fin elongate but rounded, becoming square (3)
- body heavily pigmented with star-shaped and branching spots, especially near the anal fin base (4)
- similar species: *Menticirrhus* also has heavy pigmentation, but the spots are in irregular rows; *Sciaenops* has a distinct pigment spot at origin of soft dorsal fin

---

Menticirrhus saxatilis northern kingfish 6.1 mm (2 - 8.5 mm) SL

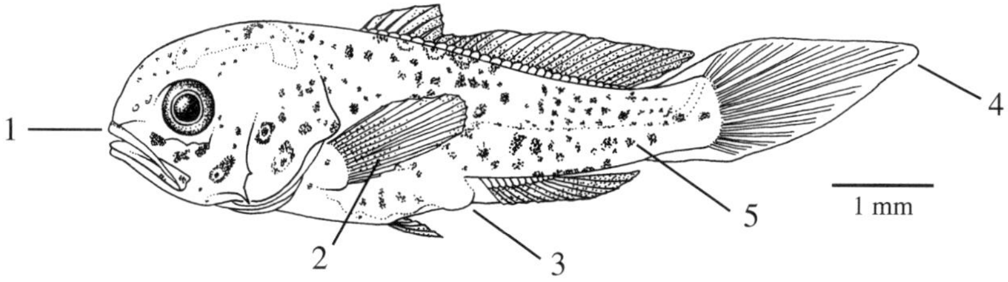

- high forehead and robust body (1)
- long pectoral fin (2)
- anus at midpoint of body close to anal fin (3)
- middle rays of caudal fin are longest, giving tail a pointed appearance (4)
- heavily pigmented with irregular rows on trunk (5)
- dark pigmentation on roof of mouth is visible from exterior (not shown)
- similar species: *Menticirrhus* spp. are not easily separated; *Pogonias* is also heavily pigmented but spots are not in rows; *Cynoscion* spp. are the only other drums with pigment on roof of mouth

---

Lophius americanus goosefish 5.3 mm (3 - 14 mm) SL

- very blunt head with low mouth (1)
- long independent dorsal rays form early (2)
- large pectoral fin becomes fan-like (3)
- long pelvic rays form early (4)
- 3 large distinct pigment spots are along trunk (5)

Lagodon rhomboides (pinfish)

Occurrence. Pinfish are among the most abundant fish larvae in winter and spring from North Carolina to Florida. They are uncommon as far north as southern New Jersey. Pinfish larvae occur throughout the Gulf from October through April, peaking during the coldest months. Ingress of larvae to South Carolina estuaries is usually from December to March.

Biology and Ecology. Pinfish spawn offshore during fall and winter, and the larvae move into estuaries toward the later stages of development. They are particularly attracted to vegetated areas in high to mid-salinities. The apparent scarcity of pinfish larvae in nearshore waters of the western Gulf may be due to a lower sampling effort during cool seasons, but it is also possible that larvae pass through larval stages offshore and enter nearshore waters as juveniles. Larvae feed on copepods.

Reference. Caldwell 1957; Muncy 1984b.

Stenotomus chrysops (scup)

Occurrence. Scup larvae occur from the Hudson River estuary north and are most abundant from May through September. Larvae are more common along the coast and in selected bays and sounds of southern New England in salinities >15.

Biology and Ecology. Scup reproduce in high-salinity areas of large sounds and estuaries and in the shallow coastal ocean when water temperatures reach about 10°C. The eggs are buoyant.

Reference. Steimle et al. 1999.

Tautogolabrus adspersus (cunner) and *Tautoga onitis* (tautog)

Occurrence. Cunner larvae occur from Massachusetts to North Carolina, with large catches noted from Rhode Island and New Jersey coastal waters. Tautog larvae are most common in embayments and nearshore coastal waters from Massachusetts to Virginia in spring and summer. Although tautog larvae have been reported as far south as North Carolina, the maximum abundance is from northern New Jersey through southern New England. In New Jersey estuaries, cunner larvae occur from June through August, and tautog larvae occur from June through September. Larvae of both species co-occur over much of this range.

Biology and Ecology. Cunner spawn in shallow ocean areas in late spring and summer in the southern part of their range and during increasingly later and shorter periods farther north. The buoyant eggs hatch in two days, and the larvae settle at a relatively early stage in about three weeks in shallow water, especially in areas with complex bottom structure, including eelgrass. Tautog spawning locations and timing are similar to those of the cunner.

References. Auster 1989; Schroedinger and Epifanio 1997; Sogard et al. 1992; Steimle and Shaheen 1999.

Lagodon rhomboides pinfish 10.7 mm (3 - 15 mm) SL

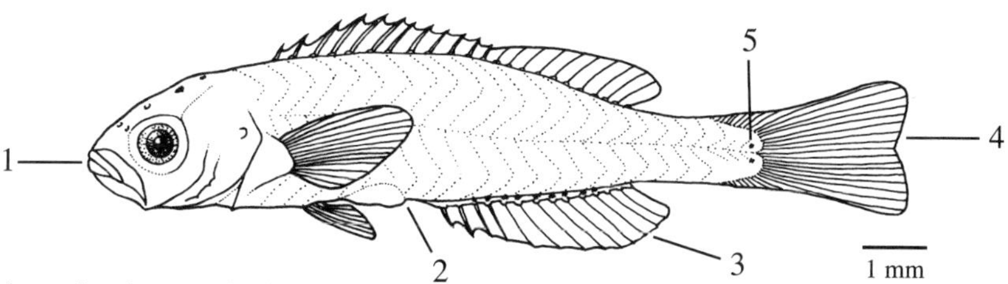

- moderately elongate body with small mouth (1)
- anus near midpoint of body close to origin of the anal fin (2)
- large anal fin (3)
- long pectoral fin almost reaches anus; long caudal fin (4)
- pigmentation absent or light; pair of spots are sometimes prominent at base of caudal fin (5)
- similar species: drums have a larger head, heavier pigmentation on the trunk, and a shorter anal fin; larvae of *Diplodus holbrooki*, which can co-occur in the coastal ocean, have more anal fin rays and pigment along the lateral line

---

Stenotomus chrysops scup 9.3 mm (3 - 20 mm) SL

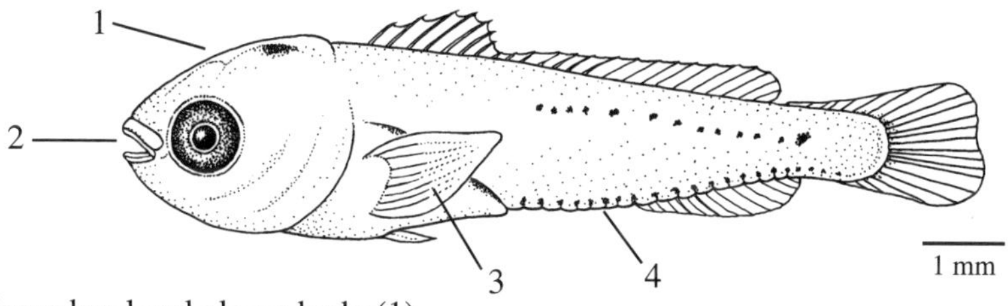

- large head and plump body (1)
- small symmetrical mouth with large eye immediately posterior to angle of jaws (2)
- pointed pectoral fin (3)
- evenly spaced rows of pigment spots along both the lateral line and ventral margin (4)

Tautogolabrus adspersus cunner 6.7 mm (3 - 10 mm) SL

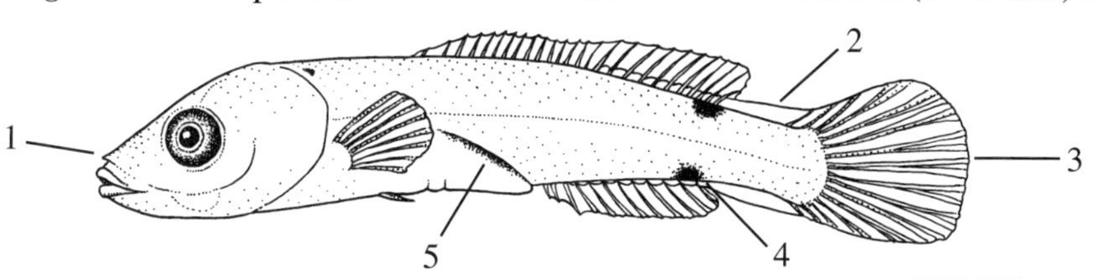

- elongate body with long sloped forehead and small mouth (1)
- dorsal and ventral finfolds persist to late in development (2)
- large, square caudal fin (3)
- large pigment spots on trunk at posterior ends of both the dorsal and anal fins (4)
- dorsal margin of gut cavity with heavy pigment (5)
- similar species: *Tautoga onitis* has a shorter snout and its body is heavily pigmented, without paired spots on posterior trunk

Pomatomus saltatrix (bluefish)

Occurrence. Bluefish larvae are most common more than 10 km off the coast but appear sporadically in nearshore waters, where they are reported in spring from North Carolina and in summer off the northern New Jersey beaches and in Narragansett Bay. In Gulf waters, larvae occur in spring and fall.

Biology and Ecology. In both the Atlantic and Gulf, most bluefish spawn well out at sea from May to July and arrive in estuaries as juveniles. Some bluefish apparently spawn in the fall and closer to shore, and thus, larvae appear in nearshore plankton collections in the fall. Larvae tend to be near the surface. Small larvae feed mostly on copepods and gradually switch to fish and crab larvae.

References. Ditty and Shaw 1995; Hare and Cowen 1996; Kendall and Walford 1979; Robillard et al. 2008.

Scomber scombrus (Atlantic mackerel)

Occurrence. Atlantic mackerel larvae occur from New Jersey northward in the surf zone and in nearshore coastal waters during spring and early summer. They are also reported in abundance from northern high-salinity bays and sounds. Several other mackerel range widely along the Atlantic and Gulf Coasts. Spanish mackerel (*Scomberomorus maculatus*) and king mackerel (*Scomberomorus cavalla*) larvae occur within 10 km of the Southeast and northern Gulf coasts from June through September.

Biology and Ecology. Spawning of the Atlantic mackerel occurs on the inner continental shelf from Chesapeake Bay to New England. The smallest larvae eat phytoplankton and copepod eggs but then switch to progressively larger copepod larvae and, finally, larger zooplankton. Larvae >6.5 mm also eat fish larvae and may be cannibalistic.

Reference. Peterson and Ausubel 1984.

Microgadus tomcod (Atlantic tomcod)

Occurrence. Early larvae may be abundant in upper reaches of selected large estuaries from the Hudson River northward from January through April. Most are found in fresh or very low salinity water. Later larvae are usually farther downstream.

Biology and Ecology. Adults occur along the entire salinity gradient and in the coastal ocean. All tomcod spawn in very low salinities or in freshwater during winter. Eggs sink and stick to the bottom. Newly hatched larvae rise to the surface and gulp air to inflate the swim bladder and then sink to near bottom, where they are usually collected. Tidally related vertical migrations facilitate export to higher-salinity areas.

References. Peterson et al. 1980; Stewart and Auster 1987.

Pomatomus saltatrix bluefish 6.6 mm (3 - 24 mm) SL

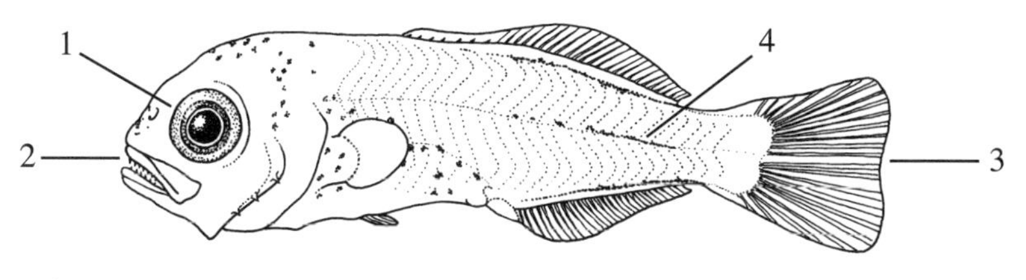

- large head and eye (1)
- stout teeth on both jaws developing early (2)
- caudal fin square becoming forked (3)
- pigment spots on top of head, posterior margin of gut cavity, and in rows along the lateral line and along the body margins of the dorsal and anal fins (4)
- similar species: similar to drum larvae, but none has the same linear pattern of pigmentation on the trunk; *Scomber* is heavily pigmented on the head and is more elongate

---

Scomber scombrus Atlantic mackerel 9.6 mm (3 - 15 mm) SL

- elongate with minimal taper (1)
- very large eye and toothed jaws (2)
- dorsal and anal fins are similar and almost opposite (3)
- heavily pigmented on head and gut (4)
- similar species: other mackerel species are similar although most have much larger, pointed mouths; *Pomatomus* has less pigmentation on the head and a linear pattern on the trunk

---

Microgadus tomcod tomcod 11.4 mm (7 - 23 mm) SL

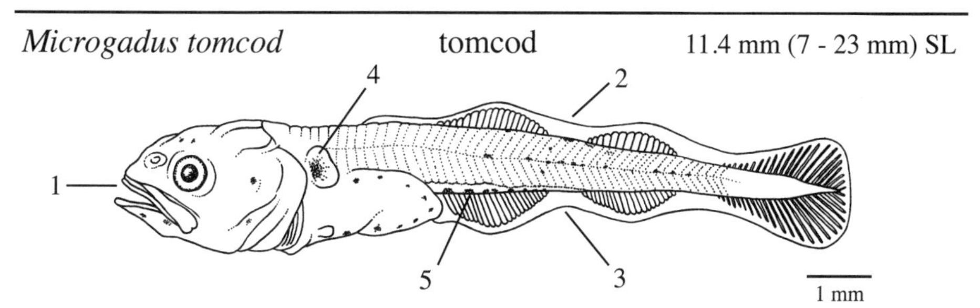

- large head and small eye; large mouth (1)
- 2 dorsal fins apparent by 8 mm SL and a third fin soon after that (2)
- 2 anal fins apparent by 8 mm SL (3)
- small pectoral fin becomes elongate (4)
- pigment spots scattered on head, gut, and trunk, but a conspicuous line of large spots at ventral edge of trunk above first anal fin (5)

LEFT-EYED FLOUNDERS

ID hint: If a flatfish is lying dark-side up with its eyes above its mouth and is facing to the left, it is called a **left-eyed flounder.**

Etropus spp.

Occurrence. *Etropus crossotus*, the fringed flounder, occurs from the Chesapeake Bay to Texas. Larvae appear in Atlantic coastal ocean and estuarine areas from May through November and in Alabama coastal waters from March through October. *Etropus microstomus*, the smallmouth flounder, ranges from at least New York to northern Florida and occurs in the northern Gulf of Mexico. Larvae occur in both nearshore and estuarine waters from February through August in Alabama, May through November in the Carolinas, and for progressively shorter periods during summer farther north. Larvae of other *Etropus* spp. occur all year in the northern Gulf.

Biology and Ecology. *Etropus microstomus* spawns during summer north of Cape Hatteras. Both species spawn during all but the coldest periods in the Gulf.

References. Scherer and Bourne 1980; Tucker 1982.

Scophthalmus aquosus (windowpane)

Occurrence. Windowpane larvae are common to abundant during the warm months from Massachusetts through the Chesapeake Bay and are reported south to North Carolina in both coastal ocean and high-salinity estuarine areas.

Biology and Ecology. A single summer spawn occurs north of New Jersey. From New Jersey south, bimodal spawning results in separate spring and fall cohorts. In the spring, larvae are also produced near inlets and estuaries. In the fall, eggs and larvae are primarily in the ocean and larvae move through inlets into estuarine areas before settlement.

Reference. Neuman and Able 2003.

Paralichthys spp. (summer, fourspot, southern, and gulf flounders)

Occurrence. Late larvae of summer flounder (*Paralichthys dentatus*) occur in nearshore waters from Massachusetts to to Florida. They are most common fall to spring. Larvae are found along the coast and in estuaries usually in salinities >10. Southern flounder (*P. lethostigma*) larvae are relatively common during winter in inshore waters and inlets from North Carolina to Florida and in the Gulf. Fourspot flounder (*Hyppoglossina oblongata*) larvae can be abundant in New England in the summer. Gulf flounder (*P. albigutta*) adults are relatively common during winter in coastal waters and inlets from North Carolina through Florida and throughout the Gulf but seldom in <20 psu, so their larvae should occur there as well. Many other common left-eyed flounders, including *Etropus* spp. and *Citharichthys* spp., produce similar larvae in shallow water during the summer.

Biology and Ecology. *Paralichthys dentatus* and the two southern species spawn offshore in winter. Later-stage larvae move inshore and arrive at inlets and mouths of estuaries in spring. Early larvae eat copepod and bivalve larvae and tintinnids. Later stages switch to larger prey: copepods, especially *Temora longicornis*, and larvaceans. Sevenspine bay shrimp feed on flounder larvae as they settle. *Paralichthys* species are difficult to distinguish, so we lack information on species-specific larval abundances where species overlap.

References. Able et al. 1990; Burke et al. 1991; Duebler 1958; Gilbert 1986; Grover 1998; Keefe and Able 1993, 1994; Moustakas et al. 2004; Smith and Fahay 1970; Taylor et al. 2010.

Etropus crossotus fringed flounder 10.3 mm (2.5 - 11 mm) SL

- left-eyed with elongate, evenly tapered body
- relatively small mouth and eyes (1); eye rotation complete by 11 mm
- second and third dorsal fin rays long; much longer in earlier stages (2)
- narrow pigment dashes along entire length of notochord (3)

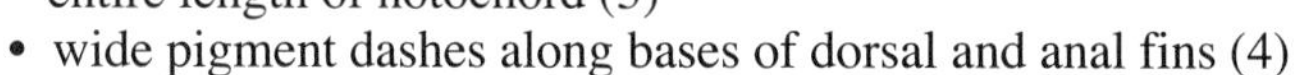

- wide pigment dashes along bases of dorsal and anal fins (4)
- similar species: other late-stage flounder larvae do not have such a symmetrical body taper; others, including *Citharichthys* spp., lack pigment dashes along notochord; other common flounders, including *Etropus microstomus*, do not have elongate second and third dorsal rays

---

Scophthalmus aquosus windowpane 6.8 mm (2 - 11 mm) SL

- left-eyed with wide body and sharply tapered trunk
- large head, mouth, and eyes; eye rotation complete by 13 mm (1)
- dark pigment spots on preopercle and surface of gut (2)
- clusters of pigment spots on dorsal and anal fins form broken bars (3)
- similar species: other left-eyed flounders are more oblong, have smaller heads, and lack vertical pigment bars across the fins

---

Paralichthys dentatus summer flounder 12 mm (3 - 14 mm) SL

- left-eyed with oblong body
- large mouth but relatively small head and eyes; eye rotation complete by 14 mm (1)
- caudal fin elongate and pointed (2)
- widely separated pigment spots on trunk and along dorsal and anal finfolds but not on fins (3)
- similar species: other *Paralichthys* spp. are similar, differing mainly in pigment patterns and fin ray counts; *Etropus* spp. and *Citharichthys* spp. are similar but occur in summer

RIGHT-EYED FLOUNDERS

ID hint: If a flatfish is lying dark-side up with its eyes above its mouth and the head is facing to the right, it is called a **right-eyed flounder.**

Trinectes maculatus (hogchoker)

Occurrence. Hogchokers occur from Massachusetts to Florida and throughout the Gulf. Larvae are particularly abundant in larger Mid-Atlantic estuaries and especially from Chesapeake Bay south. Larvae are most numerous in brackish to low-salinity reaches of estuaries during the summer, but because they are slow growing, they can occur in the fall as well.

Biology and Ecology. The hogchoker is the most abundant estuarine flatfish over much of its inshore range, but it can occur well offshore in the Gulf. Spawning takes place over a wide range of salinities from almost freshwater to nearly full-strength seawater. In some areas, adults migrate from low- to mid-salinity areas to spawn. Peak spawning in the Potomac River is in the 9–16 psu region. Larvae soon head upstream to their low-salinity nursery areas. The hogchoker is the only flatfish in our coverage area to show this unique up-estuary migration of larvae. Larvae overwinter in 0–8 psu in Chesapeake Bay.

References. Dovel et al. 1969; Hildebrand and Cable 1938; Koski 1978; Peterson 1996.

Pseudopleuronectes americanus (winter flounder)

Occurrence. Larvae are common to abundant in estuaries and coastal bays from Massachusetts to Chesapeake Bay, usually in salinities of 6–20. They may be the dominant late winter–early spring fish larvae in New England estuaries. With most spawning in New Jersey occurring from January to March, most larvae occur in April and May. Larvae are typically most abundant near the bottom.

Biology and Ecology. Some winter flounder spawn in the coastal ocean, but most spawn in estuaries. In some areas, larvae move out through inlets to settle in coves along the coast. In other areas, young larvae occur in upper estuaries and then move down-estuary as they develop. Larval development takes at least six weeks followed by a rapid metamorphosis.

References. Buckley 1989; Chant et al. 2000; Collette and Klein-MacPhee 2002; Crawford and Carey 1985; Sogard et al. 2001.

Trinectes maculatus hogchoker 4.9 mm (3 - 9 mm) SL

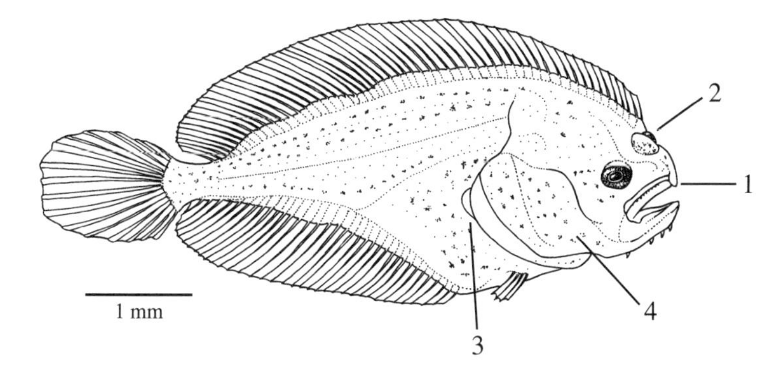

- right-eyed with robust ovate body; sharply tapered trunk
- large head with contorted upper jaw giving a hooked appearance (1)
- relatively small, flat eyes; eye rotation complete by 8 mm SL (2)
- pectoral fin very reduced after 4.9 mm (3)
- pigment spots scattered and heavily pigmented by 4.5 mm SL (4)
- similar species: all other right-eyed flounders occur in high-salinity systems and have different pigment patterns

---

Pseudopleuronectes americanus winter flounder 7.3 mm (3 - 7.7 mm) SL

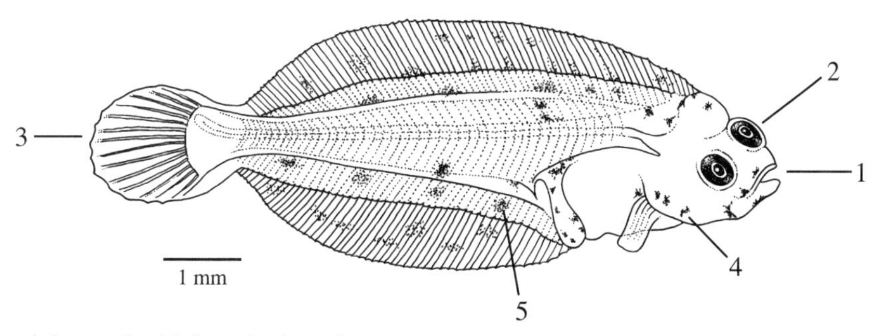

- right-eyed with long body; minor taper
- small head and mouth (1)
- oval eyes with rotation complete by 7.5 mm SL (2)
- long, round caudal fin (3)
- diagnostic pigment spots on the lower head at the corner of jaws and along edge of operculum (4)
- large scattered spots on trunk, finfolds, and fins (5)
- similar species: other right-eyed flatfishes occur in northern waters, but their larvae rarely occur nearshore or in estuaries

APPENDIXES

APPENDIX 1

Collecting Zooplankton

Collecting Using Nets

Although many methods are used to collect zooplankton, most nearshore zooplankton in our area are collected using nets (Fig. 7). Of the many design features, the two that most affect your catch are the size of the mesh and the diameter of the frame. Most plankton nets are conical with diameters that vary according to the mesh of the net. Mouths of small mesh nets are often 30 cm, medium mesh nets are often 50 cm, and large mesh nets are 100 cm or larger. Conical nets are usually mounted on metal rings, and three bridle lines are often used to connect the ring to a tow rope. Net mouth-to-length ratios can range from 1:3 to 1:6, with longer nets being less prone to clogging with small particles. For general

Table A1.1 Common mesh sizes used in plankton nets and their application

Mesh opening (μm)	Silk size (#)	Comments
10	200	Retains larger "net" phytoplankton and some of the smaller zooplankton. Because of its high resistance, it must be towed very slowly and can quickly clog with phytoplankton and detritus in "dirty" water. Poor for zooplankton in general.
64 (73)	21	A relatively uncommon size but good for copepod larval stages and rotifers. Best in waters relatively free of detritus to avoid excessive clogging.
153	10	The most commonly used estuarine zooplankton net mesh. Catches most midsized zooplankton (e.g., adult copepods, crab zoeae) and some larger and slower taxa.
365	2	Used for larger zooplankton, such as chaetognaths, crab megalopae, hydromedusae, and shrimp larvae. Not recommended for copepods and smaller mesozooplankton <1.0 mm.
505–1,000	0–000	Used for larval fishes, medusae, or large mobile crustaceans, such as mysids, amphipods, and very large copepods. Frequently used on epibenthic sleds to catch animals found just above the bottom.

Note: Early plankton nets were made of silk bolting cloth, and on occasion, mesh sizes are still referred to by their old "silk size" designations such as #10 or #12. Today's nylon mesh nets have the mesh specified in micrometers (=microns, or μm) or in millimeters (mm).

class use, we recommend using a 30–50 cm diameter net with 153–200 μm mesh. Sources of plankton nets for purchase are listed at the end of this appendix. Many marine field stations will have nets for class use but be sure to check on their availability.

You don't need a boat to collect plankton. If there are no obstructions, nets can be towed by hand while walking along a bridge or pier or while wading in shallow water. If the tide or current is moving, nets can simply be suspended from a dock or bridge. In strong currents, suspending a weight from the bottom of the net will help keep the net mouth facing into the tide. Open-water samples require a boat. Powerboats are routinely used, but small nets can be towed from a rowboat or even from a canoe in inshore areas. Take care that the net does not hit the bottom where it may snag or fill with sediment.

The following guidelines rely on our own experience and suggestions from Gerber (2000). Tow nets only long enough to collect sufficient specimens for study. If tows are too long, the meshes clog with animals and debris, and the net becomes increasingly inefficient. Worse, animals become crowded, damaged, and entangled, which makes for difficult observation. A dilute sample is preferable for live observation. Try a one- or two-minute tow first and adjust the duration as needed.

Oblique tows (from the surface to the bottom and back) sample the entire water column and provide the greatest diversity of zooplankton. The easiest and safest technique is to let the net sink, collection jar end first, from an idle boat before towing. If there is a strong current, vertical hauls using the same technique from an anchored boat will also provide a collection representative of the entire water column. Combinations of weights and floats permit tows at discrete depths for studies of vertical distribution. Sampling of the near-bottom plankton is best done with an epibenthic sled and often yields zooplankton not common higher in the water column. When you remove the net from the water, wash the plankton down to the cod end by splashing water from the outside. Drain off the excess water, and pour the concentrated sample into a large jar.

Accurate estimates of the density or number of zooplankton per unit volume of water (e.g., a cubic meter) can be determined with the use of commercially available flowmeters. These small instruments are mounted inside the mouth of the net, and numbers of propeller revolutions during the tow can be used to determine the length of the tow (in meters). By multiplying the length of the tow by the area of the mouth of the net, the volume of water filtered can be determined. Counts of the number of animals in the collection divided by the volume of that tow will yield density. This is the only way tows made at different times and places can be directly compared to determine trends in relative abundance.

Samples, Labels, and Data

Label the samples. Label paper should be heavy, with high rag content. Lightweight notebook paper disintegrates rapidly, but labels made from herbarium paper are excellent. New plastic papers are also fine but expensive. Use pencil. Pencils write in the rain or on wet paper, and the notes last indefinitely in solution. Only India ink is so durable. Add the paper directly into the sample jar. Never rely on labels applied only on the lid.

- Record date, time, weather, depth, and tide (high or low, flooding or ebbing).
- Record physical and chemical data at each station. Where possible, note temperature, salinity, oxygen levels, and other parameters of interest from the surface or from specific depths.

Care of Samples

If the object is to view living zooplankton, dilute the sample immediately in a gallon container and keep it cool or on ice and in the shade. Otherwise, add preservative immediately. Suggestions on preservatives and procedures follow.

REFERENCES

Brandt, A., Barthel, D. 1995. An improved supra- and epibenthic sledge for catching Peracarida (Crustacea, Malacostraca). *Ophelia* 43:15–23.

Fleminger, A., Clutter, R. I. 1965. Avoidance of towed nets by zooplankton. *Limnology and Oceanography* 10:96–104.

Harris, R., Wiebe, P., Lenz, J., et al., eds. 2000. *ICES Zooplankton Methodology Manual*. Academic Press, New York. 684 pp. (Except for the excellent treatment of protozooplankton, most of the treatment is for specific research techniques intended for specialists.)

Heath, M. R. 1992. Field investigations of the early life history stages of marine fish. *Advances in Marine Biology* 28:1–174.

Hernández, F. J., Jr., Lindquist, D. G. 1999. A comparison of two light-trap designs for sampling larval and presettlement juvenile fish above a reef in Onslow Bay, North Carolina. *Bulletin of Marine Science* 64:173–184.

Kane, J., Anderson, J. L. 2007. Effect of towing speed on retention of zooplankton in bongo nets. *Fishery Bulletin* 105:440–444.

Kankaala, P. 1984. A quantitative comparison of two zooplankton sampling methods, a plankton trap and a towed net, in the Baltic. *Internationale Revue für Gesamten Hydrobiologie* 69:277–287.

Minello, T. J., Matthews, G. A. 1981. Variability of zooplankton tows in a shallow estuary. *Contributions in Marine Science* 24:81–92.

Omori, M., Hamner, W. M. 1982. Patchy distribution of zooplankton: Behavior, population assessment and sampling problems. *Marine Biology* 72:193–200.

Omori, M., Ikeda, T. 1984. *Methods in Marine Zooplankton Ecology*. Wiley & Sons, New York. 332 pp.

Park, C., Wormuth, J. H., Wolff, G. A. 1989. Sample variability of zooplankton in the nearshore off Louisiana with consideration of sampling design. *Continental Shelf Research* 9:165–179.

Rey, J. R., Crossman, R. A., Kain, T. R., et al. 1987. Sampling zooplankton in shallow marsh and estuarine habitats. *Estuaries* 10:61–67.

Sameoto, D., Wiebe, P., Runge, J. 2000. Collecting zooplankton. In: Harris, R., Wiebe, P., Lenz, J., et al., eds. *ICES Zooplankton Methodology Manual*. Academic Press, San Diego, CA, 55–81. (A current review of all methods including traps, pumps, and nets. Primarily for professionals.)

Taggart, C. T., Leggett, W. C. 1984. Efficiency of large-volume plankton pumps, and evaluation of a design suitable for deployment from small boats. *Canadian Journal of Fisheries and Aquatic Science* 41:1428–1435.

Tranter, D. J., ed. 1968. *Zooplankton Sampling*. Monographs on Oceanographic Methodology 2.

Unesco Press, Paris. 174 pp. (See sections on net performance, avoidance of samplers, and filtration efficiencies.)

Youngbluth, M. J. 1982. Sampling demersal zooplankton: A comparison of field collections using three different emergence traps. *Journal of Experimental Marine Biology and Ecology* 61:111–124.

SUPPLIERS OF FIELD GEAR

Most of these suppliers have excellent Web pages. Because URLs change often, we suggest you use a Web browser to find them. *All suppliers checked online July 20, 2011.*

Aquatic Research Instruments
P.O. Box 98
Lemhi, ID 83836
Phone: 800-320-9482
Everything from student plankton nets to plankton splitters. One of few sources of light traps.

Aquatic Sampling Company
266 Elmwood Avenue, Suite 277
Buffalo, NY 14222
Phone: 800-881-3410
A selection of plankton nets and devices for plankton processing and counting.

Ben Meadows Company
P.O. Box 5277
Janesville, WI 53547-5277
Phone: 800-241-6401

Carolina Biological Supply
2700 York Court
Burlington, NC 27215
Phone: 800-334-5551
They have marine plankton samples (collected with 153 μm mesh net) that they will send for class use, and they carry a full line of water-sampling gear and water-analysis kits.

Forestry Suppliers
P.O. Box 8397
Jackson, MS 39284-8397
Phone: 800-647-5368; 601-354-3565
A selection of nets for class use and a variety of field gear, including Secchi disks, water samplers, and water chemistry kits.

General Oceanics, Inc.
1295 N.W. 163rd Street
Miami, FL 33169
Phone: 305-621-2882 Fax: 305-621-1710
Manufacture a range of plankton nets with mesh sizes from 10 microns to 1,800 microns.

What If Scientific—Leave Only Bubbles
754 Hanson Road
Spencer, WI 54479
Phone: 715-659-5427 Fax: 715-659-5235
Carry marine science curriculum supplies, including a variety of nets and items for field use.

Research Nets, Inc.
P.O. Box 249
Bothell, WA 98041-0249
Phone: 425-821-7345 Fax: 425-775-5122
Makers of custom plankton nets and trawls.

Sea-Gear Corporation
700-B1 S John Rodes Blvd.
Melbourne, FL 32904
Phone: 321-728-9116 Fax: 321-722-0351
Carry nets suitable for class use as well as specialty research nets and trawls, flowmeters, and accessories. Custom nets available on request.

Wildlife Supply Company (Wildco)
86475 Gene Lasserre Blvd.
Yulee, FL 32097
Phone: 800-799-8301 Fax: 904-225-9889
Carry a variety of nets and supplies; geared more for research.

APPENDIX 2

Observing Zooplankton

Observation in the Field

Discovery Scopes are well suited to field use and offer an inexpensive, cursory look at zooplankton fresh from the net.

Specimen Handling: Tools and Techniques

Dishes

"Finger bowls" work fine if handy, but glass custard dishes found in most grocery stores work just as well, cost less, and take up far less cupboard space. Disposable plastic petri dishes 3–7 cm are also excellent and better for many applications. "Syracuse" watch glasses, the precursors to plastic petri dishes, have the advantage that they are stackable, even when filled with living specimens.

Removing Live Specimens for Qualitative Examination

1. First pour some of the sample into a shallow enamel or plastic pan with a white bottom for preliminary sorting. This is a little easier than getting specimens directly from a deep jar.
2. Remove samples using a turkey baster or other device. A large-bore plastic pipette can be used, but it is not as effective. A small area of bright light may concentrate specific taxa to facilitate removal. Ctenophores can be removed using a tablespoon or small measuring cup.

Observation Tips

Scan the dish under low power until you find a specimen of interest. Then adjust magnification and illumination for best results. Identify the organism and repeat until that sample dish is exhausted and then get another sample. Record your observations on movement, behavior, or distinctive features. If the dish is crowded or if movement of specimens makes critical observation difficult, isolate individual specimens.

Isolating and Slowing Specimens for Observation

Small-bore disposable pipettes (3 ml size) will allow you to catch and move most living zooplankton if they are relatively quiescent. If the critter is too big, just snip the end of the pipette to enlarge the bore.

The same movements that make the zooplankton exciting to watch may move them out of your field of view. Chasing them around the dish can be frustrating. First, look for a slower specimen of the same species. If this fails, try one of the following:

- Catch and isolate the specimen. Use a pipette of appropriate size to slurp up the animal in question. A careful approach and rapid release of the pipette bulb works best. Carefully discard excess water from the pipette drop by drop until the specimen nears the tip. Now eject your specimen and a small amount of water onto a clean petri dish or microscope slide. It is now confined to a small droplet where you can always find it. If it is still too speedy, try adding a coverslip.
- Put the dish in a refrigerator or freezer for a few minutes. Check frequently until the desired sluggishness occurs. Remember, that the specimen may revive when warmed.
- Slowly add oxygen-depleted water. Boil water collected with the sample for 5–10 minutes, cap, and cool.
- Try a narcotic (see Appendix 3 for recipes and instructions). Using narcotizing agents may require some experimentation with different taxa, and some may take a while to work. Isotonic magnesium chloride (Epsom salts) is a fine and inexpensive general-purpose anesthetic suitable for mixed samples. Try setting the sample aside for 5 to 15 minutes while the anesthetic takes effect. Clove oil (toothache drops) also works well when added to samples.
- Add a viscous material such as Protoslo. Use Protoslo as a last resort because it may distort delicate specimens and can "gum up" others in the dish. Experiment carefully for best results.

Staining Live Animals

Although not usually needed for identification, vital (nontoxic) stains add contrast to enhance transparent specimens. Neutral red, methylene blue, and rose bengal are most often used. Usually, these are made to a 1% concentration and dispensed from a dropper bottle.

Quick Measures of Size and Relative Size (if the microscope lacks an ocular micrometer)

Direct Measurement. For a quick and easy approximation, simply slip a 10 cm ruler under a thin, plastic petri dish and line it up beside your specimen. Use a translucent ruler if you are using transmitted light from beneath the specimen. The 1 mm divisions may well be all that you need. Thicker glass dishes or a lot of water over the specimen may distort the measurement.

Relative Size. Use the copepod *Acartia* as a yardstick. Most samples from the Atlantic or Gulf will have either *Acartia tonsa* or *A. hudsonica* as a common species. *Acartia* is about 1–1.5 mm long. Place your unknown near an *Acartia* and estimate its size.

Viewing Tips Using Dissecting Microscopes

Appreciation and identification of most zooplankton requires at least moderate magnification. More magnification and resolution may be needed for identification. Dissecting microscopes offer magnification from 7× to 30× or 50×, and this is ideal for most purposes. Some come with built-in illumination, but separate light sources offer more lighting options.

- Always begin with the lowest power for each specimen and then increase the magnification.
- Use patience and experimentation to adjust the light for maximum effectiveness for each specimen to enhance the image and facilitate identification. Failure to adjust the illumination is a frequent and serious problem when students use dissecting microscopes.
- Adjust the light intensity and experiment with transmitted versus reflected light.
- Transparent or translucent specimens are best viewed using transmitted light from below. Light from above may cause unwanted reflections. Try turning it off.
- Opaque specimens require reflected light. Fixed lights from above may be adequate but separate "bullet" or, better, fiber-optic light sources are both brighter and more versatile. Try using these lights from a variety of angles rather than fixed to the top of the microscope.
- Incandescent bulbs heat the specimens, often causing death or disintegration in minutes. Remove specimens from the light when you are not viewing them and add cool water as needed. This is not a problem with the "cool" light from fiber-optic illuminators.
- Try to keep saltwater in your sample dish and off the microscope stage. Clean the inevitable spills from affected areas with freshwater and dry them as soon as possible.
- Cover the specimen completely with water to avoid excessive reflections. For the sharpest image, use the minimum depth of water needed to cover the specimen.

Viewing Tips Using Compound Microscopes

Compound microscopes are not essential for general zooplankton observation, but they provide detail on individual small specimens, including phytoplankton.

- First try adding the specimen to a slide in a single drop of water, without a coverslip.
- View under the lowest power.
- If the specimen is too active, or if you need more magnification, gently add a coverslip.

- For thick or delicate specimens, put a small amount of modeling clay under each corner of the coverslip to support it so that the specimen is not compressed and distorted.
- Remember to adjust the focus as you view.

Large-Screen Viewing: Camera/Video Systems

Small, high-definition video cameras that attach directly to dissecting and compound microscopes are extremely useful for classes and are now relatively affordable and widely available. Students who find particularly interesting creatures can share them with their colleagues by projecting them on the video screen, especially larger screens. See comments about equipment and suppliers at the end of this appendix.

REFERENCES

Gerber, R. P. 2000. *An Identification Manual to the Coastal and Estuarine Zooplankton of the Gulf of Maine Region.* Acadia Productions, Brunswick & Freeport, ME. (See the section on specimen handling and microscope use.)

Huys, R., Boxshall, G. A. 1991. *Copepod Evolution.* Ray Society, London. 468 pp. (Reviews observation and preservation techniques for copepods that are applicable to most zooplankton.)

Lee, J. J., Soldo, A. T. 1992. *Protocols in Protozoology.* Vol. 1, *Society of Protozoology.* Lawrence, KS. 240 pp. (Covers the special techniques needed for Protozoa.)

Montagnes, D. J., Lynn, D. H. 1987. A quantitative protargol stain (QPS) for ciliates: Method description and a test of its quantitative nature. *Marine Microbial Food Webs* 2:83–93.

Omori, M., Ikeda, T. 1984. *Methods in Marine Zooplankton Ecology.* Wiley & Sons, New York. 332 pp. (Consult for information on staining and sorting samples.)

SUPPLIERS OF ITEMS USEFUL FOR HANDLING AND OBSERVING ZOOPLANKTON

Most of these suppliers have excellent Web pages. Because URLs change often, we suggest you use a Web browser to find them. (Listing does not imply endorsement.)

Aquatic Research Instruments
P.O. Box 98
Hope, ID 83836
Phone: 800-320-9482
They carry an extensive selection equipment for sorting and counting plankton, including sample splitters, Hensen-Stemple pipettes, and counting cells.

Carolina Biological Supply
2700 York Court
Burlington, NC 27215
Phone: 800-334-5551
They have marine plankton samples (collected with 153 μm mesh net) that they will send for class use. They carry 35 mm plastic slides for projecting living zooplankton.

What If Scientific–Leave Only Bubbles
754 Hanson Road
Spencer, WI 54479
Phone: 715-659-5427 Fax: 715-659-5235

Wildlife Supply Company (Wildco)
95 Botsford Place
Buffalo NY 14216
Phone: 800-799-8301 Fax: 800-799-8115
Wildco can supply plankton splitters, counting cells, counting wheels, Henson-Stemple pipettes, and more.

APPENDIX 3

Relaxing, Fixing, and Preserving Zooplankton

Fixation and Preservation

We encourage examining living or fresh zooplankton for most class uses because living plankton are so much more interesting. However, advanced classes with objectives that require later examination will need to preserve specimens. Most zooplankton are delicate. They die quickly after capture and rapidly deteriorate once dead. Samples to be saved should be promptly fixed. Note: Most taxa lose color when preserved, and some become distorted.

Fixation provides the initial stiffening and toughening of specimens. Most fixatives include formalin. Once fixed for one to several weeks, other preservatives may be used to replace the fixative. Next are some general recommendations suitable for class use. There is no single method for fixing and preserving all taxa (see Table A3.1). These recipes and suggestions should serve most uses, but see Steedman (1976) for a comprehensive treatment.

Before fixing field-collected samples, you might want to remove large pieces of algae, plant stalks, shells, sand, or other material that you do not want to dispose of later. If ctenophores are thick, strain the sample through a coarse mesh (1 cm hardware cloth) to remove them before preserving the sample. Otherwise, ctenophores will turn your sample into a gooey mess. If you know the volume of the sample container, it is easy to prepackage formalin-based concentrates in the volumes needed. If samples are not going to be fixed immediately, placing the jar on ice in a cooler or refrigerator will slow degradation until you are ready to add the chemicals.

Always treat fixatives and preservatives with care. Some are toxic or have noxious vapors. This is equally true of the preserved samples.

Safety and health hazards should be researched before using any chemicals. MSDS (Material Safety Data Sheet) statements are excellent sources of information. Most are readily available online. Scientific supply company web sites provide search options for specific chemicals, and there is often a link to the MSDS. Care in the use and proper disposal of fixatives and preservatives is very important.

Buffered Formalin

Buffered formalin is the traditional fixative and preservative used for zooplankton collections. Full-strength formalin solution contains about 40% formaldehyde gas. Add enough full-strength buffered formalin to the sample to make a 2%–5% formalin solution. A 5% strength solution is typically used, but some researchers now use 2%, and this should be adequate for most short-term uses. If the biomass of the plankton is greater than 20% of the total volume of the container, more formalin is needed. *Wear gloves and eye protection when handling concentrated formalin.*

Advantages. Formalin is cheap and easy to use, especially in the field. It is a good general-purpose fixative for most taxa.

Disadvantages. Formalin contains toxic, irritating fumes. Before observing preserved samples, take samples to a fume hood and gently rinse them with water using an appropriate mesh to retain them. They may then be left in water or seawater for hours without harm. Samples fixed with formalin in the field can be transferred (under a fume hood) to a formalin-free solution for long-term storage. Formalin is not appropriate for samples used for DNA analysis. Some delicate taxa such as rotifers contract when put into formalin unless they are relaxed first. Because even buffered formalin is known to make fine structures (e.g., setae) of crustaceans brittle over time, it is not as good as alcohol or Steedman's preservative fluid for long-term storage.

Buffers for Formalin

Commercial borax (sodium tetraborate: $Na_2B_4O_7$) is the most common and least expensive of several buffering agents. Add 20–30 g of borax (sodium tetraborate) to 1 L of full-strength formalin (40% formaldehyde solution) and invert the jar several times during 1 hour. It is best if this solution sits for at least a week before use to allow precipitates to settle. The borax from grocery stores is pure enough for most uses. Phosphate-based buffers are also used.

Rose Bengal Stain

Use Rose Bengal when excessive levels of detritus make it hard to find or sort specimens. Use a toothpick to add a few grains directly to the sample jar. Experimentation is recommended to determine the best concentration. If too much is added, the detritus may stain, too, and critical features needed for identification may be obscured in some taxa. Add Rose Bengal before the formalin for best results. Alternatively, add enough Rose Bengal to the formalin concentrate to produce the desired degree of staining. Samples preserved in alcohol cannot be stained with Rose Bengal.

Alcohol (70% Ethanol, Not Denatured, or 40%–50% Isopropanol)

Strain samples as needed to reduce the seawater content. Alcohol is often recommended for preservation of Crustacea, although it removes pigmentation quickly. Ethanol forms a precipitate with seawater, so samples should be rinsed in freshwater first. Some (but not all) workers recommend alcohol for preservation for fishes, annelids, and cnidarians (following fixation in formalin). Some (but not all) workers suggest putting crustaceans directly into alcohol without prior fixation.

Advantages. Alcohol has low toxicity and is readily available. Isopropanol (isopropyl, or rubbing alcohol) appears to be less toxic than ethanol and methanol. All are suitable for long-term storage of collections.

Disadvantages. Because of the large volumes needed, alcohol is expensive and cumbersome to use in the field. Ethanol is a controlled substance, and permits may be required for purchase. Fumes from alcohol can be unpleasant, and they may be flammable. Ethanol will leach the pigment from specimens previously stained in Rose Bengal. As with all alcohols, potential health risks should be reviewed in the MSDS sheets before beginning use of these solutions.

Optional. For long-term storage, add glycerol to make a 1% solution. This aids flexibility and retards evaporation.

Steedman's Preservative Fluid (Propylene Phenoxetol, Propylene Glycol)

After one to several weeks of formalin-based fixation, strain and rinse the sample under a fume hood to remove the fixative and then add the preservative for long-term or archival storage.

Advantages. Specimens remain flexible and in excellent condition for prolonged periods. It is less volatile than formalin, and thus it is safer and more pleasant to work with samples in this solution than those in formalin. There also appear to be fewer health and safety concerns associated with these chemicals.

Disadvantages. Chemicals needed are difficult to find (see list of suppliers) and expensive, although they go a long way.

Recipe. Make the concentrate as follows. For 100 ml:

90 ml propylene glycol—available from most general chemical suppliers
10 ml propylene phenoxetol—available as 1-phenoxy-2-propanol from Pfaltz & Bauer
Dissolve the propylene phenoxetol in the propylene glycol.
Dilute to 5% in the sample.

Relaxation

Clove Oil, or "Toothache Drops" (Benzocaine)

Clove oil is probably the best general-purpose anesthetic for class use. Clove oil, sold as toothache drops, is inexpensive, easy to transport, and readily available in pharmacies without prescription. Dose depends on the taxa of interest and whether you want the animals knocked out or merely slowed, so a little experimentation may be needed. When using a finger bowl or petri dish, try this approach: dip the tip of a dissecting needle, fine forceps, or a fine pipette into the clove oil and then touch it to the water. The small drops of clove oil float on the surface and diffuse slowly into the water. Adding a milliliter of ethanol or stirring to create smaller droplets increases dispersion. Excess droplets can be removed using a pipette. Animals knocked out briefly may be revived when returned to fresh seawater.

Isotonic Magnesium Chloride (73.2 g $MgCl_2 . 6H_2O$ per liter of freshwater)

Isotonic magnesium chloride is a good general-purpose anesthetic. The magnesium replaces the sodium in nerve and muscle cells thus immobilizing the specimen. The isotonic solution works well if you add the specimen to the solution. If you are adding the magnesium chloride solution to specimens in a dish of seawater, try to remove as much of the seawater as possible. The concentration above is isotonic with seawater (35 psu) Delicate taxa from lower salinities require that the magnesium chloride be diluted with freshwater to avoid osmotic stress. A 1:1 mixture of isotonic magnesium chloride takes longer to immobilize specimens than stronger solutions, but recovery after returning the animals to fresh seawater is much better. Set the sample aside 5–10 minutes for full relaxation, especially for crustaceans. Epsom salts (magnesium sulfate) can be used as an inexpensive alternative.

CO_2-Club Soda

Add club soda slowly to the sample. The CO_2 in solution relaxes delicate forms like rotifers.

Alcohol

Add methanol or isopropanol dropwise. (Ethanol leaves a precipitate.)

MS-222 or Finquel (Tricaine methanesulfonate)

The muscle relaxer MS-222 is widely used in aquaculture and fisheries research. Fish larvae have been sorted using concentrations of 20–70 mg L^{-1}. It may be worth a try on various other zooplankton. Try slowly adding a 1% solution.

Table A3.1 contains recommendations for specific taxa. Because methods appear frequently, readers interested in long-term archival storage should consult museum and taxonomy web sites for the current thinking on what works best for the various groups. It is a moving target.

Table A3.1 Anesthetization and preservation of specific taxa

Taxon	Relaxation and observation	Preservation	References
Phytoplankton		Lugol's iodine	Tomas 1997
Protozoa Aloricate ciliates Tintinnids		Protargol staining, Bouin's fixative, or Lugol's Iodine	Boltovskoy 1999; Gifford and Caron 2000; Lee and Soldo 1992; Montagnes and Lynn 1993; Stoecker et al. 1994
Rotifers	Club soda or Bupivacaine (Marcaine) 0.5%	Hot-water fixation, Lugol's iodine	Edmondson 1959, p. 433; Nogrady and Rowe 1993
Cnidaria	Isotonic magnesium chloride	5% buffered formalin solution or as for Ctenophora	van Impe 1992
Ctenophora	Isotonic magnesium chloride	5% formalin solution in a 1% agar gel	van Impe 1992
Polychaete larvae and mollusc veligers	0.1% propylene phenoxetol or 4% ethanol	70% ethanol or 5% Steedman's preservative	
Crustacea (most)	Isotonic magnesium chloride Clove oil	70% ethanol with 1% glycerol or 5% Steedman's preservative	Gannon and Gannon 1975
Arrow worms (Chaetognaths)	Add methylene blue for contrast	5% Steedman's preservative; avoid alcohol	
Echinoderm and small ciliated larvae		Best viewed when alive	
Lower chordates	MS-222 (Tricaine methanesulfonate)	5% Steedman's preservative	
Fish eggs		Phosphate buffered 5% formalin	Markle 1984
Fish larvae	MS-222 (Tricaine methanesulfonate) Clove Oil	70% ethanol or fix in unbuffered 5% formalin then preserve in 5% Steedman's preservative	Markle 1984

Note: Some experts may have other preferences. Also consult Boltovskoy (1999), where coverage of each taxon includes a specific recommendation for preservation.

REFERENCES

Boltovskoy, D., ed. 1999. *South Atlantic Zooplankton*. Vols. 1, 2. Backhuys, Leiden, The Netherlands. 1706 pp. (Many sections include recommendations for preservation for that group.)

Chatain, B., Carrao, D. 1992. A sorting method for eliminating fish larvae without functional swimbladders. *Aquaculture* 107:81–88.

Edmondson, W. T. 1959. *Fresh-water Biology*. Wiley & Sons, New York. 443 pp.

Gannon, J. E., Gannon, F. A. 1975. Observations on the narcotization of crustacean zooplankton. *Crustaceana* 28:220–224.

Gifford, D. J., Caron, D. A. 2000. Sampling, preservation, enumeration and biomass of protozooplankton. In: Harris, R., Wiebe, P., Lenz, J., et al., eds. *ICES Zooplankton Methodology Manual*. Academic Press, New York, 193–221. (Detailed instructions for all groups.)

Lee, J. J., Soldo, A. T. 1992. *Protocols in Protozoology*. Society of Protozoology, Lawrence, KS.

Lynn, D. H. 2008. *The Ciliated Protozoa: Characterization, Classification, and Guide to the Literature*. Springer, New York. 606 pp.

Markle, D. F. 1984. Phosphate buffered formalin for long term preservation of formalin fixed ichthyoplankton. *Copeia* 1984:525–528.

Montagnes, D. J., Lynn, D. H. 1987. A quantitative protargol stain (QPS) for ciliates: Method description and a test of its quantitative nature. *Marine Microbial Food Webs* 2:83–93.

Nogrady, T., Rowe, T. L. A. 1993. Comparative laboratory studies of narcosis in *Brachionus plicatilis*. *Hydrobiologia* 255/256:51–56.

Steedman, H. F., ed. 1976. *Zooplankton Fixation and Preservation*. Unesco Press, Paris. 350 pp. (Outstanding practical guide with recipes.)

Stoecker, D. K., Gifford, D. J., Putt, M. 1994. Preservation of marine planktonic ciliates: Losses and cell shrinkage during fixation. *Marine Ecology Progress Series* 110:293–299.

Unesco. 1968. *Zooplankton Sampling*. Reviews on Oceanographic Methodology 2. Unesco Press, Paris. 174 pp.

Van Impe, E. 1992. A method for the transportation, long term preservation and storage of gelatinous planktonic organisms. In: Bouillon, J., Boero, F., Cicogna, F., eds. *Aspects of Hydrozoan Biology*. *Scientia Marina* 56(2–3):237–238.

CHEMICAL SUPPLIERS

All suppliers listed below have Web sites, easily found using a browser.

Scientific Supply Houses. Edmond Scientific, Fisher Scientific, Thomas Scientific, and VWR Scientific have most of the general chemicals and stains.

Educational Suppliers. Carolina Biological, Turtox Biological Supply, Sargent-Welch Scientific, and Wards Natural Science have a selection of preservatives and stains.

Specialty Suppliers. Pfaltz & Bauer provides the ingredients for Steedman's fixatives and preservatives. Argent Laboratories manufactures the anesthetic MS-222. Also consult Sigma/Aldrich.

APPENDIX 4

Sample Processing and Data Analysis

Simple observations of field-collected and preserved zooplankton collections may be all that is desired, but with some planning and extra steps, additional information about the composition and changing nature of zooplankton can be determined. For instance, by counting individuals in collections, comparisons of the relative numbers of different taxa at different locations or times can be generated. Follow these guidelines to collect quantitative data from your samples.

List all of the taxa identified, and then separate them according to phylum, holoplankton or meroplankton, or other criteria of interest. An eyeball estimate of the relative abundance of the variety of animals (abundant, common, uncommon, rare) may be enough to provide an approximation of community composition. Similarly, the percent composition of the major taxa can be estimated by quick inspection (e.g., about half copepods and about quarter medusa). However, precise counts are necessary to be quantitative and the methods presented here will enable you to accomplish that.

Subsampling Technique

If the sample contains many thousands of organisms, it will be necessary to take a subsample to quantify abundance and composition. For live samples, be sure to get subsamples from different levels in the jar. Some groups sink, while others swim near the surface. The animals should be dead or anesthetized for accurate counts. In either case, a thorough mixing before subsampling is required.

For class use, the 5 or 10 ml widemouthed dropper or pipette is usually adequate for quantitative subsamples. A specially designed subsampler called a Hensen-Stemple pipette is available from supply companies. When samples contain large animals and volumes of detritus, more elaborate rotating splitters are appropriate. Regardless of the method, subsamples must be taken from a well-mixed area in the container while the water is still in motion. The volume of water in the container must be known. Take as many subsamples as needed to get enough plankton to count. Put them in a 3–7 cm plastic petri dish.

How many animals should be in a dish for counting? Estimates vary, but about 100–200 total animals are sufficient for the more abundant taxa. Accurate estimates of rare taxa may require looking at larger volumes while ignoring the more common species. Keep track of the volume of water processed to obtain counts of individuals for each type of animal.

How do you avoid recounting the same individuals? Try this. Use a razor blade or penknife to score the back of a plastic petri dish to make a grid with squares of about 5 × 5 mm (adjust as needed). A fine-tipped permanent black marker also works but may not be compatible with alcohol. Use the lowest power on the dissecting scope that will allow you to recognize most species, and simply move the dish back and forth along the rows while recording the counts for each square. Because some specimens straddle the lines, be careful to avoid double counting. Take extra care to move the dish slowly to avoid changing the positions of animals to be counted. Record the totals on your data sheet.

Data Management

Keeping the data well organized is the key to maintaining accuracy. Set up data sheets before starting to process the samples and record information consistently. Keeping a separate sheet to record decisions you have to make along the way often becomes very important later.

Examples of data sheets that can be used follow:

Observational or Qualitative Information on the Zooplankton. Additional columns can be added to record, size, color, relative abundance (abundant, average, rare), and other information.

TAXON, STAGE, . . .	PHYLUM, GROUP, . . .	Observations on behavior or swimming . . .
a		
b		

Numerical or Quantitative Data. After deciding which taxa will be counting categories (a, b, c, . . . in the table below), subsamples can be counted. Often, the last counting category is named "others" to include all remaining individuals that do not fit into the established taxon categories. Enter the number of individuals (#) for each taxon in subsample 1 in the "Count 1" column. Repeat for subsamples 2 and 3. Calculate a "Total" number for each taxon by adding subsample counts across rows. Divide that by the total number of subsamples to determine a "Mean" for each taxon. Then add down the columns to determine the totals of "all taxa" for each subsample. Add down the "Mean" column to determine the mean number of individuals of all organisms in the total catch. Add down the "Total" column and use this number to calculate the "Percent of Total" (last column) for each taxon.

Taxon	Count 1	Count 2	Count 3	Total	Mean	Percent of total
	#	#	#	#	#	%
a						
b						
c						
TOTALS	All taxa	All Taxa	All taxa	All Taxa	Of all subsamples	100%

Comparing Collections

To determine differences between zooplankton abundance at multiple locations or between times (tides, day/night, seasons, years), densities must be calculated. Density is simply the number of individuals per unit volume (usually as number/cubic meter). One way to estimate the volume of water filtered in a tow is to calculate the volume of the imaginary cylinder through which the round framed net was towed. In tidal situations, use of a flowmeter mounted inside the net provides more accurate data for this calculation. The equation is

$$\text{volume} = (\text{distance towed}) \times 3.14\ (\text{radius of net opening in meters})^2$$

Multiply the mean number of animals in subsamples processed times the total number of subsamples to estimate the total number of individuals in each taxon in your sample jar and then divide by the volume of water filtered to determine densities. By establishing values for the numbers of individuals per unit volume of water, your zooplankton catches can be directly (quantitatively) compared with those reported by others.

Data Analyses

Enter the data into a spreadsheet or into a statistics and graphics program. One of the simplest and most informative exercises is to make bar graphs of relative abundance. Because the values that you have determined are independent, bar graphs are more appropriate for comparing samples than lines that connect values.

The following statistics may be calculated:

Means. Means for percents, raw numbers, or densities can be used.

Relative standard deviation (coefficient of variation). This statistic is intuitive and underused. It is an excellent way to assess variability.

Standard error. Standard error is useful in putting error bars on graphs. Some software programs will plot them automatically.

Similarity indices and **dendrograms** are used to compare multiple samples for more advanced classes. Students can calculate simple indices, such as percent similarity, by hand or by using a spreadsheet. Ecology statistical packages will calculate many such indices, but only a few packages construct the dendrograms, which can be constructed by by hand. See Southward and Henderson (2000) or a field ecology manual for instructions.

REFERENCES

Alden, R. W, Dahiya, R. C., Young, R. J. 1982. A method for the enumeration of zooplankton subsamples. *Journal of Experimental Marine Biology and Ecology* 59:185–206. (Good treatment on how to subsample for rarer taxa.)

Gifford, D. J., Caron, D. A. 2000. Sampling, preservation, enumeration and biomass of protozooplankton. In: Harris, R., Wiebe, P., Lenz, J., et al., eds. *ICES Zooplankton Methodology* . Academic Press, New York, 193–221. (Excellent, detailed instructions for all groups.)

Lee, W. Y., McAlice, B. J. 1979. Sampling variability of marine zooplankton in a tidal estuary. *Estuarine and Coastal Marine Science* 8:565–582. (Gives recommendations for the number and timing of samples needed for different purposes.)

Southwood, T. R. E., Henderson, P. A. 2000. *Ecological Methods.* 3rd ed. Blackwell Science, Oxford. 575 pp. (Excellent introduction to community similarity measures and to similarity dendrograms.)

SUPPLIERS OF FIELD GEAR

Aquatic Research Instruments
P.O. Box 98
Hope, ID 83836
Phone: 800-320-9482
Everything from student plankton nets to plankton splitters.

Carolina Biological Supply
2700 York Court
Burlington, NC 27215
Phone: 800-334-5551
Living marine plankton samples (collected with 153 μm mesh net) that they will send for class use. Also a full line of water-sampling gear and water-analysis kits.

Forestry Suppliers
P.O. Box 8397
Jackson, MS 39284-8397
Phone: 800-647-5368 Fax: 601-345-4565
A selection of nets for class use and a variety of field gear, including Secchi disks, water samplers, and water chemistry kits.

General Oceanics, Inc.
1295 N.W. 163rd Street
Miami, Florida 33169
Phone: 305-621-2882 Fax: 305-621-1710
Manufacture a range of plankton nets with mesh sizes from 10 to 1,800 μm.

MarineLab's Leave Only Bubbles, Inc.
P.O. Box 2397
Key Largo, FL 33037
Phone: 800-890-0134 Fax: 775-414-1126
Carry a variety of nets and items for field use. They are also one of the few suppliers of deep-well slides for slide projectors.

Research Nets, Inc.
P.O. Box 249
Bothell, WA 98041-0249
Phone: 425-821-7345 Fax: 425-775-5122
Makers of custom plankton nets and trawls.

SEA-GEAR Corporation
700-B1 S. John Rodes Blvd.
Melbourne, FL 32904
Phone: 321-728-9116 Fax: 321-722-0351
Carry nets for class use but also fabricate custom nets for research.

Wildlife Supply Company (Wildco)
95 Botsford Place
Buffalo, New York 14216
Phone: 800-799-8301 Fax: 800-799-8115
All kinds of nets and supplies, plankton splitters, counting cells, counting wheels, Henson-Stemple pipettes. Geared more for research.

VIDEO CAMERAS AND IMAGING SYSTEMS

The initial cost is steep, but the systems can be used for almost any classroom microscope application. Either dissecting or compound microscopes can be used with video to good advantage. Most larger supply houses (e.g., Carolina Biological) have micro/macro video systems. Many are under $2,000, complete with monitor.

APPENDIX 5

Regional Zooplankton Surveys

Many of the more comprehensive published surveys are from older literature. Many are excellent, but beware of changes in scientific names. Also consult government agencies, especially state departments of fish and game or natural resources and local marine laboratories, for unpublished reports.

Surveys Covering Multiple Taxa

Allen, D. M., Ogburn-Matthews, V., Buck, T., et al. 2008. Mesozooplankton responses to climate change and variability in a southeastern U.S. estuary (1981–2003). Special issue, *Journal of Coastal Research* 55:95–110.

Buskey, E. J. 1993. Annual cycle of micro- and mesozooplankton abundance and biomass in a subtropical estuary. *Journal of Plankton Research* 15:907–924. (Nueces Estuary, TX. A high-salinity coastal lagoon.)

Cronin, L. E., Daiber, J. C., Hulbert, E. M. 1962. Quantitative seasonal aspects of zooplankton in the Delaware River estuary. *Chesapeake Science* 3:63–93.

Deevey, G. B. 1956. Oceanography of Block Island Sound, 1952–1954. 5. Zooplankton. *Bulletin of the Bingham Oceanographic Collection* 15:113–155.

Fish, C. J. 1925. Seasonal distribution of plankton of the Woods Hole region. *Bureau of the US Bureau of Fisheries* 41:91–179.

Greenwood, M. F. D., Peebles, E. B., MacDonald, T. C., et al. 2006. *Freshwater Inflow Effects on Fishes and Invertebrates in the Anclote River Estuary*. Prepared for the Southwest Florida Water Management District, Brooksville. 190 pp.

Hawes, S. R., Perry, H. M. 1978. Effects of 1973 floodwaters on plankton populations in Louisiana and Mississippi. *Gulf Research Reports* 6:109–124. (Brackish and coastal areas east of the Mississippi River.)

Herman, S. S., D'Apolito, L. M. 1985. Zooplankton of the Hereford Inlet estuary, southern New Jersey. *Hydrobiologia* 124:229–236.

Herman, S. S., Mihursky, J. S., McErlean, A. M. 1968. Zooplankton and environmental characteristics of the Patuxent River estuary 1963–1965. *Chesapeake Science* 9:67–82.

Holt, J., Strawn, K. 1983. Community structure of macrozooplankton in Trinity and Upper Galveston Bays. *Estuaries* 6:66–75.

Houser, D. S., Allen, D. M. 1996. Zooplankton dynamics in an intertidal salt-marsh basin. *Estuaries* 19:659–673. (South Carolina)

Hulsizer, E. E. 1976. Zooplankton of lower Narragansett Bay. *Chesapeake Science* 17:260–270.

Jeffries, H. P. 1964. Comparative studies on estuarine zooplankton. *Limnology and Oceanography* 9:348–358. (Compares Raritan Bay, NJ, with Narragansett Bay, RI.)

Lonsdale, D. J., Coull, B. C. 1977. Composition and seasonality of zooplankton in North Inlet, South Carolina. *Chesapeake Science* 18:272–283.

Mallin, M. A. 1991. Zooplankton abundance and community structure in a mesohaline North Carolina estuary. *Estuaries* 14:481–488.

Maurer, D., Watling, L., Lambert, L., et al. 1978. Seasonal fluctuations of zooplankton populations in lower Delaware Bay. *Hydrobiologia* 6:149–160.

Nelson, D. M. 1992. *Distribution and Abundance of Fishes and Invertebrates in Gulf of Mexico Estuaries*. Vol. 1, *Data Summaries*. ELMR Report No. 10 NOAA/NOS Strategic Environmental Assessments Division, Silver Spring, MD. 273 pp.

Nelson, D. M., Monaco, M. E., Irlandi, E. S., et al. 1991. *Distribution and Abundance of Fishes and Invertebrates in Southeast Estuaries*. ELMR Report No. 9 NOAA/NOS Strategic Environmental Assessments Division, Silver Spring, MD. 167 pp.

Pattillo, M. E., Czapla, T. E., Nelson, D. M., et al. 1997. *Distribution and Abundance of Fishes and Invertebrates in Gulf of Mexico Estuaries*. Vol. 2, *Species Life History Summaries*. ELMR Report No. 11. NOAA/NOS Strategic Environmental Assessments Division, Silver Spring, MD. 377 pp. (Includes larval distributions.)

Peebles, E. B. 2005. *An Analysis of Freshwater Inflow Effects on the Early Stages of Fish and Their Invertebrate Prey in the Alafia River Estuary*. Prepared for the Southwest Florida Water Management District, Brooksville. 147 pp.

Sage, L. E., Herman, S. 1972. Zooplankton of the Sandy Hook Bay area, NJ. *Chesapeake Science* 13:29–39.

Smith, D. E., Jossi, J. W. 1984. *Net Phytoplankton and Zooplankton in the New York Bight, January 1976 to February 1978, with Comments on the Effects of Wind, Gulf Stream Eddies, and Slope Water Intrusions*. NOAA Technical Report NMFS 5.

Steinberg, D. K., Condon, R. H. 2009. Zooplankton of the York River. Special issue, *Journal of Coastal Research* 57:66–79.

Surveys of Specific Taxa

Phytoplankton

Badylak, S., Phlips, E. J. 2004. Spatial and temporal patterns of phytoplankton composition in subtropical coastal lagoon, the Indian River Lagoon, Florida, USA. *Journal of Plankton Research* 26:1229–1247.

Badylak, S., Phlips, E. J. 2008. Spatial and temporal distributions of zooplankton in Tampa Bay, Florida, including observations during a HAB event. *Journal of Plankton Research* 30:449–465.

Bledsoe, E. L., Phlips, E. J. 2000. Relationships between phytoplankton standing crop and physical, chemical, and biological gradients in the Suwannee River and plume region, USA. *Estuaries* 23:458–473.

Fahnensteil, G. L., McCormick, M. J., Lang, G. A., et al. 1995. Taxon-specific growth and loss rates for dominant phytoplankton populations from the northern Gulf of Mexico. *Marine Ecology Progress Series* 117:229–239.

Freese, L. R. 1952. Marine diatoms of the Rockport, Texas, bay area. *Texas Journal of Science* 4(3):331–386.

Karentz, D., Smayda, T. J. 1984. Temperature and seasonal occurrence patterns of 30 dominant phytoplankton species in Narragansett Bay over a 22-year period (1959–1980). *Marine Ecology Progress Series* 18:277–293.

Mallin, M. 1994. Phytoplankton ecology in North Carolina estuaries. *Estuaries* 17:561–574.

Marshall, H. G. 1982. The composition of phytoplankton within the Chesapeake Bay plume and adjacent waters off the Virginia coast, USA. *Estuarine, Coastal and Shelf Science* 15:29–44.

Marshall, H. G., Burchardt, L., Lacouture, R. 2005. A review of phytoplankton composition within Chesapeake Bay and its tidal estuaries. *Journal of Plankton Research* 27:1083–1102.

Marshall, H. G., Lacouture, R. 1986. Seasonal patterns of growth and composition of phytoplankton in the lower Chesapeake Bay and Vicinity. *Estuarine, Coastal and Shelf Science* 23:115–130.

Murrell, M. C., Lores, E. M. 2004. Phytoplankton and zooplankton seasonal dynamics in a subtropical estuary: Importance of cyanobacteria. *Journal of Plankton Research 26*:371–382. (Pensacola Bay, Florida)

Pickney, J. L., Paerl, H. W., Harrington, M., et al. 1998. Annual cycles of phytoplankton community structure and bloom dynamics in the Neuse River Estuary, NC. *Marine Biology* 131:371–381.

Quinlan, E. L., Phlips, E. J. 2007. Phytoplankton assemblages across the marine to low-salinity transition zone in a blackwater dominated estuary. *Journal of Plankton Research* 29:401–416. (Suwanee River and Florida Gulf Coast)

Simmons, E. G., Thomas, W. H. 1962. Phytoplankton of the eastern Mississippi delta. *Publications of the Institute of Marine Science, University of Texas* 8:269–298.

Smith, D. E., Jossi, J. W. 1984. *Net Phytoplankton and Zooplankton in the New York Bight, January 1976 to February 1978, with Comments on the Effects of Wind, Gulf Stream Eddies, and Slope WATER INTRUSIONS*. NOAA Technical Report NMFS 5.

Watling, L., Bottom, D., Pembroke, A., et al. 1979. Seasonal variations in Delaware Bay phytoplankton community structure. *Marine Biology* 52:207–215.

Protozooplankton

Balech, E. 1967. Dinoflagellates and tintinnids in the northeastern Gulf of Mexico. *Bulletin of Marine Science of the Gulf and Caribbean* 17:280–298.

Borror, A. C. 1962. Ciliate protozoa of the Gulf of Mexico. *Bulletin of Marine Science of the Gulf and Caribbean* 12:333–349.

Capriulo, G. M., Carpenter, E. J. 1983. Tintinnid abundance, species composition, and feeding impact in central Long Island Sound. *Marine Ecology Progress Series* 10:277–288.

Dolan, J. R. 1991. Guilds of ciliate microzooplankton in the Chesapeake Bay. *Estuarine, Coastal and Shelf Science* 33:137–152.

Dolan, J. R., Coats, D. W. 1990. Seasonal abundances of planktonic ciliates and microflagellates in mesohaline Chesapeake Bay waters. *Estuarine, Coastal and Shelf Science* 31:157–175.

Dolan, J. R., Gallegos, C. L. 2001. Estuarine diversity of tintinnids (planktonic ciliates). *Journal of Plankton Research* 23:1009–1027.

Sherr, B. F., Sherr, E. B., Newell, S. Y. 1984. Abundance and productivity of heterotrophic nanoplankton in Georgia coastal waters. *Journal of Plankton Research* 6:195–202.

Verity, P. G. 1987. Abundance, community composition, size distribution, and production rates of tintinnids in Narragansett Bay, Rhode Island. *Estuarine, Coastal and Shelf Science* 24:671–690.

Rotifers

Limited information exists for this group.

Dolan, J. R., Gallegos, C. C. 1992. Trophic role of planktonic rotifers in the Rhode River Estuary, spring–summer 1991. *Marine Ecology Progress Series* 85:187–199.

Park, G. S., Marshall, G. S. 2000. The trophic contributions of rotifers in tidal freshwater and estuarine habitats. *Estuarine, Coastal and Shelf Science* 51:729–742. (Chesapeake Bay)

Cnidaria: Hydrozoa

Also check identification references in the Cnidaria chapters for regional distribution data.

Purcell, J. E., Malej, A., Benović, A. 1999. Potential links of jellyfish to eutrophication and fisheries. *Coastal and Estuarine Studies* 55:241–263. (Seasonal distributions of hydro- and scyphomedusae in the Chesapeake Bay.)

Cnidaria: Scyphomedusae and Cubomedusae

Allwein, J. 1967. North American hydromedusae from Beaufort, N.C. *Videnskabelige Meddeleser fra Dansk Naturhistorisk Forening i Kobenhavn* 130:117–136.

Burke, W. D. 1975. Pelagic Cnidaria of Mississippi Sound and adjacent waters. *Gulf Research Reports* 5(1):23–38.

Burke, W. D. 1976. Biology and distribution of the macrocoelenterates of the Mississippi Sound and adjacent waters. *Gulf Research Reports* 5(2):17–28.

Calder, D. R. 2008. An illustrated key to the Scyphozoa and Cubozoa of the South Atlantic Bight (from Calder 2009) and a glossary of terms. Available through SERTC at <www.dnr.sc.gov/marine/sertc/info.htm>. Accessed July 19, 2011.

Calder, D. R. 2009. Cubozoan and scyphozoan jellyfishes of the Carolinian biogeographic province, southeastern, USA. *Royal Ontario Museum Contributions in Marine Science* 3:1–58.

Costello, J. H., Mathieu, H. W. 1995. Seasonal abundance of medusae in Eel Pond, Massachusetts, USA during 1990–1991. *Journal of Plankton Research* 17:199–204.

Kraeuter, J. N., Setzler, E. M. 1975. The seasonal cycle of Scyphozoa and Cubozoa in Georgia estuaries. *Bulletin of Marine Science* 25:66–74.

Phillips, P. J., Burke, W. D. 1970. The occurrence of sea wasps (Cubomedusae) in Mississippi Sound and the northern Gulf of Mexico. *Bulletin of Marine Science* 20:853–859.

Cirripedes (Barnacles)

Lang, W. H., Ackenhusen-Johns, A. 1981. Seasonal species composition of barnacle larvae (Cirripedia: Thoracica) in Rhode Island waters, 1977–1978. *Journal of Plankton Research* 3:567–575.

Cladocera (Water Fleas)

Bosch, H. F., Taylor, W. R. 1968. Marine cladocerans in the Chesapeake Bay Estuary. *Crustaceana* 15:161–164.

Egloff, D. A., Fofonoff, P. W., Onbé, T. 1997. Reproductive biology of marine cladocerans. *Advances in Marine Biology* 31:79–167. (Contains information on salinity distributions.)

Lippson, A. J., Haire, M. S., Holland, A. F., et al. 1979. *Environmental Atlas of the Potomac Estuary*. Martin Marietta Corporation for Maryland Power Plant Siting Program, Annapolis, MD. 279 pp.

Copepods

There are relatively few recent published surveys, so the data in the sources below may omit recent range extensions or name changes.

Bowman, T. E. 1971. The distribution of calanoid copepods off the southeastern U.S. between Cape Hatteras and southern Florida. *Smithsonian Contributions to Zoology* 96:1–58.

Buskey, E. J. 1993. Annual cycle of micro and mesozooplankton abundance and biomass in a subtropical estuary. *Journal of Plankton Research* 15:907–924. (Nueces Estuary, TX. A high-salinity coastal lagoon.)

Cronin, L. E., Daiber, J. C., Hulbert, E. M. 1962. Quantitative seasonal aspects of zooplankton in the Delaware River estuary. *Chesapeake Science* 3:63–93.

Deevey, G. B. 1960. The zooplankton of the surface waters of the Delaware Bay region. *Bulletin of the Bingham Oceanographic Collection* 17:5–53.

Fulton, R. S., III. 1984. Distribution and community structure of estuarine copepods. *Estuaries* 7:38–50. (Seasonal distribution near Beaufort, NC.)

Gillespie, M. C. 1971. Analysis and treatment of zooplankton of estuarine waters of Louisiana. In: *Cooperative Gulf of Mexico Estuarine Inventory and Study, Louisiana*. Phase 4, *Biology*. Louisiana Wildlife and Fisheries Commission, New Orleans, 108–175.

Grice, G. D. 1960. Calanoid and cyclopoid copepods from the Florida Gulf Coast and Florida Keys in 1954 and 1955. *Bulletin of Marine Science of the Gulf and Caribbean* 10:217–226. (Four stations north of Tampa. Transects begin in inlets or estuaries, then continue offshore.)

Hawes, S. R., Perry, H. M. 1978. Effects of 1973 floodwaters on plankton populations in Louisiana and Mississippi. *Gulf Research Reports* 6:109–124. (Brackish and coastal areas east of the Mississippi River.)

Herman, S. S., Mihursky, J. S., McErlean, A. M. 1968. Zooplankton and environmental characteristics of the Patuxent River estuary 1963–1965. *Chesapeake Science* 9:67–82.

Holt, J., Strawn, K. 1983. Community structure of macrozooplankton in Trinity and Upper Galveston Bays. *Estuaries* 6:66–75. (Collections focused on larger species retained in a 500-μm mesh net.)

Hopkins, T. L. 1966. The plankton of the St. Andrew Bay System, Florida. *Publications of the Institute of Marine Science, University of Texas* 11:12–64.

Hopkins, T. L. 1977. Zooplankton distributions in the surface waters of Tampa Bay. *Bulletin of Marine Science* 27:467–478.

Lippson, A. J., Haire, M. S., Holland, A. F., et al. 1979. *Environmental Atlas of the Potomac Estuary*. Martin Marietta Corporation for Maryland Power Plant Siting Program, Annapolis, MD. 279 pp.

Lonsdale, D. J., Coull, B. C. 1977. Composition and seasonality of zooplankton in North Inlet, South Carolina. *Chesapeake Science* 18:272–283.

Mallin, M. A. 1991. Zooplankton abundance and community structure in a mesohaline North Carolina estuary. *Estuaries* 14:481–488.

McIlwain, T. D. 1968. Seasonal occurrence of the pelagic Copepoda in Mississippi Sound. *Gulf Research Reports* 2(3):257–270.

Park, C., Wormuth, J. H., Wolff, G. A. 1989. Sample variability of zooplankton in the nearshore off Louisiana with consideration of sampling design. *Continental Shelf Research* 9:165–179.

Perry, H. M., Christmas, J. Y. 1973. Estuarine zooplankton, Mississippi. In: Christmas, J. Y., ed. *Gulf of Mexico Estuarine Inventory and Study, Mississippi*. Gulf Coast Research Laboratory, Ocean Springs, MS, 198–254.

Sage, L. E., Herman, S. 1972. Zooplankton of the Sandy Hook Bay area, N.J. *Chesapeake Science* 13:29–39.

Stancyk, S. E., Ferrell, T. L. 1982. Zooplankton. In: Allen, D. M., Stancyk, S. E., Michener, W. K., eds. *Ecology of Winyah Bay, SC and Potential Impacts of Energy* . Baruch Institute Special Publication No. 82-1. University of South Carolina, Columbia, 5-1–5-42.

Stepien, J. C., Malone, T. C., Chervin, M. B. 1981. Copepod communities in the estuary and coastal plume of the Hudson River. *Estuarine, Coastal and Shelf Science* 13:185–195.

Turner, J. T. 1982. The annual cycle of zooplankton in a Long Island estuary. *Estuaries* 5:261–274.

Turner, J. T. 1994. Planktonic copepods of Boston Harbor, Massachusetts Bay and Cape Cod Bay, 1992. *Hydrobiologia* 292/293:405–413.

Van Engel, W. A., Tan, E.-C. 1965. Investigations of inner continental shelf waters off lower Chesapeake Bay. 4. The copepods. *Chesapeake Science* 6:183–189. (Two to 40 miles offshore.)

Wilson, C. B. 1932a. The copepod crustaceans of Chesapeake Bay. *Proceedings of the U.S. National Museum* 80:1–54.

Wilson, C. B. 1932b. The copepods of the Woods Hole region, Massachusetts. *U.S. National Museum Bulletin* 158:1–635.

Mysids (Opossum Shrimp)

Grabe, S. A. 1996. Composition and seasonality of nocturnal peracarid zooplankton from coastal New Hampshire (USA) waters, 1978–1980. *Journal of Plankton Research* 18:881–894.

Hopkins, T. L. 1965. Mysid shrimp abundance in the surface waters of Indian River Inlet, Delaware. *Chesapeake Science* 6:86–91.

Maurer, D., Wigley, R. L. 1982. Distribution and ecology of mysids in Cape Cod Bay, Massachusetts. *Biological Bulletin* 163:477–491.

Price, W. W. 1976. The abundance and distribution of Mysidacea in the shallow waters of Galveston Island, TX. PhD diss., Department of Biology, Texas A&M University. 182 pp.

Stuck, K. C., Perry, H. M., Heard, R. W. 1979. Records and range extensions of Mysidacea from coastal and shelf waters of the eastern Gulf of Mexico. *Gulf Research Reports* 6:239–248.

Wigley, R. L., Burns, B. R. 1971. Distribution and biology of mysids (Crustacea, Mysidacea) from the Atlantic Coast of the United States on the NMFS Woods Hole Collection. *Fish and Wildlife Service, Fishery Bulletin* 69:717–746.

Williams, A. B. 1972. A ten-year study of the meroplankton of North Carolina estuaries: Mysid shrimps. *Chesapeake Science* 13:254–262.

Amphipods, Isopods, and Cumaceans

Primarily records from the plankton.

Grabe, S. A. 1996. Composition and seasonality of nocturnal peracarid zooplankton from coastal New Hampshire (USA) waters, 1978–1980. *Journal of Plankton Research* 18:881–894.

Modlin, R. F., Dardeau, M. 1987. Seasonal and spatial distributions of cumaceans in the Mobile Bay estuarine system, Alabama. *Estuaries* 10:291–297.

Stuck, K. C., Perry, H. M., Fish, A. G. 1980. New records of Hyperiidea (Crustacea: Amphipoda) from the north central Gulf of Mexico. *Gulf Research Reports* 6:359–370.

Williams, A. B., Bynum, K. H. 1972. A ten-year study of the meroplankton in North Carolina estuaries: Amphipods. *Chesapeake Science* 13:175–192.

Decapods

Brookins, K., Epifanio, C. 1985. Abundance of brachyuran larvae in a small coastal inlet over six consecutive tidal cycles. *Estuaries* 8:60–67. (Indian River, FL)

Bullard, S. G. 2003. Larvae of anomuran and brachyuran crabs of North Carolina. *Crustaceana Monographs* 1:1–142.

Clancy, M., Epifanio, C. E. 1989. Distribution of crab larvae in relation to tidal fronts in Delaware Bay, USA. *Marine Ecology Progress Series* 57:77–82.

DeVries, M. C., Tankersley, R. A., Forward, R. B., et al. 1994. Abundance of estuarine crab larvae is associated with hydrologic tidal variables. *Marine Biology* 118:403–413. (North Carolina)

Dittel, A. D., Epifanio, C. E. 1982. Seasonal abundance and vertical distribution of crab larvae in Delaware Bay. *Estuaries* 5:197–202.

Franks, J. S., Christmas, J. Y., Siler, W. L., et al. 1972. A study of nektonic and benthic faunas of the shallow Gulf of Mexico off the state of Mississippi as related to some physical, chemical and geological factors. *Gulf Research Reports* 4:1–148.

Hawes, S. R., Perry, H. M. 1978. Effects of 1973 floodwaters on plankton populations in Louisiana and Mississippi. *Gulf Research Reports* 6:109–124.

Hillman, N. S. 1964. Studies on the distribution and abundance of decapod larvae in Narragansett Bay, Rhode Island, with consideration of morphology and mortality. MS thesis, University of Rhode Island. 74 pp.

Houser, D. S., Allen, D. M. 1996. Zooplankton dynamics in an intertidal salt-marsh basin. *Estuaries* 19:659–673. (South Carolina)

Johnson, D. F. 1985. The distribution of brachyuran crustacean megalopae in waters of the York River, lower Chesapeake Bay and adjacent shelf: Implications for recruitment. *Estuarine, Coastal and Shelf Science* 20:693–705.

Kelley, J. A., Jr., Dragovich, A. 1967. Occurrence of macrozooplankton in Tampa Bay, Florida, and the adjacent Gulf of Mexico. *Fishery Bulletin* 66:209–221.

Kurata, H. 1970. *Studies on the Life Histories of Decapod Crustacea of Georgia*. Part 3, *Larvae of decapod Crustacea of Georgia*. Final Report. University of Georgia Marine Institute, Sapelo Island. 274 pp.

Martin, J. W. 1984. Notes and bibliography on the larvae of xanthid crabs, with a key to the known xanthid zoeas of the western Atlantic and the Gulf of Mexico. *Bulletin of Marine Science* 34:220–239.

Mense, D. J., Posey, M. H., West, T., et al. 1995. Settlement of Brachyuran postlarvae along the North Carolina Coast. *Bulletin of Marine Science* 57:793–806.

Sandifer, P. A. 1973. Distribution and abundance of decapod crustacean larvae in the York River estuary and adjacent lower Chesapeake Bay, Virginia, 1968–1969. *Chesapeake Science* 14:235–257.

Subrahmanyam, C. B. 1971. The relative abundance and distribution of penaeid shrimp larvae off the Mississippi Coast. *Gulf Research Reports* 3:291–345.

Truesdale, F. M., Andryszak, B. L. 1985. Occurrence and distribution of reptant decapod crustacean larvae in neritic Louisiana waters: July 1976. *Contributions in Marine Science* 26:37–53.

Williams, A. B. 1971. A ten-year study of meroplankton in North Carolina estuaries: Annual occurrence of some brachyuran developmental stages. *Chesapeake Science* 12:53–61.

Annelids

The limited data on distribution are contained in the identification references listed in the annelid section.

Mollusc Larvae

Mann, R. 1988. Distribution of bivalve larvae at a frontal system in the James River, Virginia. *Marine Ecology Progress Series* 50:29–44.

Thiriot-Quievreux, C. 1983. Summer meroplanktonic prosobranch larvae occurring off Beaufort, North Carolina. *Estuaries* 6:387–398.

Chaetognaths (Arrow worms)

Bigelow, H. B., Sears, M. 1939. Studies of the waters of the Continental Shelf from Cape Cod to Chesapeake Bay. 3. A volumetric study of the zooplankton. *Memoirs of the Harvard Museum of Comparative Zoology* 54:183–378.

Cowles, R. P. 1930. A biological study of the offshore waters of the Chesapeake Bay. *Bulletin of the US Bureau of Fisheries* 46:277–381.

Cronin, L. E., Daiber, J. C., Hulbert, E. M. 1962. Quantitative seasonal aspects of zooplankton in the Delaware River estuary. *Chesapeake Science* 3:63–93. (From the Atlantic to near freshwater at 20-foot-depth intervals.)

Deevey, G. B. 1960. The zooplankton of the surface waters of the Delaware Bay region. *Bulletin of the Bingham Oceanographic Collection* 17:5–53. (Samples from 1929 to 1935; Lower Delaware Bay and nearby Atlantic.)

Grant, G. C. 1963. Chaetognatha from inshore coastal waters off Delaware, and a northward extension of the known range of *Sagitta tenuis*. *Chesapeake Science* 4:38–42.

Grant, G. C. 1991. Chaetognatha from the central and southern Middle Atlantic Bight: Species composition, temperature-salinity relationships, and interspecific associations. *Fishery Bulletin* 89:33–40.

McLelland, J. A. 1984. Observations on chaetognath distributions in the northeastern Gulf of Mexico during the summer of 1974. *Northeast Gulf Science* 7(1):49–59.

Pierce, E. L. 1951. The Chaetognatha of the west coast of Florida. *Biological Bulletin* 100:403–436.

Pierce, E. L. 1958. The Chaetognatha of inshore waters of North Carolina. *Limnology and Oceanography* 3:166–170.

Pierce, E. L. 1962. Chaetognatha from the Texas coast. *Publications of the Institute of Marine Science, University of Texas* 8:147–152.

Pierce, E. L., Wass, M. L. 1962. Chaetognatha from the Florida Current and coastal water of the southeastern Atlantic states. *Bulletin of Marine Science* 12:401–436.

Lower Chordates

Boschung, H. T., Shaw, R. F. 1988. Occurrence of planktonic lancelets from Louisiana's continental shelf, with a review of pelagic *Branchiostoma* (order Amphioxi). *Bulletin of Marine Science* 43:229–240.

Cory, R. L., Pierce, E. L. 1967. Distribution and ecology of lancelets (order Amphioxi) over the continental shelf of the southeastern United States. *Limnology and Oceanography* 12:650–656.

Fishes

Allen, D. M., Barker, D. L. 1990. Interannual variation in larval fish recruitment to estuarine epibenthic habitats. *Marine Ecology Progress Series* 63:113–125.

Bozeman, E. L., Jr., Dean, J. M. 1980. Abundance of estuarine larval and juvenile fish in a South Carolina intertidal creek. *Estuaries* 3:89–97.

Cowan, J. H., Jr., Birdsong, R. S. 1985. Seasonal occurrence of larval and juvenile fishes in a Virginia Atlantic Coast Estuary with emphasis on drums (Family Sciaenidae). *Estuaries* 8:48–59.

Ditty, J. G. 1986. Ichthyoplankton in neritic waters of the northern Gulf of Mexico off Louisiana: Composition, relative abundance, and seasonality. *Fishery Bulletin* 84:935–946.

Dokken, Q. R., Matlock, G. C., Cornelius, S. 1984. Distribution and composition of larval fish populations within Alazan Bay, Texas. *Contributions in Marine Science* 27:205–222.

Dovel, W. L. 1971. *Fish Eggs and Larvae of the Upper Chesapeake Bay*. Natural Resources Institute Contribution No. 460. University of Maryland, College Park. 71 pp.

Dovel, W. L. 1981. Ichthyoplankton of the lower Hudson Estuary, New York. *New York Fish and Game Journal* 28(1):21–39.

Grothues, T. M., Cowen, R. K., Pietrafesa, L. J., et al. 2002. Flux of larval fish around Cape Hatteras. *Limnology and Oceanography* 47:165–175.

Hettler, W. F., Jr., Barker, D. L. 1993. Distribution and abundance of larval fishes at two North Carolina inlets. *Estuarine, Coastal and Shelf Science* 37:161–179.

Hettler, W. F., Jr., Hare, J. A. 1998. Abundance and size of larval fishes outside the entrance to Beaufort Inlet, North Carolina. *Estuaries* 21:476–499.

Holt, J., Strawn, K. 1983. Community structure of macrozooplankton in Trinity and upper Galveston Bays. *Estuaries* 6:66–75.

Jennings, C. A., Weyers, R. S. 2002. *Temporal and Spatial Distribution of Estuarine-Dependent Species in the Savannah River Estuary*. Annual Report. Prepared for the Georgia Ports Authority, Savannah.

Keller, A. A., Klein-MacPhee, G., St. Onge Burns, J. 1999. Abundance and distribution of ichthyoplankton in Narragansett Bay, Rhode Island, 1989–1990. *Estuaries* 22:149–163.

Monteleone, D. M. 1992. Seasonality and abundance of ichthyoplankton in Great South Bay, New York. *Estuaries* 15:230–238.

Nelson, D. M. 1992. *Distribution and Abundance of Fishes and Invertebrates in Gulf of Mexico Estuaries*. Vol. 1, *Data Summaries*. ELMR Report No. 10 NOAA/NOS Strategic Environmental Assessments Division, Silver Spring, MD. 273 pp.

Nelson, D. M., Monaco, M. E., Irlandi, E. S., et al. 1991. *Distribution and Abundance of Fishes and Invertebrates in Southeast Estuaries*. ELMR Report No. 9 NOAA/NOS Strategic Environmental Assessments Division, Rockville, MD. 167 pp.

Olney, J. E., Boehlert, G. W. 1988. Nearshore ichthyoplankton associated with seagrass beds in the lower Chesapeake Bay. *Marine Ecology Progress Series* 45:33–43.

Pattillo, M. E., Czapla, T. E., Nelson, D. M., et al. 1997. *Distribution and Abundance of Fishes and Invertebrates in Gulf of Mexico Estuaries.* Vol. 2, *Species Life History Summaries.* ELMR Report No. 11. NOAA/NOS Strategic Environmental Assessments Division, Silver Spring, MD. 377 pp.

Pearcy, W. G., Richards, S. W. 1962. Distribution and ecology of fishes of the Mystic River Estuary, Connecticut. *Ecology* 43:248–259.

Peebles, E. B. 2005. *An Analysis of Freshwater Inflow Effects on the Early Stages of Fish and Their Invertebrate Prey in the Alafia River Estuary.* Southwest Florida Water Management District, Brooksville, FL. 147 pp.

Rakocinski, C. F., Lyczkowski-Shultz, J., Richardson, S. L. 1996. Ichthyoplankton assemblage structure in Mississippi Sound as revealed by canonical correspondence analysis. *Estuarine, Coastal and Shelf Science* 43:237–257.

Ray, G., Clarke, D. 2001. *Surfzone and nearshore ichthyoplankton.* In: U.S. Army Corps of Engineers, ed. *The New York District's Biological Monitoring Program for the Atlantic Coast of New Jersey, Asbury Park to Manasquan Section Beach Erosion Control Project* (Chap. 3). Final Report. U.S. Army Corps of Engineers, Vicksburg, MS.

Raynie, R. C., Shaw, R. F. 1994. Ichthyoplankton abundance along a recruitment corridor from offshore spawning to estuarine nursery ground. *Estuarine, Coastal and Shelf Science* 39:421–450. (Louisiana)

Reyier, E. A., Shenker, J. M. 2007. Ichthyoplankton community structure in a shallow subtropical estuary of the Florida Atlantic Coast. *Bulletin of Marine Science* 80:267–293.

Ruple, D. L. 1984. Occurrence of larval fishes in the surf zone of a northern Gulf of Mexico barrier Island. *Estuarine, Coastal and Shelf Science* 18:191–208.

Tolan, J. M. 2008. Larval fish assemblage response to freshwater inflows: A synthesis of five years of ichthyoplankton monitoring within Nueces Bay, Texas. *Bulletin of Marine Science* 82: 275–296.

Wang, J. C. S., Kernehan, R. J. 1979. *Fishes of the Delaware Estuaries: A Guide to the Early Life Histories.* Ecological Analysts, Towson, MD. 410 pp.

Warlen, S. M., Burke, J. S. 1990. Immigration of larvae of fall/winter spawning marine fishes into a North Carolina estuary. *Estuaries* 13:453–461.

Witting, D. A., Able, K. W., Fahay, M. P. 1999. Larval fish assemblages of a Middle Atlantic Bight estuary: Assemblage structure and temporal stability. *Canadian Journal of Fisheries and Aquatic Science* 56:222–230.

Glossary

Most of these terms are used in at least several places in the book. Omitted are terms used in only one place and defined in that location. Definitions of some general terms are modified to reflect their application to zooplankton or usage in this book.

Actinopods (=axopods). Thin cytoplasmic projections, or rods, stiffened with microtubules that radiate outward from radiolarian, acantharian, and heliozoan cells.

Anoxia. Absence of dissolved oxygen.

Apical. At the tip, terminal.

Autotrophic. Photosynthetic; using sunlight to convert inorganic carbon into organic material or "food."

Benthic. Found on the bottom or in the sediments.

Benthos. Flora and fauna living on or in the bottom.

Binary fission. A form of asexual reproduction via cell division seen in some protozoa and phytoplankton, resulting in two separate individuals.

Bottom-up control. Indicates that primary (plant) productivity is the dominant influence in food web dynamics and structure.

Brackish. Waters with salinities between 0.5 and 30 psu.

Brown tide. A bloom of specific phytoplankton species with a brownish pigment that discolors the water.

Carapace. Shieldlike skeletal cover over the front section of decapod and shrimplike crustaceans, often covering the entire head and thorax.

Chaetae. Bristles (usually of chitin) found in annelids and previously widely referred to as setae. They are conspicuous on the parapodia of polychaetes and are embedded in the integument of oligochaetes.

Chitin. A tough structural polysaccharide found in many animal phyla. Chitin is a major component of the skeletons of arthropods, the beaks of cephalopods, and internal structures in many other groups.

Ciliary-mucoid feeding. A common type of suspension feeding in which food particles are caught in sticky mucus and brought to the mouth via cilia; particularly well suited to catching minute particles.

Cladistics. A particular method used in classifying organisms that uses multiple morphological characteristics to determine evolutionary relationships.

Copepodid (copepodite). A postnaupliar stage in copepod development that bears a general resemblance to the adult but has fewer appendages and/or segments.

Coriolis "force." The apparent deflection of winds and water currents to the right in the Northern Hemisphere caused by the earth's rotation.

Dactylozoids. Specialized polyps (zooids) on colonial hydrozoans heavily armed with cells containing stinging nematocysts and used for defense and prey capture.

Demersal eggs. Eggs that typically occur near bottom or attached to the bottom.

Demersal zooplankton. Zooplankton (mostly crustaceans) associated with the bottom that undergo periodic excursions into the water column.

Detritus. Small bits of organic material and associated microorganisms; a food source for suspension feeders.

Diel. Occurring over a 24-hour period that often covers a day and adjoining night.

DOC. Dissolved organic carbon.

Ekman circulation, or Ekman spiral. Progressive deflection of deeper-ocean currents to the right. A special application of Coriolis force.

Epibenthic plankton [=Suprabenthic =hyperbenthic]. Animals (primarily crustaceans) occupying the deepest portion of the water column; some have a direct association with the bottom.

Epibiont. An organism that lives on another organism.

Epitoke. A planktonic reproductive form of polychaete worms primarily in the Families Nereidae and Syllidae. Epitokes arise through a metamorphosis of benthic worms to produce planktonic forms with enlarged parapodia for swimming. The posterior of the worms fills with either eggs or sperm. Some authorities favor reserving the term epitoke for the posterior of the reproductive worm in contrast to the anterior part of the worm, the atoke.

Estuarine turbidity maximum (ETM). A region of high turbidity sometimes found in estuaries. Tides pushing salinity upriver beneath the outflowing river water produce turbulence, resulting in resuspension of sediment and particulate organic material in upper reaches of estuaries near the freshwater–saltwater interface. The ETM can be an important nursery area for shad, striped bass, and white perch.

Euhaline. High-salinity waters, typically 30–35 psu, and the animals that occur in these areas.

Euryhaline. A wide range of salinities and organisms with a broad salinity tolerance.

Eutrophication. An excess of nutrients that often results in algal blooms and sometimes low-dissolved oxygen levels on the bottom due to algal decomposition.

Filter feeding. See **suspension** feeding.

First crab stage. The first truly crablike stage in crab development is usually the product of a metamorphic molt from the megalops stage.

Gastrozooids. Specialized feeding polyps (zooids) on colonial hydrozoans. Gastrozooids typically have tentacles and a mouth.

Girdle. In dinoflagellates, a transverse groove containing a flagellum.

Gonozooids. Specialized reproductive polyps (zooids) on colonial hydrozoans that produce medusae by budding. They lack mouths or tentacles.

HAB (harmful algal bloom). A bloom of phytoplankton associated with toxic or other harmful effects.

Halocline. A sharp salinity gradient with depth, resulting in water-column stratification.

Heterotrophic. Feeding on organic material, such as other organisms, detritus, or dissolved organic material (DOM).

Holoplankton. Zooplankton that spend their entire life cycles in the plankton. Some groups (e.g., copepods) have resting (demersal) egg stages, but all active stages are planktonic.

Hypoxia. Low-dissolved oxygen conditions with less than 2.0 mg L^{-1} of water (=2 ppm).

Juvenile. Young individuals with morphology closely resembling that of the adult of the species.

Larvae. Prejuvenile developmental stages morphologically distinct from the adult of the species.

Lecithotrophic. Where larvae have sufficient yolk reserves to complete much or all of their planktonic phases without feeding.

Leptocephalus. The flattened, transparent larva of most eels and a few other fishes, including tarpon, ladyfish, and bonefish. Leptocephali are larger and have a longer larval development than most fish larvae.

Meroplankton. "Temporary" plankton, spending only a portion of their life cycle in the plankton, often as larval forms.

Mesohaline. Intermediate to low-salinity waters, typically 5–19 psu, and the animals that occur in these intermediate salinity areas.

Microbial loop. The part of the planktonic food web involving microorganisms. In general, this begins with bacteria using dissolved organic material and their subsequent consumption by other microorganisms, including ciliates, mixotrophic and heterotrophic dinoflagellates, and copepod nauplii.

Microheterotrophs. Small, nonphotosynthetic organisms that consume organic material to nourish themselves. The planktonic microheterotrophs include some dinoflagellates and most protozooplankton. Rotifers and copepod nauplii might also be considered microheterotrophs.

Microphagous. Feeding on minute particles, usually <20 micrometers (μm).

Mixotrophic. The ability to supplement photosynthetic nutrition by uptake of dissolved or particulate organic matter. Common in protozoans, especially dinoflagellates.

Nanoplankton. Plankton 2 to 20 micrometers (μm) in diameter.

Nauplius. The first larval stages in crustacean development, often passed within the egg. Free nauplii of copepods and barnacles may be abundant in the plankton.

Neap tides. The biweekly tides with lowest range between high and low tide; they occur between the full and new moons. See spring tides.

Nekton. Animals with sufficient swimming capacity to determine their distributions.

Neuston. Organisms specifically associated with the uppermost layer of the water column, either at or just below the surface.

Nutrients. Essential elements needed for plant or phytoplankton growth, including nitrogen, phosphorus, and silica (for diatoms). These nutrients are usually found in various inorganic compounds in marine or estuarine waters.

Ocelli (*sing.* ocellus). Small, simple eyes, sometimes called eyespots, found in some invertebrates.

Oligohaline. Low-salinity waters, typically <5 psu, and organisms that occur in these areas.

Our geographic region. Areas of the Atlantic and Gulf Coasts that comprise our area of coverage in this book: from Cape Cod, Massachusetts, to Cape Canaveral, Florida, on the Atlantic Coast and in the Gulf of Mexico from Fort Myers, Florida, to the Rio Grande River, Texas (see Fig. 3).

Ovigerous. Carrying or bearing eggs. Ovigerous as used in this book refers to female mysids, isopods, amphipods, or cumaceans when brooding their eggs or developing embryos in a ventral brood pouch.

Parapodia. A pair of fleshy, lateral extensions tipped with chaetae that occur on most segments of polychaete worms.

Parthenogenesis. A mode of "asexual" reproduction seen in rotifers and some cladocerans in which females produce diploid eggs that develop without fertilization.

Phagocytosis. The process of engulfing food particles whole as seen in some protozooplankton and heterotrophic "phytoplankton."

Picoplankton. Plankton 0.2 to 2.0 micrometers (μm) in diameter.

Planktotrophic. Where larvae feed in the plankton throughout their planktonic larval phase (as opposed to lecithotrophic).

Pleopods. Paired appendages attached to abdominal segments of decapods, stomatopods, mysids, amphipods, isopods, and cumaceans. In these groups, the pleopods are generally used for swimming, except in crabs. They appear only in later larval stages in decapods.

POC. Particulate organic carbon, including detritus.

Polyhaline. Intermediate salinity waters, typically 20–30 psu, and the animals that occur in these areas.

Protozooplankton. Small (usually <0.1 mm) nonphotosynthetic zooplankton, including protozoans, such as foraminiferans, ciliates, and radiolarians.

Pseudopod. A retractable protoplasmic protrusion employed by amoeboid protozooplankton, primarily in feeding.

psu (practical salinity units). A measure of salinity. Practical salinity units express salinity as measured by conductivity. Practical salinity units and the traditional salinity unit, parts per thousand (ppt or ‰), are numerically equivalent for biological applications. When a numerical salinity value follows the actual word salinity, the value refers to psu although psu is often omitted.

Pycnocline. A sharp, vertical density gradient produced by the joint effects of temperature and salinity, resulting in water-column stratification.

Raptorial feeding. Feeding by grasping prey or individual particles.

Recruitment. Ecological process in which larval forms arrive in a new habitat after being transported from another.

Red (or mahogany) tides. A bloom of phytoplankton with reddish pigments that give the water a reddish or reddish-brown color.

Resting eggs. Resting or diapause eggs produced by some copepods, cladocerans, and rotifers. Typically, these eggs are produced seasonally when the adults disappear. Resting eggs sink to the bottom and then hatch when conditions become more favorable to initiate a new generation in the same location.

Reynolds number (Re). The Reynolds number is the ratio of the impact of inertial versus viscous forces acting on a body moving through a fluid. Re decreases with body size as viscosity becomes dominant. Re for a fish might be 30,000 and Re for a small copepod around 0.3.

Rostrum. Anterior extension of the carapace between the antennae of crustaceans.

Salinity. The saltiness (amount of dissolved salts) in seawater, traditionally expressed as the weight of salt in a kilogram (about a liter) of seawater and reported as parts per thousand (abbreviated as ppt or ‰). Thus, a value of 35 ppt, typical of open-ocean water, would represent 35 grams of dissolved salt per 1,000 g of seawater. Today, oceanographers use conductivity to measure salinity and report salinity in terms of practical salinity units, or **psu**, which are numerically equivalent to parts per thousand, at least for biologists. When a numerical salinity value follows the actual word "salinity," the value refers to psu.

Segments (annelids). The major body divisions. In polychaetes, most segments bear a pair of parapodia.

Segments (decapods and other larger crustaceans). Segments is the preferred term for the articulations of the limbs. In these crustaceans, divisions of the body are also called segments although some specialists prefer the term somites.

Selective tidal transport. Using vertical positioning to migrate on favorable tidal currents.

Semibenthic. Animals typically residing on or in the bottom that also enter the water column in contrast to epibenthic organisms often found just above the bottom.

Setae (in annelids). See **chaetae**.

Setae (in crustaceans). Small, spinelike chitinous projections from the exoskeleton that have a flexible joint at their base (basal articulation) as opposed to spines, which lack a flexible joint.

Somite (in crustaceans). The preferred term for the body divisions, often bearing paired appendages.

sp., spp. These standard abbreviations often follow a genus name; "sp." refers to a single unspecified or unknown species in the genus, and "spp." refers to two or more unspecified species in that genus.

Spines (in crustaceans). Small, chitinous projections from the exoskeleton lacking a flexible joint at their base (basal articulation) as opposed to setae, which have a flexible joint.

Spring tides. Twice a month (biweekly) tides that occur on full and new moons and produce maximum vertical differences between tides; high tides are higher and low tides lower than on other moon phases.

Statocyst. An organ to sense orientation with respect to gravity that is found in many invertebrates.

Stratification (of the water column). A vertical layering of water masses of different densities. Stratification due to differences in temperature and/or salinity inhibits vertical mixing.

Submerged aquatic vegetation (SAV). Subtidal vegetation, usually seagrasses, growing in shallow protected areas. Eelgrass and turtlegrass beds are examples.

Suspension feeding. Feeding on small particles suspended in the water column. Sometimes loosely referred to as filter feeding.

Telson. A median, flat terminal projection from the last segment of some crustaceans, especially decapods and peracarideans; sometimes flanked by uropods to form a "tail fan."

Thermocline. A sharp difference in temperature in the water column that sometimes occurs in summer without mixing.

Top-down control. When the top predators have a dominant influence on the dynamics and structure of the food web.

Trochophore larva. First larval stage of both polychaete annelids and molluscs.

Trophic levels. A somewhat artificial but convenient hierarchy of levels in the food chain, for example, plant, herbivore, and various levels of predators.

Upwelling. Circulation of bottom waters to the surface, usually caused by a particular wind pattern.

Uropods. Two pairs of lateral extensions projecting from the last segment of many crustaceans; sometimes referred to as "tail fins," these usually flank the telson, or tail fan.

Velum. A thin flap of tissue lying just inside the margin of the bell of hydrozoan and cubozoan medusae or the ciliated swimming structures of mollusc veliger larvae.

Water column. The vertical dimension of bodies of water, often used when describing physical conditions or biotic distributions associated with depth.

Literature Cited

Able, K. W., Allen, D. M., Bath-Martin, G., et al. 2011. Life history and habitat use of the speckled worm eel, *Myrophis punctatus*, along the east coast of the United States. In: *Environmental Biology of Fishes*. Published Online First: June 15, 2011. *Environmental Biology of Fishes* 92:237–259.

Able, K. W., Cowen, R. K. 2003. Seasonal distributions and abundance of lobster (*Homarus americanus*) postlarvae in the New York Bight and an adjacent estuary. *Bulletin of the New Jersey Academy of Sciences* 47:15–20.

Able, K. W., Fahay, M. P. 1998. *The First Year in the Life of Estuarine Fishes in the Middle Atlantic Bight*. Rutgers University Press, New Brunswick, NJ. 342 pp.

Able, K. W., Fahay, M. P., Witting, D., et al. 2006. Fish settlement in the ocean vs. estuary: Comparison of pelagic larval and settled juvenile composition and abundance from southern New Jersey, USA. *Estuarine Coastal Shelf Science* 66:280–290.

Able, K. W., Matheson, R. E., Morse, W. W., et al. 1990. Patterns of summer flounder *Paralichthys dentatus* early life history in the mid-Atlantic Bight and New Jersey estuaries. *Fishery Bulletin* 88:1–12.

Abraham, B. J., Dillon, P. L. 1986. *Species Profiles: Life Histories and Environmental Requirements of Coastal Fishes and Invertebrates (Mid-Atlantic)—Softshell Clam*. US Fish and Wildlife Service Biological Report FWS/OBS-82 11.68. 18 pp.

Acosta, C. A., Butler, M. J. 1999. Adaptive strategies that reduce predation on Caribbean spiny lobster postlarvae during onshore transport. *Limnology and Oceanography* 44:494–501.

Acosta, C. A., Matthews, T. R., Butler, M. J., IV. 1997. Temporal patterns and transport processes in recruitment of spiny lobster (*Panulirus argus*) postlarvae to south Florida. *Marine Biology* 129:79–85.

Acuña, J. L., Deibel, D., Morris, C. C. 1996. Particle capture mechanisms of the pelagic tunicate *Oikopleura vanhoeffeni*. *Limnology and Oceanography* 41:1800–1814.

Adamik, P., Gallager, S. M., Horgan, E., et al. 2006. Effects of turbulence on the feeding rate of a pelagic predator: The planktonic hydroid *Clytia gracilis*. *Journal of Experimental Marine Biology and Ecology* 333:159–165.

Akimoto, H., Wu, C.-H., Kinumi, T., et al. 2004. Biological rhythmicity in expressed proteins of the marine dinoflagellate *Lingulodinium polyedrum* demonstrated by chronological proteomics. *Biochemical and Biophysical Research Communications* 315:306–312.

Alcaraz, M., Saiza, E., Calbeta, A. 2007. *Centropages* behaviour: Swimming and vertical migration. *Progress in Oceanography* 72:121–136.

Alekseev, V. R., Souissi, A. 2011. A new species within the *Eurytemora affinis* complex (Copepoda: Calanoida) from the Atlantic Coast of USA, with observations on eight morphologically different European populations. *Zootaxa* 2767:41–56.

Allen, D. M. 1982. Autecology of the cryptic mysid crustacean, *Heteromysis formosa* S. I. Smith 1873, in a temperate estuary. *Hydrobiologia* 93:1–7.

Allen, D. M. 1984. Population dynamics of the mysid shrimp *Mysidopsis bigelowi*, W. M. Tattersall in a temperate estuary. *Journal of Crustacean Biology* 4:25–34.

Alvarez-Cadena, J. N. 1993. Feeding of the chaetognath *Sagitta elegans* Verrill. *Estuarine, Coastal and Shelf Science* 36:195–206.

Amaratunga, T., Corey, S. 1975. Life history of *Mysis stenolepis* Smith (Crustacea: Mysidacea). *Canadian Journal of Zoology* 53:942–952.

Amaratunga, T., Corey, S. 1979. Marsupium and release of young in *Mysis stenolepis* Smith (Crustacean Mysidacea). *Crustaceana* 37:79–84.

Ambler, J. W., Frost, B. W. 1974. The feeding behavior of a predatory planktonic copepod, *Tortanus discaudatus*. *Limnology and Oceanography* 19:446–451.

Amend, M., Shanks, A. 1999. Timing of larval release in the mole crab *Emerita talpoida*. *Marine Ecology Progress Series* 183:295–300.

Ana, T., Winshella, I. J., Scorzettib, G., et al. 2010. Identification of okadaic acid production in the marine dinoflagellate *Prorocentrum rhathymum* from Florida Bay. *Toxicon* 55:653–657.

Andersen, V. 1986. Effect of temperature on the filtration rate and percentage assimilation of *Salpa fusiformis* Cuvier (Tunicata: Thaliacea). *Hydrobiologia* 137:135–140.

Anderson, C. R., Sapiano, M. R. P., Prasad, M. B. K., et al. 2010. Predicting potentially toxigenic *Pseudo-nitzschia* blooms in the Chesapeake Bay. *Journal of Marine Systems* 83:127–140.

Anderson, D. M., Coats, D. W., Tyler, M. A. 1985. Encystment of the dinoflagellate *Gyrodinium uncatenum*: Temperature and nutrient effects. *Journal of Phycology* 21:200–206.

Anderson, G. 1985. *Species Profiles: Life Histories and Environmental Requirements of Coastal Fishes and Invertebrates (Gulf of Mexico)—Grass Shrimp*. US Fish and Wildlife Service Biological Report SSWS/BR-82(11.35). 19 pp.

Anderson, W. W., King, J. E., Lindner, M. J. 1949. Early life stages in the life history of the common marine shrimp, *Penaeus setiferus* (Linneaus). *Biological Bulletin* 96:168–172.

Andrews, W. R., Targett, N. M., Epifanio, C. E. 2001. Isolation and characterization of the metamorphic inducer of the common mud crab, *Panopeus herbstii*. *Journal of Experimental Marine Biology and Ecology* 261:121–134.

Anger, K., Spivak, E., Luppi, T. 1998. Effects of reduced salinities on development and bioenergetics of early larval shore crab, *Carcinus maenas*. *Journal of Experimental Marine Biology and Ecology* 220:287–304.

Anger, K., Valentin, C. 1976. In situ studies on the diurnal activity pattern of *Diastylis rathkei* (Cumacea. Crustacea) and its importance for the "hyperbenthos." *Helgoländer Meeresuntersuchungen* 28:138–144.

Anning, T., MacIntyre, H. L., Pratt, S. M., et al. 2000. Photoacclimation in the marine diatom *Skeletonema costatum*. *Limnology and Oceanography* 45:1807–1817.

Anraku, M. 1964. Influence of the Cape Cod Canal on the hydrography and on the copepods in Buzzards Bay and the Cape Cod Bay, Massachusetts. 1. Hydrography and distribution of copepods. *Limnology and Oceanography* 9:46–60.

Atienza, D., Saiz, E., Calbet, A. 2008. Feeding ecology of the marine cladoceran *Penilia avirostris*: Natural diet, prey selectivity and daily ration. *Marine Ecology Progress Series* 315:211–220.

Attrill, M. J., Rundle, S. D., Fraser, A., et al. 2009. Oligochaetes as a possible entry route for terrigenous organic carbon into estuarine benthic food webs. *Marine Ecology Progress Series* 384:147–157.

Aurand, D., Daiber, F. C. 1979. Further observations on the occurrence of *Halicyclops fosteri* Wilson (Copepoda, Cyclopoida) in the Delaware Bay region, U.S.A. *Crustaceana* 36:155–165.

Auster, P. J. 1989. *Species Profile: Life Histories and Environmental Requirements of Coastal Fishes and Invertebrates (North and Mid-Atlantic)—Tautog and Cunner*. US Fish and Wildlife Service Biological Report 82(11.105). 13 pp.

Avery, D. E. 2005. Induction of embryonic dormancy in the calanoid copepod *Acartia hudsonica*: Proximal cues and variation among individuals. *Journal of Experimental Marine Biology and Ecology* 314:203–214.

Bacescu, M. 1961. *Taphromysis bowmani*, n. sp., a new brackish water mysid from Florida. *Bulletin of Marine Science of the Gulf and Caribbean* 11:517–524.

Badylak, S., Phlips, E. J. 2008. Spatial and temporal distributions of zooplankton in Tampa Bay, Florida, including observations during a HAB event. *Journal of Plankton Research* 30:449–465.

Baek, S. H., Shimode, S., Shin, K., et al. 2009. Growth of dinoflagellates, *Ceratium furca* and *Ceratium fusus* in Sagami Bay, Japan: The role of vertical migration and cell division. *Harmful Algae* 8:843–856.

Bagøien, E., Kiørboe, T. 2005. Blind dating: Mate finding in planktonic copepods. III. Hydromechanical communication in *Acartia tonsa*. *Marine Ecology Progress Series* 300:129–133.

Baier, C. T., Purcell, J. E. 1997. Trophic interactions of chaetognaths, larval fish, and zooplankton in the South Atlantic Bight. *Marine Ecology Progress Series* 146:43–53.

Bailey, R. J. E., Dick, J. T. A., Elwood, R. W., et al. 2006. Predatory interactions between the invasive amphipod *Gammarus tigrinus* and the native opossum shrimp *Mysis relicta*. *Journal of the North American Benthological Society* 25:393–405.

Bainbridge, V. 1958. Some observations on *Evadne nordmanni* Lovén. *Journal of the Marine Biological Association of the UK* 37:349–370.

Baird, D., Ulanowicz, R. E. 1989. The seasonal dynamics of the Chesapeake Bay ecosystem. *Ecological Monographs* 59:329–364.

Balcer, M. D., Korda, N. L., Dobson, S. I. 1984. *Zooplankton of the Great Lakes: A Guide to the Identification and Ecology of the Common Crustacean Species*. University of Wisconsin Press, Madison. 175 pp.

Baldwin, B. S., Newell, R. I. E. 1995. Relative importance of different size food particles in the natural diet of oyster larvae (*Crassostrea virginica*). *Marine Ecology Progress Series* 120:135–145.

Båmstedt, U. 1998. Trophodynamics of *Pleurobrachia pileus* (Ctenophora, Cydippida) and ctenophore summer occurrence off the Norwegian north-west coast. *Sarsia* 83:169–181.

Båmstedt, U., Martinussen, M. B., Matsakis, S. 1994. Trophodynamics of the two scyphozoan jellyfishes, *Aurelia aurita* and *Cyanea capillata*, in western Norway. *ICES Journal of Marine Science* 51:369–382.

Banks, M. A., Holt, G. J., Wakeman, J. M. 1991. Age-linked changes in salinity tolerance of larval spotted seatrout (*Cynoscion nebulosus*, Cuvier). *Journal of Fish Biology* 39:505–514.

Banner, A. H. 1953. On a new genus and species of mysid from southern Louisiana. *Tulane Studies in Zoology* 1:3–8.

Barnes, R. D. 1980. *Invertebrate Zoology*. 4th ed. W. B. Saunders, Philadelphia. 1089 pp.

Barse, A. M. 1998. Gill parasites of mummichogs, *Fundulus heteroclitus* (Teleostei: Cyprinodontidae): Effects of season, locality, and host sex and size. *Journal of Parasitology* 84:236–244.

Bartol, I. K., Mann, R., Vecchione, M. 2002. Distribution of the euryhaline squid *Lolliguncula brevis* in the Chesapeake Bay: Effects of selected abiotic factors. *Marine Ecology Progress Series* 226:235–247.

Bartol, I. K., Patterson, M., Mann, R. 2001. Swimming mechanics and behavior of the shallow-water brief squid *Lolliguncula brevis*. *Journal of Experimental Biology* 204:3655–3682.

Bartram, W. S. 1981. Experimental development of a model for the feeding of neritic copepods on phytoplankton. *Journal of Plankton Research* 3:25–51.

Bass, N. R., Brafield, A. E. 1972. The life-cycle of the polychaete *Nereis virens*. *Journal of the Marine Biological Association of the UK* 52:701–726.

Bauer, R. T., Delahoussaye, J. 2008. Life history migrations of the amphidromous river shrimp *Macrobrachium ohione* from a continental large river system. *Journal of Crustacean Biology* 28: 622–632.

Baxter, K. N., Renfro, W. C. 1967. Seasonal occurrence and size distribution of postlarval brown and white shrimp near Galveston, Texas, with notes on species identification. *Fishery Bulletin* 66:149–158.

Bayer, F. M. 1963. Observations on pelagic mollusks associated with the siphonophores. *Velella* and *Physalia*. *Bulletin of Marine Science* 13:454–466.

Bayne, B. L. 1976. The biology of mussel larvae. In: Baynes, B. L., ed. *Marine Mussels: Their Ecology and Physiology*. Cambridge University Press, New York, 81–121.

Bedo, A. W., Acuña, C. L., Robins, D., et al. 1993. Grazing in the micron and sub-micron particle size range: The case of *Oikopleura dioica* (Appendicularia). *Bulletin of Marine Science* 53:2–14.

Benfield, M. C., Downer, R. G. 2001. Spatial and temporal variability in the nearshore distributions of postlarval *Farfantepenaeus aztecus* along Galveston Island, Texas. *Estuarine, Coastal and Shelf Science* 52:445–456.

Bere, R. 1936. Parasitic copepods from Gulf of Mexico fish. *American Midland Naturalist* 17:577–625.

Berntsson, K. M., Jonsson, P. R., Lejhall, M., et al. 2000. Analysis of behavioral rejection of micro-textured surfaces and implications for recruitment by the barnacle *Balanus improvisus*. *Journal of Experimental Marine Biology and Ecology* 251:59–83.

Bielsa, L. M., Murdich, H., Labisky, R. F. 1983. *Species Profiles: Life Histories and Environmental Requirements of Coastal Fishes and Invertebrates (South Florida)—Pink Shrimp*. US Fish and Wildlife Service Biological Report FWS/OBS-82/11.17. 21 pp.

Bieri, R. 1970. The food of *Porpita* and niche separation in three neuston coelenterates. *Publications of the Seto Marine Biological Laboratory* 17:305–307.

Bigelow, H. B., Schroeder, W. C. 1953. Fishes of the Gulf of Maine. *Fishery Bulletin of the US* 53:1–577.

Bird, J. L., Kitting, C. L. 1982. Laboratory studies of a marine copepod (*Temora turbinata* Dana) tracking dinoflagellate migrations in a miniature water column. *Contributions in Marine Science* 25:27–44.

Blackmon, D. C., Eggleston, D. B. 2001. Factors influencing planktonic, post-settlement dispersal of early juvenile blue crabs. (*Callinectes sapidus* Rathbun). *Journal of Experimental Marine Biology and Ecology* 257:83–203.

Blades, P. I. 1977. Mating behavior of *Centropages typicus* (Copepoda: Calanoida). *Marine Biology* 40:57–64.

Blades, P. I., Youngbluth, M. J. 1979. Mating behavior of *Labidocera aestiva* (Copepoda: Calanoida). *Marine Biology* 51:339–355.

Blake, J. A. 1969. Reproduction and larval development of *Polydora* from northern New England (Polychaeta: Spionidae). *Ophelia* 7:1–63.

Blaxter, J. H. S., Fuiman, L. A. 1990. The role of the sensory systems of herring larvae in evading predatory fishes. *Journal of the Marine Biological Association of the UK* 70:413–427.

Bledsoe, E. L., Phlips, E. J. 2000. Relationships between phytoplankton standing crop and physical, chemical, and biological gradients in the Suwannee River and plume region, U.S.A. *Estuaries* 23:458–473.

Bochdansky, A. B., Bollens, S. M. 2004. Relevant scales in zooplankton ecology: Distribution, feeding, and reproduction of the copepod *Acartia hudsonica* in response to thin layers of the diatom *Skeletonema costatum*. *Limnology and Oceanography* 49:62–636.

Bockstahler, K. R., Coats, D. W. 1993a. Grazing of the mixotrophic dinoflagellate *Gymnodinium sanguineum* on ciliate populations of Chesapeake Bay. *Marine Biology* 116:477–487.

Bockstahler, K. R., Coats, D. W. 1993b. Spatial and temporal aspects of mixotrophy in Chesapeake Bay dinoflagellates. *Journal of Eukaryotic Microbiology* 40:49–60.

Bollens, S. M., Frost, B. W., Cordell, J. R. 1994. Chemical, mechanical, and visual cues in the vertical migration behavior of the marine planktonic copepod *Acartia hudsonica*. *Journal of Plankton Research* 16:555–564.

Boltkovskoy, D., ed. 1999. *South Atlantic Zooplankton*. Backhuys, Leiden, The Netherlands, 869–1098.

Bone, Q., Gorsky, G., Pulsford, A. L. 1979. On the structure and behaviour of *Fritillaria* (Tunicata: Larvacea). *Journal of the Marine Biological Association of the United Kingdom* 59:399–411.

Borkman, D. G., Smayda, T. 2009. Multidecadal (1959-1997) changes in *Skeletonema* abundance and seasonal bloom patterns in Narragansett Bay, Rhode Island, USA. *Journal of Sea Research* 61:84–94.

Bosch, H. F., Taylor, W. R. 1973a. Distribution of the cladoceran *Podon polyphemoides* in Chesapeake Bay. *Marine Biology* 19:161–171.

Bosch, H. F., Taylor, W. R. 1973b. Diurnal vertical migration of an estuarine cladoceran *Podon polyphemoides* in Chesapeake Bay. *Marine Biology* 19:172–181.

Boschung, H. T., Shaw, R. F. 1988. Occurrence of planktonic lancelets from Louisiana's continental shelf, with a review of pelagic *Branchiostoma* (Order Amphioxi). *Bulletin of Marine Science* 43:229–240.

Bouillon, J. 1999. Hydromedusae. In: Boltkovskoy, D., ed. *South Atlantic Zooplankton*. Backhuys Publishers, Leiden, The Netherlands, 385–465.

Bouillon, J., Werner, B. 1965. Production of medusae buds by the polyps of *Rathkea octopunctata* (M. Sars) (Hydroida Athecata). *Helgoländer Wissenschaftliche Meeresuntersuchungen* 12:137–148.

Bousfield, E. L. 1973. *Shallow-Water Gammaridean Amphipoda of New England*. Cornell University Press, Ithaca, NY. 312 pp.

Bowman, T. E. 1957. A new species of *Mysidopsis* (Crustacea: Mysidacea) from the southeastern coast of the United States. *Proceedings of the US National Museum* 107(3378):1–7.

Bowman, T. E. 1971. The distribution of calanoid copepods off the southeastern U.S. between Cape Hatteras and southern Florida. *Smithsonian Contributions to Zoology* 96:1–58.

Bradford-Grieve, J. M., Markhaseva, E. L., Rocha, C. E. F., et al. 1999. Copepoda. In: Boltkovskoy, D., ed. *South Atlantic Zooplankton*. Backhuys Publishers, Leiden, The Netherlands, 869–1098.

Brand, L. E., Compton, A. 2007. Long-term increase in *Karenia brevis* abundance along the Southwest Florida Coast. *Harmful Algae* 6:232–252.

Brandl, Z., Fernando, C. H. 1978. Prey selection by the cyclopoid copepods *Mesocyclops edax* and *Cyclops vicinus*. *Verhandlungen Internationale Vereinigung für Theoretische und Angewandte Limnologie* 20:2505–2510.

Branstrator, D. K. 2005. Contrasting life histories of the predatory cladocerans *Leptodora kindtii* and *Bythotrephes longimanus*. *Journal of Plankton Research* 27:569–585.

Breitburg, D. L. 1990. Demersal schooling prior to settlement by larvae of the naked goby. *Environmental Biology of Fishes* 26:97–103.

Breitburg, D. L. 1991. Settlement patterns and presettlement behavior of the naked goby, *Gobiosoma bosci,* a temperate oyster reef fish. *Marine Biology* 109:213–221.

Breitburg, D. L., Steinberg, N., DuBeau, S., et al. 1994. Effects of low dissolved oxygen on predation on estuarine fish larvae. *Marine Ecology Progress Series* 104:235–246.

Brewer, R. H. 1989. The annual pattern of feeding, growth, and sexual reproduction in *Cyanea* (Cnidaria, Scyphozoa) in the Niantic River estuary: Connecticut. *Biological Bulletin* 176:272–281.

Brinkman, A. 1964. Structure and development of *Velella. Videnskabelige Meddeleser fra Dansk Naturhistorisk Forening* 126:327–336.

Broad, A. C. 1957a. Larval development of *Palaemonetes pugio* Holthius. *Biological Bulletin* 112:144–161.

Broad, A. C. 1957b. Larval development of the crustacean *Thor floridanus* Kingsley. *Journal of the Elisha Mitchell Scientific Society* 73:317–328.

Brookins, K., Epifanio, C. 1985. Abundance of brachyuran larvae in a small coastal inlet over six consecutive tidal cycles. *Estuaries* 8:60–67.

Brooks, W. K. 1882. *Lucifer,* a study in development. *Philosophical Transactions of the Royal Society of London B, Biological Sciences* 173(1):57–137.

Brooks, W. K. 1886. The life-history of the hydromedusae: A discussion of the origin of the medusae, and of the significance of metagenesis. *Memoirs of the Boston Society of Natural History* 3:359–430.

Brossi-Garcia, A. L., Rodrigues, M. D. 1997. Zoeal morphology of *Pachygrapsus transversus* (Gibbes) (Decapoda, Grapsidae) reared in the laboratory. *Revista Brasileira de Zoologia* 14:803–819.

Browman, H. I., Kruse, S., O'Brien, W. J. 1989. Foraging behavior of the predaceous cladoceran, *Leptodora kindti*, and escape responses of their prey. *Journal of Plankton Research* 11:1075–1088.

Brown, S. E., Bert, T. M., Tweedale, W. A., et al. 1992. The effects of temperature and salinity on survival and development of early life stage Florida stone crabs (*Menippe mercenaria* Say). *Journal of Experimental Marine Biology and Ecology* 157:115–136.

Brown-Peterson, N. J., Peterson, M. S., Nieland, D. L., et al. 2002. Reproductive biology of female spotted seatrout, *Cynoscion nebulosus*, in the Gulf of Mexico: Differences among estuaries? *Environmental Biology of Fishes* 63:405–415.

Brusca, R. C., Brusca, G. J. 2003. *Invertebrates*. Sinauer Associates, Sunderland, MA. 936 pp.

Brusca, R. C., Brusca, G. J. 1990. *Invertebrates*. Sinauer Associates, Sunderland, MA. 922 pp.

Bryan, B. B. 1979. The diurnal reproductive cycle of *Evadne tergestina* Claus (Cladocera, Podonidae) in Chesapeake Bay, USA. *Crustaceana* 36:229–236.

Buckley, J. 1989. *Species Profiles: Life Histories and Environmental Requirements of Coastal Fishes and Invertebrates (North Atlantic)—Winter Flounder*. US Fish and Wildlife Service Biological Report 82(11.87). 12 pp.

Bucklin, A., Guarnieri, M., McGillicuddy, D. J., et al. 2001. Spring evolution of *Pseudocalanus* spp. abundance on Georges Bank based on molecular discrimination of *P. moultoni* and *P. newmani*. *Deep-Sea Research* 48:589–608.

Buffan-Dubau, E., Carman, K. R. 2000. Diel feeding behavior of meiofauna and their relationships with microalgal resources. *Limnology and Oceanography* 45:381–395.

Bulit, C., Díaz-Ávalos, C., Montagnes, J. S. 2004. Assessing spatial and temporal patchiness of the autotrophic ciliate *Myrionecta rubra*: A case study in a coastal lagoon. *Marine Ecology Progress Series* 268:55–67.

Bullard, S. G., Whitlach, R. B. 2004. *A Guide to the Larval and Juvenile Stages of Common Long Island Sound Ascidians and Bryozoans*. CTSG-04-07. Connecticut Sea Grant, Groton. 39 pp.

Bundy, M. H., Gross, T. F., Coughlin, D. J., Strickler, J. R. 1993. Quantifying copepod searching efficiency using swimming patterns and perceptive ability. *Bulletin of Marine Science* 53: 15–28.

Burbanck, W. D. 1962. An ecological study of the distribution of the isopod *Cyathura polita* (Stimpson) from brackish waters of Cape Cod, Massachusetts. *American Midland Naturalist* 67:449–476.

Burgess, S. C., Hart, S. P., Marshall, D. J. 2009. Pre-settlement behavior in larval bryozoans: The roles of larval age and size. *Biological Bulletin* 216:344–354.

Burke, J. S., Miller, J. M., Hoss, D. E. 1991. Immigration and settlement pattern of *Paralichthys dentatus* and *P. lethostigma* in an estuarine nursery ground, North Carolina. *Netherlands Journal of Sea Research* 27:393–405.

Burrell, V. G., Jr., Van Engel, W. A. 1976. Predation by and distribution of a ctenophore *Mnemiopsis leidyi* A. Agassiz, in the York River estuary. *Estuarine and Coastal Marine Science* 4:235–242.

Burreson, E. M., Allen, D. M. 1978. Morphology and biology of *Mysidobdella borealis* (Johansson) comb. n. (Hirudinea: Piscicolidae), from mysids in the western North Atlantic. *Journal of Parasitology* 64:1082–1091.

Burreson, E. M., Zwerner, D. E. 1982. The role of host biology, vector biology, and temperature in the distribution of *Trypanoplasma bullocki* infections in the lower Chesapeake Bay. *Journal of Parasitology* 68:306–313.

Bushing, M., Feigenbaum, D. 1984. Feeding by an expatriate population of *Sagitta enflata*. *Bulletin of Marine Science* 34:240–243.

Buskey, E. J. 1997. Behavioral components of feeding selectivity of the heterotrophic dinoflagellate *Protoperidinium pellucidum*. *Marine Ecology Progress Series* 153:77–89.

Buskey, E. J., Coulter, C. J., Brown, S. L. 1994. Feeding, growth and bioluminescence of the heterotrophic dinoflagellate *Protoperidinium huberi*. *Marine Biology* 121:373–380.

Buskey, E. J., Lenz, P. J., Hartline, D. K. 2002. Escape behavior of planktonic copepods in response to hydrodynamic disturbances: High speed video analysis. *Marine Ecology Progress Series* 235:135–146.

Buskey, E. J., Stroma, S., Coultera, C. 1992. Bioluminescence of heterotrophic dinoflagellates from Texas coastal waters. *Journal of Experimental Marine Biology and Ecology* 169:37–49.

Cahoon, L. B., Tronzo, C. R., Howe, J. C. 1986. Notes on the occurrence of "*Hyperoche medusarum*" (Kroyer) (Amphipoda, Hyperiidae) with Ctenophora off North Carolina, U.S.A. *Crustaceana* 51:95–96.

Caicci, F., Zaniolo, G., Burighela, P., et al. 2010. Differentiation of papillae and rostral sensory neurons in the larva of the ascidian *Botryllus schlosseri* (Tunicata). *Journal of Comparative Neurology* 518:547–566.

Calado, R., Bartilotti, C., Narciso, L., et al. 2004. Redescription of the larval states of *Lysmata seticaudata* (Russo, 1816) (Crustacea, Decapoda, Hyppolytidae) reared under laboratory conditions. *Journal of Plankton Research* 26:737–752.

Calbet, A., Carlotti, F., Gaudy, R. 2007. The feeding ecology of the copepod *Centropages typicus* (Kröyer). *Progress in Oceanography* 72:137–150.

Calder, D. R. 1970a. North American record of the hydroid *Probosidactyla ornata* (Hydrozoa, Proboscidactylidae). *Chesapeake Science* 11:130–132.

Calder, D. R. 1970b. Hydroid and young medusa stages of *Dipurena strangulata* (Hydrozoa, Corynidae). *Biological Bulletin* 138:109–114.

Calder, D. R. 1971. Hydroids and hydromedusae of southern Chesapeake Bay. *Virginia Institute of Marine Science, Special Papers in Marine Science* 1:1–125.

Calder, D. R. 1972. Development of the sea nettle *Chrysaora quinquecirrha* (Scyphozoa, Semaeostomeae). *Chesapeake Science* 13:40–44.

Calder, D. R. 1973. Laboratory observations on the life history of *Rhopilema verrilli* (Scyphozoa: Rhizostomeae). *Marine Biology* 21:109–114.

Calder, D. R. 1982. Life history of the cannonball jellyfish, *Stomolophus meleagris* L. Agassiz, 1860 (Scyphozoa, Rhizostomida). *Biological Bulletin* 162:149–162.

Calder, D. R. 1983. Nematocysts of stages in the life cycle of *Stomolophus meleagris*, with keys to scyphistomae and ephyrae of some western Atlantic Scyphozoa. *Canadian Journal of Zoology* 61:1185–1192.

Calder, D. R. 1988. Shallow-water hydroids of Bermuda: The Athecatae. *Royal Ontario Museum, Life Sciences Contributions* 148:1–107.

Calder, D. R. 2009. Cubozoan and scyphozoan jellyfishes of the Carolinian biogeographic province, southeastern, USA. *Royal Ontario Museum Contributions in Marine Science* 3:1–58.

Calder, D. R., Burrell, V. C., Jr. 1967. Occurrence of *Moerisia lyonsi* (Limnomedusae, Moerisiidae) in North America. *American Midland Naturalist* 78:540–541.

Calder, D. R., Burrell, V. G., Jr. 1969. Brackish water hydromedusa *Maeotias inexpectata* in North America. *Nature* 222:694–695.

Caldwell, D. K. 1957. The biology and systematics of the pinfish, *Lagodon rhomboides* (Linnaeus). *Bulletin of the Florida State Museum, Biological Sciences* 2:77–173.

Calinski, M. D., Lyons, W. G. 1983. Swimming behavior of the puerulus of the spiny lobster *Panulirus argus*. *Journal of Crustacean Biology* 3:329–335.

Calman, W. T. 1911. The Crustacea of the order Cumacea in the collections of the United States National Museum. *Proceedings of the US National Museum* 41:603–676.

Campbell, B. C., Able, K. W. 1998. Life history of the northern pipefish, *Syngnathus fuscus,* in southern New Jersey. *Estuaries* 21:470–475.

Canino, M. F., Grant, G. C. 1985. The feeding and diet of *Sagitta tenuis* (Chaetognatha) in the lower Chesapeake Bay. *Journal of Plankton Research* 7:175–188.

Caparroy, P., Pérez, M. T., Carlotti, F. 1998 Feeding behaviour of *Centropages typicus* in calm and turbulent conditions. *Marine Ecology Progress Series* 168:109–118.

Capone, D. G., Zehr, J. P., Paerl, H. W., et al. 1997. *Trichodesmium*, a globally significant marine cyanobacterium. *Science* 276:1221–1229.

Cargo, D. G. 1971. The sessile states of a scyphozoan identified as *Rhopilema verrilli*. *Tulane Studies in Zoology and Botany* 17(2):31–34.

Carre, C., Carre, D. 1991. A complete life cycle of the calycophoran siphonophore *Muggiaea kochi* (Will) in the laboratory, under different temperature conditions: Ecological implications. *Philosophical Transactions: Biological Sciences* 334:27–32.

Carriker, M. R. 2001. Functional morphology and behavior of veligers and early juveniles. In: Kraeuter, J., Castagna, M., eds. *Biology of the Hard Clam*. Elsevier, New York, 283–303.

Carvalho, W. F., Minnhagena, S., Granélia, E. 2007. *Dinophysis norvegica* (Dinophyceae), more a predator than a producer? *Harmful Algae* 7:174–183.

Casanova, J.-P. 1999. Chaetognatha. In: Boltovskoy, D., ed. *South Atlantic Zooplankton*. Vol. 2. Backhuys, Leiden, The Netherlands, 1353–1374.

Castellani, C., Irigoien, X., Harris, R. P., et al. 2005. Feeding and egg production of *Oithona similis* in the North Atlantic. *Marine Ecology Progress Series* 288:173–182.

Castellani, C., Irigoien, X., Mayor, D. J., et al. 2008. Feeding of *Calanus finmarchicus* and *Oithona similis* on the microplankton assemblage in the Irminger Sea, North Atlantic. *Journal of Plankton Research* 30:1095–1116.

Castro, L. R., Cowen, R. K. 1991. Environmental factors affecting the early life history of the bay anchovy, *Anchoa mitchilli,* in Great South Bay. *Marine Ecology Progress Series* 76:235–247.

Chambers, J. R., Musick, J. A., Davis, J. 1976. Methods of distinguishing larval alewife from larval blueback herring. *Chesapeake Science* 17:93–100.

Chanley, P., Andrews, J. D. 1971. Aids for identification of bivalve larvae of Virginia. *Malacologia* 11:45–119.

Chant, R. J., Curran, M. C., Able, K. W., et al. 2000. Delivery of winter flounder (*Pseudopleuronectes americanus*) larvae to settlement habitats in coves near tidal inlets. *Estuarine, Coastal and Shelf Science* 51:529–541.

Chatelain, E. H., Breton, S., Lemieux, H., et al. 2008. Epitoky in *Nereis* (*Neanthes*) *virens* (Polychaeta: Nereididae): A story about sex and death. *Comparative Biochemistry and Physiology Part B: Biochemistry and Molecular Biology* 149:202–208.

Checkley, D. M., Jr. 1982. Selective feeding by Atlantic herring (*Clupea harengus*) larvae on zooplankton in natural assemblages. *Marine Ecology Progress Series* 9:245–253.

Checkley, D. M., Jr., Dagg, M. J., Uye, S. 1992. Feeding, excretion, and egg production by individuals and populations of the marine planktonic copepods *Acartia* spp. and *Centropages furcatus*. *Journal of Plankton Research* 14:71–96.

Chen, D. S., Dykhuizen, G. V., Hodge, J., et al. 1996. Ontogeny of copepod predation in juvenile squid (*Loligo opalescens*). *Biological Bulletin* 190:69–81.

Chen, F., Marcus, N. H. 1997. Subitaneous, diapause, and delayed-hatching eggs of planktonic copepods from the northern Gulf of Mexico: Morphology and hatching success. *Marine Biology* 127:587–597.

Chen, Y.-H., Shaw, P. T., Wolcott, T. G. 1997. Enhancing estuarine retention of planktonic larvae by tidal currents. *Estuarine, Coastal and Shelf Science* 45:525–533.

Chesney, E. J. 2008. Foraging behavior of bay anchovy larvae, *Anchoa mitchilli*. *Journal of Experimental Marine Biology and Ecology* 362:117–124.

Choudhury, P. C. 1970. Complete larval development of the palaemonid shrimp *Macrobrachium acanthurus* (Wiegmann, 1836). *Crustaceana* 21:113–126.

Christensen, A. M., McDermott, J. J. 1958. Life-history and biology of the oyster crab, *Pinnotheres ostreum* Say. *Biological Bulletin* 114:146–179.

Christy, J. H. 1982. Adaptive significance of semilunar cycles of larval release in fiddler crabs (Genus *Uca*): Test of an hypothesis. *Biological Bulletin* 163:251–263.

Clancy, M., Cobb, J. S. 1997. Effect of wind and tidal advection on distribution patterns of rock crab *Cancer irroratus* megalopae in Block Island Sound, Rhode Island. *Marine Ecology Progress Series* 152:217–225.

Cloney, R. A., Young, C. M., Svane, I. 2002. Phylum Chordata. In: Young, C. M., ed. *Atlas of Marine Invertebrate Larvae*. Academic Press, New York, 563–605.

Cobb, J. S., Wahle, R. A. 1994. Early life history and recruitment processes of clawed lobsters. *Crustaceana* 67:1–25.

Cohen, J. H., Forward, R. B., Jr. 2005a. Diel vertical migration of the marine copepod *Calanopia americana*. I. Twilight DVM and its relationship to the diel light cycle. *Marine Biology* 147:387–398.

Cohen, J. H., Forward, R. B., Jr. 2005b. Diel vertical migration of the marine copepod *Calanopia americana*. II. Proximate role of exogenous light cues and endogenous rhythms. *Marine Biology* 147:399–410.

Colin, S. P., Costello, J. H. 2007. Functional characteristics of nematocysts found on the scyphomedusa *Cyanea capillata*. *Journal of Experimental Marine Biology and Ecology* 351:114–120.

Colin, S. P., Costello, J. H., Hansson, L. J., et al. 2010. Stealth predation and the predatory success of the invasive ctenophore *Mnemiopsis leidyi*. *Proceedings of the National Academy of Sciences* 107:17223–17227.

Colin, S. P., Kremer, P. 2002. Population maintenance of the scyphozoan *Cyanea* sp. settled planulae and the distribution of medusae in the Niantic River, Connecticut, USA. *Estuaries* 25:70–75.

Collette, B. B., Klein-MacPhee, G., eds. 2002. *Bigelow and Schroeder's Fishes of the Gulf of Maine*. 3rd ed. Smithsonian Institution Press, Washington, DC. 748 pp.

Compton, C. E., Jr., Price, W. W. 1979. Range extension to Texas for *Taphromysis bowmani* Bacescu (Crustracea: Mysidacea) with notes on its ecology and generic distribution. *Contributions in Marine Science* 22:121–125.

Condon, R. H., Steinberg, D. K. 2008. Development, biological regulation, and fate of ctenophore blooms in the York River estuary, Chesapeake Bay. *Marine Ecology Progress Series* 369:153–168.

Conley, W. J., Turner, J. T. 1985. Omnivory by the coastal marine copepods *Centropages hamatus* and *Labidocera aestiva*. *Marine Ecology Progress Series* 21:113–120.

Connaughton, V. P., Epifanio, C. E. 1993. Influence of previous experience on the feeding habits of larval weakfish *Cynoscion regalis*. *Marine Ecology Progress Series* 101:237–241.

Connaughton, V. P., Epifanio, C. E., Thomas, R. 1994. Effects of varying irradiance on feeding in larval weakfish (*Cynoscion regalis*). *Journal of Experimental Marine Biology and Ecology* 180:151–163.

Cook, H. L., Murphy, M. A. 1971. Early developmental stages of the brown shrimp, *Penaeus aztecus* Ives, reared in the laboratory. *Fishery Bulletin* 69:223–239.

Corkett, C. J., McLaren, I. A. 1978. The biology of *Pseudoclanus*. *Advances in Marine Biology* 15:1–231.

Cornelius, P. F. S. 1982. Hydroids and medusae of the family Campanulariidae recorded from the eastern North Atlantic, with a world synopsis of genera. *Bulletin of the British Museum of Natural History* 42:37–148.

Costello, J. H., Bayha, K. M., Mianzan, H. W., et al. 2012. The ctenophore *Mnemiopsis leidyi*—transitions from a native to an exotic species. *Hydrobiologia*. Published online: doi: 10.1007/s10750-012-1037-9.

Costello, J. H., Colin, S. P. 1994. Morphology, fluid motion and predation by the scyphomedusa *Aurelia aurita*. *Marine Biology* 121:327–334.

Costello, J. H., Coverdale, R. 1998. Planktonic feeding and evolutionary significance of the lobate body plan within the Ctenophora. *Biological Bulletin* 195:247–248.

Costello, J. H., Loftus, R., Waggett, R. 1999. Influence of prey detection on capture success for the ctenophore *Mnemiopsis leidyi* feeding upon adult *Acartia tonsa* and *Oithona colcarva* copepods. *Marine Ecology Progress Series* 191:207–216.

Costello, J. H., Stancyk, S. 1983. Tidal influence upon appendicularian distribution and abundance in North Inlet, South Carolina. *Journal of Plankton Research* 5:263–277.

Costello, J. H., Strickler, J. R., Marrase, C., et al. 1990. Grazing in a turbulent environment: Behavoral responses of a calanoid copepod, *Centropages hamatus*. *Proceedings of the National Academy of Sciences* 87:1648–1652.

Costello, J. H., Sullivan, B. K., Gifford, D. J., et al. 2006. Seasonal refugia, shoreward thermal amplification and metapopulation dynamics of the ctenophore *Mnemiopsis leidyi* in Narragansett Bay, Rhode Island. *Limnology and Oceanography* 52:1819–1831.

Costlow, J. D., Jr., Bookhout, C. G. 1959. The larval development of *Callinectes sapidus* Rathbun reared in the laboratory. *Biological Bulletin* 116:373–396.

Costlow, J. D., Jr., Bookhout, C. G. 1960. The complete larval development of *Sesarma cinereum* (Bosc.) reared in the laboratory. *Biological Bulletin* 118:203–214.

Costlow, J. D., Jr., Bookhout, C. G. 1961a. The larval stages of *Eurypanopeus depressus* (Smith) reared in the laboratory. *Crustaceana* 2:6–15.

Costlow, J. D., Jr., Bookhout, C. G. 1961b. The larval stages of *Panopeus herbstii* Milne-Edwards reared in the laboratory. *Journal of the Elisha Mitchell Scientific Society* 77:33–42.

Costlow, J. D., Jr., Bookhout, C. G. 1962. The larval development of *Sesarma reticulatum* Say reared in the laboratory. *Crustaceana* 4:281–294.

Costlow, J. D., Jr., Bookhout, C. G. 1966a. The larval development of *Ovalipes ocellatus* (Herbst) under laboratory conditions. *Journal of the Elisha Mitchell Scientific Society* 82:160–171.

Costlow, J. D., Jr., Bookhout, C. G. 1966b. Larval development of the crab *Hexopanopeus angustifrons*. *Chesapeake Science* 7:148–156.

Costlow, J. D., Jr., Bookhout, C. G. 1966c. Larval stages of the crab, *Pinnotheres maculatus,* under laboratory conditions. *Chesapeake Science* 7:157–163.

Costlow, J. D., Jr., Bookhout, C. G., Monroe, R. 1962. Salinity-temperature effects on the larval development of the crab, *Panopeus herbstii* Milne-Edwards, reared in the laboratory. *Physiological Zoology* 35:79–93.

Costlow, J. D., Jr., Bookhout, C. G., Monroe, R. 1966. Studies on the larval development of the crab, *Rhithropanopeus harrisii* (Gould). 1. The effect of salinity and temperature on larval development. *Physiological Zoology* 39:81–100.

Coston-Clements, L., Waggett, R. J., Tester, P. A. 2009. Chaetognaths of the United States South Atlantic Bight: Distribution, abundance and potential interactions with newly spawned larval fish. *Journal of Experimental Marine Biology and Ecology* 373:111–123.

Cowan, J. H., Jr., Birdsong, R. S., Houde, E. D., et al. 1992. Enclosure experiments on survival and growth of black drum eggs and larvae in lower Chesapeake Bay. *Estuaries* 15:392–402.

Cowles, R. P. 1903. Notes on the rearing of the larvae of *Polygordius appendiculatus* and on the occurrence of the adult on the Atlantic Coast of America. *Biological Bulletin* 4:125–128.

Crawford, R. E., Carey, C. G. 1985. Retention of winter flounder larvae within a Rhode Island salt pond. *Estuaries* 8:217–227.

Criales, M. M., Browder, J. A., Mooers, C. N. K., et al. 2007. Cross-shelf transport of pink shrimp larvae: Interactions of tidal currents, larval vertical migrations and internal tides. *Marine Ecology Progress Series* 345:167–184.

Cupp, E. E. 1943. Marine plankton diatoms of the west coast of North America. *Bulletin of the Scripps Institution of Oceanography* 5:1–238.

Dabiri, J. O., Colin, S. P., Costello, J. H. 2006. Fast-swimming hydromedusae exploit velar kinematics to form an optimal vortex wake. *Journal of Experimental Biology* 209:2026–2033.

Dahlberg, M. D., Conyers, J. C. 1973. An ecological study of *Gobiosoma bosci* and *G. ginsburgi* (Pisces, Gobiidae) on the Georgia coast. *Fishery Bulletin* 71:279–287.

Dales, R. P. 1950. The reproduction and larval development of *Nereis diversicolor* O. F Müller. *Journal of the Marine Biological Association of the UK* 29:321–360.

Dam, H. G., Peterson, W. T. 1991. In situ feeding behavior of the copepod *Temora longicornis* effects of seasonal changes in chlorophyll size fractions and female size. *Marine Ecology Progress Series* 71:113–123.

Dam, H. G., Peterson, W. T. 1993. Seasonal contrasts in the diel vertical distribution, feeding behavior, and grazing impact of the copepod *Temora longicornis* in Long Island Sound. *Journal of Marine Research* 51:561–594.

D'Ambra, I., Costello, J. H., Bentivegna, F. 2001. Flow and prey capture by the scyphomedusa *Phyllorhiza punctata* von Lendenfeld, 1884. *Hydrobiologia* 451(1/3):223–227.

Daniels, B. A., Sawyer, R. T. 1975. The biology of the leech *Myzobdella lugubris* infesting blue crabs and bluefish. *Biological Bulletin* 148:193–198.

Darby, D. G. 1965. *Ecology and Taxonomy of Ostracoda in the vicinity of Sapelo Island, Georgia.* Project GB-26. Report No. 2. National Science Foundation, Washington, DC. 76 pp.

Darnell, R. M. 1961. Trophic spectrum of an estuarine community, based on studies of Lake Pontchartrain, Louisiana. *Ecology* 42:553–568.

Davidovich, N. I., Bates, S. S. 1999. Sexual reproduction in the pennate diatoms *Pseudo-nitzschia multiseries* and *P. pseudodelicatissima* (Bacillariophyceae). *Journal of Phycology* 34:126–137.

Davis, C. C. 1955. *The Marine and Freshwater Plankton.* Michigan State University Press, East Lansing, 562 pp.

Davis, C. C. 1984. Planktonic Copepoda (including Monstrilloida). In: Steidinger, K. M., Walter, L. M., eds. *Marine Plankton Life Cycle Strategies.* CRC Press, Boca Raton, FL, 67–91.

Davis, C. O., Hollibaugh, J. T., Siebert, D. L. R., et al. 1980. Formation of resting spores by *Leptocylindrus danicus* (Bacillariophyceae) in a controlled experimental ecosystem. *Journal of Phycology* 16:296–302.

Dawe, E. G., Beck, P. C. 1985. Distribution and size of short-finned squid (*Illex illecebrosus*) larvae in the Northwest Atlantic. *Journal of Northwest Atlantic Fisheries Science* 6:43–55.

Dean, D. 1965. Larval development of *Streblospio benedicti. Biological Bulletin* 128:67–76.

Decker, M. B., Brown, C. W., Hood, R. R., et al. 2007. Predicting the distribution of the scyphomedusa *Chrysaora quinquecirrha* in Chesapeake Bay. *Marine Ecology Progress Series* 329:99–113.

Deevey, G. B. 1968. Pelagic ostracods of the Sargasso Sea off Bermuda. *Bulletin of the Peabody Museum of Natural History (Yale University)* 26:1–125.

Deibel, D., Lee, S. H. 1992. Retention efficiency of sub-micrometer particles by the pharyngeal filter of the pelagic tunicate *Oikopleura vanhoeffeni. Marine Ecology Progress Series* 81:25–30.

Deibel, D., Paffenhöfer, G.-A. 2009. Predictability of patches of neritic salps and doliolids (Tunicata, Thaliacea). *Journal of Plankton Research* 31:1571–1579.

DeManche, J. M., Curl, H. C., Lundy, D. W., et al. 1979. The rapid response of the marine diatom *Skeletonema costatum* to changes in external and internal nutrient concentration. *Marine Biology* 53:323–333.

De Melo, R., Hebert, P. D. N. 1994. Taxonomic reevaluation of North American Bosminidae. *Canadian Journal of Zoology* 72:1808–1825.

de Schweinitz, E. H., Lutz, R. A. 1976. Larval development of the northern horse mussel, *Modiolus modiolus* (L.), including a comparison with the larvae of *Mytilus edulis* L. as an aid in planktonic identification. *Biological Bulletin* 150:348–360.

Devreker, D., Souissi, S., Seuront, L. 2004. Development and mortality of the first naupliar stages of *Eurytemora affinis* (Copepoda, Calanoida) under different conditions of salinity and temperature. *Journal of Experimental Marine Biology and Ecology* 303:31–46.

DeVries, M. C., Rittschof, D., Forward, R. B. 1991. Chemical mediation of larval release behaviors in the crab *Neopanope sayi*. *Biological Bulletin* 180:1–11.

Díaz, G. A. 1998. Description of the last seven pelagic larval stages of *Squilla* sp. (Crustacea Stomatopoda). *Bulletin of Marine Science* 62:753–762.

Díaz, G. A., Manning, R. B. 1998. The last pelagic stage and juvenile of *Lysiosquilla scabricauda* (Lamarck, 1818) (Crustacea, Stomatopoda). *Bulletin of Marine Science* 63:453–457.

Diaz, H., Costlow, J. D., Jr. 1972. Larval development of *Ocypode quadrata* under laboratory conditions. *Marine Biology* 15:120–131.

Dineen, J. F., Hines, A. F. 1994. Larval settlement of the polyhaline barnacle *Balanus eburneus* (Gould): Cue interactions and comparisons with two estuarine congeners. *Journal of Experimental Marine Biology and Ecology* 179:223–234.

Dittel, A., Epifanio, C. E., Natunewicz, C. 1996. Predation on mud crab megalopae, *Panopeus herbstii* H. Milne Edwards: Effect of habitat complexity, predator species and postlarval densities. *Journal of Experimental Marine Biology and Ecology* 198:191–202.

Dittrich, B. 1987. Postembryonic development of the parasitic amphipod *Hyperia galba*. *Helgoländer Wissenschaftliche Meeresuntersuchengen* 41:217–232.

Ditty, J. G. 1989. Separating early larvae of sciaenids from the western North Atlantic: A review and comparison of larvae off Louisiana and Atlantic coasts of the U.S. *Bulletin of Marine Science* 44:1083–1105.

Ditty, J. G., Shaw, R. F. 1995. Seasonal occurrence, distribution, and abundance of larval bluefish, *Pomatomus saltatrix* (Family: Pomatomidae), in the northern Gulf of Mexico. *Bulletin of Marine Science* 56:592–601.

Doall, M. H., Colin, S. P., Strickler, J. R., et al. 1998. Locating a mate in 3D: The case of *Temora longicornis*. *Philosophical Transactions of the Royal Society of London B, Biological Sciences* 353:681–689.

Dobkin, S. 1961. Early developmental stages of pink shrimp, *Penaeus duorarum,* from Florida waters. *Fishery Bulletin of the US* 61:321–349.

Dobkin, S. 1962. Abbreviated larval development of a species of *Thor* (Decapoda: Caridea). *American Zoologist* 2:404–405.

Dobkin, S. 1968. The larval development of a species of *Thor* (Caridea, Hippolytidae) from south Florida, U.S.A. *Crustaceana*. Suppl. 2, *Studies on Decapod Larval Development*, 1–18.

Dobkin, S. 1971. A contribution to knowledge of the larval development of *Macrobrachium acanthurus* (Wiegmann, 1836) (Decapoda, Palaemonidae). *Crustaceana* 21:294–297.

Doherty, M., Tamura, M., Costas, B. A., et al. 2010. Ciliate diversity and distribution across an environmental and depth gradient in Long Island Sound, USA. *Environmental Microbiology* 12:886–898.

Dolan, J. R. 1991. Guilds of ciliate microzooplankton in the Chesapeake Bay. *Estuarine, Coastal and Shelf Science* 33:137–152.

Dolan, J. R., Gallegos, C. L. 1991. Trophic coupling of rotifers, microflagellates, and bacteria during the fall months in the Rhode River Estuary. *Marine Ecology Progress Series* 77:147–156.

Dolan, J. R., Gallegos, C. L. 1992. Trophic role of planktonic rotifers in the Rhode River Estuary, spring–summer 1991. *Marine Ecology Progress Series* 85:187–199.

Dorsey, S. E., Houde, E. D., Gamble, J. C. 1996. Cohort abundances and daily variability in mortality of eggs and yolk-sac larvae of bay anchovy, *Anchoa mitchilli*, in Chesapeake Bay. *Fishery Bulletin of the US* 94:257–267.

Dortch, Q., Robichaux, R., Pool, S., et al. 1997. Abundance and vertical flux of *Pseudo-nitzschia* in the northern Gulf of Mexico. *Marine Ecology Progress Series* 146:249–264.

Dovel, W. L., Mihursky, J. A., McErlean, A. J. 1969. Life history aspects of the hogchoker, *Trinectes maculatus,* in the Patuxent River estuary, Maryland. *Chesapeake Science* 10:104–119.

Duebler, E. E., Jr. 1958. A comparative study of the postlarvae of three flounders (*Paralichthys*) in North Carolina. *Copeia* 1958(2):112–116.

Duffy, J. T., Epifanio, C. E., Fuiman, L. A. 1997. Mortality rates imposed by three scyphozoans on red drum (*Sciaenops ocellatus* Linnaeus) larvae in field enclosures. *Journal of Experimental Marine Biology and Ecology* 212:123–131.

Dugger, D. M., Dobkin, D. 1975. A contribution to the knowledge of the larval development of *Macrobrachium olfersii* (Wiegmann, 1836) (Decapoda, Palaemonidae). *Crustaceana* 29:1–30.

Durante, K. M. 1991. Larval behavior, settlement preference, and induction of metamorphosis in the temperate solitary ascidian *Molgula citrina* Alder & Hancock. *Journal of Experimental Marine Biology and Ecology* 145:175–187.

Durbin, E., Kane, J. 2007. Seasonal and spatial dynamics of *Centropages typicus* and *C. hamatus* in the western North Atlantic. *Progress in Oceanography* 72:249–258.

Duval, E. J., Able, K. W. 1998. Life history of the seaboard goby, *Gobiosoma ginsburgi,* in New Jersey waters. *Bulletin of the New Jersey Academy of Science* 43:5–10.

Edmondson, W. T. 1959. *Fresh-water Biology.* 2nd ed. Wiley & Sons, New York, 1248 pp.

Egloff, D. A. 1988. Food and growth relations of the marine microzooplankter *Synchaeta cecilia* (Rotifera). *Hydrobiologia* 157:129–141.

Egloff, D. A., Fofonoff, P. W., Onbé, T. 1997. Reproductive biology of marine cladocerans. *Advances in Marine Biology* 31:79–167.

Elbourne, P. D., Clare, A. S. 2010. Ecological relevance of a conspecific, waterborne settlement cue in *Balanus amphitrite* (Cirripedia). *Journal of Experimental Marine Biology and Ecology* 392:99–106.

Elbrächter, M., Qi, Y.-Z. 1998. Aspects of *Noctiluca* (Dinophyceae) population dynamics. In: Anderson, D. M., Cembella, A. D., Hallegraeff, G. M., eds. *Physiological Ecology of Harmful Algal Blooms.* Vol. G 41. NATO ASI Series. Springer, Berlin, 315–335.

Ennis, G. P. 1995. Larval and postlarval ecology. In: Factor, J. R., ed. *Biology of the Lobster Homarus americanus.* Academic Press, New York, 23–46.

Epifanio, C. E. 1987. The role of tidal fronts in maintaining patches of brachyuran zoeae in estuarine waters. *Journal of Crustacean Biology* 7:513–517.

Epifanio, C. E. 1995. Transport of blue crab (*Callinectes sapidus*) larvae in waters off mid-Atlantic states. *Bulletin of Marine Science* 57:713–725.

Epifanio, C. E. 2007. Biology of larvae. In: Kennedy, V. S., Cronin, L. E., eds. *The Blue Crab Callinectes sapidus.* Maryland Sea Grant, College Park, MD, 513–533.

Epifanio, C. E., Dittel, A. I., Park, S., et al. 1998. Early life history of *Hemigrapsus sanguineus,* a non-indigenous crab in the Middle Atlantic Bight (USA). *Marine Ecology Progress Series* 170:231–238.

Esnal, G. B., Daponte, M. C. 1999. Salpida. In: Boltovskoy, D., ed. *South Atlantic Zooplankton.* Vol. 2. Backhuys, Leiden, The Netherlands, 1423–1444.

Esser, M., Greve, W., Boersma, M. 2004. Effects of temperature and the presence of benthic predators on the vertical distribution of the ctenophore *Pleurobrachia pileus. Marine Biology* 145:595–601.

Etherington, L. L., Eggleston, D. B. 2003. Spatial dynamics of large-scale, multistage crab *Callinectes sapidus* dispersal: Determinants and consequences for recruitment. *Canadian Journal of Fisheries and Aquatic Sciences* 60:873–887.

Eversole, A. G. 1987. *Species Profiles: Life Histories and Environmental Requirements of Coastal Fishes and Invertebrates (South Atlantic)—Hard Clam.* US Fish and Wildlife Service Biological Report FWS/OBS-82/11.75. 33 pp.

Ewald, J. J. 1965. The laboratory rearing of pink shrimp, *Penaeus duorarum* Burkenroad. *Bulletin of Marine Science* 15:436–449.

Facey, C. E., Van Den Avyle, M. J. 1987. *Species Profiles: Life Histories and Environmental Requirements of Coastal Fishes and Invertebrates (North Atlantic)—American Eel.* US Fish and Wildlife Service Biological Report FWS/OBS-82/11.45. 28 pp.

Factor, J. R., Dexter, B. L. 1993. Suspension feeding in larval crabs (*Carcinus maenas*). *Journal of the Marine Biological Association of the UK* 73:207–211.

Fahay, M. P. 1983. Guide to the early stages of marine fishes occurring in the western North Atlantic Ocean, Cape Hatteras to the southern Scotian Shelf. *Journal of the Northwest Atlantic Fisheries Science* 4:1–423.

Fahay, M. P., Obenchain, C. L. 1978. Leptocephali of the Ophichthid Genera *Ahlia, Myrophis, Ophichthus, Pisodonophis, Callechelys, Letharchus,* and *Apterichtus* on the Atlantic Continental Shelf of the United States. *Bulletin of Marine Science* 28:442–486.

Faimali, M., Garaventa, F., Terlizzi, A., et al. 2004. The interplay of substrate nature and biofilm formation in regulating *Balanus amphitrite* Darwin, 1854 larval settlement. *Journal of Experimental Marine Biology and Ecology* 306:37–50.

Faria, A. M., Ojanguren, A. F., Fuimam, L. A., et al. 2009. Ontogeny of critical swimming speed of wild-caught and laboratory-reared red drum *Sciaenops ocellatus* larvae. *Marine Ecology Progress Series* 384:221–230.

Fay, C. W., Neves, R. J., Pardue, G. B. 1983a. *Species Profiles: Life Histories and Environmental Requirements of Coastal Fishes and Invertebrates (Mid-Atlantic)—Striped Bass.* US Fish and Wildlife Service Biological Report FWS/OBS-82/11.8. 36 pp.

Fay, C. W., Neves, R. J., Pardue, G. B. 1983b. *Species Profiles: Life Histories and Environmental Requirements of Coastal Fishes and Invertebrates (Mid-Atlantic)—Atlantic Silverside.* US Fish and Wildlife Service Biological Report FWS/OBS-82/11.10. 15 pp.

Fay, C. W., Neves, R. J., Pardue, G. B. 1983c. *Species Profiles: Life Histories and Environmental Requirements of Coastal Fishes and Invertebrates (Mid-Atlantic)—Surf Clam.* US Fish and Wildlife Service Biological Report FWS/OBS-82 11.13. 23 pp.

Fenchel, T., Hansen, J. P. 2006. Motile behaviour of the bloom-forming ciliate *Mesodinium rubrum*. *Marine Biology Research* 2:33–40.

Figueroa, R. I., Bravo, I. 2005. Sexual reproduction and two different encystment strategies of *Lingulodinium polyedrum* (dinophyceae) in culture. 1. *Journal of Phycology* 41:370–379.

Fitzgerald, T. P., Forward, R. B., Tankersley, R. A. 1998. Metamorphosis of the estuarine crab *Rhithropanopeus harrisii:* Effect of water type and adult odor. *Marine Ecology Progress Series* 165:217–223.

Fitzhugh, G. R., Nixon, S. W., Arenholz, D. W., et al. 1997. Temperature affects on otolith microstructure and birth month estimation from otolith increment patterns in Atlantic menhaden. *Transactions of the American Fisheries Society* 126:579–593.

Flaherty, K. E., Landsberg, J. H. 2010. Effects of a persistent red tide (*Karenia brevis*) bloom on community structure and species-specific relative abundance of nekton in a Gulf of Mexico estuary. *Estuaries and Coasts* 34:417–439.

Flood, P. R. 2003. House formation and feeding behaviour of *Fritillaria borealis* (Appendicularia: Tunicata). *Marine Biology* 143:467–475.

Flores, A. A. V., Mazzuco, A. C. A., Bueno, M. 2007. A field study to describe diel, tidal and semilunar rhythms of larval release in an assemblage of tropical rocky shore crabs. *Marine Biology* 151:1989–2002.

Flores, A. A. V., Negreieros-Fransozo, M. L., Fransozo, A. 1998. The megalopa and juvenile development of *Pachygrapsus transversus* (Gibbes, 1850) (Decapoda, Brachyura) compared with other grapsid crabs. *Crustaceana* 71:197–222.

Flores-Coto, C., Warlen, S. M. 1993. Spawning time, growth, and recruitment of larval spot, *Leiostomus xanthurus,* into a North Carolina estuary. *Fishery Bulletin* 91:8–22.

Forward, R. B., Jr. 1989. Behavioral responses of crustacean larvae to rates of salinity change. *Biological Bulletin* 176:229–238.

Forward, R. B., Jr. 2009. Larval biology of the crab *Rhithropanopeus harrisii* (Gould): A synthesis. *Biological Bulletin* 216:243–256.

Forward, R. B., Jr., Bourla, M. H. 2008. Entrainment of the larval release rhythm of the crab *Rhithropanopeus harrisii* (Brachyura: Xanthidae) by cycles in hydrostatic pressure. *Journal of Experimental Marine Biology and Ecology* 357:128–133.

Forward, R. B., Jr., Burke, J. S., Rittschof, D., et al. 1996. Photoresponses of larval Atlantic menhaden (*Brevoortia tyrannus* Latrobe) in offshore and estuarine waters: Implications for transport. *Journal of Experimental Marine Biology and Ecology* 199:123–135.

Forward, R. B., Rittschof, D. 1994. Photoresponses of crab megalopae in offshore and estuarine waters: Implications for transport. *Journal of Experimental Marine Biology and Ecology* 182:183–192.

Forward, R. B., Tankersley, R. A., Rittschof, D. 2001. Cues for metamorphosis of brachyuran crabs: An overview. *American Zoologist* 41:1108–1122.

Forward, R. B., Jr., Tankersley, R. A., Welch, J. M. 2003. Selective tidal-stream transport of the blue crab *Callinectes sapidus*: An overview. *Bulletin of Marine Science* 72:347–365.

Foster, J. M., Heard, R. W. 2002. *Ameroculodes miltoni*, a new species of estuarine amphipod (Crustacea: Malacostraca: Peracarida: Oedicerotidae) from the southeastern United States. *Zootaxa* 28:1–12.

Francis, L. 1991. Sailing downwind: Aerodynamic performance of the *Velella* sail. *Journal of Experimental Biology* 158:117–132.

Frank, K. T. 1986. Ecological significance of the ctenophore *Pleurobrachia pileus* off southwestern Nova Scotia. *Canadian Journal of Fisheries and Aquatic Science* 43:211–222.

Franke, H.-D. 1999. Reproduction of the Syllidae (Annelida: Polychaeta). *Hydrobiologia* 402:39–55.

Frankenberg, D., Burbanck, W. D. 1963. A comparison of the physiology and ecology of the estuarine isopod *Cyathura polita* in Massachusetts and Georgia. *Biological Bulletin* 125:81–95.

Frankenberg, D., Menzies, R. J. 1966. A new species of Asellote marine isopod, *Munna* (*Uromunna*) *reynoldsi*, Crustacea (Isopoda). *Bulletin of Marine Science* 16:200–208.

Fredette, T. J., Diaz, R. J. 1986. Life history of *Gammarus mucronatus* Say (Amphipoda: Gammaridae) in warm temperate estuarine habitats, York River, Virginia. *Journal of Crustacean Biology* 6:57–78.

French, H. W., Hargraves, P. E. 1986. Population dynamics of the spore-forming diatom *Leptocylindrus danicus* in Narragansett Bay, Rhode Island. *Journal of Phycology* 22:411–420.

Fritzsche, R. A. 1978. *Development of Fishes of the Mid-Atlantic Bight: An Atlas of Egg, Larval, and Juvenile Stages.* Vol. 5, *Chaetodontidae through Ophidiidae*. Biological Services Program. US Fish and Wildlife Service. FWS/OBS-78/12. 340 pp.

Frost, B. W. 1989. A taxonomy of the marine calanoid copepod genus *Pseudocalanus*. *Canadian Journal of Zoology* 67:525–551.

Frost, B. W., Bollens, S. M. 1992. Variability of diel vertical migration in the marine planktonic copepod *Pseudocalanus newmani* in relation to its predators. *Canadian Journal of Fisheries and Aquatic Science* 49:1137–1141.

Frost, J. R., Jacoby, C. A., Youngbluth, M. J. 2010. Behavior of *Nemopsis bachei* L. Agassiz, 1849 medusae in the presence of physical gradients and biological thin layers. *Hydrobiologia* 645:97–111.

Fuiman, L. A., Cowan, J. H., Jr. 2003. Behavior and recruitment success in fish larvae: Repeatability and covariation of survival skills. *Ecology* 84:53–67.

Fuiman, L. A., Gamble, J. C. 1988. Predation by Atlantic herring, sprat, and sand eels on herring larvae in large enclosures. *Marine Ecology Progress Series* 44:1–6.

Fuiman, L. A., Rose, K. A., Cowan, J. H., Jr., et al. 2006. Survival skills required for predator evasion by fish larvae and their relation to laboratory measures of performance. *Animal Behaviour* 71:1389–1399.

Fulton, R. S., III. 1984. Effects of chaetognath predation and nutrient enrichment on enclosed estuarine communities. *Oecologia* 62:97–101.

Gallager, S. M. 1988. Visual observations of particle manipulation during feeding in larvae of a bivalve mollusc. *Bulletin of Marine Science* 43:344–365.

Galt, C. P., Fenaux, R. 1990. Urochordata-Larvacea. In: Adyiodi, K. G., Adyiodi, R. G., eds. *Reproductive Biology of Invertebrates*. Oxford & IBH, New Delhi, 471–500.

Galtsoff, P. S. 1964. The American oyster *Crassostrea virginica* (Gmelin). *Fishery Bulletin of the US* 64:1–480.

Gannon, J. E., Gannon, F. A. 1975. Observations on the narcotization of crustacean zooplankton. *Crustaceana* 28:220–224.

Garcia-Cuetos, L., Moestrup, O., Hansen, P. J., et al. 2010. The toxic dinoflagellate *Dinophysis acuminata* harbors permanent chloroplasts of cryptomonad origin, not kleptochloroplasts. *Harmful Algae* 9:25–38.

Garrison, L. P., Morgan, J. A. 1999. Abundance and vertical distribution of drifting, post-larval *Macoma* spp. (Bivalvia: Tellinidae) in the York River, Virginia, USA. *Marine Ecology Progress Series* 182:175–185.

Gasca, R., Manzanilla, H., Suárez-Morales, E. 2009. Distribution of hyperiid amphipods (Crustacea) of the southern Gulf of Mexico, summer and winter, 1991. *Journal of Plankton Research* 31:1493–1504.

Gehringer, J. W. 1959. Early development and metamorphosis of the ten-pounder, *Elops saurus* Linnaeus. *US Fish and Wildlife Service Fishery Bulletin* 59:619–647.

Gentile, J. H., Gentile, S. M., Hairston, N. G., Jr., et al. 1982. The use of life-tables for evaluating the chronic toxicity of pollutants to *Mysidopsis bahia*. *Hydrobiologia* 93:179–187.

Gentsch, G., Kreibich, T., Hagen, W., et al. 2009. Dietary shifts in the copepod *Temora longicornis* during spring: Evidence from stable isotope signatures, fatty acid biomarkers and feeding experiments. *Journal of Plankton Research* 31:45–60.

Gerber, R. P. 2000. *An Identification Manual to the Coastal and Estuarine Zooplankton of the Gulf of Maine Region*. Part 1: Text and identification keys (80 pp.). Part 2: Figures (98 pp.). Freeport Village Press, Brunswick, ME.

Gershwin, L.-A. 2006. Comments on *Chiropsalmus* (Cnidaria: Cubozoa: Chirodropida): A preliminary revision of the Chiropsalmidae, with descriptions of two new genera and two new species. *Zootaxa* 1231:P1–P42.

Gibson, V. R., Grice, G. D. 1977a. The developmental stages of *Labidocera aestiva* A. Wheeler, 1900 (Copepoda, Calanoida). *Crustaceana* 32:7–20.

Gibson, V. R., Grice, G. D. 1977b. The developmental stages of *Pontella meadi* Wheeler (Copepoda: Calanoida). *Journal of the Fisheries Research Board of Canada* 33:847–854.

Gidholm, L. 1965. Reproduction in *Autolytus*. *Zoologiska Bidrag från Uppsala* 37:1–44.

Gifford, D. J., Caron, D. A. 2000. Sampling, preservation, enumeration and biomass of protozooplankton. In: Harris, R., Wiebe, P., Lenz, J., et al., eds. *ICES Zooplankton Methodology Manual*. Academic Press, New York, 193–221.

Gilbert, C. R. 1986. *Species Profiles: Life Histories and Environmental Requirements of Coastal Fishes and Invertebrates (South Florida)—Southern, Gulf, and Summer Flounders*. US Fish and Wildlife Service Biological Report FWS/OBS-82/11. 54 pp.

Gilbert, J. J. 2007. Induction of mictic females in the rotifer *Brachionus*: Oocytes of amictic females respond individually to population-density signal only during oogenesis shortly before oviposition. *Freshwater Biology* 52:1417–1426.

Gillespie, M. C. 1971. *Cooperative Gulf of Mexico Estuarine Inventory and Study, Louisiana*. Phase 4, *Biology, Analyses and Treatment of Zooplankton of estuarine Waters of Louisiana*. Louisiana Wildlife and Fisheries Commission, New Orleans. 175 pp.

Gilmer, R. W., Harbison, G. R. 1986. Morphology and field behavior of pteropod molluscs: Feeding methods in the families Cavoliniidae, Limacinidae and Peraclididae (Gastropoda: Thecosomata). *Marine Biology* 91:47–57.

Gleason, T. R., Bengtson, D. A. 1996. Growth, survival and size-selective predation mortality of larval and juvenile inland silversides, *Menidia beryllina* (Pisces: Atherinidae). *Journal of Experimental Marine Biology and Ecology* 199:165–177.

Gnewuch, W. T., Croker, R. A. 1973. Macrofauna of northern New England marine sand. I. The biology of *Mancocuma stellifera* Zimmer, 1943 (Crustacea: Cumacea). *Canadian Journal of Zoology* 51:1011–1020.

Go, Y.-B., Oh, B.-C., Terazaki, M. 1998. Feeding behavior of the poecilostomatoid copepods *Oncaea* spp. on chaetognaths. *Journal of Marine Systems* 15:475–482.

Gomes, C. L., Marazzo, A., Valentin, J. L. 2004. The vertical migration behaviour of two calanoid copepods, *Acartia tonsa* Dana, 1849 and *Paracalanus parvus* (Claus, 1863) in a stratified tropical bay in Brazil. *Crustaceana* 77:941–954.

Gonor, S. L., Gonor, J. J. 1973. Feeding, cleaning, and swimming behavior in larval stages of porcellanid crabs (Crustacea: Anomura). *Fishery Bulletin* 71:225–234.

González, J. G., Bowman, T. E. 1965. Planktonic copepods from Bahia Fosforescente, Puerto Rico, and adjacent waters. *Proceedings of the United States National Museum* 117:241–303.

Gooding, R. U. 1988. The *Saphirella* problem. *Hydrobiologia* 167/168:363–366.

Gore, R. H. 1972. *Petrolisthes armatus* (Gibbes, 1850): The development under laboratory conditions of larvae from a Pacific specimen (Decapoda, Pocellanidae). *Crustaceana* 22:67–83.

Goshorn, D. M., Epifanio, C. E. 1991. The diet and prey abundance of weakfish larvae in Delaware Bay. *Transactions of the American Fisheries Society* 120:684–692.

Govoni, J. J. 1993. Flux of larval fishes across frontal boundaries: Examples from the Mississippi River Plume front and the western Gulf Stream front in winter. *Bulletin of Marine Science* 53:538–566.

Govoni, J. J. 2010. Feeding on protists and particulates by the leptocephali of the worm eels *Myrophis* spp. (Teleostei: Anguilliformes: Ophichthidae), and the potential energy contribution of large aloricate protozoa. *Scientia Marina (Barcelona)* 74:339–344.

Govoni, J. J., Chester, A. J. 1990. Diet composition of larval *Leiostomous xanthurus* in and about in the Mississippi River Plume. *Journal of Plankton Research* 12:819–830.

Govoni, J. J., Hoss, D. E., Chester, A. J. 1983. Comparative feeding of three species of larval fishes in the Northern Gulf of Mexico: *Brevoortia patronus, Leiostomus xanthurus,* and *Micropogonias undulatus. Marine Ecology Progress Series* 13:189–199.

Govoni, J. J., Merriner, J. V. 1978. The occurrence of ladyfish, *Elops saurus*, larvae in low salinity waters and another record for Chesapeake Bay. *Estuaries* 1:205–206.

Govoni, J. J., Ortner, P. B., Al-Yamani, F., et al. 1986. Selective feeding of spot *Leiostomus xanthurus,* and Atlantic croker, *Micropogonias undulatus,* larvae in the northern Gulf of Mexico. *Marine Ecology Progress Series* 28:175–183.

Govoni, J. J., Stoecker, D. K. 1984. Food selection by young gulf menhaden (*Brevoortia patronus*). *Marine Biology* 80:299–306.

Goy, J. W., Provenzano, A . J., Jr. 1978. Larval development of the rare burrowing mud shimp, *Naushonia crangonoides* Kingsley (Decapoda; Thalassinidae: Laomediidae). *Biological Bulletin* 154:241–261.

Graham, K. R., Sebens, K. P. 1996. The distribution of marine invertebrate larvae near vertical surfaces in the rocky subtidal zone. *Ecology* 77:933–949.

Graham, W. M., Kroutil, R. M. 2001. Size-based prey selectivity and dietary shifts in the jellyfish, *Aurelia aurita. Journal of Plankton Research* 23:67–74.

Graham, W. M., Martin, D. L., Felder, D. L., et al. 2003. Ecological and economic implications of a tropical jellyfish invader in the Gulf of Mexico. *Biological Invasions* 5:53–69.

Grammer, G. L., Brown-Peterson, N. J., Peterson, M. S., et al. 2009. Life history of silver perch (*Bairdiella chrysoura*, Lacepède 1803) in north-central Gulf of Mexico estuaries. *Gulf of Mexico Science* 27:62–73.

Grant, G. C. 1988. Seasonal occurrence and dominance of *Centropages* congeners in the Middle Atlantic Bight, USA. *Hydrobiologia* 167/168:227–237.

Grant, G. C., Olney, J. E. 1991. Distribution of striped bass, *Morone saxitilis* (Walbaum) eggs and larvae in major Virginia rivers. *Fishery Bulletin* 89:187–193.

Granum, E., Kirkvold, S., Myklestad, S. M. 2002. Cellular and extracellular production of carbohydrates and amino acids by the marine diatom *Skeletonema costatum*: Diel variations and effects of N depletion. *Marine Ecology Progress Series* 242:83–94.

Greene, C. H., Landry, M. R., Monger, B. C. 1986. Foraging behavior and prey selection by the ambush entangling predator *Pleurobrachia bachei. Ecology* 67:1493–1501.

Greenwood, J. G. 1965. The larval development of *Petrolisthes elongatus* (H. Milne-Edwards) and *Petrolisthes novaezelandiae* Filhol (Anomura, Porcellanidae) with notes on breeding. *Crustaceana* 8:285–307.

Gribble, K. E., Anderson, D. M., Coats, D. W. 2009. Sexual and asexual processes in *Protoperidinium steidingerae* Balech (Dinophyceae), with observations on life-history stages of *Protoperidinium depressum* (Bailey) Balech (Dinophyceae). *Journal of Eukaryotic Microbiology* 56:88–103.

Grice, G. D. 1969. The developmental stages of *Pseudodiaptomus coronatus* Williams (Copepoda, Calanoida). *Crustaceana* 16:291–301.

Grice, G. D. 1971. The developmental stages of *Eurytemora americana* Williams, 1906, and *Eurytemora herdmani* Thompson & Scott, 1897 (Copepoda, Calanoida). *Crustaceana* 20:145–158.

Grice, G. D., Gibson, V. R. 1977. Resting eggs in *Pontella meadi* (Copepoda: Calanoida). *Journal of the Fisheries Research Board of Canada* 34:410–412.

Grimes, B. H., Huish, M. T., Kerby, J. H., et al. 1989. *Species Profiles: Life Histories and Environmental Requirements of Coastal Fishes and Invertebrates (Mid-Atlantic)—Atlantic Marsh Fiddler*. US Fish and Wildlife Service Biological Report 82 (11.114). 18 pp.

Grosberg, R. K. 1987. Limited dispersal and proximity-dependent mating success in the colonial ascidian *Botryllus schlosseri*. *Evolution* 41:372–38.

Grover, J. J. 1998. Feeding habits of pelagic summer flounder, *Paralichthys dentatus,* larvae in oceanic and estuarine habitats. *Fishery Bulletin* 96:248–257.

Guillory, V., Perry, H. M., Leard, R. L. 1995. A profile of the western gulf stone crab, *Menippe adina*. *Gulf States Marine Fisheries Commission Publications* 31:1–49.

Guo, C., Tester, P. A. 1994. Toxic effect of the bloom-forming *Trichodesmium* sp. (cyanophyta) to the copepod *Acartia tonsa*. *Natural Toxins* 2:222–227.

Gupta, S. J., Lonsdale, D. J., Wang, D.-P. 1994. The recruitment patterns of an estuarine copepod: A biological-biophysical model. *Journal of Marine Research* 52:687–710.

Gustafson, D. E., Jr., Stoecker, D. K., Johnson, M. D., et al. 2000. Cryptophyte algae are robbed of their organelles by the marine ciliate *Mesodinium rubrum*. *Nature (London)* 405:1049–1052.

Gutow, L., Franke, H.-D. 2003. Metapopulation structure of the marine isopod *Idotea metallica*, a species associated with drifting habitat patches. *Helgoland Sea Research* 56:259–264.

Gutow, L., Strahl, J., Wienckel, C., et al. 2006. Behavioural and metabolic adaptations of marine isopods to the rafting life style. *Marine Biology* 149:821–828.

Hadfield, M. G., Meleshkevitch, E. A., Boudko, D. Y. 2000. The apical sensory organ of a gastropod veliger is a receptor for settlement cues. *Biological Bulletin* 198:67–76.

Hagiwara, A., Lee, C.-S., Miyamoto, G., et al. 1989. Resting egg formation and hatching of the S-type rotifer *Brachionus plicatilis* at varying salinities. *Marine Biology* 103:327–332.

Halsband-Lenk, C., Pierson, J. J., Leising, A. W. 2005. Reproduction of *Pseudocalanus newmani* (Copepoda: Calanoida) is deleteriously affected by diatom blooms: A field study. *Progress in Oceanography* 67:332–348.

Hansen, P. J. 1991. *Dinophysis:* A planktonic dinoflagellate genus which can act both as a prey and a predator of a ciliate. *Marine Ecology Progress Series* 69:201–204.

Hansen, P. J. 1992. Prey size selection, feeding rates and growth dynamics of marine heterotrophic dinoflagellates, with special emphasis on *Gyrodinium spirale*. *Marine Biology* 114:327–334.

Hansen, P. J., Fenchel, T. 2006. The bloom-forming ciliate *Mesodinium rubrum* harbours a single permanent endosymbiont. *Marine Biology Research* 2:169–177.

Hansen, P. J., Miranda, L., Azanza, R. 2004. Green *Noctiluca scintillans*: A dinoflagellate with its own greenhouse. *Marine Ecology Progress Series* 275:79–87.

Hansson, L. J. 1997. Capture and digestion of the scyphozoan jellyfish *Aurelia aurita* by *Cyanea capillata* and prey response to predator contact. *Journal of Plankton Research* 19:195–208.

Hansson, L. J. 2006. A method for in situ estimation of prey selectivity and predation rate in large plankton, exemplified with the jellyfish *Aurelia aurita* (L.). *Journal of Experimental Marine Biology and Ecology* 328:113–126.

Harbison, G. R., Biggs, D. C., Madin, L. P. 1977. The associations of Amphipoda Hyperiidea with gelatinous zooplankton. 2. Associations with Cnidaria, Ctenophora and Radiolaria. *Deep-Sea Research* 24:465–488.

Harding, J. M. 1999. Selective feeding behavior of larval naked gobies *Gobiosoma bosc* and blennies *Chasmodes bosquianus* and *Hypsoblennius hentzi:* Preferences for bivalve veligers. *Marine Ecology Progress Series* 179:145–153.

Harding, J. M., Mann, R. 2000. Estimates of naked goby (*Gobiosoma bosc*), striped blenny (*Chasmodes bosquianus*) and eastern oyster (*Crassostrea virginica*) larval production around a restored Chesapeake Bay oyster reef. *Bulletin of Marine Science* 66:29–45.

Harding, L. W., Jr., Coats, D. W. 1988. Photosynthetic physiology of *Prorocentrum mariae-lebour-*

iae (Dinophyceae) during its subpycnocline transport in Chesapeake Bay. *Journal of Phycology* 24:77–89.

Harding, L. W., Jr., Mallonee, M. E., Henderson, K. W. 1991. Spectral distribution and species specific photosynthetic performance of natural populations of *Prorocentrum mariae-lebouriae* (Dinophyceae) in the Chesapeake Bay, 1989. *Marine Ecology Progress Series* 52:261–272.

Harding, S. M., Chittenden, M. E. 1987. *Reproduction, Movements, and Population Dynamics of the Southern Kingfish, Menticirrhus americanus, in the Northwestern Gulf of Mexico*. NOAA Technical Report NMFS 49. 21 pp.

Hardy, J. D., Jr. 1978. *Development of Fishes of the Mid-Atlantic Bight: An Atlas of Egg, Larval and Juvenile Stages*. Vol. 2, *Anguillidae through Syngnathidae*. US Fish and Wildlife Service, Biological Services Program FWS/OBS-78/12. 458 pp.

Hare, J. A., Cowen, R. K. 1996. Transport mechanisms of larval and pelagic juvenile bluefish (*Pomatomus saltatrix*) from south Atlantic Bight spawning grounds to mid-Atlantic nursery habitats. *Limnology and Oceanography* 41:1264–1280.

Harms, J. 1992. Larval development and delayed metamorphosis in the hermit crab *Clibanarius erythropus* Latreille) (Crustacea, Diogenidae). *Journal of Experimental Marine Biology and Ecology* 156:151–160.

Harper, D. E., Jr. 1968. Distribution of *Lucifer faxoni* (Crustacea: Decapoda: Sergestidae) in neritic waters off the Texas coast, with notes on the occurrence of *Lucifer typicus*. *Contributions in Marine Science* 13:1–16.

Harper, D. E., Jr., Runnels, R. J. 1990. The occurrence of *Rhopilema verrilli* (Cnidaria: Scyphozoa: Rhizostomeae) on Galveston Island, Texas, and a discussion of its distribution in U.S. waters. *Northeast Gulf Science* 11:19–27.

Harris, C. E., Vogelbein, W. K. 2006. Parasites of mummichogs, *Fundulus heteroclitus*, from the York River, Virginia, U.S.A., with a checklist of parasites of Atlantic Coast *Fundulus* spp. *Comparative Parasitology* 73:72–110.

Harris, R. P. 1977. Some aspects of the biology of the harpacticoid copepod, *Scottolana canadensis* (Willey), maintained in laboratory culture. *Chesapeake Science* 18:245–252.

Harvey, A. W. 1993. Larval settlement and metamorphosis in the sand crab *Emerita talpoida* (Crustacea: Decapoda: Anomura). *Marine Biology* 117:575–581.

Harvey, A. W. 1996. Delayed metamorphosis in Florida hermit crabs: Multiple cues and constraints (Crustacea: Decapoda: Paguridae and Diogenidae). *Marine Ecology Progress Series* 141:27–36.

Harvey, E. A., Epifanio, C. E. 1997. Prey selection by larvae of the common mud crab *Panopeus herbstii* Milne-Edwards. *Journal of Experimental Marine Biology and Ecology* 217:79–91.

Hashizume, K. 1999. Larval development of seven species of *Lucifer* (Dendrobranchiata, Sergestoidea), with a key for the identification of their larval forms. In: Schram, F. R., von Vaupel Klein, C. E., eds. *Crustaceans and the Biodiversity Crisis*. Vol. 1. Koninklijke Brill NV, Leiden, The Netherlands, 753–804.

Hasle, G. R. 1972. The distribution of *Nitzschia seriata* Cleve and allied species. *Beiheft zur Nova Hedwigia* 39:171–190.

Hasle, G. R. 1995. *Pseudo-nitzschia pungens* and *P. multiseries* (Bcillariophyceae): Nomenclatural history, morphology, and distribution. *Journal of Phycology* 31:428–435.

Hasle, G. R., Syvertsen, E. E. 1980. The diatom genus *Cerataulina*: Morphology and taxonomy. *Bacillaria* 3:79–113.

Hatschek, B. 1878. Studien über Entwicklungsgeschichte der Anneliden. *Arbeiten aus den Zoologischen Instituten der Universität Wien und der Zoologischen Station in Triest* 3:177–404.

Hausmann, K., Patterson, D. J. 1982. Pseudopod formation and membrane production during prey capture by a heliozoon (feeding by *Actinophrys*, II). *Cell Motility* 2:9–24.

Hawser, S. P., O'Neil, M. J., Roman, M. R., et al. 1992. Toxicity of blooms of the cyanobacterium *Trichodesmium* to zooplankton. *Journal of Applied Phycology* 4:79–86.

Head, E. J. H., Harris, L. R., Campbell, R. W. 2000. Investigations on the ecology of *Calanus* spp. in the Labrador Sea. 1. Relationship between the phytoplankton bloom and reproduction and development of *Calanus finmarchicus* in spring. *Marine Ecology Progress Series* 193:3–73.

Heard, R. W. 1982. *Guide to the Tidal Marsh Invertebrates of the Northeast Gulf Of Mexico*. Mississippi-Alabama Sea Grant Cosortium. MASGP-79-004. 82 pp.

Heard, R. W., Roccatagliata, D., Petrescu, I. 2007. *An Illustrated Guide to Cumacea (Crustacea: Malacostraca: Peracarida) from Florida Coastal and Shelf Waters to Depths of 100m*. Florida Department of Environmental Protection, Tallahassee. 175 pp.

Heinbokel, J. F., Coats, D. W., Henderson, K. W., et al. 1988. Reproduction rates and secondary production of three species of the rotifer genus *Synchaeta* in the estuarine Potomac River. *Journal of Plankton Research* 10:659–672.

Hendon, J. R., Mark S., Peterson, M. S., et al. 2000. Spatio-temporal distribution of larval *Gobiosoma bosc* in waters adjacent to natural and altered marsh-edge habitats of Mississippi coastal waters. *Bulletin of Marine Science* 66:143–156.

Henley, C. 1974. Platyhelminthes. In: Giese, A. C., Pearse, J. S., eds. *Reproduction in Marine Invertebrates*. Vol. 1, *Acoecomate and Pseudocoelomate Metazoans*. Academic Press, New York, 267–344.

Herman, S. S. 1963. Vertical migration of the opossum shrimp, *Neomysis americana* Smith. *Limnology and Oceanography* 8:228–238.

Herman, S. S., Coull, B. C., Brickman, L. M. 1971. Infestation of harpacticoid copepods (Crustacea) with ciliate protozoans. *Journal of Invertebrate Pathology* 17:141–142.

Herman, S. S., Mihursky, J. A. 1964. Infestation of the copepod *Acartia tonsa* with the stalked ciliate *Zoothamnium*. *Science* 146:543–544.

Hermans, C. O., Satterlie, R. A. 1992. Fast-strike feeding behaviour in a pteropod mollusk, *Clione limacina* Phipps. *Biological Bulletin* 182:1–7.

Heron, A. C. 1972. Population ecology of a colonizing species: The pelagic tunicate *Thalia democratica*. 1. Individual growth rate and generation time. *Oecologia* 10:269–293.

Herrick, F. H. 1978/1911. Natural history of the American lobster. Arno Press, New York. (Reprint of the classic 1911 publication from *Bulletin of the US Bureau of Fisheries* 29:149–408.)

Herrnkind, W. F. 1968. The breeding of *Uca pugilator* (Bosc) and mass rearing of the larvae with comments on the behavior of the larval and early crab stages (Brachyura, Ocypodidae). *Crustaceana*. Suppl. 2, *Studies on Decapod Larval Development*, 214–224.

Hettler, W. F., Jr. 1984. Description of eggs, larvae and early juveniles of gulf menhaden, *Brevoortia patronus,* and comparisons with Atlantic menhaden, *Brevoortia tyrannus,* and yellowfin menhaden, *Brevoortia smithi*. *Fishery Bulletin* 82:85–95.

Hildebrand, S. F., Cable, L. E. 1934. Reproduction and development of whitings or kingfishes, drums, spot, croaker, and weakfishes or seatrouts family Sciaenidae of the Atlantic coast of the United States. *Bulletin of the US Bureau of Fisheries* 48:41–117.

Hildebrand, S. F., Cable, L. E. 1938. Further notes on the development and life history of some teleosts at Beaufort, N.C. *US Bureau of Fisheries Bulletin* 48:505–642.

Hiller, A., Kraus, H., Almon, M., et al. 2006. The *Petrolisthes galathinus* complex: Species boundaries based on color pattern, morphology and molecules, and evolutionary interrelationships between this complex and other Porcellanidae (Crustacea: Decapoda: Anomura). *Molecular Phylogenetics and Evolution* 40:547–569.

Hlawa, S., Heerkloss, R. 1995. Feeding biology of two brachionoid rotifers: *Brachionus quadridentalis* and *Brachionus plicatilis*. *Hydrobiologia* 313/314:219–221.

Holt, G. J., Holt, S. A. 2000. Vertical distribution and the role of physical processes in the feeding dynamics of two larval sciaenids *Sciaenops ocellatus* and *Cynoscion nebulosus*. *Marine Ecology Progress Series* 193:181–190.

Holt, J., Strawn, K. 1983. Community structure of macrozooplankton in Trinity and Upper Galveston Bays. *Estuaries* 6:66–75.

Hood, M. R. 1962. Studies on the larval development of *Rhithropanopeus harrisii* (Gould) of the family Xanthidae (Brachyura). *Gulf Research Reports* 1:122–130.

Houde, E. D., Schekter, R. C. 1980. Feeding by marine fish larvae: Developmental and functional responses. *Environmental Biology of Fishes* 5:315–334.

Hough, A. R., Naylor, E. 1991. Field studies on retention of the planktonic copepod *Eurytemora affinis* in a mixed estuary. *Marine Ecology Progress Series* 76:115–122.

Hovel, K. A., Morgan, S. G. 1997. Planktivory as a selective force for reproductive synchrony and larval migration. *Marine Ecology Progress Series* 157:79–95.

Hudon, C. 1983. Selection of unicellular algae by the littoral amphipods *Gammarus oceanicus* and *Calliopius laeviusculus* (Crustacea). *Marine Biology* 78:59–67.

Hughes, D. A., Richard, J. D. 1973. Some current-directed movements of *Macrobrachium acanthurus* (Wiegmann 1836) (Decapoda, Palaemonidae) under laboratory conditions. *Ecology* 54:927–929.

Hulburt, E. M. 1957a. The distribution of *Neomysis americana* in the estuary of the Delaware River. *Limnology and Oceanography* 2:1–11.

Hulburt, E. M. 1957b. The taxonomy of unarmored Dinophyceae of shallow embayments on Cape Cod, Massachusetts. *Biological Bulletin* 112:196–219.

Hwang, J.-S., Strickler, R. 2001. Can copepods differentiate prey from predator hydromechanically? *Zoological Studies* 40:1–6.

Hwang, J.-S., Turner, J. T., Costello, J. H., et al. 1993. A cinematographic comparison of behavior by the calanoid copepod *Centropages hamatus* Lilljeborg: Tethered versus free-swimming animals. *Journal of Experimental Marine Biology and Ecology* 167:277–288.

Hwang, S. G., Lee, C., Kim, C. H. 1993. Complete larval development of *Hemigrapsus sanguineus* (Decapoda, Brachyura, Grapsidae) reared in the laboratory. *Korean Journal of Systematic Zoology* 9(2):69–86.

Ianora, A., Santella, L. 1991. Diapause embryos in the neustonic copepod *Anomalocera patersoni*. *Marine Biology* 108:387–394.

Itoh, H., Nishida, S. 1995. Copepodid stages of *Hemicyclops japonicus* Itoh and Nishida (Poecilostomatoida: Clausidiidae) reared in the laboratory. *Journal of Crustacean Biology* 15:134–155.

Jackson, D. J., MacMillan, D. L. 2000. Tailflick escape behavior in larval and juvenile lobsters (*Homarus americanus*) and crayfish (*Cherax destructor*). *Biological Bulletin* 198:307–318.

Jacobs, J. 1961. Laboratory cultivation of the marine copepod *Pseudodiaptomus coronatus* Williams. *Chesapeake Science* 6:443–446.

Jagger, R. A., Kimmerer, W. J., Jenkins, G. P. 1988. Food of the cladoceran *Podon intermedius* in a marine embayment. *Marine Ecology Progress Series* 43:245–250.

Jahn, T. L., Jahn, F. F. 1949. *How to Know the Protozoa*. William C. Brown, Dubuque, IA. 234 pp.

Jakobsen, H. H., Halvorsen, E., Hansen, B. W., et al. 2005. Effects of prey motility and concentration on feeding in *Acartia tonsa* and *Temora longicornis*: The importance of feeding modes. *Journal of Plankton Research* 27:775–785.

Janicki, A. J., Jr., DeCosta, J. 1990. An analysis of prey selection by *Mesocyclops edax*. *Hydrobiologia* 198:133–139.

Janson, S., Siddiqui, P. J. A., Walsby, A. E., et al. 1995. Cytomorphological characterization of the planktonic diazotrophic cyanobacteria *Trichodesmium* spp. from Indian Ocean and Caribbean and Sargasso Seas. *Journal of Phycology* 31:463–477.

Jeffries, H. P. 1962. Salinity-space distribution of the estuarine copepod genus *Eurytemora*. *International Revue für Gesamten Hydrobiologie* 47:291–300.

Jenkins, R. L. 1983. Observations on the commensal relationship of *Nomeus gronovii* with *Physalia physalis*. *Copeia* 1983:250–252.

Jeong, H. J. 1994a. Predation effect of the calanoid copepod *Acartia tonsa* on a population of the heterotrophic dinoflagellate *Protoperidinium* cf. *divergens* in the presence of co-occurring red-tide dinoflagellate prey. *Marine Ecology Progress Series* 111:87–97.

Jeong, H. J. 1994b. Predation by the heterotrophic dinoflagellate *Protoperidinium* cf. *divergins* on copepod eggs and early naupliar stages. *Marine Ecology Progress Series* 114:203–208.

Jeong, H. J., Kim, S. K., Kim, J. S., et al. 2001. Growth and grazing rates of the heterotrophic dinoflagellate *Polykrikos kofoidii* on red-tide and toxic dinoflagellates. *Journal of Eukaryotic Microbiology* 48:298–308.

Jiang, M., Brown, M. W., Turner, J. T., et al. 2007. Springtime transport and retention of *Calanus finmarchicus* in Massachusetts and Cape Cod Bays, USA, and implications for right whale foraging. *Marine Ecology Progress Series* 349:183–197.

Jiménez-Cueto, S., Suárez-Morales, E. 1999. Tomopterids (Polychaeta: Tomopteridae) of the western Caribbean Sea. *Bulletin de l'Institut Royal des Sciences Naturelles de Belgique Biologie* 69:5–14.

Johansson, M., Coats, D. W. 2002. Ciliate grazing on the parasite *Amoebophrya* sp. decreases infection of the red-tide dinoflagellate *Akashiwo sanguinea*. *Aquatic Microbial Ecology* 28:69–78.

Johns, D. M., Lang, W. H. 1977. Larval development of the spider crab, *Libinia emarginata* (Majidae). *Fishery Bulletin* 75:831–841.

Johnson, D. R., Perry, J. M. 1999. Blue crab larval dispersion and retention in the Mississippi Bight. *Bulletin of Marine Science* 65:129–149.

Johnson, G. D. 1978. *Development of Fishes of the Mid-Atlantic Bight: An Atlas of Egg, Larval, And Juvenile Stages*. Vol. 4, *Carangidae through Ephippidae*. Biological Services Program. US Fish and Wildlife Service. FWS/OBS-78/12.

Johnson, M. D., Oldach, D., Delwiche, C. F., et al. 2007. Retention of transcriptionally active cryptophyte nuclei by the ciliate *Myrionecta rubra*. *Nature* 445:426–428.

Johnson, M. D., Rome, M., Stoecker, D. K. 2003. Microzooplankton grazing on *Prorocentrum minimum* and *Karlodinium micrum* in Chesapeake Bay. *Limnology and Oceanography* 48:238–248.

Johnson, M. D., Stoecker, D. K. 2005. The role of feeding in growth and the photophysiology of *Myrionecta rubra*. *Aquatic Microbial Ecology* 39:303–312.

Johnson, M. W. 1939. The correlation of water movement and dispersal of planktonic larval stages of certain littoral animals, especially the sand crab, *Emerita*. *Journal of Marine Research* 2:236–245.

Johnson, W. S., Allen, D. M., Ogburn, M. V., et al. 1990. Short-term predation responses of adult bay anchovies *Anchoa mitchilli* to estuarine zooplankton availability. *Marine Ecology Progress Series* 64:55–68.

Juinio, M. R., Cobb, J. S. 1992. Natural diet and feeding habits of the postlarval lobster *Homarus americanus*. *Marine Ecology Progress Series* 85:83–91.

June, F. C., Carlson, F. T. 1971. Food of young Atlantic menhaden, *Brevoortia tyrannus,* in relation to metamorphosis. *Fishery Bulletin* 68:493–512.

Kankaala, P. 1983. Resting eggs, seasonal dynamics, and production of *Bosmina longispina maritima* (P. E. Müller) (Cladocera) in the northern Baltic proper. *Journal of Plankton Research* 5:53–69.

Katechakis, A., Stibor, H. 2004. Feeding selectivities of the marine cladocerans *Penilia avirostris, Podon intermedius* and *Evadne nordmanni*. *Marine Biology* 145:529–539.

Katona, S. K. 1971. The developmental stages of *Eurytemora affinis* (Poppe, 1880) (Copepoda, Calanoida) raised in laboratory cultures, including a comparison with the larvae of *Eurytemora americana* Williams, 1906, and *Eurytemora herdmani* Thompson and Scott, 1897. *Crustaceana* 21:5–20.

Katz, C. H., Cobb, J. S., Spaulding, M. 1994. Larval behavior, hydrodynamic transport, and potential offshore-to-inshore recruitment in the American lobster *Homarus americanus*. *Marine Ecology Progress Series* 103:265–273.

Keefe, M., Able, K. W. 1993. Patterns of metamorphosis in summer flounder, *Paralichthys dentatus*. *Journal of Fish Biology* 42:713–728.

Keefe, M., Able, K. W. 1994. Contributions of abiotic and biotic factors to settlement in summer flounder, *Paralichthys dentatus*. *Copeia* 1994:458–465.

Kelly, K. H., Moring, J. R. 1986. *Species Profiles: Life Histories and Environmental Requirements of Coastal Fishes and Invertebrates (North Atlantic)—Atlantic Herring*. US Fish and Wildlife Biological Report 82(11.38). 22 pp.

Kendall, A. W., Jr., Walford, L. A. 1979. Sources and distribution of bluefish, *Pomatomus saltatrix*, larvae and juveniles off the east coast of the United States. *Fishery Bulletin of the US* 77:213–227.

Kennedy, V. S. 1996. Biology of larvae and spat. In: Kennedy, V. S., Newell, R. I. E., Eble, A. F., eds. *The Eastern Oyster, Crassostrea virginica*. Maryland Sea Grant, College Park, MD, 371–421.

Kensley, B., Schotte, M. 1999. *Guide to the Isopod Crustaceans of the Caribbean*. Smithsonian Institution Press. Washington, DC. 308 pp.

Kim, I. H., Ho, J. S. 1992. Copepodid stages of *Hemicyclops ctenidis* Ho and Kim, 1990 (Clausidiidae), a poecilostomatiod copepod associated with a polychaete. *Journal of Crustacean Biology* 12:631–646.

Kim, J. -S., Jeong, H.-J. 2004. Feeding by the heterotrophic dinoflagellates *Gyrodinium dominans* and *G. spirale* on the red-tide dinoflagellate *Prorocentrum minimum*. *Marine Ecology Progress Series* 280:85–94.

Kimmel, D. G., Miller, W. D., Roman, M. R. 2006. Regional climate forcing of mesozooplankton dynamics in Chesapeake Bay. *Estuaries and Coasts* 2:375–387.

Kiørboe, T., Saiz, E., Visser, A. W. 1999. Hydrodynamic signal perception in the copepod *Acartia tonsa*. *Marine Ecology Progress Series* 179:97–111.

Kiørboe, T., Titelman, J. 1998. Feeding, prey selection and prey encounter mechanisms in the heterotrophic dinoflagellate *Noctiluca scintillans*. *Journal of Plankton Research* 20:1615–1636.

Kivi, K., Setälä, O. 1995. Simultaneous measurement of food particle selection and clearance rates of planktonic oligotrich ciliates (Ciliophora: Oligotrichina). *Marine Ecology Progress Series* 119:125–137.

Kjelson, M. A., Peters, D. S., Thayer, G. W., et al. 1975. The general feeding ecology of postlarval fishes in the Newport River Estuary. *Fishery Bulletin* 73:137–144.

Knowlton, R. E. 1973. Larval development of the snapping shrimp *Alpheus heterochaelis* Say, reared in the laboratory. *Journal of Natural History* 7:273–306.

Koehl, M. A. R., Strickler, J. R. 1981. Copepod feeding currents: Food capture at low Reynolds number. *Limnology and Oceanography* 26:1062–1073.

Kofoid, C. A., Swezy, O. 1921. The free-living unarmored dinoflagellates. *Memoirs of the University of California* 5:1–562.

Kokinos, J. P., Anderson, D. M. 1995. Morphological development of resting cysts in cultures of the marine dinoflagellate *Lingulodinium polyedrum* (=*L. machaerophorum*). *Palynology* 19:143–166.

Kolesar, S. E., Breitburg, D. L., Purcell, J. E., et al. 2010. Effects of hypoxia on *Mnemiopsis leidyi*, ichthyoplankton and copepods: Clearance rates and vertical habitat overlap. *Marine Ecology Progress Series* 411:173–188.

Kopin, C. Y., Epifanio, C. E., Nelson, S., et al. 2001. Effects of chemical cues on metamorphosis of the Asian shore crab *Hemigrapsus sanguineus,* an invasive species of the Atlantic Coast of North America. *Journal of Experimental Marine Biology and Ecology* 265:141–151.

Koski, R. T. 1978. Age, growth, and maturity of the hogchoker, *Trinectes maculatus,* in the Hudson River, New York. *Transactions of the American Fisheries Society* 107:449–453.

Kraeuter, J. N., Setzler, E. M. 1975. The seasonal cycle of Scyphozoa and Cubozoa in Georgia estuaries. *Bulletin of Marine Science* 25:66–74.

Kramp, P. L. 1959. The Hydromedusae of the Atlantic Ocean and adjacent waters. *Dana Reports* 46:1–283.

Kremer, P., Madin, L. P. 1992. Particle retention efficiency of salps. *Journal of Plankton Research* 14:1009–1015.

Kreps, T. A., Purcell, J. E., Heidelberg, K. B. 1997. Escape of the ctenophore *Mnemiopsis leidyi* from the scyphozoan predator *Chrysaora quinquecirrha. Marine Biology* 128:441–446.

Krimsky, L. S., Gravinese, P. M., Tankersley, R. A., et al. 2009. Patterns of larval release in the Florida stone crab, *Menippe mercenaria. Journal of Experimental Marine Biology and Ecology* 373:96–101.

Kruczynski, W. L., Subrahmanyam, C. B. 1978. Distribution and breeding cycle of *Cyathura polita* (Isopoda: Anthuridae) in a *Juncus roemerianus* marsh of Northern Florida. *Estuaries* 1:93–100.

Kurata, H. 1970. *Larvae of Decapod Crustacea of Georgia.* Technical Report. Parts 1 & 2. University of Georgia Marine Institute, Sapelo Island. 274 pp.

LaFrance, K., Ruber, E. 1985. The life cycle and productivity of the amphipod *Gammarus mucronatus* on a northern Massachusetts salt marsh. *Limnology and Oceanography* 30:1067–1077.

Lagersson, N. C., Hoeg, J. T. 2002. Settlement behavior and antennulary biomechanics in cypris larvae of *Balanus amphitrite* (Crustacea: Cirripedia). *Marine Biology* 141:513–526.

Lalli, C. M. 1970. Structure and functioning of the buccal apparatus of *Clione limacina* (Phipps) with a review of feeding in gymnosomatous pteropods. *Journal of Experimental Marine Biology and Ecology* 4:101–118.

Lalli, C. M., Gilmer, R. W. 1989. *Pelagic Snails: The Biology of Holoplanktonic Gastropod Mollusks.* Stanford University Press, Palo Alto, CA. 259 pp.

Landry, M. R., Fagerness, V. L. 1988. Behavioral and morphological influences on predatory interactions among marine copepods. *Bulletin of Marine Science* 43:509–529.

Lang, E. T., Peterson, E. T., Slack, W. T. 2011. Comparative development of five sympatric coastal fundulid species from the northern Gulf of Mexico. *Zootaxa* 2901:1–18.

Lang, W. H. 1979. *Larval Development of Shallow Water Barnacles of the Carolinas (Cirripedia: Thoracica) with Keys to Naupliar Stages*. NOAA Technical Report NMFS Circular 421. 39 pp.

Larsen, J., Moestrup, Ø. 1989. *Guide to Toxic and Potentially Toxic Marine Algae*. Fish Inspection Service, Ministry of Fisheries, Copenhagen. 61 pp.

Larson, R. J. 1986. The feeding and growth of the sea nettle, *Chrysaora quinquecirrha* (Desor), in the laboratory. *Estuaries* 9:376–379.

Larson, R. J. 1987. A note on the feeding, growth, and reproduction of the epipelagic scyphomedusa *Pelagia noctiluca* (Forskal). *Biological Oceanography* 4:447–454.

Larson, R. J. 1988. Feeding and functional morphology of the lobate ctenophore *Mnemiopsis mccradyi*. *Estuarine, Coastal and Shelf Science* 27:495–502.

Larson, R. J. 1991. Diet, prey selection and daily ration of *Stomolophus meleagris,* a filter feeding scyphomedusa from NE Gulf of Mexico. *Estuarine, Coastal and Shelf Science* 332:511–525.

Larson, S. C., Van Den Avgyle, M. J., Bozeman, E. L., Jr. 1989. *Species Profiles: Life Histories and Environmental Requirements of Coastal Fishes and Invertebrates (South Atlantic)—Brown Shrimp*. US Fish and Wildlife Service Biological Report 82(11.90). 14 pp.

LaSalle, M. W., de la Cruz, A. A. 1985. *Species Profiles: Life Histories and Environmental Requirements of Coastal Fishes and Invertebrates (Gulf of Mexico)—Common Rangia*. US Fish and Wildlife Biological Report 82 (11.31). 18 pp.

Lassuy, D. R. 1983a. *Species Profiles: Life Histories and Environmental Requirements (Gulf of Mexico)—Brown Shrimp*. US Fish and Wildlife Publication FWS/OBS-82/11.1. 15 pp.

Lassuy, D. R. 1983b. *Species Profiles: Life histories and Environmental Requirements of Coastal Fishes and Invertebrates (Gulf of Mexico)—Gulf Menhaden*. US Fish and Wildlife Publication FWS/OBS-82/11.2. 13 pp.

Lassuy, D. R. 1983c. *Species Profiles: Life Histories and Environmental Requirements of Coastal Fishes and Invertebrates (Gulf of Mexico)—Atlantic Croaker*. US Fish and Wildlife Publication FWS/OBS-82/11.3. 12 pp.

Lassuy, D. R. 1983d. *Species Profiles: Life Histories and Environmental Requirements of Coastal Fishes and Invertebrates (Gulf of Mexico)—Spotted Seatrout*. US Fish and Wildlife Service Biological Report FWS/OBS-82/11.4. 14 pp.

Latz, M. I., Jeong, H. L. 1996. Effect of red tide dinoflagellate diet on the bioluminescence of *Protoperidinium* spp. *Marine Ecology Progress Series* 132:275–285.

Laughlin, R. A., Livingston, R. J. 1982. Environmental and trophic determinants of the spatial/temporal distribution of the brief squid (*Lolliguncula brevis*) in the Apalachicola estuary (North Florida, USA). *Bulletin of Marine Science* 32:489–497.

Laval, P. 1980. Hyperiid amphipods as crustacean parasitoids associated with gelatinous zooplankton. *Oceanography and Marine Biology: An Annual Review* 18:11–56.

Laval-Peuto, M. 1981. Construction of the lorica in Ciliata Tintinnina. In vivo study of *Favella ehrenbergii:* Variability of the phenotypes during the cycle, biology, statistics, biometry. *Protistologica* 17:249–272.

Lawrence, D., Valiela, I., Tomasky, G. 2004. Estuarine calanoid copepod abundance in relation to season, salinity, and land-derived nitrogen loading, Waquoit Bay, MA. *Estuarine, Coastal and Shelf Science* 61:547–557.

Lawson, T. S., Grice, G. D. 1973. The developmental stages of *Paracalanus crassirostris* Dahl, 1894 (Copepoda, Calanoida). *Crustaceana* 24:43–56.

Lebour, M. V. 1928. The larval stages of the Plymouth Brachyura. *Proceedings of the Zoological Society of London* 32:473–560.

Lee, J. J., Hugo, D., Freudenthal, H. D., et al. 2007. Cytological observations on two planktonic Foraminifera, *Globigerina bulloides* d'Orbigny, 1826, and *Globigerinoides ruber* (d'Orbigny, 1839) Cushman, 1927. *Journal of Eukaryotic Microbiology* 12:531–542.

Lee, J. J., Soldo, A. T., eds. 1992. *Protocols in Protozoology*. Society of Protozoology, Lawrence, KS. 240 pp.

Lee, W. Y. 1978. The cyclopoid copepods, *Hemicyclops adhaerens* and *Saphirella* sp., in the Damariscotta River Estuary, Maine, with a note to their possible relationship. *Estuaries* 1:200–202.

Leiby, M. M. 1979. Morphological development of the eel, *Myrophis punctatus* (Ophichthidae) from hatching to metamorphosis, with emphasis on the developing head skeleton. *Bulletin of Marine Science* 29:509–521.

Lekan, D. K., Tomas, C. R. 2010. The brevetoxin and brevenal composition of three *Karenia brevis* clones at different salinities and nutrient conditions. *Harmful Algae* 9:39–47.

Lenz, P. H., Hower, A. E., Hartline, D. K. 2004. Force production during pereiopod power strokes in *Calanus finmarchicus*. *Journal of Marine Systems* 49:133–144.

Lester, K. M., Heil, C., Neeley, M. B., et al. 2008. Zooplankton and *Karenia brevis* in the Gulf of Mexico. *Continental Shelf Research* 28:99–111.

Leverone, J. R., Shumway, S. E., Blake, N. J. 2007. Comparative effects of the toxic dinoflagellate *Karenia brevis* on clearance rates in juveniles of four bivalve molluscs from Florida, USA. *Toxicon* 49:634–645.

Levin, L. A., Caswell, H., DePatra, K. D., et al. 1987. Demographic consequences of larval development mode: Planktotrophy vs. lecithotrophy in *Streblospio benedicti*. *Ecology* 68:1877–1886.

Levin, L. A., Creed, E. L. 1986. Effect of temperature and food availability on reproductive responses of *Streblospio benedicti* (Polychaeta: Spionidae) with planktotrophic or lecithotrophic development. *Marine Biology* 92:103–113.

Levine, D. M., Sulkin, S. D. 1984. Ingestion and assimilation of microencapsulated diets by brachyuran crab larvae. *Marine Biology Letters* 5:147–153.

Lewis, J. B. 1951. The phyllosoma larvae of the spiny lobster *Panulirus argus*. *Bulletin of Marine Science of the Gulf and Caribbean* 1:89–103.

Li, A., Stoecker, D. K., Coats, D. W. 2000a. Spatial and temporal aspects of *Gyrodinium galatheanum* in Chesapeake Bay: Distribution and mixotrophy. *Journal of Plankton Research* 22:2105–2124.

Li, A., Stoecker, D. K., Coats, D. W. 2000b. Mixotrophy in *Gyrodinium galatheanum* (Dinophyceae): Grazing responses to light intensity and inorganic nutrients. *Journal of Phycology* 36:33–45.

Li, Y., Swift, E., Buskey, E. J. 1996. Photoinhibition of mechanically stimulable bioluminescence in the heterotrophic dinoflagellate *Protoperidinium depressum*. *Journal of Phycology* 32:974–982.

Limburg, K. E., Pace, M. L., Arend, K. K. 1999. Growth, mortality, and recruitment of larval *Morone* spp. in relation to food availability and temperature in the Hudson River. *Fishery Bulletin* 97:80–91.

Limburg, K. E., Pace, M. L., Fischer, D., et al. 1997. Consumption, selectivity, and use of zooplankton by larval striped bass and white perch in a seasonally pulsed estuary. *Transactions of the American Fisheries Society* 126:607–621.

Lin, J., Zhang, D. 2001. Reproduction in a simultaneous hermaphroditic shrimp, *Lysmata wurdemanni:* Any two will do? *Marine Biology* 139:919–922.

Lindholm, T. 1985. *Mesodinium rubrum:* A unique photosynthetic ciliate. *Advances in Aquatic Microbiology* 3:1–48.

Lindner, A., Migotto, A. E. 2002. The life cycle of *Clytia linearis* and *Clytia noliformis*: Metagenic campanulariids (Cnidaria: Hydrozoa) with contrasting polyp and medusa stages. *Journal of the Marine Biological Association of the UK* 82:541–553.

Lindsay, J. A., Radle, E. R., Wang, J. C. S. 1978. A supplemental sampling method for estuarine ichthyo-plankton with emphasis on the Atherinidae. *Estuaries* 1:61–64.

Lippson, A. J., Haire, M. S., Holland, A. F., et al. 1979. *Environmental Atlas of the Potomac Estuary.* Martin Marietta Corporation for Maryland Power Plant Siting Program, Annapolis, MD. 279 pp.

Lippson, A. J., Moran, R. L. 1974. *Manual for Identification of Early Developmental Stages of Fishes of the Potomac River Estuary.* Report PPSP-MP-13. Martin Marietta Corporation, Baltimore. 282 pp.

Litaker, R. W., Tester, P. A., Duke, C. S., et al. 2002. Seasonal niche strategy of the bloom-forming dinoflagellate *Heterocapsa triqueta. Marine Ecology Progress Series* 232:45–62.

Little, K. T., Epifanio, C. E. 1991. Mechanism for the re-invasion of an estuary by two species of brachyuran megalopae. *Marine Ecology Progress Series* 68:235–242.

Lochhead, J. H. 1936. On the feeding mechanism of a ctenopod cladoceran, *Penilia avirostris. Proceedings of the Zoological Society of London* 1936:335–355.

Lochhead, J. H. 1954. On the distribution of a marine cladoceran, *Penilia avirostris* Dana (Crustacea, Branchiopoda), with a note on its reported bioluminescence. *Biological Bulletin* 107:92–105.

Lonsdale, D. J., Hassett, R. P., Dobbs, F. C., et al. 1998. Physiological traits associated with a reproductive-resting stage in *Coullana canadensis* (Copepoda: Harpacticoida). *Marine Biology* 131:123–131.

Lonsdale, D. J., Levinton, J. S. 1985. Latitudinal differentiation in embryonic duration, egg size, and newborn survival in a harpacticoid copepod. *Biological Bulletin* 168:419–431.

Lopez, H. G., Peterson, M. S., Walker, J., et al. 2011. Distribution, abundance, and habitat characterization of the saltmarsh topminnow, *Fundulus jenkinsi* (Everman 1892). *Estuaries and Coasts* 34:148–158.

López-Urrutia, Á., Harris, R. P., Smith, T. 2004. Predation by calanoid copepods on the appendicularian *Oikopleura dioica. Limnology and Oceanography* 49:303–307.

Lozano, C. 2011. Dynamics of ingress, hatch dates, growth, and feeding of Atlantic menhaden, *Brevoortia tyrannus*, larvae at the Chesapeake Bay mouth. Master's thesis, University of Maryland, College Park. 175 pp.

Lucas, C. H. 2001. Reproduction and life history strategies of *Aurelia aurita* in relation to its ambient environment. In: Purcell, J. E., Graham, W. M., Dumont, H. J., eds. Jellyfish blooms: Ecological and societal importance. *Hydrobiologia* 451(1–3) (*Developments in Hydrobiology*) 155:229–246.

Luckenbach, M. W., Orth, R. J. 1990. A chemical defense in Crustacea? *Journal of Experimental Marine Biology and Ecology* 137:79–87.

Luckenbach, M. W., Orth, R. J. 1992. Swimming velocities and behavior of blue crab (*Callinectes sapidus* Rathbun) megalopae in still and flowing water. *Estuaries* 15:186–192.

Lyczkowski-Shultz, J., Ruple, D. L., Richardson, S. L., et al. 1990. Distribution of fish larvae relative to time and tide in a Gulf of Mexico barrier island pass. *Bulletin of Marine Science* 46:563–577.

Maar, M., Visser, A., Nielsen, T., et al. 2006. Turbulence and feeding behaviour affect the vertical distributions of *Oithona similis* and *Microsetella norwegica. Marine Ecology Progress Series* 313:157–172.

MacGregor, J. M., Houde, E. D. 1996. Onshore-offshore pattern and variability in distribution and abundance of bay anchovy *Anchoa mitchilli* eggs and larvae in Chesapeake Bay. *Marine Ecology Progress Series* 138:15–25.

MacKenzie, B. R., Kiørboe, T. 1995. Encounter rates and swimming behavior of the pause-travel and cruise larval fish predators in calm and turbulent laboratory environments. *Limnology and Oceanography* 40:1278–1289.

Mackie, G. O., Marx, R. M., Meech, R. W. 2003. Central circuitry of the jellyfish *Aglantha digitale*. IV. Pathways coordinating feeding behaviour. *Journal of Experimental Biology* 206:2487–2505.

Madin, L. P. 1974. Field observations on the feeding behavior of salps (Tunicata: Thaliacea). *Marine Biology* 25:143–147.

Madin, L. P. 1990. Aspects of the jet propulsion in salps. *Canadian Journal of Zoology* 68:765–777.

Madin, L. P., Bollens, S. M., Horgan, E., et al. 1996. Voracious planktonic hydroids: Unexpected predatory impact on a coastal marine ecosystem. *Deep-Sea Research* 43:1823–1829.

Madin, L. P., Harbison, G. R. 1977. The association of Hyperiidea with gelatinous zooplankton. 1. Associations with Salpidae. *Deep-Sea Research* 24:449–463.

Maki, J. S., Rittschof, D., Schmidt, A. R., et al. 1989. Factors controlling attachment of bryozoan larvae: A comparison of bacterial films and unfilmed surfaces. *Biological Bulletin* 177:295–335.

Mallin, M. A. 1991. Zooplankton abundance and community structure in a mesohaline North Carolina estuary. *Estuaries* 14:481–488.

Mansueti, R. 1963. Symbiotic behavior between small fishes and jellyfishes, with new data on that between the stromateid, *Peprilus alepidotus* and the scyphomedusa, *Chrysaora quinquecirrha*. *Copeia* 1963(1):40–80.

Mansueti, R. J. 1964. Eggs, larvae and young of the white perch, *Roccus americanus,* with comments on its ecology in the estuary. *Chesapeake Science* 5:3–45.

Maris, R. C. 1983. A key to the porcellanid crab zoeae (Crustacea: Decapoda: Anomura) of the north central Gulf of Mexico and a comparison of meristic characters of four species. *Gulf Research Reports* 7:237–246.

Markle, D. F. 1984. Phosphate buffered formalin for long term preservation of formalin fixed ichthyoplankton. *Copeia* 1984:525–528.

Marks, R. E., Juanes, F., Hare, J. A., et al. 1996. Occurrence and effect of the parasitic isopod *Lironeca ovalis* (Isopods, Cymothoidae) on young-of-the-year bluefish (*Pomatomus saltatrix*) (Pices: Pomatomidae). *Canadian Journal of Fisheries and Aquatic Science* 53:2052–2057.

Marshall, S. M., Orr, A. P. 1955. *The Biology of a Marine Copepod Calanus finmarchicus (Gunnerus).* Oliver & Boyd, Edinburgh. 188 pp.

Marshalonis, D., Pinckney, J. L. 2008. Grazing and assimilation rate estimates of hydromedusae from a temperate tidal creek system. *Hydrobiologia* 606:203–211.

Martin, R. B., Truesdale, F. M., Felder, D. L. 1988. The megalopa stage of the Gulf stone crab, *Menippe adina* Williams and Felder, 1986, with a comparison of megalopae in the genus *Menippe*. *Fishery Bulletin* 86:289–297.

Martindale, M. Q. 2002. Phylum Ctenophora. In: Young, C. M., ed. *Atlas of Marine Invertebrate Larvae*. Academic Press, New York, 107–122.

Matanoski, J. C., Hood, R. R. 2006. An individual-based numerical model of medusa swimming behavior. *Marine Biology* 149:595–608.

Matanoski, J. C., Hood, R. R., Purcell, J. E. 2001. Characterizing the effect of prey on swimming and feeding efficiency of the scyphomedusa *Chrysaora quinquecirrha*. *Marine Biology* 139:191–200.

Matsuoka, K., Cho, H.-J., Jacobson, D. M. 2000. Observations of the feeding behavior and growth rates of the heterotrophic dinoflagellate *Polykrikos kofoidii* (Polykrikaceae, Dinophyceae). *Phycologia* 39:82–86.

Matsuyama, Y., Miyamoto, M., Kotani, Y. 1999. Grazing impacts of the heterotrophic dinoflagellate *Polykrikos kofoidii* on a bloom of *Gymnodinium catenatum*. *Aquatic Microbial Ecology* 17:91–98.

Mayer, A. G. 1912. Ctenophores of the Atlantic coast of North America. *Publications of the Carnegie Institution of Washington* 162:1–58.

McBride, R. S., MacDonald, T. C., Matheson, R. E., Jr., et al. 2001. Nursery habitats for ladyfish, *Elops saurus*, along salinity gradients in two Florida estuaries. *Fishery Bulletin* 99:443–458.

McCleave, J. D., Kleckner, R. C. 1982. Selective tidal stream transport in the estuarine migration of glass eels of the American eel (*Anguilla rostrata*). *Journal du Conseil International pour l'Exploration de la Mer* 40:262–271.

McConaugha, J. 2002. Alternate feeding mechanisms in megalopae of the blue crab *Callinectes sapidus*. *Marine Biology* 140:1227–1233.

McDermott, J. J. 2005. Biology of the brachyuran crab *Pinnixa chaetopterana* Stimpson (Decapoda: Pinnotheridae) symbiotic with tubicolous polychaetes along the Atlantic coast of the United States, with additional notes on other polychaete associations. *Proceedings of the Biological Society of Washington* 118:742–764.

McDermott, J. J. 2009. Notes on the unusual megalopae of the ghost crab *Ocypode quadrata* and related species (Decapoda: Brachyura: Ocypodidae). *Northeastern Naturalist* 16:637–646.

McDermott, J. J., Zubkoff, P., Lin, A. L. 1982. The occurrence of the anemone *Peachia parasitica* as a symbiont in the scyphozoan *Cyanea capillata* in the lower Chesapeake Bay. *Estuaries* 5:319–321.

McEdward, L. R., Strathmann, R. R. 1987. The body plan of the cyphonautes larva of bryozoans prevents high clearance rates: Comparison with pluteus and a growth model. *Biological Bulletin* 172:30–45.

McGillicuddy, D., Bucklin, A. 2002. Intermingling of two *Pseudocalanus* species on Georges Bank. *Journal of Marine Research* 60:583–604.

McGovern, J., Olney, J. 1996. Factors affecting survival of early life stages and subsequent recruitment of striped bass in the Pamunkey River, Virginia. *Canadian Journal of Fisheries and Aquatic Science* 53:1713–1726.

McManus, G. B., Zhang, H. Lin, S. 2004. Marine planktonic ciliates that prey on macroalgae and enslave their chloroplasts. *Limnology and Oceanography* 49:308–313.

McMichael, R. H., Jr., Peters, K. M. 1989. Early life history of spotted seatrout, *Cynoscion nebulosus* (Pices: Sciaenidae), in Tampa Bay, Florida. *Estuaries* 12:98–110.

Mense, D. J., Wenner, E. L. 1989. Distribution and abundance of early life history stages of the blue crab, *Callinectes sapidus,* in tidal marsh creeks near Charleston, South Carolina. *Estuaries* 12:157–168.

Menzies, R. J., Frankenberg, D. 1966. *Handbook on the Common Marine Isopod Crustacea of Georgia*. University of Georgia Press, Athens. 93 pp.

Mercer, L. P. 1989. *Species Profiles: Life Histories and Environmental Requirements of Coastal Fishes and Invertebrates (Mid-Atlantic)—Weakfish*. US Fish and Wildlife Service Biological Report FWS/OBS-82/11.109. 17 pp.

Merrell, J. R., Stoecker, D. K. 1998. Differential grazing on protozoan microplankton by developmental stages of the calanoid copepod *Eurytemora affinis* Poppe. *Journal of Plankton Research* 20:289–304.

Meyer-Harms, B., Irigoien, X., Head, R., et al. 1999. Selective feeding on natural phytoplankton by *Calanus finmarchicus* before, during, and after the 1997 spring bloom in the Norwegian Sea. *Limnology and Oceanography* 44:154–165.

Mianzan, H. W., Cornelius, P. F. S. 1999. Cubomedusae and Scyphomedusae. In: Boltovskoy, D., ed. *South Atlantic Zooplankton*. Backhuys, Leiden, The Netherlands, 513–559.

Michalec, F.-G., Souissi, A., Dur, G., et al. 2010. Differences in behavioral responses of *Eurytemora affinis* (Copepoda, Calanoida) reproductive stages to salinity variations. *Journal of Plankton Research* 32:805–813.

Middaugh, D. P. 1981. Reproductive ecology and spawning periodicity of the Atlantic silverside, *Menidia menidia* (Pices: Atherinidae). *Copeia* 1981:766–776.

Miglietta, M. P., Piraino, S., Kubota, S., et al. 2007. Species in the genus *Turritopsis* (Cnidaria, Hydrozoa): A molecular evaluation. *Journal of Zoological Systematics and Evolutionary Research* 45:11–19.

Mikheev, V., Mikheev, A., Pasternak, A., et al. 2000. Light mediated host-searching strategies in a fish ectoparasite, *Argulus foliaceus* L. (Crustacea: Branchiura). *Parasitology* 120:409–416.

Miller, M. J., Rowe, P. M., Able, K. W. 2002. Occurrence and growth rates of young-of-year northern kingfish, *Menticirrhus saxatilis*, on ocean and estuarine beaches in Southern New Jersey. *Copeia* 2002:815–823.

Miller, M. M., Burbanck, W. D. 1961. Systematics and distribution of an estuarine isopod crustacean, *Cyathura polita* (Stimpson, 185) new comb., from the Gulf and Atlantic seaboard of the United States. *Biological Bulletin* 120:62–84.

Miner, R. W. 1950. *Field Book of Seashore Life*. G. P. Putnam's Sons, New York. 888 pp.

Modlin, R. F. 1992. Population structure, distribution, life cycle, and reproductive strategy of *Spilocuma watlingi* Omholt and Heard, 1979 and *S. salomani* Watling, 1977 (Cumacea: Bodotriidae) from coastal waters of the Gulf of Mexico. *Northeast Gulf Science* 12:83–91.

Modlin, R. F., Dardeau, M. 1987. Seasonal and spatial distributions of cumaceans in the Mobile Bay Estuarine System, Alabama. *Estuaries* 10:291–297.

Modlin, R. F., Harris, P. A. 1989. Observations on the natural history and experiments on the reproductive strategy of *Hargeria rapax* (Tanaidacea). *Journal of Crustacean Biology* 9:578–586.

Molenock, J. 1969. *Mysidopsis bahia*, a new species of mysid (Crustacea: Mysidacea) from Galveston Bay, Texas. *Tulane Studies in Zoology and Botany* 15:113–116.

Møller, L. F., Canon, J. M., Tiselius, P. 2010. Bioenergetics and growth in the ctenophore *Pleurobrachia pileus*. *Hydrobiologia* 645:167–178.

Møller, O. S., Olesen, J., Waloszek, D. 2007. Swimming and cleaning in the free-swimming phase of *Argulus* larvae (Crustacea, Branchiura): Appendage adaptation and functional morphology. *Journal of Morphology* 268:1–11.

Montagnes, D. J. S. 1996. Growth responses of planktonic ciliates in the genera *Strobilidium* and *Strombidium*. *Marine Ecology Progress Series* 130:241–254.

Montagnes, D. J. S., Lynn, D. H. 1993. A quantitative protargol stain (QPS) for ciliates and other protists. In: Kemp, P. F., Sherr, B. F., Sherr, E. B., et al. eds. *Aquatic Microbial Ecology*. Lewis Publishers, Boca Raton, FL, 229–240.

Monteleone, D. M., Houde, E. D. 1992. Vulnerability of striped bass *Morone saxatilis* Waldbaum eggs and larvae to predation by juvenile white perch *Morone americana* Gmelin. *Journal of Experimental Marine Biology and Ecology* 158:93–104.

Monteleone, D. M., Peterson, W. T. 1986. Feeding ecology of the American sand lance *Ammodytes americanus* larvae from Long Island Sound. *Marine Ecology Progress Series* 30:133–143.

Monteleone, D. M., Peterson, W. T., Williams, G. C. 1987. Interannual fluctuations in density of sand lance, *Ammodytes americanus,* larvae in Long Island Sound, 1951-1983. *Estuaries* 10:246–254.

Morand, P., Carré, C., Biggs, C. 1987. Feeding and metabolism of the jellyfish *Pelagia noctiluca* (Scyphomedusae, Semaeostomeae). *Journal of Plankton Research* 9:651–665.

Moreira, F. T., Harari, J., Flores, A. A. V. 2007. Neustonic distribution of decapod planktonic stages and competence of brachyuran megalopae in coastal waters. *Marine and Freshwater Research* 58:519–530.

Morgan, S. G. 1980. Aspects of larval ecology of *Squilla empusa* (Crustacea, Stomatopoda) in Chesapeake Bay. *Fishery Bulletin of the US* 78:693–700.

Morgan, S. G., Christy, J. H. 1996. Survival of marine larvae under the countervailing selective pressures of photodamage and predation. *Limnology and Oceanography* 41:498–504.

Morgan, S. G., Provenzano, A. J., Jr. 1979. Development of pelagic larvae and postlarva of *Squilla empusa* (Crustacea, Stomatopoda), with an assessment of larval characters within the Squillidae. *Fishery Bulletin* 77:61–90.

Morton, T. 1989. *Species Profiles: Life histories and Environmental Requirements of Coastal Fishes and Invertebrates (Mid-Atlantic)—Bay Anchovy.* US Fish and Wildlife Service Biological Report FWS/OBS-82/11.97. 13 pp.

Moustakas, C. T., Watanabe, W. O., Copeland, K. A. 2004. Combined effects of photoperiod and salinity on growth, survival and osmoregulatory ability of larval southern flounder *Paralichthys lethostigma. Aquaculture* 229:159–179.

Mullin, M. M. 1979. Differential predation by the carnivorous marine copepod *Tortanus discaudatus. Limnology and Oceanography* 24:774–777.

Mullin, M. M., Onbé, T. 1992. Diel reproduction and vertical distribution of the marine cladocerans, *Evadne tergestina* and *Penilia avirostris,* in contrasting coastal environments. *Journal of Plankton Research* 14:41–59.

Mullineaux, L. S., Butman, C. A. 1991. Initial contact, exploration and attachment of barnacle (*Balanus amphitrite*) cyprids settling in flow. *Marine Biology* 110:93–103.

Muncy, R. J. 1984a. *Species Profiles: Life histories and Environmental Requirements of Coastal Fishes and Invertebrates (Gulf of Mexico)—White Shrimp.* US Fish and Wildlife Service Biological Report FWS/OBS-82/11.20. 19 pp.

Muncy, R. J. 1984b. *Species Profiles: Life Histories and Environmental Requirements of Coastal Fishes and Invertebrates (Gulf of Mexico)—Pinfish.* US Fish and Wildlife Service Publication FWS/OBS-82/11.26. 18 pp.

Muncy, R. J. 1984c. *Species Profiles: Life Histories and Environmental Requirements of Coastal Fishes and Invertebrates (South Atlantic)—White Shrimp.* US Fish and Wildlife Service Biological Report FWS/OBS-82/11.27. 19 pp.

Nakamura, Y., Turner, J. 1997. Predation and respiration by the small cyclopoid copepod *Oithona similis:* How important is feeding on ciliates and heterotrophic flagellates? *Journal of Plankton Research* 19:1275–1288.

Nassogne, A. 1970. Influence of food organisms on the development and culture of pelagic copepods. *Helgoländer Wissenschaftliche Meeresuntersuchengen* 20:333–345.

Nates, S. F., Felder, D. L., Lamaitre, R. 1997. Comparative larval development in two species of the burrowing ghost shrimp genus *Lepidophthalmus* (Decapoda: Callianassidae). *Journal of Crustacean Biology* 17:497–519.

National Oceanic and Atmospheric Administration (NOAA). 1996. *NOAA's Estuarine Eutrophica-*

tion Survey. Vol. 1, *South Atlantic Region*. Office of Ocean Resources Conservation Assessment, Silver Spring, MD. 50 pp.

National Oceanic and Atmospheric Administration (NOAA). 1997a. *NOAA's Estuarine Eutrophication Survey*. Vol. 2, *Mid-Atlantic Region*. Office of Ocean Resources Conservation Assessment, Silver Spring, MD. 51 pp.

National Oceanic and Atmospheric Administration (NOAA). 1997b. *NOAA's Estuarine Eutrophication Survey*. Vol. 3, *North Atlantic Region*. Office of Ocean Resources Conservation Assessment, Silver Spring, MD. 45 pp.

National Oceanic and Atmospheric Administration (NOAA). 1997c. *NOAA's Estuarine Eutrophication Survey*. Vol. 4, *Gulf of Mexico Region*. Office of Ocean Resources Conservation Assessment, Silver Spring, MD. 77 pp.

Needham, J. G., Needham, P. R. 1962. *A Guide to the Study of Fresh-Water Biology*. 5th ed. Holden-Day, San Francisco. 108 pp.

Nelson, T. C. 1925. On the occurrence and food habits of ctenophores in New Jersey inland coastal waters. *Biological Bulletin* 48:92–111.

Neuman, M. J., Able, K. W. 2003. Inter-cohort differences in spatial and temporal settlement patterns of young-of-the-year windowpane (*Scophthalmus aquosus*) in southern New Jersey. *Estuarine and Coastal Shelf Science* 56:527–538.

Newell, G. E., Newell, R. C. 1963. *Marine Plankton: A Practical Guide*. Hutchinson Educational, London. 244 pp.

Newell, R. I. E. 1989. *Species Profiles: Life Histories and Environmental Requirements of Coastal Fishes and Invertebrates (North And Mid-Atlantic)—Blue Mussel*. US Fish and Wildlife Service Biological Report FWS/OBS-82/11.102. 25 pp.

Newell, R. I. E., Langdon, C. J. 1996. Mechanisms and physiology of larval and adult feeding. In: Kennedy, V. S., Newell, R. I. E., Eble, A., eds. *The Eastern Oyster, Crassostrea virginica*. Maryland Sea Grant, College Park, MD, 185–230.

Ngoc-Ho, N. 1981. A taxonomic study of the larvae of four thalassinid species (Decapoda, Thalassinidea) from the Gulf of Mexico. *Bulletin of the British Museum of Natural History* 40:237–273.

Nicholls, A. G. 1935. The larval stages of *Longipedia coronata* Claus, *L. scotti* G. O. Sars, and *L. minor* T. and A. Scott, with a description of the male of *L. scotti*. *Journal of the Marine Biological Association of the United Kingdom (New Series)* 20:29–45.

Nielsen, T. G., Sabatini, M. 1996. Role of cyclopoid copepods *Oithona* spp. in North Sea plankton. *Marine Ecology Progress Series* 139:79–93.

Nogrady, T., Rowe, T. L. A. 1993. Comparative laboratory studies of narcosis in *Brachionus plicatilis*. *Hydrobiologia* 255/256:51–56.

Noij, T. T., Bathmann, U. V., von Budungen, B., et al. 1997. Clearance of picoplankton-sized particles and formation of sinking aggregates by the pteropod *Limacina retroversa*. *Journal of Plankton Research* 19:863–875.

Norcross, B. L. 1991. Estuarine recruitment mechanisms of larval Atlantic croakers. *Transactions of the American Fisheries Society* 120:673–683.

Norcross, J. J., Massmann, W. H., Joseph, E. B. 1961. Investigations of inner continental shelf waters off lower Chesapeake Bay. Part 2. Sand lance larvae, *Ammodytes americanus*. *Chesapeake Science* 2:49–59.

North, E. W., Hood, R. R., Chao, S.-Y., et al. 2005. The influence of episodic events on transport of striped bass eggs to the estuarine turbidity maximum nursery area. *Estuaries* 28:108–123.

North, E. W., Houde, E. D. 2001. Retention of white perch and striped bass larvae: Biological-physical interactions in Chesapeake Bay estuarine turbidity maximum. *Estuaries* 24:756–769.

North, E. W., Houde, E. D. 2004. Distribution and transport of bay anchovy (*Anchoa mitchilli*) eggs and larvae in the Chesapeake Bay. *Estuarine, Coastal and Shelf Science* 60:409–429.

North, E. W., Houde, E. D. 2006. Retention mechanisms of white perch (*Morone americana*) and striped bass (*Morone saxatilis*) early-life stages in an estuarine turbidity maximum: An integrative fixed-location and mapping approach. *Fisheries Oceanography* 15:429–450.

Nyblade, C. E. 1970. Larval development of *Pagurus annulipes* (Stimpson 1862) and *Pagurus pollicaris* Say, 1817 reared in the laboratory. *Biological Bulletin* 139:557–573.

Ocaña-Luna, A., Sánchez-Ramirez, M. 1998. Feeding of sciaenid (Pisces: Sciaenidae) larvae in two coastal lagoons of the Gulf of Mexico. *Gulf Research Reports* 10:1–10.

O'Connell, A. M., Angermeier, P. L. 1997. Spawning location and distribution of early life stages of alewife and blueback herring in a Virginia stream. *Estuaries* 20:779–791.

O'Connor, N. J., Van, B. T. 2006. Adult fiddler crabs *Uca pugnax* (Smith) enhance sediment-associated cues for molting of conspecific megalopae. *Journal of Experimental Marine Biology and Ecology* 335:123–130.

Ogburn, M. B., Jackson, J. T., Forward, R. B., Jr. 2007. Comparison of low salinity tolerance in *Callinectes sapidus* Rathbun and *Callinectes similis* Williams postlarvae upon entry into an estuary. *Journal of Experimental Marine Biology and Ecology* 352:343–350.

Ohman, M. D., Runge, J. A. 1994. Sustained fecundity when phytoplankton resources are in short supply: Omnivory by *Calanus finmarchicus* in the Gulf of St. Lawrence. *Limnology and Oceanography* 39:21–36.

Ohman, M. D., Snyder, R. A. 1991. Growth kinetics of the omnivorous oligotrich ciliate *Strombidium* sp. *Limnology and Oceanography* 36:922–935.

Ohtsuka, S., Kubo, N., Okada, M., et al. 1993. Attachment and feeding of pelagic copepods on larvacean houses. *Journal of Oceanography* 49:115–120.

Olesen, J., Richter, S., Scholtz, G. 2003 On the ontogeny of *Leptodora kindtii* (Crustacea, Branchiopoda, Cladocera), with notes on the phylogeny of the Cladocera. *Journal of Morphology* 256:235–259.

Olesen, N. J., Purcell, J. E., Stoecker, D. K. 1996. Feeding and growth by ephyrae of the scyphomedusae *Chrysaora quinquecirrha. Marine Ecology Progress Series* 137:149–159.

Olney, J. E., Boehlert, G. W. 1988. Nearshore ichthyoplankton associated with seagrass beds in the lower Chesapeake Bay. *Marine Ecology Progress Series* 45:33–43.

Olson, M. B., Lessarda, E. J. 2010. The influence of the *Pseudo-nitzschia* toxin, domoic acid, on microzooplankton grazing and growth: A field and laboratory assessment. *Harmful Algae* 9:540–547.

Omholt, P. E., Heard, R. W. 1979. A new species of *Spilocuma* (Cumacea: Bodotriidae: Mancocuminae) from the Gulf of Mexico. *Proceedings of the Biological Society of Washington* 92:184–194.

Onbé, T. 1984. The developmental stages of *Longipedia americana* (Copepoda: Harpacticoida) reared in the laboratory. *Journal of Crustacean Biology* 4:615–631.

O'Neil, J. M. 1998. The colonial cyanobacterium *Trichodesmium* as a physical and nutritional substrate for the harpacticoid copepod *Macrosetella gracilis. Journal of Plankton Research* 20:43–59.

O'Neil, J. M., Roman, M. R. 1994. Ingestion of the cyanobacterium *Trichodesmium* spp. by pelagic harpacticoid copepods *Macrosetella, Miracia* and *Oculosetella. Hydrobiologia* 292/293:235–240.

Ong, K. S., Costlow, J. D., Jr. 1970. The effect of salinity and temperature on the larval development of the stone crab *Menippe mercenaria* reared in the laboratory. *Chesapeake Science* 11:16–29.

Oshiro, L. M. Y., Omori, M. 1996. Larval development of *Acetes americanus* (Decapoda: Sergestidae) at Paranaguá and Laranjeiras Bays, Brazil. *Journal of Crustacean Biology* 16:709–729.

Osman, R. W., Whitlatch, R. B. 1995. The influence of resident adults on larval settlement: Experiments with four species of ascidians. *Journal of Experimental Marine Biology and Ecology* 190:199–220.

Pace, M. C., Carman, K. R. 1996. Interspecific differences among meiobenthic copepods in the use of microalgal resources. *Marine Ecology Progress Series* 143:77–86.

Paffenhöfer, G.-A., Knowles, S. C. 1980. Omnivorousness in marine planktonic copepods. *Journal of Plankton Research* 2:355–365.

Paffenhöfer, G.-A., Orcutt, J. D., Jr. 1986. Feeding, growth and food conversion of the marine cladoceran *Penilia avirostris*. *Journal of Plankton Research* 8:741–754.

Paffenhöfer, G.-A., Stearns, D. E. 1988. Why is *Acartia tonsa* (Copepoda: Calanoida) restricted to nearshore environments? *Marine Ecology Progress Series* 42:33–38.

Pagès, F., González, H. E., Gonzáles, S. R. 1996. Diet of the gelatinous zooplankton in Hardangerfjord (Norway) and potential predatory impact by *Aglantha digitale* (Trachymedusae). *Marine Ecology Progress Series* 139:69–77.

Park, M. G., Cooney, S. K., Yih, W., et al. 2002. Effects of two strains of the parasitic dinoflagellate *Amoebophyra* on growth, photosynthesis, light absorption, and quantum yield of bloom-forming dinoflagellates. *Marine Ecology Progress Series* 227:281–292.

Pasternak, A. F., Mikheev, V. N., Valtonen, E. T. 2000. Life history characteristics of *Argulus foliaceus* L. (Crustacea: Branchiura) populations in Central Finland. *Annales Zoologici Fennici* 37:25–35.

Patterson, D. J. 1979. On the organization and classification of the protozoon, *Actinophrys sol* Ehrenberg, 1830. *Microbios* 26:65–208.

Pattillo, M. E., Czapla, T. E., Nelson, D. M., et al. 1997. *Distribution and Abundance of Fishes and Invertebrates in Gulf of Mexico Estuaries*. Vol. 2, *Species Life History Summaries*. ELMR Report No. 11. NOAA/NOS Strategic Environmental Assessments Division, Silver Spring, MD. 377 pp.

Paz, B., Riobó, P., Luisa Fernández, M., Fraga, J. M. 2004. Production and release of yessotoxins by the dinoflagellates *Protoceratium reticulatum* and *Lingulodinium polyedrum* in culture. *Toxicon* 44:251–258.

Pearre, S., Jr. 1973. Vertical migration and feeding in *Sagitta elegans* Verrill. *Ecology* 54:300–314.

Pearson, J. C. 1939. The early life histories of some American Penaeidae, chiefly the commercial shrimp, *Penaeus setiferus* (Linn.). *Bulletin of the US Bureau of Fisheries* 49:1–73.

Pearson, J. C. 1941. The young of some marine fishes taken in lower Chesapeake Bay, Virginia, with special reference to the gray sea trout, *Cynoscion regalis* (Block). *Fishery Bulletin of the US* 36:77–102.

Pechenik, J. A., Rittschof, D., Schmidt, A. R. 1993. Influence of delayed metamorphosis on survival and growth of juvenile barnacles *Balanus amphitrite*. *Marine Biology* 115:287–294.

Peebles, E. B. 2002. Temporal resolution of biological and physical influences on bay anchovy *Anchoa mitchilli* egg abundance near a river-plume frontal zone. *Marine Ecology Progress Series* 237:257–269.

Peebles, E. B., Tolley, S. G. 1988. Distribution, growth, and mortality of larval spotted seatrout *Cy-*

noscion nebulosus: A comparison between two adjacent estuarine areas of southwest Florida. *Bulletin of Marine Science* 42:387–410.

Pérez-Domínguez, D., Holt, S. A., Holt, G. J. 2006. Environmental variability in seagrass meadows: Effects of nursery environment cycles on growth and survival in larval red drum *Sciaenops ocellatus*. *Marine Ecology Progress Series* 321:41–53.

Pérez-Farfante, I. 1969. Western Atlantic shrimps of the genus *Penaeus*. *Fishery Bulletin* 67:461–591.

Perry, H., Johnson, D. R., Larsen, K., et al. 2003. Blue crab larval dispersion and retention in the Mississippi Bight: Testing the hypothesis. *Bulletin of Marine Science* 72:331–346.

Perry, H. M., Eleuterius, C., Trigg, C., et al. 1995. Settlement patterns of *Callinectes sapidus* megalopae in Mississippi Sound: 1991, 1992. *Bulletin of Marine Science* 57:821–833.

Perry, H. M., Graham, W. M. 2000. *The Spotted Jelly-Fish: Alien Invader.* Mississippi-Alabama Sea Grant Publication 00-007. 2 pp.

Perry, H. M., McIlwain, T. D. 1986. *Species Profiles: Life Histories and Environmental Requirements of Coastal Fishes and Invertebrates (Gulf of Mexico)—Blue Crab.* US Fish and Wildlife Biological Report 82 (11.55). 11 pp.

Perry, H. M., McLelland, J. A. 1981. First recorded observance of the dinoflagellate *Prorocentrum minimum* (Pavillard) Schiller 1933 in Mississippi sound and adjacent waters. *Gulf Research Reports* 7:83–85.

Petersen, C. W., Salinas, S., Preston, R. L., et al. 2010. Spawning periodicity and reproductive behavior of *Fundulus heteroclitus* in a New England Salt Marsh. *Copeia* 2010:203–210.

Peterson, R. H., Johansen, P. H., Metcalfe, J. L. 1980. Observations on early life stages of Atlantic tomcod, *Microgadus tomcod*. *Fishery Bulletin* 78:147–158.

Peterson, T. L. 1996. Seasonal migration in the southern hogchoker, *Trinectes maculatus fasciatus* (Achiridae). *Gulf Reseach Reports* 9:169–176.

Peterson, W. T., Ausubel, S. J. 1984. Diets and selective feeding by larvae of Atlantic mackerel *Scomber scombrus* on zooplankton. *Marine Ecology Progress Series* 17:65–75.

Petrone, C., Jancaitis, L. B., Jones, M. B., et al. 2005. Dynamics of larval patches: Spatial distribution of fiddler crab larvae in Delaware Bay and adjacent waters. *Marine Ecology Progress Series* 293:177–190.

Pettibone, M. H. 1963. *Marine Polychaete Worms of the New England Region*. Part 1. *Families Aphroditidae to Trochochaetidae*. Smithsonian Institution, Museum of Natural History, Washington, DC. 340 pp.

Pfeiler, E. 1986. Towards an explanation of the developmental strategy in leptocephalus larvae of marine teleost fishes. *Environmental Biology of Fishes* 15:3–13.

Phillips, B. F., Sastry, A. N. 1980. Larval ecology. In: Cobb, J. S., Phillips, B. H., eds. *The Biology and Management of Lobsters*. Vol. 2. Academic Press, New York, 11–57.

Phillips, J. M., Huish, M. T., Kerby, J. H., et al. 1989. *Species Profiles: Life Histories and Environmental Requirements of Coastal Fishes and Invertebrates (Mid-Atlantic)—Spot.* US Fish and Wildlife Service Biological Report. 82(11.98). 13 pp.

Phillips, P. J., Burke, W. D. 1970. The occurrence of sea wasps (Cubomedusae) in Mississippi Sound and the northern Gulf of Mexico. *Bulletin of Marine Science* 20:853–859.

Piasecki, W. 1996. The developmental stages of *Caligus elongatus* von Nordmann, 1832 (Copepoda: Caligidae). *Canadian Journal of Zoology* 74:1459–1478.

Piasecki, W., MacKinnon, B. M. 1995. Life-cycle of a sea louse, *Caligus elongatus* von Nordmann,1832 (Copepoda, Siphonostomatoida, Calgidae). *Canadian Journal of Zoology* 73:74–82.

Pierce, R. W., Coats, D. W. 1999. The feeding ecology of *Actinophrys sol* (Sarcodina: Heliozoa) in Chesapeake Bay. *Journal of Eukaryotic Microbiology* 46:451–457.

Pohle, G., Santana, W., Jansen, G., et al. 2011. Plankton-caught zoeal stages and megalopa of the lobster shrimp *Axius serratus* (Decapoda: Axiidae) from the Bay of Fundy, Canada, with a summary of axiidean and gebiidean literature on larval descriptions. *Journal of Crustacean Biology* 31:82–99.

Poling, K. R., Fuiman, L. A. 1997. Sensory development and concurrent behavioural changes in Atlantic croaker larvae. *Journal of Fish Biology* 51:402–421.

Pollock, L. W. 1997. *A Practical Guide to Marine Animals of Northeastern North America*. Rutgers University Press, New Brunswick, NJ. 367 pp.

Porter, H. J. 1960. Zoeal stages of the stone crab, *Menippe mercenaria Say. Chesapeake Science* 1:168–177.

Poulet, S. A. 1976. Feeding of *Pseudocalanus minutus* on living and non-living particles. *Marine Biology* 34:117–125.

Powell, A. B. 1993. A comparison of the early life-history traits in Atlantic menhaden *Brevoortia tyrannus* and Gulf menhaden *B. patronus*. *Fishery Bulletin* 91:119–128.

Powles, P. M., Warlen, S. M. 2002. Recruitment season, size, and age of young American eels (*Anguilla rostrata*) entering an estuary near Beaufort, North Carolina. *Fishery Bulletin* 100:299–306.

Pratt, H. S. 1935. *A Manual of the Common Invertebrate Animals*. Blakiston, Philadelphia. 854 pp.

Price, W. W., Heard, R. W., Stuck, L. 1994. Observations on the genus *Mysidopsis* Sars, 1864 with the designation of a new genus, *Americamysis,* and the descriptions of *Americamysis alleni* and *A. stucki* (Peracarida: Mysidacea: Mysidae), from the Gulf of Mexico. *Proceedings of the Biological Society of Washington* 107:680–698.

Pryor, V. K., Epifanio, C. E. 1993. Prey selection by larval weakfish (*Cynoscion regalis*): The effects of prey size, speed, and abundance. *Marine Biology* 116:31–37.

Purcell, J. E. 1982. Feeding and growth of the siphonophore *Muggiaea atlantica* (Cunningham 1893). *Journal of Experimental Marine Biology and Ecology* 62:39–54.

Purcell, J. E. 1984. Predation on fish larvae by *Physalia physalis,* the Portuguese man of war. *Marine Ecology Progress Series* 19:189–191.

Purcell, J. E. 1992. Effects of predation by the scyphomedusan *Chrysaora quinquecirrha* on zooplankton populations in Chesapeake Bay, USA. *Marine Ecology Progress Series* 87:65–76.

Purcell, J. E., Båmstedt, U., Båmstedt, A. 1999. Prey, feeding rates, and asexual reproduction rates of the introduced oligohaline hydrozoan *Moerisia lyonsi*. *Marine Biology* 134:317–325.

Purcell, J. E., Nemazie, D. A. 1992. Quantitative feeding ecology of the hydromedusan *Nemopsis bachei* in Chesapeake Bay. *Marine Biology* 113:305–311.

Purcell, J. E., Nemazie, D. A., Dorsey, S. E., et al. 1994. Predation mortality of bay anchovy *Anchoa mitchilli* eggs and larvae due to scyphomedusae and ctenophores in Chesapeake Bay. *Marine Ecology Progress Series* 114:47–58.

Purcell, J. E., Sturdevant, M. V., Galt, C. P. 2005. A review of appendicularians as prey of invertebrate and fish. In: Gorsky, G., Youngbluth, M. J., Deibel, D., eds. *Response of Marine Ecosystems to Global Change: Ecological Impact of Appendicularians*. Contemporary Publishing International, Paris, 359–435.

Pyne, R. R. 1972. Larval development and behaviour of the mantis shrimp *Squilla armata* Milne Edwards (Crustacea: Stomatopoda). *Journal of the Royal Society of New Zealand* 2:121–146.

Qiu, J. W., Gosselin, L. A., Qian, P. Y. 1997. Effects of short-term variation in food availability on larval development in the barnacle *Balanus amphitrite amphitrite*. *Marine Ecology Progress Series* 161:83–91.

Rakusa-Suszczewski, S. 1968. Predation of Chaetognatha by *Tomopteris helgolandica* Greff. *ICES Journal of Marine Science* 32:226–231.

Ram, J. L., Xubo, F., Danaher, S. M., et al. 2008. Finding females: Pheromone-guided reproductive tracking behavior by male *Nereis succinea* in the marine environment. *Journal of Experimental Biology* 211:757–765.

Ramey, P. A. 2008. Life history of a dominant polychaete, *Polygordius jouinae*, in inner continental shelf sands of the Mid-Atlantic Bight, USA. *Marine Biology* 154:443–452.

Rapoza, R., Novak, D., Costello, J. H. 2005. Life-stage dependent, *in situ* dietary patterns of the lobate ctenophore *Mnemiopsis leidyi* A. Agassiz from Woods Hole, Massachusetts, USA. *Journal of Plankton Research* 27:951–956.

Rees, G. H. 1959. Larval development of the sand crab *Emerita talpoida* (Say) in the laboratory. *Biological Bulletin* 117:356–370.

Regan, R. E. 1985. *Species Profiles: Life Histories and Environmental Requirements of Coastal Fishes and Invertebrates (Gulf of Mexico)—Red Drum.* US Fish and Wildlife Service Biological Report FWS/OBS-82 11.36. 16 pp.

Reimer, R. D., Strawn, K., Dixon, A. 1974. Notes on the river shrimp, *Macrobrachium ohione* (Smith) 1874, in the Galveston Bay system of Texas. *Transactions of the American Fisheries Society* 103:120–126.

Richmond, C., Marcus, N. H., Sedlacek, C., et al. 2006. Hypoxia and seasonal temperature: Short-term effects and long-term implications for *Acartia tonsa* Dana. *Journal of Experimental Marine Biology and Ecology* 328:177–196.

Richoux, N. B., Deibel, D., Thompson, R. J. 2004. Population biology of hyperbenthic crustaceans in a cold water environment (Conception Bay, Newfoundland). I. *Mysis mixta* (Mysidacea). *Marine Biology* 144:881–894.

Rico-Martinez, R., Snell, T. W. 1995. Mating behavior and mate recognition pheromone blocking of male receptors in *Brachionus plicatilis* Muller (Rotifera). *Hydrobiologia* 313/314: 105–110.

Riisgård, H. U., Svane, I. 1999. Filter feeding in lancelets (amphioxus), *Branchiostoma lanceolatum*. *Invertebrate Biology* 118:423–432.

Rilling, G. C., Houde, E. D. 1999. Regional and temporal variability in growth and mortality of bay anchovy, *Anchoa mitchilli*, larvae in Chesapeake Bay. *Fishery Bulletin* 97:555–569.

Rines, J. E. B., Hargraves, P. E. 1988. The *Chaetoceros* Ehrenberg (Bacillariophyceae) flora of Narragansett Bay, Rhode Island, USA. *Bibliotheca Phycologia* 79:1–196.

Ringo, R. D., Zamora, G., Jr. 1968. A penaeid postlarval character of taxonomic value. *Bulletin of Marine Science* 18:471–476.

Riser, N. W. 1974. Nemertinea. In: A. C. Giese, A. C., Pearse, J. S. eds. *Reproduction in Marine Invertebrates*. Vol. 1, *Acoecomate and Pseudocoelomate Metazoans*. Academic Press, New York, 359–390.

Roberts, M. H., Jr. 1968. Larval development of the decapod *Euceramus praelongus* in laboratory culture. *Chesapeake Science* 9:121–130.

Roberts, M. H., Jr. 1970. Larval development of *Pagurus longicarpus* Say reared in the laboratory. 1. Description of larval instars. *Biological Bulletin* 139:188–202.

Roberts, M. H., Jr. 1971a. Larval development of *Pagurus longicarpus* Say reared in the laboratory. 2. Effects of reduced salinity on larval development. *Biological Bulletin* 140:104–116.

Roberts, M. H., Jr. 1971b. Larval development of *Pagurus longicarpus* Say reared in the laboratory. 4. Aspects of the ecology of the megalopa. *Biological Bulletin* 141:162–166.

Roberts, M. H., Jr. 1975. Larval development of *Pinnotheres chamae* reared in the laboratory. *Chesapeake Science* 16:242–252.

Robertson, P. B. 1968. The complete larval development of the sand lobster, *Scyllarus americanus* (Smith) (Decapoda, Scyllaridae) in the laboratory, with notes on larvae from the plankton. *Bulletin of Marine Science* 18:294–342.

Robillard, E., Reiss, C. S., Jones, C. M. 2008. Reproductive biology of bluefish (*Pomatomus saltatrix*) along the East Coast of the United States. *Fisheries Research* 90:198–208.

Robinette, H. R. 1983. *Species Profiles: Life Histories and Environmental Requirements of Coastal Fishes and Invertebrates (Gulf of Mexico)—Bay Anchovy and Striped Anchovy.* US Fish and Wildlife Service Biological Report FWS/OBS-82/11. 15 pp.

Roccatagliata, D., Heard, R. W. 1995. Two species of *Oxyurostylis* (Crustacea: Cumacea: Diastylidae), *O. smithi* Calman, 1912 and *O. lecroyae*, a new species from the Gulf of Mexico. *Proceedings of the Biological Society of Washington* 108:596–612.

Rodrigues, S. I., Manning, R. B. 1992. The first stage larva of *Coronis scolopendra* Latreille (Stomatopoda: Nannosquillidae). *Journal of Crustacean Biology* 12:79–82.

Rodriguez, R. A., Epifanio, C. E. 2000. Multiple cues for induction of metamorphosis in larvae of the common mud crab *Panopeus herbstii*. *Marine Ecology Progress Series* 195:221–229.

Rogers, B. D., Shaw, R. F., Herke, W. H., et al. 1993. Recruitment of postlarval and juvenile brown shrimp (*Penaeus aztecus* Ives) from offshore to estuarine waters of the northwestern Gulf of Mexico. *Estuarine, Coastal and Shelf Science* 36:377–394.

Rogers, C. A., Biggs, D. C., Cooper, R. A. 1978. Aggregation of the siphonophore *Nanomia cara* in the Gulf of Maine: Observations from a submersible. *Fishery Bulletin* 76:281–284.

Rogers, S. G., Van Den Avyle, M. J. 1983. *Species Profiles: Life Histories and Environmental Requirements of Coastal Fishes and Invertebrates (Mid-Atlantic)—Atlantic Menhaden.* US Fish and Wildlife Service Publication FWS/OBS-82/11.11. 20 pp.

Roman, M. R. 1978. Ingestion of the blue-green algae *Trichodesmium thiebauti* by the harpacticoid copepod *Macrosetella gracilis*. *Limnology and Oceanography* 23:1245–1248.

Roman, M. R. 1984. Utilization of detritus by the copepod, *Acartia tonsa*. *Limnology and Oceanography* 29:949–959.

Roman, M. R., Holliday, D. V., Sanford, L. P. 2001. Temporal and spatial patterns of zooplankton in the Chesapeake Bay turbidity maximum. *Marine Ecology Progress Series* 213:215–227.

Rome, N. E., Conner, S. L., Bauer, R. T. 2009. Delivery of hatching larvae to estuaries by an amphidromous river shrimp: Tests of hypotheses based on larval moulting and distribution. *Freshwater Biology* 54:1924–1932.

Rooker, J. R., Holt, S. A., Soto, M. A., et al. 1998. Postsettlement patterns of habitat use by sciaenid fishes in subtropical seagrass meadows. *Estuaries* 21:318–327.

Rooney, P., Cobb, J. S. 1991. Effects of time of day, water temperature, and water velocity on swimming by postlarvae of the American lobster, *Homarus americanus*. *Canadian Journal of Fisheries and Aquatic Science* 48:1944–1950.

Rottini Sandrini, L., Avian, M. 1989. Feeding mechanism of *Pelagia noctiluca* (Scyphozoa: Semaeostomeae): Laboratory and open sea observations. *Marine Biology* 102:49–55.

Ruppert, E. E., Fox, R. S. 1988. *Seashore Animals of the Southeast: A Guide to Common Shallow-Water Invertebrates of the Southeastern Atlantic Coast.* University of South Carolina Press, Columbia. 429 pp.

Ruppert, E. E., Fox, R. S., Barnes, R. D. 2004. *Invertebrate Zoology. A Functional Evolutionary Approach.* 7th ed. Thompson Learning, Belmont, CA. 963 pp.

Ruppert, R. E. 1978. A review of metamorphosis of turbellarian larvae. In: Chia, F.-S., Rice, M. E., eds. *Settlement and Metamorphosis of Marine Invertebrate Larvae* Elsevier / North-Holland Biomedical Press, New York, 65–81.

Rushton-Mellor, S. K., Boxshall, G. A. 1994. The developmental sequence of *Argulus foliaceus. Journal of Natural History* 28:763–785.

Russell-Hunter, W. D., Apley, M. L., Hunter, R. D. 1972. Early life-history of *Melampus* and the significance of semilunar synchrony. *Biological Bulletin* 143:623–656.

Ryland, J. S. 1976. Physiology and ecology of marine bryozoans. *Advances in Marine Biology* 14:285–443.

Saage, A., Vadstein, O., Sommer, U. 2009. Feeding behaviour of adult *Centropages hamatus* (Copepoda, Calanoida): Functional response and selective feeding experiments. *Journal of Sea Research* 62:16–21.

Sabatini, M. E. 1990. The developmental stages (copepodids I to VI) of *Acartia tonsa* Dana, 1849 (Copepoda, Calanoida). *Crustaceana* 59:53–61.

Saito, H., Kiørboe, T. 2001. Feeding rates in the chaetognath *Sagitta elegans:* Effects of prey size, prey swimming behaviour and small-scale turbulence. *Journal of Plankton Research* 23:1385–1398.

Saiz, E., Kiørboe, T. 1995. Predatory and suspension feeding of the copepod *Acartia tonsa* in turbulent environments. *Marine Ecology Progress Series* 122:147–158.

Sanders, R. W. 1995. Seasonal distributions of the photosynthesizing ciliates *Laboea strobila* and *Myrionecta rubra* (=*Mesodinium rubrum*) in an estuary of the Gulf of Maine. *Aquatic Microbial Ecology* 9:237–242.

Sandifer, P. A. 1971. The first two phyllosomas of the sand lobster, *Scyllarus depressus* (Smith) (Decapoda, Scyllaridae). *Journal of the Elisha Mitchell Scientific Society* 87:183–187.

Sandifer, P. A. 1972a. Effects of diet on larval development of *Thor floridanus* (Decapoda, Caridea) in the laboratory. *Virginia Journal of Science* 23:5–8.

Sandifer, P. A. 1972b. *Morphology and Ecology of Chesapeake Bay Decapod Larvae.* Ph.D diss. University of Virginia, Charlottesville. 531 pp.

Sandifer, P. A. 1973a. Larvae of the burrowing shrimp, *Upogebia affinis* (Crustacea, Decapoda, Upogebiidae) from Virginia plankton. *Chesapeake Science* 14:98–104.

Sandifer, P. A. 1973b. Mud shrimp (*Callianassa*) larvae (Crustacea, Decapoda, Callianassidae) from Virginia plankton. *Chesapeake Science* 14:149–159.

Sandifer, P. A. 1973c. Effects of temperature and salinity on larval development of grass shrimp, *Palaemonetes vulgaris* (Decapoda, Caridea). *Fishery Bulletin* 71:115–123.

Sandifer, P. A., Kerby, J. H. 1983. Early life history and biology of the common fish parasite, *Lironeca ovalis* (Say) (Isopoda, Cymothoidae). *Estuaries* 6:420–425.

Sandifer, P. A., Van Engel, W. A. 1970. *Modiolus demissus:* A new host for oyster crab, *Pinnotheres ostreum* in Virginia. *Veliger* 13:145–146.

Sandifer, P. A., Van Engel, W. A. 1971. Larval development of the spider crab, *Libinia dubia* H. Milne Edwards (Brachyura, Majidae, Pisinae) reared in the laboratory. *Chesapeake Science* 12:18–25.

Sandifer, P. A., Van Engel, W. A. 1972. *Lepidopa* larvae (Crustacea, Decapoda, Albuneidae) from Virginia plankton. *Journal of the Elisha Mitchell Scientific Society* 88:220–225.

Sandoz, M., Hopkins, S. H. 1947. Early life history of the oyster crab *Pinnotheres ostreum* (Say). *Biological Bulletin* 93:250–258.

Sastry, A. N. 1965. The development and external morphology of pelagic larvae and post larval stages of the bay scallop *Aequipecten irradians concentricus* Say, reared in the laboratory. *Bulletin of Marine Science* 15:417–435.

Sastry, A. N. 1977a. The larval development of the Jonah crab, *Cancer borealis* Stimpson, 1859, under laboratory conditions (Decapoda Brachyura). *Crustaceana* 32:290–303.

Sastry, A. N. 1977b. The larval development of the rock crab, *Cancer irroratus* Say, 1817, under laboratory conditions (Decapoda Brachyura). *Crustaceana* 32:155–168.

Sato, M., Jumars, P. A. 2008. Seasonal and vertical variations in emergence behaviors of *Neomysis americana*. *Limnology and Oceanography* 53:1665-1677

Sawyer, R. T., Hammond, D. H. 1973. Observations on the biology of *Calliobdella carolinensis* (Hirudinea: Piscicolidae): Parasitic on the Atlantic menhaden in epizootic proportions. *Biological Bulletin* 145:373–388.

Sawyer, R. T., Lawler, A. R., Overstreet, R. M. 1975. Marine leeches of the eastern United States and the Gulf of Mexico with a key to the species. *Journal of Natural History* 9:633–667.

Schaffler, J. J., Reiss, C. S., Jones, C. M. 2009. Patterns of larval Atlantic croaker ingress into Chesapeake Bay, USA. *Marine Ecology Progress Series* 378:187–197.

Scheltema, R. S. 1961. Metamorphosis of the veliger larvae of *Nassarius obsoletus* (Gastropoda) in response to bottom sediment. *Biological Bulletin* 120:92–109.

Scherer, M. D., Bourne, D. W. 1980. Eggs and early larvae of smallmouth flounder, *Etropus microstomus*. *Fishery Bulletin* 77:708–712.

Schiebel, R., Bijma, J., Hemleben, C. 1997. Population dynamics of the planktonic foraminiferan *Globigerina bulloides* from the eastern North Atlantic. *Deep-Sea Research* 44:1701–1713.

Schiedges, K. L. 1979a. Reproduction in *Autolytus*. *International Journal of Invertebrate Reproduction* 1:359–370.

Schiedges, K. L. 1979b. Reproductive biology and ontogenesis in the polychaete genus *Autolytus* (Annelida: Syllidae): Observations of laboratory-cultured individuals. *Marine Biology* 54:239–250.

Schroedinger, S. E., Epifanio, C. E. 1997. Growth, development and survival of larval *Tautoga onitis* (Linnaeus) in large laboratory containers. *Journal of Experimental Marine Biology and Ecology* 210:143–155.

Schultz, E. T., Cowen, R. K., Lwiza, K. M. M., et al. 2000. Explaining advection: Do larval bay anchovy (*Anchoa mitchilli*) show selective tidal stream transport? *ICES Journal of Marine Science* 57:360–371.

Schuyler, Q., Sullivan, B. K. 1997. Light responses and diel migration of the scyphomedusa *Chrysaora quinquecirrha* in mesocosms. *Journal of Plankton Research* 19:1417–1428.

Sellner, K. G. 1997. Physiology, ecology, and toxic properties of marine cyanobacteria blooms. *Limnology and Oceanography* 42:1089–1104.

Sellner, K. G., Brownlee, D. C. 1990. Dinoflagellate-microzooplankton interactions in Chesapeake Bay. In: Graneli, E., Sondstrom, B., Edler, L., Anderson, D. M., eds. *Toxic Marine Phytoplankton*. Elsevier, New York, 221–226.

Setälä, O., Autio, R., Kuosa, H. 2005. Predator-prey interactions between a planktonic ciliate *Strombidium* sp. (Ciliophora, Oligotrichida) and the dinoflagellate *Pfiesteria piscicida* (Dinamoebiales, Pyrrophyta). *Harmful Algae* 4:235–247.

Setzler-Hamilton, E. M., Jones, P. W., Martin, F. D., et al. 1982. *Comparative Feeding Habits of White Perch and Striped Bass Larvae*. Maryland Sea Grant Publication UM-SGRS-82-08. 19 pp.

Seuront, L. 2006. Effect of salinity on the swimming behaviour of the estuarine calanoid copepod *Eurytemora affinis*. *Journal of Plankton Research* 28:805–813.

Shanks, A. L. 1998. Abundance of post-larval *Callinectes sapidus, Penaeus* spp., *Uca* spp., and *Libinia* spp. collected at an outer coastal site and their cross-shelf transport. *Marine Ecology Progress Series* 168:57–69.

Shanks, A. L., Graham, W. M. 1987. Oriented swimming in the jellyfish *Stomolophus meleagris* L. Agassiz (Scyphozoa: Rhizostomida). *Journal of Experimental Marine Biology and Ecology* 108:159–170.

Sharpe, R. W. 1911. Notes on the marine Copepoda and Cladeocera of Woods Hole and adjacent regions including a synopsis of the genera of the Harpacticoida. *Proceedings of the U.S. National Museum* 38:405–436.

Shenker, J. M., Hepner, D. J., Frere, P. E., et al. 1983. Upriver migration and abundance of naked goby (*Gobiosoma bosci*) larvae in the Patuxent River Estuary, Maryland. *Estuaries* 6:36–42.

Shield, P. 1973. The chromatophores of *Emerita talpoida* (Say) zoeae considered as a diagnostic character. *Chesapeake Science* 14:41–47.

Shield, P. 1978. Larval development of the caridean shrimp, *Hyppolyte pleuracanthus* (Stimpson), reared in the laboratory. *Estuaries* 1:1–16.

Shoji, J., Matsuda, R., Yamashita, Y., et al. 2008. Predation on fish larvae by moon jellyfish *Aurelia aurita* under low dissolved oxygen concentrations. *Fisheries Science* 71:748–753.

Shoji, J., North, E., Houde, E. 2005. The feeding ecology of *Morone americana* larvae in the Chesapeake Bay estuarine turbidity maximum: The influence of physical conditions and prey concentrations. *Journal of Fish Biology* 66:1328–1341.

Shuvayev, Y. D. 1979. Movements of some planktonic copepods. *Hydrobiological Journal* 14:32–36.

Smalley, G. W., Coats, D. W. 2002. Ecology of the red-tide dinoflagellate *Ceratium furca:* Distribution, mixotrophy, and grazing impact on ciliate populations of Chesapeake Bay. *Journal of Eukaryotic Microbiology* 49:64–74.

Smayda, T. J., Boleyn, B. J. 1966. Experimental observations on the flotation of marine diatoms. II. *Skeletonema costatum* and *Rhizosolenia setigera. Limnology and Oceanography* 11:18–34.

Smigielski, A. S., Halavik, T. A., Buckley, L. J., et al. 1984. Spawning, embryo development and growth of the American sand lance *Ammodytes americanus* in the laboratory. *Marine Ecology Progress Series* 14:287–292.

Smith, D. G. 2001. *Pennak's Freshwater Invertebrates of the United States: Porifera to Crustacea.* 4th ed. Wiley & Sons, New York. 638 pp.

Smith, R. E., Kernehan, R. J. 1981. Predation by the free-living copepod, *Cyclops bicuspidatus thomasi,* on larvae of the striped bass and white perch. *Estuaries* 4:81–83.

Smith, R. I., ed. 1964. *Keys to the Marine Invertebrates of the Woods Hole Region*. Contribution No. 11. Systematics-Ecology Program, Marine Biological Laboratory, Woods Hole, MA. 208 pp. (Updated keys available online.)

Smith, W. G., Fahay, M. P. 1970. *Description of Eggs and Larvae of the Summer Flounder, Paralichthys dentatus*. US Fish and Wildlife Service Report No. 75. 21 pp.

Snell, T. W., Burke, B. E., Messur, S. D. 1983. Size and distribution of resting eggs in a natural population of the rotifer *Brachionus plicatilis*. *Gulf Research Reports* 7:285–287.

Sogard, S. M., Able, K. W., Fahay, M. P. 1992. Early life history of the tautog, *Tautoga onitis,* in the Mid-Atlantic Bight. *Fishery Bulletin* 90:529–539.

Sogard, S. M., Able, K. W., Hagan, S. M. 2001. Long-term assessment of settlement and growth of juvenile winter flounder (*Pseudopleuronectes americanus*) in New Jersey estuaries. *Journal of Sea Research* 45:189–204.

Sogard, S. M., Hoss, D. E., Govoni, J. J. 1987. Density and depth distribution of larval gulf menhaden, *Brevoortia patronus*. Atlantic croaker, *Micropogonias undulatus,* and spot, *Leiostomus xanthurus,* in the northern Gulf of Mexico. *Fishery Bulletin* 85:601–609.

Sorensen, P. W. 1986. Origins of the freshwater attractant(s) of migrating elvers of the American eel, *Anguilla rostrata. Environmental Biology of Fishes* 17:185–200.

Stanley, J. G., Danie, D. S. 1983. *Species Profiles: Life Histories and Environmental Requirements of Coastal Fishes and Invertebrates (Mid-Atlantic)—White Perch.* US Fish and Wildlife Service Publication FWS/OBS-82/11.7. 12 pp.

Stanley, J. G., DeWitt, R. 1983. *Species Profiles: Life Histories and Environmental Requirements of Coastal Fishes and Invertebrates (North Atlantic)—Hard Clam.* US Fish Wildlife Service Publication FWS/OBS-82/11.18. 19 pp.

Staton, J. L., Sulkin, S. D. 1991. Nutritional requirements and starvation resistance in larvae of the brachyuran crabs *Sesarma cinereum* (Bosc) and *S. reticulatum* (Say). *Journal of Experimental Marine Biology and Ecology* 152:271–284.

Stearns, D. E., Dardeau, M. R. 1990. Nocturnal and tidal vertical migrations of "benthic" crustaceans in an estuarine system with diurnal tides. *Northeast Gulf Science* 11:93–104.

Steimle, F. W., Morse, W. W., Johnson, D. L. 1999. *Essential Fish Habitat Source Document: Goose-Fish, Lophius americanus, Life History and Habitat Characteristics.* NOAA Technical Memorandum NMFS-NE-118. 23 pp.

Steimle, F. W., Shaheen, P. A. 1999. *Tautog (Tautoga onitis) Life History and Habitat Requirements.* NOAA Technical Memorandum NMFS-NE-127. 31 pp.

Steimle, F. W., Zetlin, C. A., Berrien, P. L., et al. 1999. *Scup, Stenotomus chrysops, Life History* and *Habitat Characteristics.* NOAA Technical Memorandum NMFS-NE-149. 39 pp.

Steinberg, M. K., Epifanio, C. E., Andon, A. 2007. A highly specific chemical cue for the metamorphosis of the Asian shore crab, *Hemigrapsus sanguineus. Journal of Experimental Marine Biology and Ecology* 347:1–7.

Steppe, C. N., Epifanio, C. E. 2006. Synoptic distribution of crab larvae near the mouth of Delaware Bay: Influence of nearshore hydrographic regimes. *Estuarine, Coastal and Shelf Science* 70:654–662.

Stewart, L., Auster, P. 1987. *Species Profiles: Life Histories and Environmental Requirements of Coastal Fishes and Invertebrates (North Atlantic)—Atlantic Tomcod.* US Fish and Wildlife Service Biological Report 82(11.76). 8 pp.

Stoecker, D. K., Guillard, R. R. L., Kavee, R. M. 1981. Selective predation by *Favella ehrenbergii* (Tintinia) on and among dinoflagellates. *Biological Bulletin* 160:136–145.

Stoecker, D. K., Adolf, J. E., Place, A. R., et al. 2008. Effects of the dinoflagellates *Karlodinium veneficum* and *Prorocentrum minimum* on early life history stages of the eastern oyster (*Crassostrea virginica*). *Marine Biology* 154:81–90.

Stoecker, D. K., Egloff, D. A. 1987. Predation by *Acartia tonsa* Dana on planktonic ciliates and rotifers. *Journal of Experimental Marine Biology and Ecology* 110:53–68.

Stoecker, D. K., Gallager, S. M., Langdon, C. J., et al. 1995. Particle capture by *Favella* sp. (Ciliata, Tintinnina) *Journal of Plankton Research* 17:1105–1124.

Stoecker, D. K., Johnson, M. D., de Vargas, C. 2009. Acquired phototrophy in aquatic protists *Aquatic Microbial Ecology.* 57:279–310.

Stoecker, D. K., Li, A., Coats, D. W., et al. 1997. Mixotrophy in the dinoflagellate *Prorocentrum minimum. Marine Ecology Progress Series* 152:1–12.

Stoecker, D. K., Gifford, D. J., Putt, M. 1994. Preservation of marine planktonic ciliates: Losses and cell shrinkage during fixation. *Marine Ecology Progress Series* 110:293–299.

Stoecker, D. K., Michaels, A. E., Davis, L. H. 1987. Grazing by the jellyfish, *Aurelia aurita* on microzooplankton. *Journal of Plankton Research* 9:901–915.

Stoecker, D. K., Michaels, A. E., Davis, L. H. 1988. Obligate mixotrophy in *Laboea strobila,* a ciliate which retains chloroplasts. *Marine Biology* 99:415–423.

Stokes, M. D. 1996. Larval settlement, post-settlement growth and secondary production of the Florida lancelet (=Amphioxus) *Branchiostoma floridae. Marine Ecology Progress Series* 130:71–84.

Stokes, M. D. 1997. Larval locomotion of the lancelet. *Journal of Experimental Biology* 200:1661–1680.

Stokes, M. D., Holland, N. D. 1995a. Ciliary hovering in larval lancelets. *Biological Bulletin* 188:231–233.

Stokes, M. D., Holland, N. D. 1995b. Embryos and larvae of a lancelet (*Branchiostoma floridae*), from hatching through metamorphosis: Growth in the laboratory and external morphology. *Acta Zoologica* 76(2):105–120.

Strasser, K. M., Felder, D. L. 1999a. Larval development of two populations of the ghost shrimp *Callichirus major* (Decapoda: Thalassinidea) under laboratory conditions. *Journal of Crustacean Biology* 19:844–878.

Strasser, K. M., Felder, D. L. 1999b. Settlement cues in an Atlantic coast population of the ghost shrimp *Callichirus major* (Crustaces: Decapoda: Thalassinidea). *Marine Ecology Progress Series* 183:217–225.

Strasser, K. M., Felder, D. L. 2000. Larval development of the ghost shrimp *Callichirus islagrande* (Decapoda: Thalassinidea: Callianassidae) under laboratory conditions. *Journal of Crustacean Biology* 20:100–117.

Strathmann, R. R., Bonar, D. 1976. Ciliary feeding of tornaria larvae of *Ptychodera flava* (Hemichordata: Enteropneusta). *Marine Biology* 34:317–324.

Strathmann, R. R., McEdward, L. R. 1986. Cyphonautes' ciliary sieve breaks a biological rule of inference. *Biological Bulletin* 171:694–700.

Stuck, K. C., Perry, H. M. 1992. Life history characteristics of *Menippe adina* in Mississippi waters. In: Bert, T. M., ed. *Proceedings of a Symposium on the Stone Crab (Genus Menippe) Biology and Fisheries*. Florida Marine Research Publication No. 50, 82–98.

Stuck, K. C., Perry, H. M., Heard, R. W. 1979. An annotated key to the Mysidacea of the north central Gulf of Mexico. *Gulf Research Reports* 6:225–238.

Suárez-Morales, E., Ramírez-Fernando, C., Derisio, C. 2008. Monstrilloida (Crustacea: Copepoda) from the Beagle Channel, South America. *Contributions to Zoology* 77:217–226.

Suchman, C. L., Sullivan, B. K. 2000. Effect of prey size on vulnerability of copepods to predation by the scyphomedusae *Aurelia aurita* and *Cyanea* sp. *Journal of Plankton Research* 22:2289–2306.

Sugie, K., Kuma, K. 2008. Resting spore formation in the marine diatom *Thalassiosira nordenskioeldii* under iron- and nitrogen-limited conditions. *Journal of Plankton Research* 30:1245–1255.

Sulkin, S., Strom, J., Hutchinson, D. 1998. Nutritional role of protists in the diet of first stage larvae of the Dungeness crab *Cancer magister. Marine Ecology Progress Series* 169:237–242.

Sulkin, S. D., Van Heukelem, W., Kelly, P. 1983. Behavioral basis of depth regulation in the hatching and post-larval states of the mud crab *Eurypanopeus depressus* Hay and Shore. *Marine Ecology Progress Series* 11:157–164.

Sullivan, B. K. 1980. In situ feeding behavior of *Sagitta elegans* and *Eukrohnia hamata* (Chaetognatha) in relation to the vertical distribution and abundance of prey at Ocean Station "P." *Limnology and Oceanography* 25:317–326.

Sullivan, B. K., Costello, J. H., Van Keuren, D. 2007. Seasonality of the copepods *Acartia hudsonica* and *Acartia tonsa* in Narragansett Bay, RI, USA during a period of climate change. *Estuarine, Coastal and Shelf Science* 73:259–267.

Sullivan, B. K., McManus, L. T. 1986. Factors controlling seasonal succession of the copepods *Acartia hudsonica* and *A. tonsa* in Narragansett Bay, Rhode Island: Temperature and resting egg production. *Marine Ecology Progress Series* 28:121–128.

Sullivan, B. K., Suchman, C. L., Costello, J. H. 1997. Mechanics of prey selection by ephyrae of the scyphomedsua *Aurelia aurita*. *Marine Biology* 130:213–222.

Sullivan, L. J., Gifford, D. J. 2007. Growth and feeding rates of the newly hatched larval ctenophore *Mnemiopsis leidyi* A. Agassiz (Ctenophora, Lobata). *Journal of Plankton Research* 29:949–965.

Sun, J., Feng, Y., Zhang, Y., et al. 2007. Fast microzooplankton grazing on fast-growing, low-biomass phytoplankton: A case study in spring in Chesapeake Bay, Delaware Inland Bays and Delaware Bay. *Hydrobiologia* 589:127–139.

Sutter, F. C., Waller, R. S., McIlwain, T. D. 1986. *Species Profiles: Life Histories and Environmental Requirements of Coastal Fishes and Invertebrates (Gulf of Mexico)—Black Drum*. US Fish and Wildlife Service Biological Report FWS/OBS-82/11.51. 10 pp.

Svensen, C., Kiørboe, T. 2000. Remote prey detection in *Oithona similis:* Hydromechanical versus chemical cues. *Journal of Plankton Research* 22:1155–1166.

Swanberg, N. 1974. The feeding behavior of *Beröe ovata*. *Marine Biology* 24:69–76.

Sweatt, J. A., Forward, R. B., Jr. 1985. Diel vertical migration and photoresponses of the chaetognath *Sagitta hispida* Conant. *Biological Bulletin* 168:18–31.

Tackx, M. L. M., Herman, P. J. M., Gasparini, S., et al. 2003. Selective feeding of *Eurytemora affinis* (Copepoda, Calanoida) in temperate estuaries: Model and field observations. *Estuarine and Coastal Marine Science* 56:305–311.

Tamburri, M. N., Finelli, C. M., Wethey, D. S., et al. 1996. Chemical induction of larval settlement behavior in flow. *Biological Bulletin* 191:367–373.

Tamburri, M. N., Zimmerfaust, R. K., Tampi, M. I. 1992. Natural sources and properties of chemical inducers mediating settlement of oyster larvae. *Biological Bulletin* 183:327–338.

Tamm, S. L., Tamm, S. 1991. Reversible epithelial adhesion closes the mouth of *Beroe,* a carnivorous marine jelly. *Biological Bulletin* 181:463–473.

Tankersley, R. A., McKelvey, L. M., Forward, R. B., Jr. 1995. Response of estuarine crab megalopae to pressure, salinity and light: Implications for flood-tide transport. *Marine Biology* 122:391–400.

Taylor, J. C., Miller, J. M., Pietrafesa, L. J., et al. 2010. Winter winds and river discharge Determine juvenile southern flounder (*Paralichthys lethostigma*) recruitment and distribution in North Carolina estuaries. *Journal of Sea Research* 64:15–25.

Taylor, N. G. H., Wootten, R., Sommerville, C. 2009. Using length-frequency data to elucidate the population dynamics of *Argulus foliaceus* (Crustacea: Branchiura). *Parisitology* 136:1023–1032.

Tesmer, C. A., Broad, A. C. 1964. The larval development of *Crangon septemspinosa* (Say) (Crustacea: Decapoda). *Ohio Journal of Science* 64:239–250.

Tester, P. A., Cohen, J. H., Cervetto, G. 2004. Reverse vertical migration and hydrographic distribution of *Anomalocera ornata* (Copepoda: Pontellidae) in the U.S. South Atlantic Bight. *Marine Ecology Progress Series* 268:195–204.

Tester, P. A., Steidinger, K. A. 1997. *Gymnodinium breve* red tide blooms: Initiation, transport, and consequences of surface circulation. *Limnology and Oceanography* 42:1039–1051.

Thessen, A. E., Stoecker, D. K. 2008. Distribution, abundance and domoic acid analysis of the toxic diatom *Pseudo-nitzschia* from the Chesapeake Bay. *Estuaries and Coasts* 31:664–672.

Thiyagarajan, V. 2010. A review on the role of chemical cues in habitat selection by barnacles: New insights from larval proteomics. *Journal of Experimental Marine Biology and Ecology* 392:22–36.

Thomas, D. L., Smith, B. A. 1973. Studies of young black drum, *Pogonias cromis,* in low salinity waters of the Delaware estuary. *Chesapeake Science* 14:124–130.

Thorp, J. H., Covich, A. P., ed. 2010. *Ecology and Classification of North American Freshwater Invertebrates*. 3rd ed. Academic Press, New York. 1021 pp.

Thorson, G. 1946. Reproductive and larval development of Danish marine bottom invertebrates. *Meddelelser fra Kommissionen for Danmarks Fiskeri-Og Havundersøgelser, Kobenhaven* 4:1–523.

Tilburg, C. E., Seay, J. E., Bishop, T. D., et al. 2010. Distribution and retention of *Petrolisthes armatus* in a coastal plain estuary: The role of vertical movement in larval transport. *Estuarine, Coastal and Shelf Science* 88:60–266.

Tomas, C. R. ed. 1997. *Identifying Marine Phytoplankton*. Academic Press, New York. 858 pp.

Tonnesson, K., Tiselius, P. 2005. Diet of the chaetognaths *Sagitta setosa* and *S. elegans* in relation to prey abundance and vertical distribution. *Marine Ecology Progress Series* 289:177–190.

Tremblay, R., Olivier, F., Bourget, E., et al. 2007. Physiological condition of *Balanus amphitrite* cyprid larvae determines habitat selection success. *Marine Ecology Progress Series* 340: 1–8.

Troedsson, C., Bouquet, J.-M., Skinnes, R., et al. 2009. Regulation of filter-feeding house components in response to varying food regimes in the appendicularian, *Oikopleura dioica*. *Journal of Plankton Research* 31:1453–1463.

Truesdale, F. M., Andryszak, B. L. 1983. Occurrence and distribution of reptant decapod crustacean larvae in neritic Louisiana waters: July 1976. *Contributions in Marine Science* 26:37–53.

Truesdale, F. M., Mermilliod, W. J. 1979. The river shrimp *Macrobrachium ohione* (Smith) (Decapoda, Palaemonidae): Its abundance, reproduction, and growth in the Atchafalaya River Basin of Louisiana, U.S.A. *Crustaceana* 36:61–73.

Tucker, J. W., Jr. 1982. Larval development of *Citharichthys cornutus, C. gymnorhinus, C. spilopterus,* and *Etropus crossotus* (Bothidae), with notes on larval occurrence. *Fishery Bulletin of the US* 80:35–73.

Turner, J. T. 1978. Scanning electron microscope investigations of feeding habits and mouthpart structures of three species of copepods in the family Pontellidae. *Bulletin of Marine Science* 28:487–500.

Turner, J. T. 1984a. Zooplankton feeding ecology: Contents of fecal pellets of the copepods *Temora turbinata* and *T. stylifera* from continental shelf and slope waters near the mouth of the Mississippi River. *Marine Biology* 82:73–83.

Turner, J. T. 1984b. Zooplankton feeding ecology: Contents of fecal pellets of the copepods *Eucalanus pileatus* and *Paracalanus quasimodo* from continental shelf waters of the Gulf of Mexico. *Marine Ecology Progress Series* 15:27–46.

Turner, J. T. 1985. Zooplankton feeding ecology: Contents of fecal pellets of the copepod *Anomalocera ornata* from continental shelf and slope waters of the Gulf of Mexico. *P.S.Z.N.I Marine Ecology* 6:285–298.

Turner, J. T. 1986. Zooplankton feeding ecology: Contents of fecal pellets of the cyclopoid copepods

Oncaea venusta, Corycaeus amazonicus, Oithona plumifera and *O. simplex* from the northern Gulf of Mexico. *P.S.Z.N.I Marine Ecology* 7:289–302.

Turner, J. T. 1987. Zooplankton feeding ecology: Contents of fecal pellets of the copepod *Centropages velificatus* from waters near the mouth of the Mississippi River. *Biological Bulletin* 173:377–386.

Turner, J. T., Tester, P. A. 1989. Zooplankton feeding ecology: Nonselective grazing by the copepods *Acartia tonsa* Dana, *Centropages velificatus* De Oliveira, and *Eucalanus pileatus* Giesbrecht in the plume of the Mississippi River. *Journal of Experimental Marine Biology and Ecology* 126:21–43.

Turner, J. T., Tester, P. A., Ferguson, R. L. 1988. The marine cladoceran *Penilia avirostris* and the "microbial loop" of pelagic food webs. *Limnology and Oceanography* 33:245–255.

Turner, J. T., Tester, P. A., Hettler, W. F. 1985. Zooplankton feeding ecology: A laboratory study of predation on fish eggs and larvae by the copepods *Anomalocera ornata* and *Centropages typicus*. *Marine Biology* 90:1–8.

Turner, J. T., Tester, P. A., Strickler, J. R. 1993. Zooplankton feeding ecology--a cinematographic study of animal-to-animal variability in the feeding behavior of *Calanus finmarchicus*. *Limnology and Oceanography* 38:255–264.

Uchida, T., Kamiyama, T., Matsuyama, Y. 1997. Predation by a photosynthetic dinoflagellate *Gyrodinium instriatum* on loricated ciliates. *Journal of Plankton Research* 19:603–608.

Uchima, M., Hirano, R. 1988. Swimming behavior of the marine copepod *Oithona davisae:* Internal control and search for environment. *Marine Biology* 99:47–56.

Utz, L. R. P., Coats, D. W. 2005. Spatial and temporal patterns in the occurrence of peritrich ciliates as epibionts on calanoid copepods in the Chesapeake Bay, USA. *Journal of Eukaryotic Microbiology* 52:236–244.

Utz, L. R. P., Coats, D. W. 2008. Telotroch formation, survivorship, attachment success, and attachment patterns of the epibiotic peritrich *Zoothamnium intermedium* (Ciliophora, Oligohymenophorea). *Invertebrate Biology* 127:237–248.

van Duren, L. A., Videler, J. J. 2003. Escape from viscosity: The kinematics and hydrodynamics of copepod foraging and escape swimming. *Journal of Experimental Biology* 206:269–279.

van Impe, E. 1992. A method for the transportation, long term preservation and storage of gelatinous planktonic organisms. *Scientia Marina* 56:237–238.

van Montfrans, J., Peery, C. A., Orth, R. J. 1990. Daily, monthly, and annual settlement patterns by *Callinectes sapidus* and *Neopanope sayi* megalopae on artificial collectors deployed in the York River, Virginia: 1985-1988. *Bulletin of Marine Science* 46:214–229.

Van Wassenbergh, S., Roos, G., Genbrugg, A., et al. 2009. Suction is kid's play: Extremely fast suction in newborn seahorses. *Biology Letters* 5:200–203.

Vaque, D., Pace, M. L., Findlay, S., et al. 1992. Fate of bacterial production in a heterotrophic ecosystem: Grazing by protists and metazoans in the Hudson Estuary. *Marine Ecology Progress Series* 89:155–163.

Vargas, C. A., Madin, L. P. 2004. Zooplankton feeding ecology: Clearance and ingestion rates of the salps *Thalia democratica*, *Cyclosalpa affinis* and *Salpa cylindrica* on naturally occurring particles in the Mid-Atlantic Bight. *Journal of Plankton Research* 26:827–833.

Vargo, G. A. 2009. A brief summary of the physiology and ecology of *Karenia brevis* Davis (G. Hansen and Moestrup comb. nov.) red tides on the West Florida Shelf and of hypotheses posed for their initiation, growth, maintenance, and termination. *Harmful Algae* 8:73–584.

Vázquez, E., Young, C. M. 2000. Effects of low salinity on metamorphosis in estuarine colonial ascidians. *Invertebrate Biology* 119:433–444.

Vecchione, M. 1981. Aspects of early life history of *Loligo pealei* (Cephalopoda: Myopsida). *Journal of Shellfish Research* 1:171–180.

Vecchione, M. 1982. Development and morphology of planktonic *Lolliguncula brevis*. *Proceedings of the Biological Society of Washington* 95:601–608.

Vecchione, M. 1991a. Dissolved oxygen and the distribution of the euryhaline squid *Lolliguncula brevis* (Cephalopoda: Loliginidae). *Bulletin of Marine Science* 49:668–669.

Vecchione, M. 1991b. Observations on the paralarval ecology of a euryhaline squid *Lolliguncula brevis* (Cephalopoda: Loliginidae). *Fishery Bulletin* 89:515–521.

Vecchione, M., Roper, C. F. E., Sweeney, M. J., Lu, C. C. 2001. *Distribution, Relative Abundance and Developmental Morphology of Paralarval Cephalopods in the Western North Atlantic Ocean*. NOAA Technical Report 152. US Department of Commerce, Seattle, WA. 54 pp.

Vecchione, M., Shea, E. K. 2002. Phylum Mollusca: Cephalopoda. In: Young, C. M., ed. *Atlas of Marine Invertebrate Larvae*. Academic Press, New York, 327–335.

Verity, P. G. 2010. Expansion of potentially harmful algal taxa in a Georgia Estuary (USA). *Harmful Algae* 9:144–152.

Verity, P. G., Villareal, T. A. 1986. The relative value of diatoms, dinoflagellates, flagellates, and cyanobacteria for tintinnid ciliates. *Archiv für Protistenkünde* 131:71–84.

Waggett, R., Buskey, E. 2006. Copepod sensitivity to flow fields: Detection by copepods of predatory ctenophores. *Marine Ecology Progress Series* 323:205–211.

Waggett, R., Buskey, E. 2007. Calanoid copepod escape behavior in response to a visual predator. *Marine Biology* 150:599–607.

Waggy, G. L., Peterson, M. S., Comyns, B. H. 2007. Feeding habits and mouth morphology of young silver perch, *Bairdiella chrysoura*, from the north-central Gulf of Mexico. *Southeastern Naturalist* 6:743–751.

Walker, P. D., Flik, G., Wendelaar Bongs, S. E. 2004. The biology of parasites from the genus *Argulus* and a review of the interactions with its host. In: Weigertjes, G., Flik, G., eds. *Host-Parasite Interactions. Symposia of the Society for Experimental Biology*. Vol. 55. Garland / BIOS Scientific Publishers, Abingdon, UK, 107–129.

Wang, J. C. S., Kernehan, R. J. 1979. *Fishes of the Delaware Estuaries: A Guide to the Early Life Histories*. Ecological Analysts, Towson, MD. 410 pp.

Wang, W. X., Widdows, J. 1991. Physiological responses of mussel larvae *Mytilus edulis* to environmental hypoxia and anoxia. *Marine Ecology Progress Series* 70:223–236.

Ward, H. B., Whipple, G. C. 1918. *Freshwater Biology*. Wiley & Sons, New York. 1111 pp.

Warlen, S. M. 1988. Age and growth of larval Gulf menhaden, *Brevoortia patronus*, in the northern Gulf of Mexico. *Fishery Bulletin of the US* 86:77–90.

Warlen, S. M., Able, K. W., Laban, E. 2002. Recruitment of larval Atlantic menhaden (*Brevoortia tyrannus*) to North Carolina and New Jersey estuaries: Evidence for larval transport northward along the east coast of the United States. *Fishery Bulletin* 100:609–623.

Webb, J. E. 1969. On the feeding and behaviour of the larva of *Branchiostoma lanceolatum*. *Marine Biology* 3:58–72.

Webber, H. H. 1977. Gastropoda: Prosobranchia. In: Giese, A. C., Pearse, J. S., eds. *Reproduction in Marine Invertebrates*. Vol. 4, *Molluscs: Gastropods and Cephalopods*. Academic Press, New York, 1–97.

Wehrtmann, I. S. 1991. How important are starvation periods in early larval development for survival of *Crangon septemspinosa* larvae? *Marine Ecology Progress Series* 73:183–190.

Weiss, H. M. 1995. *Marine Animals of Southern New England and New York: Identification Keys to Common Nearshore and Shallow Water Macrofauna*. Department of Environmental Protection Bulletin 115. State Geological and Natural History Survey of Connecticut, Hartford. 344 pp.

Weissburg, M. J., Doall, M. H., Yen, J. 1998. Following the invisible trail: Kinematic analysis of mate-tracking in the copepod *Temora longicornis*. *Philosophical Transactions of the Royal Society of London B, Biological Sciences* 353:701–712.

Weissman, P., Lonsdale, D. J., Yen, J. 1993. The effect of peritrich ciliates on the production of *Acartia hudsonica* in Long Island Sound. *Limnology and Oceanography* 38:613–622.

Wendt, D. E. 1998. Effect of larval swimming duration on growth and reproduction of *Bugula neritina* (Bryozoa) under field conditions. *Biological Bulletin* 195:126–135.

Werner, B. 1955. On the development and reproduction of the anthomedusan *Margelopsis haeckeli* Hartlaub. *Annals of the New York Academy of Sciences* 62:3–29.

Wiadnyana, N. N., Rassoulzadegan, F. 1989. Selective feeding of *Acartia tonsa* and *Centropages typicus* on microzooplankton. *Marine Ecology Progress Series* 53:37–45.

Wickstead, J. H. 1959. A predatory copepod. *Journal of Animal Ecology* 28:69–72.

Wikfors, G. H. 2005. A review and new analysis of trophic interactions between *Prorocentrum minimum* and clams, scallops, and oysters. *Harmful Algae* 4:585–592.

Williams, A. B. 1972. A ten-year study of the meroplankton of North Carolina estuaries: Mysid shrimps. *Chesapeake Science* 13:254–262.

Williams, A. B. 1984. *Shrimps, Lobsters, and Crabs of the Atlantic Coast of the Eastern United States, Maine to Florida*. Smithsonian Institution Press, Washington, DC. 550 pp.

Williams, A. B., Bowman, T. E., Damkaer, D. M. 1974. Distribution, variation, and supplemental description of the opossum shrimp, *Neomysis americana* (Crustacea: Mysidacea). *Fishery Bulletin* 72:835–842.

Williams, B. G. 1968. Laboratory rearing of the larval stages of *Carcinus maenas* (Crustacea: Decapoda). *Journal of Natural History* 2:121–126.

Williams, P. J., Brown, J. A. 1992. Development changes in the escape response of larval winter flounder *Pleuronectes americanus* from hatch through metamorphosis. *Marine Ecology Progress Series* 8:185–193.

Williamson, C. E. 1986. The swimming and feeding behavior of *Mesocyclops*. *Hydrobiologia* 134:11–19.

Wilson, C. B. 1905. North American parasitic copepods belonging to the family Caligidae. 1. Caliginae. *Proceedings of the United States National Museum* 31:479–672.

Wilson, C. B. 1932a. The copepod crustaceans of Chesapeake Bay. *Proceedings of the US National Museum* 80:1–54.

Wilson, C. B. 1932b. The copepods of the Woods Hole region, Massachusetts. *US National Museum Bulletin* 158:1–635.

Wilson, M. S. 1958. The copepod genus *Halicyclops* in North America, with a description of a new species from Lake Pontchartrain, Louisiana and the Texas coast. *Tulane Studies in Zoology* 6:176–189.

Wilson, W. H., Jr., Ruff, R. E. 1988. *Species Profiles: Life Histories and Environmental Requirements of Coastal Fishes and Invertebrates (North Atlantic)—Sandworm and Bloodworm.* US Fish and Wildlife Service Biological Report FWS/OBS-82/11.80. 23 pp.

Wimpenny, R. S. 1966. *The Plankton of the Sea*. Elsevier, New York. 456 pp.

Winkler, G., Martineau, C., Dodson, J. J., et al. 2007. Trophic dynamics of two sympatric mysid species in an estuarine transition zone. *Marine Ecology Progress Series* 332:171–187.

Wittmann, K. J. 2009. Revalidation of *Chlamydopleon aculeatum* Ortmann, 1893, and its consequences for the taxonomy of Gastrosaccinae (Crustacea: Mysida: Mysidae) endemic to coastal waters of America. *Zootaxa* 2115:21–23.

Wong, C. K., Li, V. C. Y., Chan, A. 2008. Diel cycles of reproduction and vertical migration in the marine cladocerans *Pseudevadne tergestina* and *Penilia avirostris*. *Journal of Plankton Research* 30:65–73.

Wood, E. J. F. 1968. *Dinoflagellates of the Caribbean Sea and Adjacent Areas*. University of Miami Press, Coral Gables, FL. 142 pp.

Wood, R. D., Lutes, J. 1967. *Guide to the Phytoplankton of Narragansett Bay, Rhode Island*. University of Rhode Island Printing, Kingston. 65 pp.

Woodmansee, R. A. 1966. Daily vertical migration of *Lucifer:* Planktonic numbers in relation to solar and tidal cycles. *Ecology* 47:847–850.

Wortham-Neal, J. L., Price, W. W. 2002. Marsupial developmental stages in *Americamysis bahia* (Mysida: Mysida). *Journal of Crustacean Biology* 22:98–112.

Wu, C.-H., Dahms, H.-U., Buskey, E. J., et al. 2010. Behavioral interactions of the copepod *Temora turbinata* with potential ciliate prey. *Zoological Studies* 49:157–168.

Wulff, A. 1919. Ueber das Kleinplankton der Barentssee. *Wissenschaftliche Meeresunterschungen, Neue Folge Abteilung Helgoland* 13:95–125.

Wyanski, D. M., Targett, T. E. 2000. Development of transformation larvae and juveniles of *Ctenogobius boleosoma*, *Ctenogobius shufeldti*, and *Gobionellus oceanicus* (Pisces: Gobiidae) from western North Atlantic estuaries, with notes on early life history. *Bulletin of Marine Science* 67:709–728.

Yamaguti, S. 1963. *Parasitic Copepoda and Branchiura of Fishes*. Wiley Interscience, New York. 1104 pp.

Yamaji, I. 1966. *Illustrations of the Marine Plankton of Japan*. Hoikusha, Osaka, Japan. 369 pp. (In Japanese.)

Yeung, C., Jones, D. L., Criales, M. M., et al. 2001. Influence of coastal eddies and counter-currents on the influx of spiny lobster, *Panulirus argus*, postlarvae into Florida Bay. *Marine and Freshwater Research* 52:1217–1232.

Yih, W., Coats, D. W. 2000. Infection of *Gymnodinium sanguineum* by the dinoflagellate *Amoebophrya* sp.: Effect of nutrient environment on parasite generation time, reproduction, and infectivity. *Journal of Eukaryotic Microbiology* 47:504–510.

Yih, W., Kim, H. S., Jeong, H. J., et al. 2004. Ingestion of cryptophyte cells by the marine photosynthetic ciliate *Mesodinium rubrum*. *Aquatic Microbial Ecology* 36:165–170.

Young, C. M., ed.; Rice, M. E., Sewell, M., assoc. eds. 2001. *Atlas of Marine Invertebrate Larvae*. Academic Press, New York. 656 pp.

Yule, A. B., Crisp, D. J. 1983. A study of feeding behaviour in *Temora longicornis* (Müller) (Crustacea: Copepoda). *Journal of Experimental Marine Biology and Ecology* 71:271–282.

Zagursky, G., Feller, R. J. 1985. Macrophyte detritus in the winter diet of the estuarine mysid, *Neomysis americana*. *Estuaries* 8:355–362.

Zeng, C. S., Naylor, E. 1996a. Endogenous tidal rhythms of vertical migration in field collected zoea-1 larvae of the shore crab *Carcinus maenas:* Implications for ebb tide offshore dispersal. *Marine Ecology Progress Series* 132:71–82.

Zeng, C. S., Naylor, E. 1996b. Occurrence in coastal waters and endogenous tidal swimming rhythms of late megalopae of the shore crab *Carcinus maenas:* Implications for onshore recruitment. *Marine Ecology Progress Series* 136:69–79.

Zeng, C. S., Naylor, E., Abelló, P. 1997. Endogenous control of timing of metamorphosis in megalopae of the shore crab *Carcinus maenas*. *Marine Biology* 128:299–305.

Ziegler, T. A., Forward, R. B., Jr. 2006. Larval release behaviors of the striped hermit crab, *Clibanarius vittatus* (Bosc): Temporal pattern in hatching. *Journal of Experimental Marine Biology and Ecology* 335:245–255.

Zimmer-Faust, R. K., Tamburri, M. N., Decho, A. W. 1996. Chemosensory ecology of oyster larvae: Benthic-pelagic coupling. In: Lenz, P. H., Hartline, D. K., Purcell, J. E., Macmillan, D. L., eds. *Zooplankton: Sensory Ecology and Physiology.* Gordon & Breach, Amsterdam, The Netherlands, 37–50.

Zimmerman, T. L., Felder, D. L. 1991. Reproductive ecology of an intertidal brachyuran crab, *Sesarma* sp. (NR *reticulatum*), from the Gulf of Mexico. *Biological Bulletin* 181:387–340.

Index

Page numbers of illustrations of specific taxa are in italics.

Sources of Models for Illustrations

The illustrations in this book, all drawn by Marni Fylling, were inspired by a number of sources. They were usually drawn using a combination of photographs, specimens collected in the wild, experience of the authors, advice from external reviewers, and descriptions and illustrations from the following references:

Balcer et al. (1984); Barnes (1980); Bigelow and Schroeder (1953); Bouillon (1999); Bousfield (1973); Bowman (1971); Bradford-Greive et al. (1999); Brossi Garcia and Rodrigues (1997); Brusca and Brusca (1990); Bullard and Whitlach (2004); Calman (1911); Casanova (1999); Choudhury (1970); Clark, Tammy (from specimens collected in North Inlet, SC); Cloney et al. (2002); Costlow and Bookhout (1959, 1960, 1961a, 1961b, 1962, 1966a, 1966b, 1966c); Cupp (1943); Davis (1955); Deevey (1968); Diaz and Costlow (1972); Ditty (1989); Dugger and Dobkin (1975); Egloff et al. (1997); Esnal and Daponte (1999); Fahay (1983); Flores et al. (1998); Frankenberg and Menzies (1966); Fritzsche (1978); Galtsoff (1964); Gehringer (1959); Gerber (2000); Gibson and Grice (1977a); González and Bowman (1965); Gore (1972); Goy and Provenzano (1978); Hardy (1978); Hasle (1972); Hasle and Syvertsen (1980); Hatschek (1878); Heard (1982); Heard et al. (2007); Herrick (1978/1911); Hildebrand and Cable (1934, 1983); Hood (1962); Hulburt (1957b); Hwang et al. (1993); Itoh and Nishida (1995); Jahn and Jahn (1949); Johnson (1978); Kensley and Schotte (1999); Knowlton (1973); Kofoid and Swezy (1921); Kramp (1959); Kurata (1970); Lang (1979); Larsen and Moestrup (1989); Lebour (1928); Lippson and Moran (1974); Maris (1983); Martindale (2002); Mayer (1912); Menzies and Frankenberg (1966); Mianzan and Cornelius (1999); Miller and Burbanck (1961); Miner (1950); Morgan and Provenzano (1979); Needham and Needham (1962); Newell and Newell (1963); Omholt and Heard (1979); Pearson (1939, 1941); Pettibone (1963); Pollock (1997); Pratt (1935); Price et al. (1994); Rees (1959); Rodrigues and Manning (1992); Ruppert and Fox (1988); Sandifer (1972b); Sandifer and Van Engel (1971); Sandoz and Hopkins (1947); Sastry (1977b); Schiedges (1979a, 1979b); Sharpe (1911); Smith (1964); Strüder-Kypke, et al. <www.liv.ac.uk/ciliate/>; Stuck et al. (1979); Suárez-Morales et al. (2008); Tesmer and Broad (1964); Thorp and Covich (2001); Thorson (1946); Tomas (1997); Tucker (1982); Utz, Laura, photographs; Vecchione and Shea (2002); Wang and Kernehan (1979); Ward and Whipple (1918); Weiss (1995); Williams (1984); Wilson, C. B. (1932a, 1932b); Wilson, M. S. (1958); Wimpenny (1966); Wood (1968); Wood and Lutes (1967); Wulff (1919); Yamaji (1966).